Climate Change Management

Series Editor

Walter Leal Filho, International Climate Change Information and Research Programme, Hamburg University of Applied Sciences, Hamburg, Germany

The aim of this book series is to provide an authoritative source of information on climate change management, with an emphasis on projects, case studies and practical initiatives – all of which may help to address a problem with a global scope, but the impacts of which are mostly local. As the world actively seeks ways to cope with the effects of climate change and global warming, such as floods, droughts, rising sea levels and landscape changes, there is a vital need for reliable information and data to support the efforts pursued by local governments, NGOs and other organizations to address the problems associated with climate change.

This series welcomes monographs and contributed volumes written for an academic and professional audience, as well as peer-reviewed conference proceedings. Relevant topics include but are not limited to water conservation, disaster prevention and management, and agriculture, as well as regional studies and documentation of trends. Thanks to its interdisciplinary focus, the series aims to concretely contribute to a better understanding of the state-of-the-art of climate change adaptation, and of the tools with which it can be implemented on the ground.

More information about this series at http://www.springer.com/series/8740

Walter Leal Filho · Sarah L. Hemstock
Editors

Climate Change and the Role of Education

Springer

Editors
Walter Leal Filho
HAW Hamburg
Hamburg, Germany

Sarah L. Hemstock
Department of Geography
Bishop Grosseteste University
Lincoln, UK

ISSN 1610-2002
Climate Change Management
ISBN 978-3-030-32900-6
https://doi.org/10.1007/978-3-030-32898-6

ISSN 1610-2010   (electronic)

ISBN 978-3-030-32898-6   (eBook)

© Springer Nature Switzerland AG 2019
This work is subject to copyright. All rights are reserved by the Publisher, whether the whole or part of the material is concerned, specifically the rights of translation, reprinting, reuse of illustrations, recitation, broadcasting, reproduction on microfilms or in any other physical way, and transmission or information storage and retrieval, electronic adaptation, computer software, or by similar or dissimilar methodology now known or hereafter developed.
The use of general descriptive names, registered names, trademarks, service marks, etc. in this publication does not imply, even in the absence of a specific statement, that such names are exempt from the relevant protective laws and regulations and therefore free for general use.
The publisher, the authors and the editors are safe to assume that the advice and information in this book are believed to be true and accurate at the date of publication. Neither the publisher nor the authors or the editors give a warranty, expressed or implied, with respect to the material contained herein or for any errors or omissions that may have been made. The publisher remains neutral with regard to jurisdictional claims in published maps and institutional affiliations.

This Springer imprint is published by the registered company Springer Nature Switzerland AG
The registered company address is: Gewerbestrasse 11, 6330 Cham, Switzerland

Preface

Climate change is without a doubt one of the most pressing challenges of modern times, one which affects both industrialised and developing nations. Education is known to potentially play a key role in catalysing the participation of individuals and communities in climate change mitigation and adaptation processes.

But the role of education is not only about helping people to understand the impacts of global warming. Rather, education on matters related to climate change is also about catalysing changes in attitudes and behaviour, which may encourage individuals to take a more active role in both climate change mitigation and adaptation efforts.

International experiences show that there is a perceived need to make climate change education a more central and more prominent component of responses to climate change, at the international, regional and local levels. Yet, there is a paucity of truly international publications, which address the many pedagogical, social and economic variables which characterise climate change education.

This book has been produced to address this need. It is the outcome of the "International Symposium on Climate Change and the Role of Education", organised by Bishop Grosseteste University (UK), the Hamburg University of Applied Sciences (Germany) and the International Climate Change Information and Research Programme (ICCIRP).

Papers here compiled look at matters related to the use of a variety of educational approaches to educate, inform or raise awareness about climate change across a variety of audiences. It contains experiences from empirical research, practical projects and teaching methods being deployed round the world, all with the aim of bringing the message across various audiences.

Moreover, the book also entails contributions on how to promote the climate agenda and foster adaptation efforts at the local level. Thanks to its scope, this is a truly interdisciplinary publication. We thank the many authors who contributed to this volume and for their willingness to share their knowledge and expertise.

We hope it will be useful to scholars, social movements, practitioners and members of governmental agencies, undertaking research and/or executing projects on climate change education across the world.

Hamburg, Germany
Lincoln, UK
Winter 2019/2020

Walter Leal Filho
Sarah L. Hemstock

Contents

1 Climate Change Education: An Overview of International Trends and the Need for Action 1
Walter Leal Filho and Sarah L. Hemstock

2 Fiery Spirits: Educational Opportunities for Accelerating Action on Climate Change for Sustainable Development 19
Terence Miller and Mark Charlesworth

3 Integrating Climate Change Competencies into Mechanical Engineering Education 33
Sven Linow

4 Climate Change Education Across the Curriculum 53
Rahul Chopra, Aparna Joshi, Anita Nagarajan, Nathalie Fomproix and L. S. Shashidhara

5 Researching Climate Change in Their Own Backyard—Inquiry-Based Learning as a Promising Approach for Senior Class Students 71
Sebastian Brumann, Ulrike Ohl and Carolin Schackert

6 Energy Transitions: Linking Energy and Climate Change 87
John H. Perkins

7 Delivery Mode and Learner Emissions: A Comparative Study from Botswana 107
Alexis Carr, Stanley Modesto, K. Balasubramanian, Kayla Ortlieb and John Lesperance

8 Adolescents' Perceptions of the Psychological Distance to Climate Change, Its Relevance for Building Concern About It, and the Potential for Education 129
Moritz Gubler, Adrian Brügger and Marc Eyer

9 Addressing Climate Change at a Much Younger Age Than just at the Decision-Making Level: Perceptions from Primary School Teachers in Fiji 149
Peni Hausia Havea, Apenisa Tamani, Anuantaeka Takinana, Antoine De Ramon N' Yeurt, Sarah L. Hemstock and Hélène Jacot Des Combes

10 The Benefits and Downsides of Multidisciplinary Education Relating to Climate Change 169
Lino Briguglio and Stefano Moncada

11 Climate Change, Disaster Risk Management and the Role of Education: Benefits and Challenges of Online Learning for Pacific Small Island Developing States 189
Diana Hinge Salili and Linda Flora Vaike

12 Learning *with* Idea Station: What Can Children on One Canadian Playground Teach Us About Climate Change? 201
Sarah Hennessy

13 Using a Masters Course to Explore the Challenges and Opportunities of Incorporating Sustainability into a Range of Educational Contexts 219
Alison Fox, Paula Addison-Pettit, Clare Lee and Kris Stutchbury

14 Capacity Building Itinerary on Sustainable Energy Solutions for Islands and Territories at Risk for the Effects of Climate Change 237
Lara de Diego, María Luisa Marco and Mirian Bravo

15 Taking Current Climate Change Research to the Classroom—The "Will Hermit Crabs Go Hungry in Future Oceans?" Project 255
Christina C. Roggatz, Neil Kenningham and Helga D. Bartels-Hardege

16 Why Is Early Adolescence So Pivotal in the Climate Change Communication and Education Arena? 279
Inez Harker-Schuch

17 Developing a Climate Literacy Framework for Upper Secondary Students 291
Inez Harker-Schuch and Michel Watson

18 Realities of Teaching Climate Change in a Pacific Island Nation 319
Charles Pierce

Contents ix

19 From Academia to Response-Ability 349
Raichael Lock

20 Recognition of Prior Learning (RPL) in Resilience (Climate Change Adaptation and Disaster Risk Reduction) in the Pacific: Opportunities and Challenges in Climate Change Education 363
Helene Jacot Des Combes, Amelia Siga, Leigh-Anne Buliruarua, Titilia Rabuatoka, Nixon Kua and Peni Hausia Havea

21 The Role of Informal Education in Climate Change Resilience: The Sandwatch Model 371
G. Cambers, P. Diamond and M. Verkooy

22 A Plexus Curriculum in School Geography—A Holistic Approach to School Geography for an Endangered Planet 385
Phil Wood and Steven Puttick

23 (Latent) Potentials to Incorporate and Improve Environmental Knowledge Using African Languages in Agriculture Lessons in Malawi 401
Michael M. Kretzer and Russell H. Kaschula

24 Sixty Seconds Above Sixty Degrees: Connecting Arctic and Non-Arctic Classrooms in the Age of Climate Change 419
Mary E. Short and Laura C. Engel

25 Diving Ecotourism as Climate Change Communicating Means: Greek Diving Instructors' Perceptions 435
Georgios Maripas-Polymeris, Aristea Kounani, Maria K. Seleventi and Constantina Skanavis

26 A Model to Integrate University Education Within Cultural Traditions for Climate Change Resilience 457
Keith Morrison

27 Nurturing Adaptive Capacity Through Self-regulated Learning for Online Postgraduate Courses on Climate Change Adaptation 481
Keith Morrison, Moleen Monita Nand and Heena Lal

28 Increasing Environmental Action Through Climate Change Education Programmes that Enable School Students, Teachers and Technicians to Contribute to Genuine Scientific Research 507
Elizabeth A. C. Rushton

29 Teenagers Expand Their Conceptions of Climate Change Adaptation Through Research-Education Cooperation 525
Oliver Gerald Schrot, Lars Keller, Dunja Peduzzi, Maximilian Riede, Alina Kuthe and David Ludwig

30 Engaging and Empowering Business Management Students to Support the Mitigation of Climate Change Through Sustainability Auditing 549
Kay Emblen-Perry

Chapter 1
Climate Change Education: An Overview of International Trends and the Need for Action

Walter Leal Filho and Sarah L. Hemstock

Abstract Climate change is perceived as one of the major challenges of modern times. But in order to help people to comprehend the various messages around climate change, it is important to foster climate change education. This paper reviews the state of the art on climate change education at university level, and examines the problems associated with it. Its novelty is based on the fact that it presents the results of an international analysis, which illustrates different understandings and engagements of a variety of universities with climate change across diverse audiences and geographical conditions. The paper concludes by outlining some of the lessons learned and some areas where interventions are needed.

Keywords Climate change · Education · Universities · Societies · Research · Curriculum

Introduction

Education is one of the factors to prepare societies to handle climate change (UNESCO 2017), which by itself is one of the greatest world challenges (Alberta Council for Environmental Education 2017). Climate change information in an important part of the formula (Leal Filho 2009). At the United Nations Conference on Environment and Development (UNCED) held in Rio de Janeiro, Brazil in 1992, nations agreed to bring to life the United Nations Framework Convention on Climate Change (UNFCCC), which coordinates global efforts in this key field. Among other elements, UNFCCC expressed the need to develop and implement education initiatives to raise awareness about climate change and its impacts, and to develop and share educational programmes and materials (United Nations 1992).

W. Leal Filho (✉)
European School of Sustainability Science and Research, Hamburg University of Applied Sciences, Ulmenliet 20, 21033 Hamburg, Germany
e-mail: Walter.leal2@haw-hamburg.de

S. L. Hemstock
School of Humanities, Bishop Grosseteste University, Lincoln LN1 3DY, UK
e-mail: Sarah.hemstock@bishopg.ac.uk

© Springer Nature Switzerland AG 2019
W. Leal Filho and S. L. Hemstock (eds.), *Climate Change and the Role of Education*,
Climate Change Management, https://doi.org/10.1007/978-3-030-32898-6_1

As a matter of principle, climate change education includes relevant content knowledge on the climate system, climate science, and on climate change impacts. It also requires the following components to be considered: issue analysis, community and personal decision-making, political processes, social justice, inter-cultural sensitivity and inter-cultural competence, behaviour change, stewardship, and connections between climate change and economics (Hopkins and McKeown 2010; Forrest and Feder 2011). Climate change education focuses on the institutional environment in which that content is taught, so as to ensure that education systems themselves foster climate resilient communities (Anderson 2012).

However, various studies have identified the fact that many teaching staff at both universities and at schools, still lack confidence in their personal subject knowledge, and some of them feel that they are unprepared for the integration of action and content knowledge that characterises climate change education. This is especially so for those teaching science, where subject knowledge tends to be more factual (Oversby 2015). This paper intends to reiterate the importance of climate change education, and the need to work towards a better handling of climate issues in teaching programmes.

Since the Special Report on Managing the Risks of Extreme Events and Disasters to Advance Climate Change Adaptation states that, "Inequalities influence local coping and adaptive capacity, and pose disaster risk management and adaptation challenges from the local to national levels"; it should be acknowledged here that in many of the least developed countries most affected by climate change impacts, such as the Pacific small island states, there is little formal provision of climate change education at university level. For example, studies from 15 Pacific island nations, identified a lack of human capacity and expertise as key barrier to improving national resilience to climate change impacts (Buliruarua et al. 2015). The United Nations Development Programme (2013) established that most Pacific island populations lack climate change awareness and knowledge of appropriate adaptation strategies, leaving them powerless to make informed choices about adaptation to climate change impacts affecting their livelihoods and resources—both now and in the future (UNDP 2013). This lack of human capacity resulted from the absence of sustainable accredited and quality assured formal education around social, managerial and technical climate change adaptation issues (Buliruarua et al. 2015; Hemstock et al. 2016).

Importance of Climate Change Education

Climate change education is perceived as one of the instruments which may be employed in order to create an informed society, knowledgeable workforce, and enable government officials to make decisions and preparations to help protect communities against climate change and its adverse impacts (UNESCO 2010; American Chemical Society 2016). Analysis of global policy frameworks dealing with climate change and disaster risk, for example, the Sendai Framework on Disaster Risk Reduction 2015–2013 (adopted by 187 countries), illustrates that they identify the central role of training and capacity development as being critical for meeting policy goals (Hemstock et al. 2016). Climate change mitigation and adaptation are two main strategies to address climate change. Both require education, which can provide the knowledge and skills needed for redefining lifestyles, change social structures to become more resilient, foster energy efficiency, reduce ecological footprint, change consumption and production patterns, and build adaptive capacity and resilient societies (Anderson 2012).

Education is to play one of the major roles in the transition towards a global economy characterised by low carbon emissions (Education International 2017), which gained a new momentum with the ratification of the Paris Agreement in 2015, due to its ambitious emissions reduction goals. Climate change education is also urgently needed because of the time delay between decisions that may worsen climate change (e.g. increases in CO_2 emissions) and when their full environmental and societal impacts such as droughts, floods, hunger due to crop failure or climate induced migration are felt. As a result, it would benefit society if a due emphasis could be given to the evidences gathered from scientific projections in decision-making, and more efforts could be employed in addressing the roots of the problem. Effective education could accelerate climate change mitigation by transferring scientific knowledge across societal sectors and building social will or pressure to shape climate policy (Ledley et al. 2017).

But even though climate change education is important, it is not as widely practised as it should or could be. Table 1.1 outlines some of the problems seen in pursuing climate change education.

In order to address these problems and further efforts in respect of information, communication, training and education on climate change, the International Climate Change Information and Research Programme (ICCIRP)[*] was created by the Hamburg University of Applied Sciences in 2008. Since its creation, ICCIRP has involved over one million people on its on-line climate conferences, and thousands of scientists have attended the symposia it regularly organises.

The critical role of climate change education has been also recognized in a number of international agreements, such as:

- Article 6 of the UN Framework Convention on Climate Change.
- Article 12 of the Paris Agreement, the Sustainable Development Goals (SDGs) which have on specific goal on climate change (SDG 13, Climate Action).
- The Lima Ministerial Declaration on Education and Awareness-Raising: Ministers and heads of delegations attending the UN Climate Change Conference 2014—COP20 (1–12 December 2014, Lima, Peru) adopted The Lima Ministerial Declaration on Education and Awareness-raising. This Declaration calls on governments to include climate change into school curricula, and to include climate awareness into national development and climate change plans. Marcin Korolec, President of COP 19/CMP 9 and at the time Secretary of State, Ministry of the Environment, Poland, said that "this declaration is an important step towards bringing education back into the spotlight where it belongs".

Table 1.1 Some of the problems that hinder the pursuit of climate change education

Problem	Impact
Complexity and scale	Climate change relates to a variety of atmospheric, meteorological, social and economic factors which makes it a complex issue. The scale of climate change causes and impacts can be seen as overwhelming and individual acts can be rationalised as being inconsequential
Lack of training and professional development opportunities	Not all teaching staff have the training or feel qualified to engage on climate change teaching
Limited teaching resources	The restricted availability of specific teaching resources which may clearly communicate on climate change in some countries
Curriculum constraints	Lack of flexible time-tables to allow discussions on climate change; inflexible and politically biased state curricula
Competing themes	Climate change needs to compete with a variety of themes which are equally important
Limited institutional support	In many cases climate change is on the one hand perceived as important, but institutional support to it is limited, which inhibits progress
Scepticism and controversy	Lack of interest to tackle the topic; fear of being controversial

Source Author

- Article 27 (j) of the Sendai Framework for Disaster Risk Reduction 2015–2030 asks for accredited formal qualifications for capacity development and professionalisation of the disaster risk reduction sector (UNISDR 2016, p. 18).
- The Aichi-Nagoya Declaration on Education for Sustainable Development (EDS).
- Article 6 of the UN Framework Convention on Climate Change (UNESCO 2017).

*Details at: https://www.haw-hamburg.de/en/ftz-nk/programmes/iccirp/.

In addition, the Paris Committee on Capacity-building was formed by the UNFCCC in 2015 to addresses current and emerging gaps and needs in implementing the Paris agreement by analysing climate capacity-building issues and making policy recommendations to support countries in enhancing climate action. For the years 2017–2019, its focus area is capacity-building activities for the implementation of nationally-determined contributions in the context of the Paris Agreement (UNFCCC 2019).

Some Case Studies

There have been various studies on climate change, and on education, but very few have tackled climate change education in an integrated way. Based on the perceived need to address this research gap and the paucity of first-hand information on the obstacles which hinder developments in this field, a survey was performed, with a view to selecting some case studies from universities where climate change education is being pursued. These are herewith presented.

Evidences of the increasing significance of climate change education was seen by the launch of the UN Alliance on Climate Change Education, Training and Public Awareness (Læssøe and Mochizuki 2015). Climate change, as a topic, has also been included in the U.S. National Science Standards such as the 2012 Framework for K-12 Science Education (National Research Council 2012) and Next Generation Science Standards 2013 (NGSS 2013).

In Canada, the province of Newfoundland and Labrador addressed the importance of education in its Climate Change Action Plan in 2005. The document promises continued support and funding to the Newfoundland and Labrador Climate Change Education Centre, which work focuses on educating the public about greenhouse gas emission reduction measures and encouraging actions to reduce personal emissions (Nazir et al. 2009).

A further example of action comes from China. The Chinese government highlights the importance of education as one of the necessary approaches to handle climate change in a variety of government policy documents. For instance, the document "China's Scientific and Technological Actions on Climate Change and Policies", in addition to various others, include recommendations on specific education initiatives to be integrated in schools, universities and research institutes (Yi and Wu 2009).

In Denmark and Australia, matters related climate change are explicitly incorporated in the Danish Education for Sustainable Development strategy and the Australian National Education for Sustainable Development policy, respectively (Breiting et al. 2009; Chambers 2009). Denmark followed the recommendation of the European Commission to implement climate change topics in the formal national education through its integration to Education for Sustainable Development (Milér et al. 2012). In Brazil, the Sustainable Schools Programme, which is included in the National Plan on Climate Change calls upon the Ministry of Education to introduce climate change into the curricula and learning materials (Valentin et al. 2015).

In the Pacific-African, Caribbean and Pacific (P-ACP) grouping, key national policies on climate change adaptation (CCA) and disaster risk management (DRM)/disaster risk reduction (DRR) have been surveyed (Table 1.2) with regard to policy support at national level for climate change education.

Table headers and content:

Table 1.2 Policy support for climate change education in the P-ACP region

Country	Key policy	Awareness raising	Non-formal training	Strengthen capacity (general)	Education (formal assumed)
Cook Island	Cook Islands Joint National Action Plan for DRM and CCA	✓		✓	
Fiji	Fiji National CC Policy 2012 and National DRM Plan 1995, National Disaster Management Act 1998; Climate Change Adaptation and Disaster Risk Reduction Strategies 2013	✓	✓	✓	✓
FSM	Joint State Action Plan for CC and DRM and 2nd National Communications report to the UNFCCC			✓	✓
Kiribati	Kiribati Joint Implementation Plan for CC and DRM 2014–2023	✓		✓	✓
Nauru	Republic of Nauru Framework for Climate Change Adaptation and Disaster Risk Reduction 2015			✓	✓
Niue	Niue's Joint Action Plan for DRM and CCA			✓	✓

(continued)

Table 1.2 (continued)

Country	Key policy	Awareness raising	Non-formal training	Strengthen capacity (general)	Education (formal assumed)
Palau	Palau Climate Change Policy: For Climate and Disaster Resilient Low Emissions Development 2015			✓	✓
PNG	The National Development Strategic Plan (DSP) (2010–2030)			✓	✓
RMI	RMI Joint Action Plan for CCA and DRM; Vision 2018 (2003–2018); National Climate Change Policy Framework 2011; Ministry of Education Strategic Plan (2013–2016)			✓	✓
Samoa	Samoa National Action Plan for DRM 2011–2016	✓	✓	✓	✓
Solomon Islands	National Development Strategy 2011–2020; Solomon Islands Climate Change Policy (2012); Solomon Islands National Disaster Risk Reduction Policy (NDRRP 2010)	✓	✓	✓	✓

(continued)

Table 1.2 (continued)

Country	Key policy	Awareness raising	Non-formal training	Strengthen capacity (general)	Education (formal assumed)
Timor Leste	National Adaptation Programme of Action (NAPA 2010) on Climate Change Adaptation National Disaster Risk Management Policy 2008	✓	✓	✓	
Tonga	Tonga National Climate Change Policy and Joint National Action Plan for CCA and DRM 2010–2015			✓	✓
Tuvalu	Tuvalu National Strategic Action Plan for CCA and DRM 2012–2016	✓	✓	✓✓	✓
Vanuatu	Vanuatu Climate Change and Disaster Risk Reduction Policy 2016–2030	✓	✓	✓	

Integration of Climate Change Issues at Universities: Some Examples

Many countries are at present working on improving the efficiency of their education systems towards a climate literate society (Milér et al. 2012). Higher education institutions are considered to have a critical role in preparing society to adapt to the impacts of climate change, and to help local communities to create, test, and disseminate knowledge about regional climate projections and adaptation strategies (Dyer and Andrews 2014). This is especially important in the regions where leaders and media have created confusion among the public about environmental issues such as climate change (Hess and Collins 2018). Some recent publications such as "Universities and Climate Change: Introducing Climate Change to University Programmes" (Leal Filho 2010) and "Climate Change Research at Universities: Addressing the Mitigation and Adaptation Challenges" (Leal Filho 2017), have attempted to raise awareness about the role played by universities in the climate change debate, and have documented and promoted many concrete experiences and case studies which have shown how this can be done in practice.

In practical terms, the integration of quality climate change learning into existing education systems requires reconsideration of existing approaches by developing a system that equips learners with the requisite skills, knowledge and attributes to deal with future challenges (Bangay and Blum 2010).

Monroe et al. (2017) identified a number of effective climate change education strategies:

- development of programs focused on making climate change information personally relevant and meaningful for learners;
- engagement of learners into activities or educational interventions (e.g. debates, small group discussions, laboratory investigations, and simulations, field trips etc.);
- design of programs specifically to uncover and address misconceptions about climate change;
- engagement in designing and implementing projects to address some aspect of climate change, e.g. energy saving, emissions reduction, change of environmental behaviours (Monroe et al. 2017).

There are several ways to integrate climate change education into higher education. One of them is via **curriculum modifications**. Such modifications across a diverse range of disciplines could ensure professionals understand climate change, its impacts, and the best practices for responding to them (Dyer and Andrews 2014). Hess and Collins (2018) identified three pathways toward a greater inclusion of climate science in the core curriculum:

- to have a single specified core course that explicitly requires education about climate change and taken by all students;
- to have a high volume of options of climate-related courses. Climate-change courses could be offered in multiple departments, thus increasing the proportion of courses that cover climate change in the general education curriculum;

- to have a menu of courses that would include climate-science or climate-change education for an environmental or sustainability studies programs (Hess and Collins 2018).

Naturally, there can be no prescription as to which approach is the best one. A decision about the most adequate method depends on the local context, available expertise and socio-economic settings.

In any case, experts note that climate education cannot only be limited to the STEM disciplines (science, technology, engineering, and math). It is critical to capacity building to focus on social and behavioural disciplines, including emerging interdisciplinary fields, such as ecological psychology and ecological economics. Incorporating new learning objectives into a variety of programs is necessary to ensure that tomorrow's professionals have an understanding of the new climate reality that will impact their work (Dyer and Andrews 2014).

Higher education institutions across the world have developed robust examples of integrating climate change into their curriculum. Even countries which have limited resources such as Belarus, can be active. In Belarus for instance, its universities integrated a new study course "Climate Change: Consequences, Mitigation, Adaptation", which has been developed under the UNDP-ENVSEC "Environment and Security" project framework. The four-unit study course applies innovative educational approaches to help students and young people to understand, address, mitigate, and adapt to the impacts of climate change (UNDP in Belarus 2015).

Under the USAID Lowering Emissions in Asia's Forests (USAID LEAF), program participating universities integrated a regional climate change curriculum. It was created by USAID LEAF in close collaboration with 14 universities from six countries, as well as experts from the US Forest Service and from several American universities. The climate change curriculum consists of four modules: Basic Climate Change; Social and Environmental Soundness; Low Emission Land Use Planning; Carbon Measurement and Monitoring (USAID LEAF 2016). For instance, at the Da Lat University, a leading research and technology institution in central Vietnam, 8000 students are required to take an introductory climate change course. At Kasetsart University in Thailand, new undergraduate students take a general education course on climate change that integrates all four curriculum modules. The Vietnam Forestry University (VFU) has integrated the materials into existing courses, including undergraduate courses on "Global Climate Change Impacts and Mitigation" and "Climate Change and Forestry" (USAID LEAF 2015).

A number of European Universities have also set standalone graduate programmes on climate change. For instance, the University of Copenhagen offers a 2-year interdisciplinary M.Sc. programme, that combines natural and social science approaches to the study of climate change, its causes and effects, and adaptation methods (University of Copenhagen 2017). The Lund University offers a master program in Disaster Risk Management and Climate Change Adaptation with a strong focus on group work and interaction between students and teaching staff, supported by national and international institutions, such as UN agencies, the Red Cross/Red Crescent movement, NGOs, and national authorities (Lund University 2017).

In the P-ACP region, the University of the South Pacific in partnership with the Pacific Community and national tertiary education providers were the first to offer a suit of technical, vocational education and training qualifications (TVET) in Climate Change Adaptation at levels 1–4 on the Pacific Qualifications Framework (Hausia Havea et al. 2019). This was part of a region-wide European Union funded initiative with all 15 P-ACP countries participating (the European Union Pacific Technical Vocational Education and Training in Climate Change Adaptation and Sustainable energy—EU PacTVET). For the P-ACP region, a vocational skill-sets and competencies approach worked since technical knowledge and skills are required to improve community resilience to climate change impacts. Additionally, Small Island States such as Tuvalu have around 30% of their employees working in jobs linked to climate change and disaster risk reduction, so climate change adaptation and "resilience" can be defined as an employment sector. These TVET qualifications were developed by all 15 P-ACP countries as equal partners and were regionally accredited (rather than on a national basis). The qualifications were then offered across the region at various tertiary education providers (for example, Fiji National University, Solomon Islands National University, Vanuatu Institute of Technology, etc.) (Martin et al. 2017).

It is important to note that in all cases, it should be ensured that student learning outcomes are appropriate and allow them to seek solutions through well informed activities.

Another way of integrating climate change education to universities is by means of **research**. By providing cutting-edge scientific research, higher education contributes to climate adaptation by identifying the most pressing climate impacts at different levels. Research universities have also the opportunity to provide much needed research on solutions (Dyer and Andrews 2014).

For instance, a number of South African universities are demonstrating transdisciplinary research and teaching, guided by the Global Change Grand Challenge National Research Plan (Valentin et al. 2015).

In 2008, the University of Victoria established the New Zealand Climate Change Research Institute (CCRI) to develop interdisciplinary research into all aspects of climate change. The Institute undertakes internationally significant climate change research that informs policy makers in New Zealand and other countries, develops and delivers high quality, interdisciplinary taught courses to students and practitioners (Victoria University 2017).

The Stanford School of Earth, Energy and Environmental Sciences and the Stanford Teacher Education Program launched a Climate Change Education Project in 2009. It aims to support teaching of scientifically accurate climate change curriculum in middle and high schools. The project documents in detail the full circle of curriculum development, teacher professional development, classroom implementation, analysis of student achievement data, and curriculum revision (Stanford School of Earth, Energy and Environmental Sciences 2017).

The third way to integrate climate education is by building **collaborative partnerships**. The establishment of coordinated climate change education networks would help to integrate, and synergize these diverse efforts by conducting research on effective methods, sharing best practices and educational resources (Forrest and Feder 2011). UNESCO recommends to seek collaboration and partnerships with ministries, civil society, communities, media, the private sector etc. In the specific case of African countries, the organization recommends to establish regional centres of excellence and regional climate change education network (Valentin et al. 2015).

Under the Asian University Network for Environment and Disaster Management (AUEDM) 18 universities from 13 countries have come together to share knowledge resources, advocate for policy change, and develop guidelines related to environment and disaster risk management. One of the objectives of AUEDM is to seek possibilities of mutual collaboration on field-based action research focusing on climate change adaptation (UNESCO 2012).

In the P-ACP region, the Pacific Regional Federation of Resilience Professionals was formed as an industry association for resilience to achieve sustainable outcomes in skills development, education, training and employment for climate change adaptation and disaster risk reduction and to align closely with regional and national needs and priorities. It is also intended to administer an industry certification scheme for practitioners that sets the benchmark of quality for the Resilience (climate change adaptation CCA and disaster risk reduction/management (DRR/DRM)) sectors. To ensure the sustainability of the EU PacTVET project outcomes, this federation will host a Resilience Industry Skills Advisory Committee to facilitate reviews and updates of education and training curriculum and practices in resilience for the regionally accredited qualifications (Hemstock et al. 2016, 2017).

The next approach is **community engagement**. Colleges and universities have the opportunity to serve as 'hubs' on climate adaptation issues and support their communities preparing for growing climate change impacts (Dyer and Andrews 2014). This role includes but is not limited to, conducting applied climate change research, assessment of current conditions and risks from severe weather events, translating science for lay audience and local decision makers, disseminating local-scaled climate information, providing technical support for multi-sector collaborative planning efforts, and evaluating the effectiveness of local adaptation actions (Gruber et al. 2017).

For instance, the Vietnam Forestry University has adapted climate change modules for a training program for provincial government officers titled "Green Growth and Climate Change Mitigation" and for training courses in Malaysia and Laos entitled, "Improving Forest Governance in Southeast Asia" (USAID LEAF 2015). In the United States, the Cornell Institute for Climate Smart Solutions supports local farmers with decision tools for strategic adaptation to climate change and builds their capacity to cope with potential negative effects of climate change, and to take advantage of any opportunities that it might bring (Cornell Institute for Climate Smart Solutions 2017). Also in the US, the Center for Climate Preparedness and Community Resilience of Antioch University in New England implements climate resilience at the local level through stakeholder capacity building, applied research, education and training provided to community leaders (Antioch University New England 2017).

Community engagement can be also focused on **campus operations**. Campuses are often complex physical plant and infrastructure systems that include elements such as:

(a) buildings and operations such as, power generation, heating and cooling systems, storm water management,
(b) transportation,
(c) waste management.

In some campuses, elements such as forests and agriculture are also relevant. All these sectors are vulnerable to the risks posed by climate change (Dyer and Andrews 2014). They also offer real opportunities via which universities may engage in environmental action. Indeed, many universities at present address these challenges as part of climate action strategies, blueprints or plans. For instance, the Climate Action Plan of the University of Saskatchewan, Canada includes energy awareness training, available to any department or unit on campus informing employees and students of expectations and goals for energy conservation and efficiency on campus. Furthermore, sustainability awareness is a part of new employee orientation through which new staff become aware of the university's commitment to GHG reduction and the ways in which they can be involved (University of Saskatchewan 2012).

It is import to note that climate change education has been shown to be more effective with a focus on concrete actions that can be taken by individual students (Alberta Council for Environmental Education 2017). At the same time various studies suggest that many universities and colleges are still struggling to modify the general education curriculum and to ensure that all students are exposed to education about climate science and climate change. The reluctance to engage has many reasons, one of which can be the slow pace of change reflected in different priorities of professors and administrators for core curriculum reform (Hess and Collins 2018). Another potential threat to the effective integration of climate change education might be differing views of roles and responsibilities in this process hold by professors (McGinnis et al. 2016).

Lessons Learned and Conclusions

Education is central to efforts towards informing and raising awareness about climate change. But in order to yield the expected benefits, it requires a holistic and interdisciplinary approach, which recognizes the complexities of climate change and includes scientific, economic, political, ethical and cultural dimensions of expertise (UNESCO 2011). Although higher education may contribute to climate change mitigation and adaptation in research, education, operational and community engagement activities areas, this potential has not yet been realised (Dyer and Andrews 2014) partly by the reasons here outlined.

Some of the lessons which may be drawn from the literature and from the various examples and experiences listed on this paper, and some interventions which may lead to the further development of climate change education, are follows:

1. climate change education should be equally pursued at school level and at universities, preferably in a more systematic way by means of formal policies and programmes, since these offer a sense of continuity, as opposed to being ad hoc and short-termed, as it is often the case with many of the current and past initiatives;
2. the existing international emphasis provided to climate change education by UNFCCC or the Paris Declaration, should be better used. This is particularly so if one considers the fact that over 100 countries have signed on to these documents, and are at least in principle committed to education on matters related to climate change;
3. climate change education can be planned and delivered flexibly. There is no prescription on fixed approaches or methods, much depends on the local context and reality;
4. much could be gained by combining climate change education with efforts in the field of education for sustainable development. After all, climate change is one of the major obstacles to sustainable development so that a combination of both education streams can have a far greater impact, than handling them in separate;
5. climate change education is not a prerogative of industrialised nations. On the contrary: education institutions in developing countries should actively engage and pursue initiatives to foster awareness and see local solutions.

As far as universities are concerned, the plurality of options available means that climate considerations can be given to campus operations, teaching, research and in community relations. This, combined with an increase in teachers' confidence in teaching the political, social and economic side of climate change, may lead to changes and to a better motivation among students be more engaged towards addressing a problem which is global in nature, but whose impacts are felt at the local level.

References

Alberta Council for Environmental Education (2017) What is excellent climate change education? A guidebook based on peer-reviewed research and practitioner best practices. Available at: http://www.abcee.org/sites/abcee.org/files/3-%27What%20is%20Excellent%20Climate%20Change%20Education%20(7%20Dec%202017).pdf

American Chemical Society (2016) Global climate change ACS position statement. Available at: https://www.acs.org/content/acs/en/policy/publicpolicies/sustainability/globalclimatechange.html. Accessed 29 Dec 2017

Anderson A (2012) Climate change education for mitigation and adaptation. J Educ Sustain Dev 6:191–206. https://doi.org/10.1177/0973408212475199

Antioch University New England (2017) Our mission. Available at: http://www.communityresilience-center.org/. Accessed 29 Dec 2017

Bangay C, Blum N (2010) Education responses to climate change and quality: two parts of the same agenda? Int J Educ Dev 30:335–450. Available at: http://discovery.ucl.ac.uk/1526915/1/Bangay2010Education359.pdf

Breiting S, Læssøe J, Rolls S (2009) Climate change and sustainable development: the response from education. Danish national report. Danish School of Education, University of Aarhus, Copenhagen, Denmark. Available at: http://www.hilaryinwood.ca/pdfs/researcb/ESD%20in%20Canada%202009.pdf

Buliruarua LA, Hemstock SH, Jacot Des Combes H, Kua N, Martin T, Satiki V, Manuella-Morris T (2015) P-ACP training needs and gap analysis. Reports for Fiji, Cook Islands, the Federated States of Micronesia, Kiribati, Marshall Islands, Nauru, Niue, Palau, Papua New Guinea, Samoa, Solomon Islands, Timor Leste, Tonga, Tuvalu and Vanuatu (FED/2014/347-438). The Pacific Community and The University of the South Pacific. EU PacTVET, Suva, Fiji

Chambers D (2009) Sustainable development: the response from education. Australian country report. Melbourne Graduate School of Education, Australia. Available at: http://www.hilaryinwood.ca/pdfs/researcb/ESD%20in%20Canada%202009.pdf

Cornell Institute for Climate Smart Solutions (2017) Smart solutions for a changing climate. Available at: http://climateinstitute.cals.cornell.edu/. Accessed 29 Dec 2017

Dyer G, Andrews J (2014) Higher education's role in adapting to a changing climate. American College and University. Presidents' Climate Commitment, Boston, MA. Available at: http://secondnature.org/wp-content/uploads/Higher_Education_Role_Adapting_Changing_Climate.pdf

Education International (2017) COP23: states must be more ambitious with regard to climate change education and training. Available at: https://ei-ie.org/en/detail/15526/cop23-states-must-be-more-ambitious-with-regard-to-climate-change-education-and-training. Accessed 29 Dec 2017

Forrest S, Feder MA (2011) Climate change education: goals, audiences, and strategies: a workshop summary. National Academics Press, Washington, DC. Available at: https://www.nap.edu/read/13224/chapter/2

Gruber JS, Rhoades JL, Simpson M, Stack L, Yetka L, Wood R (2017) Enhancing climate change adaptation: strategies for community engagement and university-community partnerships. J Environ Stud Sci 7:10–24. https://doi.org/10.1007/s13412-015-0232-1

Hausia Havea P, Siga A, Rabuatoka T, Tagivetaua Tamani A, Devi P, Senikula R, Hemstock SL, Jacot Des Combes H (2019) Using vocational education to support development solutions in the Pacific: an emphasis on climate change and health. Appl Environ Educ Commun. ISSN 1533-0389

Hemstock SL, Buliruarua LA, Chan EYY, Chan G, Jacot Des Combes H, Davey P, Farrell P, Griffiths S, Hansen H, Hatch T, Holloway A, Manuella-Morris T, Martin T, Renaud FG, Ronan K, Ryan B, Szarzynski J, Shaw D, Yasukawa S, Yeung T, Murray V (2016) Accredited qualifications for capacity development in disaster risk reduction and climate change adaptation. Australas J Disaster Trauma Stud 20(1):15–33

Hemstock SH, Jacot Des Combes H, Martin T, Vaike FL, Maitava K, Buliruarua L-A, Satiki V, Kua N, Marawa T (2017) A case for formal education in the technical, vocational education and training (TVET) sector for climate change adaptation and disaster risk reduction in the Pacific Islands region. In: Climate change adaptation in Pacific countries: fostering resilience and improving the quality of life. Climate change management. Springer International Publishing, Berlin, pp 309–324. ISBN 978-3-319-50093-5

Hess DJ, Collins BM (2018) Climate change and higher education: assessing factors that affect curriculum requirements. J Clean Prod 170:1451–1458. https://doi.org/10.1016/j.jclepro.2017.09.215

Hopkins C, McKeown R (2010) Rethinking climate change education. Green Teach 89:17–21. Available at: https://www.humphreyfellowship.org/system/files/Rethinking%20Climate%20Change%20Education.pdf

Læssøe J, Mochizuki Y (2015) Recent trends in national policy on education for sustainable development and climate change education. J Educ Sustain Dev 9:27–43. https://doi.org/10.1177/0973408215569112

Leal Filho W (2009) Communicating climate change: challenges ahead and action needed. Int J Clim Change Strat Manage 1(1):6–18. https://doi.org/10.1108/17568690910934363

Leal Filho W (ed) (2010) Universities and climate change: introducing climate change to university programmes. Springer, Berlin

Leal Filho W (ed) (2017) Climate change research at universities: addressing the mitigation and adaptation challenges. Springer, Berlin

Ledley TS, Rooney-Varga J, Niepold F (2017) Addressing climate change through education. In: Oxford research encyclopedia of environmental science, pp 1–40. Available at: http://dx.doi.org/10.1093/acrefore/9780199389414.013.56

Lund University (2017) Disaster risk management and climate change adaptation—master's programme. Available at: https://www.lunduniversity.lu.se/lubas/i-uoh-lu-TAKAK. Accessed 29 Dec 2017

Martin T, Hemstock SL, Jacot Des Combes H, Pierce C (2017) Capacity development and TVET: accredited qualifications for improving resilience of coastal communities: a Vanuatu case study. In: Climate change impacts and adaptation strategies for coastal communities. Climate change management. Springer International Publishing AG (Springer Nature), International, pp 119–131. ISBN 978-3-319-70702-0

McGinnis JR, McDonald C, Hestness E, Breslyn W (2016) An investigation of science educators' view of roles and responsibilities for climate change education. Sci Educ Int 27:179–192. Available at: https://files.eric.ed.gov/fulltext/EJ1104645.pdf

Milér T, Hollan J, Válek J, Sládek P (2012) Teachers' understanding of climate change. Procedia Soc Behav Sci 69:1437–1442. https://doi.org/10.1016/j.sbspro.2012.12.083

Monroe MC, Plate RR, Oxarart A, Bowers A, Chaves WA (2017) Identifying effective climate change education strategies: a systematic review of the research. Environ Educ Res 1–22. http://dx.doi.org/10.1080/13504622.2017.1360842

National Research Council (2012) A framework for K-12 science education: practices, crosscutting concepts, and core ideas. Committee on a conceptual framework for new K-12 science education standards

Nazir J, Pedretti E, Wallace J, Montemurro D, Inwood H (2009) Climate change and sustainable development: the response from education—the Canadian perspective. Centre for Science, Mathematics and Technology Education Ontario Institute for Studies in Education, University of Toronto. Available at: http://www.hilaryinwood.ca/pdfs/research/ESD%20in%20Canada%202009.pdf

NGSS (2013) Next generation science standards. Available at: https://www.nextgenscience.org/. Accessed 27 Dec 2017

Oversby J (2015) Teachers' learning about climate change education. Procedia Soc Behav Sci 167:23–27. https://doi.org/10.1016/j.sbspro.2014.12.637

Stanford School of Earth, Energy and Environmental Sciences (2017) About the climate change education project. Available at: https://pangea.stanford.edu/programs/outreach/climatechange/about-climate-change-education-project. Accessed 29 Dec 2017

UNDP (2013) Climate change and Pacific Island countries. Asia Pacific human development report. Background papers series 2012/07. HDR-2013-APHDR-TBP-07. United Nations Development programme

UNDP in Belarus (2015) International assistance helps Belarus to face challenges and explore new opportunities of the climate change. Available at: http://www.by.undp.org/content/belarus/en/home/presscenter/pressreleases/2015/12/08/international-assistance-helps-belarus-to-face-challenges-and-explore-new-opportunities-of-the-climate-change.html. Accessed 29 Dec 2017

UNESCO (2010) Climate change education for sustainable development. Paris, France. Available at: http://unesdoc.unesco.org/images/0019/001901/190101E.pdf

UNESCO (2011) Recommendations from the UNESCO expert meeting on climate change education for sustainable development and adaptation in small island developing states. Available at: http://www.unesco.org/fileadmin/MULTIMEDIA/HQ/ED/pdf/2011Bahamasrecommendations.pdf

UNESCO (2012) Education sector responses to climate change. UNESCO Bangkok, Bangkok, Thailand. Available at: http://unesdoc.unesco.org/images/0021/002153/215305e.pdf

UNESCO (2017) UNESCO at COP23. Climate change education. Available at: http://unesdoc.unesco.org/images/0026/002600/260083e.pdf

UNFCCC (2019) 8th Durban forum on capacity building. Available at: https://unfccc.int/8th-durban-forum

UNISDR (2016) The science and technology roadmap to support the implementation of the Sendai framework for disaster risk reduction 2015–2030. Retrieved from www.preventionweb.net/files/45270_unisdrscienceandtechnologyroadmap.pdf

United Nations (1992) United Nations framework convention on climate change

University of Copenhagen (2017) Master of Science (MSc) in climate change. Available at: http://studies.ku.dk/masters/climate-change/. Accessed 29 Dec 2017

University of Saskatchewan (2012) Climate action plan. Available at: http://sustainability.usask.ca/documents/UofS_Climate_Action_Plan_2012.pdf

USAID LEAF (2015) Universities in Asia-Pacific implement innovative climate change curriculum. Available at: http://www.leafasia.org/sites/default/files/public/resources/USAID%20LEAF%20SS5-Climate%20Change%20Curriculum%20and%20Universities_0.pdf

USAID LEAF (2016) Climate change curriculum development brochure. Available at: http://www.leafasia.org/library/climate-change-curriculum-development-brochure. Accessed 29 Dec 2017

Valentin B, Abreu D, Ramirez O, Bynoe P, Ferguson T, Simmons D, Gokool-Ramdoo S, Lotz-Sisitka H, Mandikonza C, Sweeney D, Pritchard M (2015) Not just hot air. Putting climate change education into practice. UNESCO. Available at: http://unesdoc.unesco.org/images/0023/002330/233083e.pdf

Victoria University (2017) About the climate change research institute. Available at: https://www.victoria.ac.nz/sgees/research-centres/ccri/about. Accessed 29 Dec 2017

Yi J, Wu P (2009) IALEI-project: climate change and sustainable development: the response from education. Report from China. School of Education, Beijing Normal University, China. Available at: http://rce-denmark.dk/sites/default/files/2017-04/Climatechangeandeducation.pdf

Chapter 2
Fiery Spirits: Educational Opportunities for Accelerating Action on Climate Change for Sustainable Development

Terence Miller and Mark Charlesworth

Abstract This chapter considers questions of how people move from being informed about climate change and unsustainability to acting to address these issues – including in a passionate way. Parts of the answer relate to an individual experience or experiences. Parts of the answer relate to education including self-consciously ecological schools often with hands-on learning by doing and immersion in addressing ecological concerns as part of the way that they operate. Pedagogical literature in this field suggests that education practices that engage heart-head-hand rather than simply head are often more effective and perhaps necessary to achieve changes in behaviour at the required speed to address the reality of climate change and unsustainability. The global movement sparked by Greta Thunberg suggests individual experiences can start widespread change very rapidly. The chapter explores the possibility of encouraging these pivotal ecological experiences via tailored activities and suggests that producing a spark can be done. However, it also indicates that often turning that spark into a wider flame is difficult because the 'damp wood' of context snuffs out enthusiasm – at least for a time or until another spark relights a fire elsewhere or embers are allowed to rekindle.

Keywords Fiery-spirit · Ecological schools · Immersion · Hands on · Learning by doing

Introduction

In her recent TED Talk Greta Thunberg (2018), the 16 year old Swedish school student, said, "Once we start to act, hope is everywhere. Instead of looking for hope, look for action. Actions will give us hope". As Greta has emphasised *now* is the time for action, for accelerating our actions for climate change. Greta has galvanised sections of a whole generation of young people to demand action, action

---

T. Miller (✉)
University of Lincoln, Lincoln, UK
e-mail: terry@terryemiller.co.uk

M. Charlesworth
Bishop Grosseteste University, Lincoln, UK

© Springer Nature Switzerland AG 2019
W. Leal Filho and S. L. Hemstock (eds.), *Climate Change and the Role of Education*, Climate Change Management, https://doi.org/10.1007/978-3-030-32898-6_2

by politicians, business leaders, teachers, every man and woman and child. How is this to be achieved? As she points out we have known about the problem for many years but we have done nothing, merely continued our dangerous behaviours. Thirty-four years ago the Lutheran theologian Jürgen Moltmann writing in 1985 in 'God and Creation' starkly summarised our situation when he said:

> What we call the environmental crisis is not merely a crisis in the natural environment of human beings. It is nothing less than a crisis in human beings themselves. It is a crisis of life on this planet, a crisis so comprehensive and so irreversible that it can not unjustly be described as apocalyptic. It is not a temporary crisis. As far as we can judge, it is the beginning of a life and death struggle for creation on this earth. (Moltmann 1985, xi)

In 1985, global CO_2 was 345.72 ppm (NASA 2019), and in the intervening period has grown to 411.75 ppm in February 2019 (ESRL 2019), averaging about 2–3 ppm per year. Yet even as CO_2 grew year on year actions have been too little to make a difference. The IPCC Special Report 15 launched from Incheon in Korea on 8th October 2018 in preparation for the Katowice COP 24 Climate change Conference in December 2018, warned that global warming must be limited to 1.5 °C above pre-industrial levels to avoid the environmental impacts of a rapidly heating world (IPCC 2018). The report has challenged the world to reduce CO_2 emissions to 45% below 2010 levels by 2030 in order to have a chance of keeping to the 1.5 °C limit. This would require an emergency level of action which is what Greta was talking about.

The challenge is a stark one, and Greta was at pains to say that it is an emergency, and asked why was nobody acting as if there is one? The high level of interest in Greta's intervention does however demonstrate a possible way forward. The mobilisation of millions of young people is an opportunity and points to the valuable role that education offers if we are quick to offer support. Education is one of the best opportunities for accelerating actions and equipping young people to take action for climate change that can bring about the radical reductions in CO_2 emissions required to avoid or minimise the extreme risks of runaway global warming.

In Sweden Greta might be called an 'eldjfäl' (pronounced eldshevl)—a 'fiery spirit', a Swedish word literally translated, 'soul', 'mind', 'fire', 'spirit' [The idea of the 'fiery spirit' was explained to me by a Swedish friend, the late Karin Silk. See also Eriksson and Ujvari (2015)]. It is a phrase stronger than 'activist' or 'pioneer', describing someone who is utterly convinced and transformed by what they believe. Recent research by the lead author of this chapter (Miller 2015) interviewed highly motivated environmental activists to explore how they were moved to take radical action. The term fiery spirit perfectly describes the research subjects in the UK, USA, and India, who are drawn from a range of backgrounds and beliefs. That study utilises a theory of action to explore the mechanism that was presented in the data for motivation for environmental action. The study in particular focuses on the pivotal experiences that influenced behaviour and in the theory of action that links together experience, belief and what is termed 'fire', the force of animation. "Experience which is mediated externally or internally through physical or mental actions, situated in a place, riding a bicycle, sitting in a garden, standing by a lake, watching a sunset, working with nature, caring for animals, watching a film, shocked by a sight, hearing a lecture, reading a book and so on, may engender an insight, a belief about the world that engages emotion, feelings and rational/reflective thought, and which precipitates into further action" (Miller 2015, p. 187).

Action and Experience

All experience affects us, but there are certain kinds of experiences that change us—that is to say change our personalities. It is the mechanism of behavioural change and motivation that is important to understand in order to design appropriate pedagogy. The psychologist Abraham Maslow focused on personality change in his work 'Motivation and Personality' saying, "When an effective stimulus, a traumatic experience let us say impinges on the personality... the experience, if it is effective, changes the whole personality. This personality, now different from what it was before, expresses itself differently and behaves differently from before" (Maslow 1987, p. 216). Here the linkage is made between personality change and behaviour. Likewise the neurologist Damasio (2004) also describes an experiential pathway that changes the way people see the world. His work is on emotional intelligence and he explores how a person's motivational worldview is rooted in experience and is thus consistent with the cognitive pathway of the 'exteroceptive' sense which can become part of our 'interoceptive' sense—an external experience that becomes an internal experience, if the experience is of such a kind as to be an 'emotionally competent stimuli', a phrase that echoes Maslow's 'effective stimulus'. Such an experience can be absorbed deep into the 'somatosensory' cortices so as to become part of our deep somatic and emotional motivation (Damasio 2004).

The experiences of the research subjects may also be described as possessing a characteristic of 'oneness' or 'at-one-ment' as William James has described in 'Varieties of Religious Experience' (James 1902) which is characteristic of the experiences described by the research subjects. Both Maslow and Damasio thus describe a mechanism whereby experience can change us, so that we see the world differently, and thus we act differently. In the research cohort the subjects shared their 'pivotal' stories, the manner in which they felt their world-view change. It is significant for this chapter that many of them described childhood experiences, which suggests childhood is a particularly fertile time for formative experiences.

Changing the Way We Do Things

Greta in her TED talk (Thunberg 2018) says we must change everything. It is difficult to imagine how our world conquering Western materialist values could now be changed, and that we revalue wealth, and learn to live by limiting our wants. Changing the way we do things and re-examining and re-imagining all aspects of our society and our behaviour is commonplace amongst environmentalists. Now the idea is being discussed openly and with some urgency, so how is this to be turned into a reality? Ulrich Beck in the 'Risk Society' says that when risks become apparent, and anxiety grows, even fear, we enter a new relationship with the world and a redefinition of nature is possible, a new scenario of change becomes possible, with new and alternative communities "whose world views, norms and certainties are grouped around the center of invisible threats" (Beck 1992, p. 95). With a growing middle class the world economy is set to treble in size by 2050. The International Energy Agency predicts that oil demand is still forecast to grow in developing countries, and there will be shortages unless more oil projects are allowed to go ahead (IEA 2018). The clarity which Greta approaches this problem, is something which she ascribes to being "on the autistic spectrum", and being able to see things in a more black and white manner, and is helpful in focussing on a solution.

Becoming a fiery spirit is to embrace dramatic change in our way of seeing the world, in our relationship of ourselves to the world, and in our motivation for action, as one of the research subjects said who was trying to work out how to look after her ADHD child, "I had to choose between a massive rearranging of my lifestyle and medicating of my child. I rearranged my life. But in the process of educating myself, I became aware of what we are doing to the next generation. *And once you know something, you can't unknow it. Once you see, you can't unsee*". And another research subject, whose partner died of cancer, said "It was living through this experience that cemented my understanding of faith, peace, justice, sustainability and health. And made me see how the very structures of our society are based on unsustainable premises.... *We are the ones that we've been waiting for. If we don't do it, it won't happen.*" The statements of the fiery spirits are radical and extreme, and represent a complete embrace of what has to be done, whatever the cost. One strong characteristic is that the personal experience ripples outwards overcoming boundaries to embrace the whole world. Freya Matthews in her work 'The Ecological Self' seeks to widen the concept of intrinsic value through the extension of Self to the whole cosmos. Matthews says:

> But the thesis that we, as human selves, stand in holistic relation - a relation of 'oneness' - with the cosmos itself, promises more than a list of ethical prescriptions. It promises a key to the perennial questions of who we are, why we are born, what is our reason for living, etc. In short it promises to throw light on the *meaning* of life. (Matthews 1991, p. 147)

Matthews goes on to talk about Deep Ecology and the connection with "a form of human 'self-realization' which springs out of 'ecological consciousness' and she refers to the work" of Arne Naess, and quotes in footnotes, "Identification as a source of deep ecological attitudes" (Matthews 1991, p. 142, footnote 24).

A wide range of experiences can trigger this transformation into a 'super-activist', a fiery-spirit, which constitutes a form of 'self-realization', from the 'moment' of at-one-ness. To recount examples from Miller (2015) a Muslim girl in her home garden in Delhi experienced "a powerful sense of the miraculous and overwhelming unity between my pulse-beat, the ants that were crawling in the grass and the wind in the trees," that gave her a sense of unity of all things, to the young man who was cycling arduously up a Pyreneean mountain, zig-zagging a way to the top of a high pass, only to become aware that an eagle was following his path directly above him tracking his every turn. He described it as an experience that changed his life, and that he was only intending to get to the cafe at the top to have a coffee. Then there was the little boy in Sri Lanka who watched the sun setting over the sea and saw the famous 'green flash' as the sun dips below the horizon and light comes through the ocean itself, and then there was the shock of a little girl who was coming home from school in Argentina and as she stepped out of a taxi saw the dead body of a bird covered in oil in the harbour, or then the young man who watched the film of the Global Sunrise of 2000 "the first dawn of the Millennium over Kiribati - an island which the narrator casually said was likely to be under water at the end of the century because of sea level rise. I had a huge sense of foreboding". Many however have 'slower' experiences, longer associations, working and playing with nature, looking after animals, tending gardens, student activities, studying, the influence of teachers, climbing a mountain, and so on. Experiences of being out in nature are frequent triggers but experiences of shock, anxiety, anger are also significant.

Educational Opportunities

To illustrates possible vehicles for change the lead author worked with both adults and children over many years in environmental education, particularly implementing the guidance of 'Sharing Nature with Children' (Cornell 1979) and its invitation to deepen our relationship with nature through play and has been able to explore opportunities that encountering nature can offer. Even before the phrase 'sustainable development' had been invented the concept was being explored at a farm in Kent called Bore Place (www.boreplace.org). Here the task embarked upon was called 'Commonwork' (CW) and the 'strap line' was 'waste is a misplaced resource'. A core tenet of CW was to invite the public to experience the farm environment and have 'hands-on' learning (Leach 2010). Part charity, part business it was what we would now call a social enterprise engaging at all levels with the wider society. It utilised all the assets on offer, about 24 identifiable habitats, grass leys, ancient woodlands, shelterbelts, copses, coppice, forestry, meandering streams, ponds, lakes, as well as clay, milk, cattle and cattle slurry for methane digesting, gardens, verges, field walks with stiles and bridges, farm buildings, meadows, crops, hedges, history and archaeology, team work, sheep, horses, all the wildlife we share these spaces with, the wider landscapes and the weather, and the skills of the people involved, for global citizenship and interdependency. Natural cycles underlay the life and work. Children and adults alike were immersed in the experience, and many of those exposed to this alternative world perspective were changed by their experiences. Living and working in a farm environment has a way of embracing the whole of life so as to become a visceral experience. Bore Place Farm continues today to welcome many children and adults and in recent years new activities have been added including healthy eating and growing food, crafts and arts, sustainable energy projects, and especially a new programme that grew out of that early exploration into introducing visitors to sustainable development and ecological awareness, called 'Leading for the Future' (LfF) partnered with WWF (Buckley Sander and Blair 2011), and based on the research of a Junior School head-teacher (Dixon 2009). This was designed to give a more structured immersion experience especially to senior teachers to enable them to reshape their school environments as a catalyst for wider social and environmental transformation.

On the Lincolnshire/Nottinghamshire border the lead author has also worked with another rural social enterprise called Hill Holt Wood (HHW) (www.hillholtwood. co.uk) which helps to imagine what a sustainable school might be like. HHW in its Study Programme offers a broad based largely vocational curriculum in a woodland environment for 11–19 year olds through active engagement with the ancient woodland management and in a social community context. The goal is 'enthusing' the young people to grow through their experiences, and to 'learn-by-doing'. Environmental study is part of all classes and many of the 'classrooms' are outside in the woodland with minimal shelter and surrounded by the woodland environment and equipment, and many pedagogical tools made from the woodland itself. The management goal of HHW is the 'three legged stool' (or the Triple Bottom Line) of the social enterprise—Social, Environmental, and Economic goals and activities, all of which have to be in balance. HHW is a self-sustaining rural social enterprise, and it has also been called a Community Co-operative (Frith et al. 2009), that runs a number of businesses in the woodland and further afield which are collectively described as 'benefit stacking' (Parrish 2007), and the education of young people is a key part of these activities. Not only does the social enterprise have a high success rate with the education of young people, many of the students subsequently seek further study and employment in environmentally related activities. HHW also has an award winning Mind project in a dedicated woodland, which also works on the ecological immersion principle. The ecological ethos and the activities at HHW ripple outward into the local communities who are invited to be part of the 'Community' and Trustees of the Enterprise, and it works in partnership with the local District Council, other partners such as the Woodland Trust, and wider afield "at both a regional and national level due to effective networking, partnership working as well as word of mouth" (O'Brien 2004, p. 5).

Both Bore Place and Hill Holt Wood are examples of the method of 'ecological immersion' in the natural environment, and of hands on 'learning by doing'. This is applicable to both children and adults as a pedagogical approach. It is the advantages of the ecological school that is the particular interest of this chapter, and the opportunity to accelerate action for climate change as a methodology of experiential learning through active immersion for sustainable development by both students and teachers. Action is at the heart of this method, and 'learning by doing' describes the practical application. Pedagogy today is still dominated in both school and college by abstract conceptual learning for environmental education, "technological solutions" and "content knowledge acquisition" (Glackin and King 2018, p. 1), and in the wider public discourse by advice on actions and encouragement to change attitudes. This makes understanding sustainable development harder to comprehend if it is not embraced as a way of life, and is not a heart-head-hand experiential process. In the academic literature on climate change the emphasis is on changing consumer behaviour using behavioural science principles, which is in itself moving on from "rigid legislation, incremental policy changes, data driven reports and fines for rule-breakers" (Gilchrist 2018). The literature nowadays abounds with recommendations on behavioural change such as a recent review headline in the New Scientist, "Saving Earth by changing us - Boldly altering human behaviour may be our best shot at saving earth." (Barnett 2019, p. 44) which refers to the list of actions given by Mike Berners-Lee in his book, 'There is No Planet B: A handbook for the make or break years'. The 'Nudge' theory of behavioural change by Thaler and Sunstein (2008) has proved controversial, some saying it is not an effective strategy to solve the big problems such as climate change (Godwin 2012). Nudge "alters people's behaviour in a predictable way" (Thaler and Sunstein 2008), and utilises emotional cues, but arguably does not alter behaviour more permanently without frequent reinforcing. This may be called externalised behavioural change which is not internalised as a personality trait and as would be the case with virtue (cf. Barry 1999; Dobson 1998; Charlesworth 2015). In exploring the challenge of positive actions for climate change the Futerra consultancy which describes itself as "a change agency", gave up the more rational approach of attitudinal change, saying "Threats of Climate Hell haven't seemed to hold us back from running headlong towards it" and the approach advocated is to see a positive vision that is desirable and exciting—"hope, a sense of progress and excitement about tomorrow" (Futerra Sustainability Communications 2012, p. 3), where 'hope' here is a 'positivity', overcoming the feelings of 'hopelessness' to engage in action. This is their idea of 'selling the sizzle' and goes some way beyond the rational self-interest theory to a more emotional emphasis, but is still not sufficient to generate the type of personality change that fundamentally sees the world differently.

So these methods still do not boldly alter human behaviour, but only scratch the surface. Our knowledge of how people change their behaviour and in what circumstances is still poorly understood. Religious traditions on the other hand do have the experience of conversion and of the change of heart, sometimes called metanoia (lit. repentance). George Marshall in his recent exploration of climate change and the difficulties of belief turned to American evangelical churches for their experiences and some understanding of 'climate conviction'. He says in conclusion that, "Learning from religions, I suggest we could find a different approach to climate change that recognizes the importance of *conviction*: the point at which the rational crosses into the emotional, the head into the heart, and we can say, 'I've heard enough, I've seen enough - now I am convinced.'" (Marshall 2014, p. 255). So theology of a range of religious traditions could help here more than is typically assumed by secular academics though clearly not perfect. We know that changing behaviour for sustainable development is one of the hardest things, and though incremental change is happening, it is too slow, and the forces of inertia and opposition, amongst the public and civil society, media, business and government, equivocate, as Greta pointed out at Davos when she said, "…you say nothing in life is black or white, but this is a lie, a very dangerous lie…". The contribution this chapter seeks to outline is how people come to change their behaviour through a 'Theory of Action' that emphasises a triangle of relationships between experience, belief and fire, where 'fire' is the force of animation. It is a process whereby experience becomes part of our somatic and emotional intelligence as belief, and so shapes the way we see the world.

Living Within the Ecological Possible

Moltmann wrote his prescient warning two years before 'Our Common Future', better known as the Brundtland Report, which laid out the scope and challenge of sustainable development, giving us the famous definition: "Sustainable development is development that meets the needs of the present without compromising the ability of future generations to meet their own needs" (Our Common Future 1987, p. 43). In emphasising needs the statement implies an equality of all people present and future, and thus also the intrinsic condition of limitation. This is made clearer when it goes on to say that, "Sustainable development requires the promotion of values that encourage consumption standards that are within the bounds of the ecological possible and to which all can reasonably aspire" (Our Common Future 1987, p. 44). It is the phrase "within the bounds of the ecological possible" that makes sense of the main definition, but unfortunately the two statements are rarely connected. It is acting "within the bounds of the ecological possible" that has so eluded us.

It was the Rio Earth Summit in 1992 that firmly put Sustainable Development on the map and flagged up the crucial significance of Climate Change, and the practical method it proposed was to be Local Agenda 21—local people acting locally but thinking globally for the twenty-first century. This did not catch on however and national governments did not make it a statutory responsibility of local government. The local–global connection was a missed opportunity. Furthermore market growth was left unrestrained and surged forwards until the Global Financial Crisis 2008, which slowed economic growth and raised questions about the viability of the whole project. At the same time there was a new awareness of the imminent dangers of climate change.

Even so, the now long story of sustainable development is more nuanced with many bold grassroots examples, people committing their lives in practical projects of learning by doing. Some of the more famous are the alternative communities of Findhorn (https://www.findhorn.org/) in Scotland and New Alchemy (https://newalchemists.net/) in Cape Cod, two of a range of experiments in living which are still active (Barnhart 1982), Findhorn now the largest single intentional community in the UK (Tinsley and George 2006), with Findhorn Ecovillage (www.ecovillagefindhorn.com) at its heart. They linked together voluntary simplicity, community living, food-sufficiency, alternative technology and ecological economics. Many of the alternative communities faded from view subsequently, and are not less important for that fact, but some have survived long enough to see the ideas becoming mainstream. Two others which have some characteristics of community have also persisted are Permaculture which was established in 1978 and CAT. "Permaculture is a word we have coined for an integrated, evolving system of perennial or self-perpetuating plant and animal species useful to man." (Mollison and Holmgren 1982, p. 1). Permaculture is essentially a design process which aims to link protecting wildlife, and integrating food production with environmentally friendly living. Permaculture is now a global movement (Veteto and Lockyer 2008). CAT, the Centre for Alternative Technology (https://www.cat.org.uk/), at Machynlleth in Wales, started in 1973, and is now both a visitor centre and a demonstration of sustainable development, and sustainable living with graduate level courses. These and many others have interpreted the phrase "within the bounds of the ecological possible" to mean having less 'stuff', sharing, reducing our human impacts on the wider environment and working with natural systems.

How can the fiery spirits help? Is it something that only happens to individuals in exceptional circumstances, or can it cascade through a population, a group, a community, a school, a college perhaps? With Greta Thunberg we are witnessing such a cascade of global dimensions. Clearly the impulse to take strong action and stand up for our beliefs can be triggered by a key experience, and can be far-reaching, rippling through our lives and changing us. George Marshall in his 2009 observations of environmental behavioural motivation amongst scientist and environmental activists concluded that the mere provision of information is not enough, there is a disconnection between individual actions and global problems. He concludes that belief and imagination are essential, and asks, "How then should we go about generating a shared belief in the reality of climate change?" He proposes better communication, but his key point he calls "the collective emotional imagination" (Marshall 2009). This could be described as turning emotion into belief. Experience then changes us as people, it is the root of conviction, the well-spring of action. There is a certain kind of pivotal experience that is more intense, that changes our personalities and develops the kind of belief which motivates action. Quite simply we need belief to act.

The Educational Opportunity: The Sustainable School

This 'Theory of Action', which brings together 'fire', experience and belief, potentially has wide educational applications for sustainable development, and education arguably has one of the best opportunities for accelerating climate change action. Schools are connected to families and families to their communities, and communities to wider society. There is the possibility of a cascade effect, or tipping point, when the insights and experience of a few committed people are adopted by others who are also likewise convinced. In the last few decades schools have become enthusiastic to promote environmental activities, but in the last decade the momentum for Education for Sustainable Development (ESD) has not been sustained. However there have been some success stories and one is Eco-Schools (www.eco-schools. org.uk) which has been operating for about 25 years in 67 countries. Eco-Schools are pupil led, largely curriculum based study enabling progression through a bronze and silver award scheme and seven steps to gain Green Flag success, with ten topics and many practical opportunities for creating school recycling, wildlife gardens and ponds, sensory and reflective spaces, reducing the school's carbon footprint and even installing alternative energy technologies. Forest Schools (www.forestschools. com), in operation for over 18 years, on the other hand is an outdoor approach to education and play and are very much hands on, catering particularly for young children. Forest Schools are learner led, nature-based learning, embracing an emotional and social approach. A UK Government led initiative to creating sustainable schools was the Sustainable Schools Alliance (http://sustainable-schools-alliance. org.uk/sustainable-schools-framework/) with a target to make all UK schools sustainable by 2020, however it has suffered from government funding cuts in education. Environmental education in England is now found to be "falling through the gaps" (Glackin and King 2018). In the other non-governmental sector UK environmental charities such as the County Wildlife Trusts with large memberships run Watch Groups for children with active outdoor programmes. The Lincolnshire Wildlife Trust Watch Group (https://www.wildlifewatch.org.uk/where-you-live/east-midlands/lincolnshire-wildlife-trust) for instance has helped secure sand dunes with Marram grass in 2017 on the vulnerable Lincolnshire coast. The UK bird charity the RSPB also has an active children's and teachers' programme of engagement with nature. There are other such nature based charitable initiatives and beyond these activity based schemes there are well devised classroom based learning programmes. One such on offer is from Oxfam, 'Climate Change for 7–11 years', and there is a wide range of literature to support learning for climate change. All these initiatives are helpfully pushing us in the right direction but do not have the ecologically immersive characteristics of truly sustainable schools. There is a growth of awareness in schools and in wider society, possibly even a surge, which is more propitious for cascading sustainable behaviours than ever before. The challenge is to take on these educational initiatives to a new intensity of ecological community living. Whilst there are learning schemes for climate change and sustainable development fitting them into the curriculum is a hurdle to get over without Government support, and going beyond that to reshaping the school environment and the school ethos will require passionate conviction.

One imaginative way forwards was developed at Bore Place, Commonwork (CW). In 2010 the experiences of welcoming people onto the farm to share in the work and activities over many years, and from the 2008 CW pilot, "Sustainable and Global Schools' Learning Network" (Buckley Sander 2013), were gathered up and an initiative was led by Jane Buckley Sander and then partnered by WWF and David Dixon, a junior school head-teacher, and based on his Ed.D. of 2009 (Dixon 2009). The first report was entitled 'Habitats for Humanity: Linking Sustainable and Global Learning for the Future' which developed "an education programme that sought to make links between the natural environment, and with the humanity that inhabits it, by offering a variety of experiential and hands-on activities" (Buckley Sander 2010, p. 3). This was piloted as the first Leading for the Future (LfF) courses with two cohorts of teachers. "The course aims to provide stimulus and inspiration for school teachers to develop both their personal and professional leadership capacities" in the peaceful, contemplative surroundings of Bore Place, away from the pressures and demands of school life. The course guides the participants on a journey that engages them intellectually, physically and emotionally, and to examine the need for a new educational paradigm in face of likely global futures (Buckley Sander 2010, p. 13). The course provides for both intellectual learning and reflection, aware there "is the ever present danger that exists in some ESD and DE work of leading students into particular actions without the necessary thinking questioning and learning. CW has sought to encourage students to have a sense of empowerment and agency rather than being overwhelmed by gloomy prognoses of the future and thus has stressed action at times" (Buckley Sander 2010, p. 6). LfF proposes a significant and radical change in schools, a deep change in ethos. The first two cohorts of teachers demonstrated the importance of taking heads and senior staff out of school for these residential courses at Bore Place farm as they would more likely be in a position to make changes in their schools.

The whole LfF process was written up in a more detailed form in 2011:

> The purpose was to explore how a combination of experiential learning, hosted space, deep reflection on values and interconnection within self, between each other and with the natural world could inspire and create leaders for sustainability. The hope is that such engagement and exploration will lead to transformational change within leaders and within the wider education systems. (Buckley Sander and Blair 2011, p. 7)

Hosted space is described as something that is created mindfully for reflection and exploration on one's own journey to sustainability. It was felt that the programmes were successful, that:

> Evidence showed that all experienced the programme as transformative and empowering. They expressed a rediscovery of their sense of self, their true values and potential agency, as well as a wish to increase reflection time for themselves, both personally and in professional practice, in order to step back regularly from the pressures of measured performance. (Buckley Sander and Blair 2011, p. 5)

The teachers themselves have to believe in the importance of SD and have that determination to take action to enable their students, and it follows that the school environment must also be supportive of ESD, and the pedagogical environment, principally through addressing the needs of the heart, the head and the hands. This may be a challenge for many schools, particularly in urban environments. Hill Holt Wood, like Commonwork, demonstrate the holistic nature of the ecological learning environment where the place and the process are interconnected, and they reflect and reinforce each other, which means that sustainable development is thereby explicitly accessible. Both leaders and students will accept ecological living as normative and are motivated to engage with the challenges of our time. Those schools that cannot transform their environments may be able to twin with others than can, or possibly to have a 'country lung'. There are numbers of community owned woodlands which could develop HHW style education programmes. The follow-up to the LfF programme is tantalizing with very positive teacher feedback. Suggestions for development include, "Arrange a whole school INSET with Commonwork at Bore Place", to "Action plan for whole school sustainability involving the wider community". It is as if the stone has been thrown into the pool and the ripples spread wider and wider.

Conclusion

The fiery spirits research underlines the great opportunity that exists to bring young people regularly into contact with the natural environment and to normalise an ecological world view. It is particularly disappointing that the LfF initiative has not moved forwards due to cuts in funding for ESD and for the "patchy and restricted" access to environmental education in England (Glackin and King 2018). Nevertheless LfF is ready to be activated should the situation change. We are in danger of missing a generation of activists which is a tragedy at a time of great challenge with climate change. However the encouraging signs of a new awareness by young people through the intervention of Greta Thunberg may well change the situation as young people themselves demand action. HHW on the other hand is very much up and running and we need to look further to its learning method, its replicability and the lessons to be learnt.

We are aware that the suggestions made here cannot be seen as definitive; however, it is clear that there is much more that schools education can do using lessons learned from past experience to address climate change and sustainable development more broadly. That this could constructively channel the enthusiasm of the movement inspired by Greta Thunberg would be a distinct benefit as long as it did not distract from the pressure for change that the movement is bringing about.

References

Barnett A (2019) There is no planet B review: how to save earth by changing humans. New Scientist, vol 3217, p 44. Retrieved from https://www.newscientist.com/article/mg24132170-400-there-is-no-planet-b-review-how-to-save-earth-by-changing-humans/

Barnhart E (1982) Food forest: an agricultural strategy for the North East, USA. In: Hill S, Ott P (eds) Basic technics in ecological farming. Birkhäuser, Basel. Retrieved from https://link.springer.com/chapter/10.1007/978-3-0348-6310-0_13

Barry J (1999) Rethinking green politics: nature, virtue and progress. Sage, London

Beck U (1992) Risk society: towards a new modernity. Sage, London

Buckley Sander J (2010) Habitats and humanity: linking sustainable and global learning for the future. Commonwork

Buckley Sander J (2013) The harvest: sustainable and global schools learning network 2008–2011. Commonwork

Buckley Sander J, Blair F (2011) Leading for the future: a collaborative project to support, inspire and engage school leaders for positive change in education. Commonwork/WWF. Retrieved from http://assets.wwf.org.uk/downloads/leadingforthefuture_finalreport.pdf

Charlesworth M (2015) Transdisciplinary solutions for sustainable development: from planetary management to stewardship. Routledge, Abingdon

Cornell J (1979) Sharing nature with children. Exley Publications Ltd, Watford

Damasio A (2004) Looking for Spinoza. Vintage, London

Dixon D (2009) Developing a green leaders model for primary schools. Ed D in education leadership, University of Lincoln

Dobson A (1998) Justice and the environment: conceptions of environmental sustainability and theories of distributive justice. Oxford University Press, Oxford

Eriksson N, Ujvari S (2015) Fiery Spirits in the context of institutional entrepreneurship in Swedish healthcare. J Health Organ Manag 29(4):515–531

ESRL (2019) Trends in atmospheric carbon dioxide. Retrieved from https://www.esrl.noaa.gov/gmd/ccgg/trends/

Frith K, McElwee G, Somerville P (2009) Building a 'community cooperative' at Hill Holt Wood. Retrieved from http://irep.ntu.ac.uk/id/eprint/8661/1/200411_6812%20McElwee%20Publisher.pdf

Futerra Sustainability Communications (2012) Sizzle, the new climate message. Retrieved from https://www.wearefuterra.com/wp-content/uploads/2018/03/Sellthesizzle.pdf

Gilchrist E (2018) Climate change needs behavior change. Yale program for climate change communication. Retrieved from http://climatecommunication.yale.edu/news-events/climate-change-needs-behavior-change/

Glackin M, King H (2018) Understanding environmental education in secondary schools in England. Kings College London environmental education research group. Retrieved from https://www.kcl.ac.uk/news/understanding-environmental-education-in-secondary-schools-in-england-1

Godwin T (2012) Why we should reject nudge. Politics 32(2):85–92

IEA (2018) World energy outlook 2018. Retrieved from https://www.iea.org/weo2018/

IPCC (2018) Summary for policymakers of IPCC special report on global warming of 1.5°C approved by governments. Retrieved from https://www.ipcc.ch/2018/10/08/summary-for-policymakers-of-ipcc-special-report-on-global-warming-of-1-5c-approved-by-governments/

James W (1902) Varieties of religious experience. Longmans, Green, and Co., New York

Leach J (2010) The commonwork vision, mission and principles. Commonwork, Edenbridge [I estimated the last bit of the reference—do tweak if you have the original documents]

Marshall G (2009) Are you a believer? Opinion Article, New Scientist, vol 2718

Marshall G (2014) Don't even think about it. Bloomsbury, New York, London

Maslow AH (1987) Motivation and personality, 3rd edn. Longman, New York

Matthews F (1991) The ecological self. Routledge

Miller T (2015) From belief to action: cross cultural perspectives for sustainable development. PhD in business management, University of Lincoln

Mollison B, Holmgren D (1982) Permaculture one: a perennial agriculture for human settlements, 2nd edn. Tagari, Tyalgum, NSW

Moltmann J (1985) God in creation: the Gifford lectures 1984–1985. SCM Press, London

NASA (2019) Global mean CO_2 mixing observations. Retrieved from https://data.giss.nasa.gov/modelforce/ghgases/Fig1A.ext.txt

O'Brien L (2004) Hill Holt Wood social enterprise and community woodland. Social Research Group, Forest Research. Retrieved from https://www.forestresearch.gov.uk/documents/1433/eliv_hhw_report.pdf

Our Common Future (1987) World commission on environment and development. Oxford University Press, Oxford

Parrish B (2007) Sustainability entrepreneurship: design principles, processes and parables. PhD Thesis, University of Leeds

Thaler R, Sunstein C (2008) Nudge, improving decisions about wealth, health and happiness. Penguin Books, London

Thunberg G (2018) Greta Thunberg: school strike for climate—save the world by changing the rules [Video File], Dec 2018. Retrieved from https://www.ted.com/talks/greta_thunberg_school_strike_for_climate_save_the_world_by_changing_the_rules/transcript?language=en

Tinsley S, George H (2006) Ecological footprint of the Findhorn foundation and community. Sustainable development research centre. Retrieved from https://www.ecovillagefindhorn.org/docs/FF%20Footprint.pdf

Veteto JR, Lockyer J (2008) Environmental anthropology engaging permaculture: moving theory and practice toward sustainability. Retrieved from https://onlinelibrary.wiley.com/doi/full/10.1111/j.1556-486X.2008.00007.x

Chapter 3
Integrating Climate Change Competencies into Mechanical Engineering Education

Sven Linow

Abstract Engineers will be needed to enable technical solutions that minimise the impact of climate change to the earth system and to humanity. Thus a basic understanding of mechanisms driving climate change to the earth system as well as means and methods to generate and evaluate technical solutions for change should be part of engineering education. Mechanical engineering as a discipline has a well-developed understanding of skills to be acquired during a bachelor degree programme—especially if graduates are intended to directly start working as an engineer. Mechanical engineering departments tend to be conservative and are hesitant to change much of their curriculum. Today's engineering curricula are already quite crammed to meet all the requirements for educating good engineers and meeting requirements set by professional bodies. Typical approaches for overcoming this are adding some mandatory courses from the social sciences with the intent to broaden the view of future engineers (e.g. the Darmstädter approach) or to invent new interdisciplinary degree programmes. Neither approach will reach to the core of engineering education. This paper instead focuses on introducing climate change skills as hard engineering tasks into the technical degree programme: today thermodynamics and fluid dynamics courses teach most of the basic competencies that would be needed, but classically without any climate change context. This paper aims at discussing the possibility to include basic understanding, relevant mitigation approaches and evaluation tools into thermodynamics without overburdening the course or endangering learning of the basic skills. Experience from an ongoing first run will be shared.

Keywords Engineering thermodynamics · Competence oriented teaching · Climate change teaching · Curriculum development

---

S. Linow (✉)
Hochschule Darmstadt, Fachbereich Maschinenbau und Kunststofftechnik, Schöfferstraße 3, 64295 Darmstadt, Germany
e-mail: sven.linow@h-da.de

© Springer Nature Switzerland AG 2019
W. Leal Filho and S. L. Hemstock (eds.), *Climate Change and the Role of Education*, Climate Change Management, https://doi.org/10.1007/978-3-030-32898-6_3

Introduction

Addressing climate change is a truly interdisciplinary undertaking, where meaningful approaches to solving or at least governing are defined by the problems addressed rather than by disciplines employed (Seager et al. 2012). Addressing climate change is part of any approach to a meaningful sustainable development, as is outlined in the UN SDGs (2015). Thus climate change in education is usually well covered in interdisciplinary courses, where transcending the limits of a peculiar discipline is part of the intended learning outcome (McCright et al. 2013 as an example), and vice versa that skills and competences deemed necessary to address climate change are placed in an interdisciplinary context (Wiek et al. 2011). Teaching such interdisciplinary courses is fun and is educating the teachers as well. More so, if interdisciplinary co-teaching is used where teachers from different departments and perspectives attend the same course at the same time and expand the discussion through their disciplinary view and through overcoming their own disciplinary limitations.

Students with a broad interest or focus on sustainable development will have no trouble in understanding the interdisciplinary approach and its necessity; they will make the connections by themselves. Such students feel empowered to look and reach beyond their usual degree programme or tend to be enrolled in interdisciplinary programmes. But there are students that do not have such a broad interest or who did choose a defined discipline and programme with a specific curriculum on purpose. An example is mechanical engineering or any other specific engineering degree programme, where the majority of students has a strong interest in technology and aims to become a specialist in a defined technical field inside the broader frame of engineering. Many universities add some mandatory interdisciplinary courses to engineering degrees, to overcome such focus, to broaden the scope, to teach non-technical skills, and to place the specific technical content in a broader interdisciplinary frame. Hochschule Darmstadt developed such additions during its transformation from an engineering school into an university of applied sciences in 1969 and is proud of its mandatory Soziales und Kulturelles Begleitstudium (accompanying studies in social and cultural aspects, SuK). Today engineering students can choose from a broad offer of courses to get the mandated 10 credit points (ECTS). An example where skills as discussed here are learned is given by Linow et al. (2017).

From the perspective of a dedicated and focused student of engineering all content that is not part of the core curriculum will be seen as superfluous. For such a student this is not a part of what makes a good engineer and usually does not readily interest him. Mandatory interdisciplinary courses outside engineering are seen as a burden and are often chosen on a base of minimising effort. Such courses do not reach their audience and do not give a lasting learning experience, especially if such courses are staffed by teacher that, as seen from an engineering perspective, lack the ability to make themselves understood: Overtly disciplinary language, use of vague terms and examples without connection into technology alienate students. Thus it is often found, that graduates have not acquired relevant and lasting competences in such courses. This is not to argue that an addition of interdisciplinary content to the curriculum is superfluous, on the contrary, but great care in addressing students from a different discipline is needed and there are other aspects better suited for such 'ad on' courses. To reach all students in an engineering program it is necessary to make competences and knowledge for sustainable development a part of what defines a good engineer and thus include it in the engineering curriculum itself. This chapter starts from this position and tries to answer to the questions: What kind of skills are relevant for engineers to respond to climate change? Where does one place this content inside the standard engineering education? How does one start today in a department, were sustainable development is yet not part of the consensus?

Is Climate Change an Issue for Mechanical Engineers?

The typical content of a bachelor programme in mechanical engineering is summarised in Table 3.1. The table provides generic intended learning outcomes for the different matters, i.e. the skill level or competency level using the classification of Bloom according to Biggs and Tang (2011). Differences between universities exist, they affect how deep the specific issues are treated or to what extent theory and application is part of a course. This content of engineering education as well as the distinct image of what makes a good engineer is developing since the first steam engines: From the beginning it were technical issues like efficiency of an engine, creating the necessary materials, calculating performance and strength, understanding and manufacturing standardised parts that drove the development of engineering as a science. Engineers through their work expand the technical space of the possible, but are always limited by technical and economic constraints.

Climate change belongs to the most challenging problems of our time, others are the energy transition, availability of resources and loss of biodiversity. These are wicked problems as defined by Rittel and Webber (1973). The concept has been further developed and clarified by Levin et al. (2012). These problems will need contributions from technology for a meaningful development of possible solutions and thus engineers able and willing to provide them. On the other hand any approach solely based on scientific expertise—especially if we consider the small scale and linear approach of most disciplinary problem solving—are 'doomed to failure' as is argued by Rittel and Webber as well as Levin et al. Based on this Seager et al. (2012) uses the concept of 'governing a wicked condition' instead of solving. Thus any approach intended to govern a wicked condition will need a tight and interdisciplinary contact between innovation in society, in politics and in technological development involving engineers. This document does not argue that such necessary technical contributions will be examples of grandiose progress, quite on the contrary, many technical contributions will need to be humble or frugal instead.

Another argument starts from the history of engineering and the ultimate machine of the fossil age, the coal fired steam engine: Newcomen's engine was the first for *raising water (with a power) made by fire* (which is the title of a well-known engraving of the machine), i.e. being able to convert the released chemical energy

Table 3.1 Usual content of a bachelor programme in mechanical engineering and possible deeper integration of climate change competences

Course cluster	Intended learning outcome	Possible inclusion of climate change
Engineering mathematic	Applies necessary mathematical tools	–
Basic physics	Applies basic physical concepts to engineering problems Chooses applicable measurement method	Basic concepts on an introductory level
Electro-technical basics	Choosing matching electric machines and supply for a specific task	–
Technical mechanics	Analyses all kinds of mechanical problems connected to machines: static and dynamic	–
Elements of machines	Chooses fitting parts and components for all kinds of machines	Durability, repairability
Engineering thermodynamics	Apply the thermodynamic method for engineering tasks	(Fossil) fuels and carbon dioxide Energy and mass flows in earth system Radiative heat transfer and greenhouse gases Nonlinear feedback
Fluid dynamics	Design simple fluid systems	Earth systems fluid dynamics, heat and mass transfer
Engineering materials	Choose materials and materials processing methods	Basic processes, (carbon) footprint
Design methodology	Designs complex machines using parts or own new design	Include regulatory and stakeholders demands, make resilient
Industrial process and control	Design simple and safe control systems for machines	Feedbacks in the earth system
Manufacturing technology	Define efficient manufacturing processes	Basics in industrial ecology, ecological footprints
Interdisciplinary studies	Social and cultural aspects, language and other cultures, ...	Connection between technical and social aspects of climate change

(heat) of burned fossil fuels into usable work. Today's main use of the coal fired steam engine is generating electricity: It is now a steam turbine, but steam is still generated by burning coal or other fossil fuels. And this is still by far society's major source of electricity, as Smil (2017) shows. It is this technology that is central in creating climate change in the first place, so it is indeed a core engineering issue, to understand and to provide solutions for climate change: Climate change is an unintended consequence of engineering itself. As Allenby (2006) discusses, it does not make sense, to seek individual responsibility for the (early) decisions taken to enable this system. These decisions are part of a macro-ethics on a large scale and over a long time of the self-organising technological system.

Different orientations toward sustainable development are found in engineering departments. Seager et al. (2012) identify three basic orientations:

- *business-as-usual*, where simplification, reduction and very high technological optimism result in neglect or denial of other perspectives; often found in combination with a disinterest in climate change or seeing the discussion and consideration of climate change as hindering monetary growth and technological progress and thus to be ignored or to be wronged.
- *systems engineering*, where risk minimisation or cost-benefit are seen as major driver for decision processes. Triple bottom line approaches (profit, people, planet, usually applied in this order) include society and climate change aspects in technical considerations: this orientation accepts the great challenges to engineering and tries to create optimised solutions in a specific frame. It is not the approach or the frame itself that is seen as problematic, the approach is just not fitting for wicked problems, as it still focuses on single engineering tasks. It acts as a small part in a large and self-organising environment. Individual decisions are taken with consideration of a professional conduct and engineering ethics, but the large scale development is not included in considerations. It may be questioned, as Seager et al. argue, if such an approach is making a measurable difference when applied to (super) wicked problems.
- *sustainable engineering science*, where the great challenges are seen as wicked problems and where technologies role is seen with scepticism. Seager et al. (2012) further defines the concept as the necessary approach for engineers to contribute in a systemic way to govern and handle climate change.

Such wide spread of conflicting orientations in an organisation will lead to a blockade: it is usually near impossible to create consensus to include wicked problems, climate change or sometimes even engineering ethics as a major aspect into the programme. If a university would be externally forced, to create mechanical engineering as *sustainable engineering science* and could get personnel with matching orientation, it would be possible. But this is usually not the case and a stalemate exists. Accepting this often leads to acquiescence to extreme positions, thus there must be a way for individuals to start including wicked problems teaching not top down, but in their own courses from the bottom up—with the hope of eventually changing the organisation.

This classification of orientations follows engineering education in a way: *Business-as-usual* without any ideology and on a small scale is what an engineer does, what he is usually trained for and where teaching starts: simplify the problem, exclude any complex considerations and solve it by optimising some given constraints, like energy efficiency. In this small scale, purely technical and otherwise detached version it is indeed good engineering; the transgression happens, when problems are addressed through this frame that have environmental impacts outside the engineering domain. *Systems engineering* seen from this perspective is the inclusion of extra constraints to a possible solution for a problem. Engineers either are informed about them, or need to seek these constraints themselves: This systems approach needs additional competences and soft skills. It is the *sustainable engineering science* orientation that leads to a different understanding of the role of engineers. The engineering tools and the technical competences learned and used are the same, but they are used with a different orientation. From this orientation engineers ask about the long term or large scale environmental impacts, they consider pro-actively complex feedbacks, and include applicable aspects of macro-ethics. Macro-ethics is defined by Seager et al. (2012) as ethical consideration at the scale of the collective; this includes, but goes beyond the VDI (2002) engineering ethics requirement, demanding from engineers to actively avoid technology lock-in or technological constraints that will hamper or disable future actions. An open question beyond this paper is, if society understands, wants, and will readily employ graduates that were educated in *sustainable engineering science*, even though this is what society urgently needs.

Experience in interdisciplinary courses and engineering courses shows that students will stick to simple *business-as-usual* approaches as long as possible. Even interdisciplinary educated students from programmes with a sustainable development focus struggle to include a meaningful *systems engineering* viewpoint in open-ended problem-oriented assignments, an approach to the problem that names or considers large scales or complex feedbacks is very uncommon. Thus *systems engineering* and even more so *sustainable engineering science* are orientations and involve competences that will need deep considerations in teaching/learning. Both can't be left to a single add-on course but must be part of the core of engineering teaching if one wants to reach most students.

What Skills/Competences Are Needed for Engineers?

From an engineering perspective that wants to address climate change it helps to grasp of the workings of the system earth. Kleidon (2017) provides a comprehensive introduction well beyond of what can be learned in an interdisciplinary course. Even though the basics of climate change seem to be all over the news or secondary education, many students starting a bachelor degree seem to have profound misconceptions, as the author's own experience or Jarrett et al. (2012) and literature cited therein illustrate. The concepts summarised in Table 3.2 and at the taxonomy level

Table 3.2 Relevant technical competences to consider climate change as an engineer

Concept	Content	To understand	To act	Fitting into
Energy, heat and work	Energy conservation, flow of energy through the system earth, equilibrium conditions	Apply and calculate amounts and flows for a system	Apply and calculate amounts and flows for a system	Thermodynamics—basics
Radiative heat transfer	Radiation and spectra, interaction of radiation with matter	Apply and calculate heat transfer rates, heat transfer in semi-transparent media gases	Apply and calculate heat transfer rates	Thermodynamics—basics
Greenhouse gases and radiation	Radiative properties Short wave transparency versus long wave absorption Effect on energy flow through atmosphere	Apply radiative forcings of different greenhouse gases	Understand and interpret	Engineering thermodynamics—usually not included
Entropy and radiation	Entropy budget of earth, availability of work in the earth system	Apply and calculate rates	Understand and interpret	Engineering thermodynamics—usually not included
Weather and climate	Convective heat transfer, dry adiabatic processes (wind) and wet adiabatic processes (clouds, heat transfer)	Apply and calculate simple cases	Understand and interpret	Thermodynamics and fluid dynamics—usually not included
Feedback in complex systems	Non-linear interaction between parameters, basic feedbacks in earths system	Make simple models	Use the concept: expect non-linear feedback	Thermodynamics and fluid dynamics—usually not included
Carbon cycle	Mass conservation, geological carbon cycle, lifetime of carbon in the system earth	Make simple models	Understand and interpret	–
System earth: flows, feedbacks, climate system	Climate system, Milankowich cycles	Make simple models	Understand and interpret	–

given are required to generate climate-change-literacy. It is distinguished between an intimate learning of the climate system itself (understand) and a level seen necessary for action as an engineer. The concepts are based on the concept inventory developed by Jarrett et al. (2012) and focus on the natural earth system. Therefore they are not strictly disciplinary from an engineering perspective. Some of the concepts are included in today's engineering thermodynamics, some of the other would fit readily into the curriculum. Acquiring these concepts will enable students to understand climate change, but not to act as engineers, as this needs additional skills.

One of the most relevant competences defining what makes a good engineer is getting to sound decisions as part of the design process (Pahl et al. 2007). This competence needs to be trained. This *design competence* is the ability to first analyse the problem and define criteria on which a decision will be based, then to create some meaningful and substantially different technical solutions based on the full possibility of the technical solution space, and only afterwards start to decide based on the criteria defined and not based on some preliminary thoughts or long held beliefs. A good engineer delays decisions to the point where he is actually able to decide based on meaningful criteria. Experience shows that projects that base on early decisions, decisions taken before a relevant analysis has been undertaken, will overrun cost and underperform on a regular basis. There exists a broad literature on bad decisions usually taken when political and monetary interests trump good engineering practice, for example Flyvbjerg (2009). Thus this skill, to delay (important) decisions is a general, a basic and an ethical competence. In teaching this skill with regard to climate change it becomes necessary to include the differentiation between simple task and complex engineering problem which are both part of the disciplinary realm and wicked problems on the other hand, where any good approach is interdisciplinary and only becomes solvable through a team effort.

A common understanding for competences needed to enable students to participate meaningful in sustainable development is summarised by Wiek et al. (2011). Sustainable development as defined in the UN SDGs is understood as the way forward for humanity as being agreed between all members of the UN. These competences are here interpreted with the focus on enabling engineers to participate and contribute to addressing climate change. Aspects from the orientations as defined by Seager et al. (2012) in view of the definition of sustainable engineering are included. Engineering ethics is well defined by many professional organisations, for example in VDI (2002), but does usually not touch all aspects of *macro-ethics*. In Mezirov (1997), the process and the ultimate Intended Learning Outcome for transformative learning is discussed in a general frame. The relevant competences are:

- *systems thinking competence*: seeing any problem in a global perspective, understanding that any problem may be wicked and that a simplistic search for one simple, single, "true" answer is futile or nonsense. This is connected to the *design competence*, but understanding the systemic aspects, the scale and possible feedbacks of a wicked problem is the core.
- *anticipatory competence*: to be able to look into the future in a meaningful way, to avoid wishful thinking or technological over-optimism, not to be lead astray

by computational bias as defined by Bridle (2018), to consider the full scale of developments, and to include feedbacks into other systems or developments. This is basic for any *postponing competence*, i.e. being able to postpone decisions to the point where sound deciding actually becomes possible and meaningful beyond wishful thinking or personal enrichment. For wicked problems, postponing is understood as overcoming the first layers of full ignorance and trying to grasp the full picture.

– *normative competence*: this includes understanding and enabling compliance but expands beyond and into being able to have a clear understanding of one's own position and decision space, to base decisions on engineering ethics and include long term repercussions of possible concepts onto one owns decisions through a *macro-ethics* perspective, where scale and feedbacks matter.

– *strategic competence*: the ability of creating transformations. This is typically well beyond the task of usual engineers at the start of their careers, but often the aim of a career. The focus is here on meaningful transformation in the direction of sustainable development. It may be argued that this kind of transformation is outside the typical motivation of engineers, but this is the competence identified as being at the core of sustainable development, as Levin et al. (2012) argue.

– *interpersonal competence*: bridging intercultural distances, the basics of good leadership, the ability to forge consensus through seeking and understanding different positions. With respect to wicked problems where any approach for a transformation is interdisciplinary, this includes respect for other disciples, the willingness to learn, to understand and to include.

Good engineers need a comprehensive and substantial set of technical competences within their field to act. These are at the core of any classical engineering education. As is argued these competences do not enable engineers to fully understand or address climate change or other wicked problems of sustainable development. Thus part of any engineering education oriented towards a sustainable future is the need, to integrate both approaches and to interconnect technical considerations with these generic competences. As engineering competences and this set intended for *sustainability science* are both complex and will need time to develop, these competences will need to find their way into the full programme.

Where Does Climate Change Fit into the Curriculum?

The curriculum of any mechanical engineering bachelor programme is crammed, the basic issues summarised in Table 3.1 need time and effort. In addition the increasing number of students attending university demands that more care and effort is needed for the average student to understand and learn the core skills. To use the picture of Biggs and Tang (2011) the number of Susan's (students that do well on their own independent of teaching quality) is limited and the number of Roberts (students that need support) increases. This situation places a higher responsibility on universities to provide an opportunity for quality learning for all students. One of the approaches to increase the learning experience is didactic reduction: not to teach each possible screw and bolt somehow connected to the course, but to carefully choose what is relevant as theory, what is necessary for getting the broad picture, what is needed to acquire the relevant competences, and what can be left out. As Table 3.1 illustrates, getting through all relevant basic skills creates a tight schedule. In addition engineers need to learn the use of many digital tools—being able to use digital tools is part of what defines an engineer today (Landfester et al. 2019). Thus the curriculum leaves no space for additional courses. There is a general understanding in industry, as argued by for example by VDMA in Germany, that there is no need for specialised new engineering programs. On the contrary, there is a clear need for the classical broad, deep and balanced engineering education. All this leaves no simple opportunity to include new interdisciplinary courses. As has been argued earlier, the add-on interdisciplinary courses are often not well fitting into the programme, amount to between 5 and 10% of the full programme and will only reach the predisposed and interested.

One could give a full course on climate change without reaching students, i.e. without the course affecting views, frames or opinions of the students. Biggs and Tang (2011) differentiate between surface and deep approaches to learning and teaching. If we take this seriously and see surface learning as waste of precious resources, the obvious question is how to present the wicked problems of climate change in a way that students will directly understand them as wicked, see the technical and non-technical aspects and are engaged by the problem in a way that enables deep learning. In case of climate change or other aspects of sustainable development the intended learning outcome is not only understanding some technical matter, but an acceptance of complex repercussions throughout society, necessary changes to problem solving and to society itself. This is transformative learning; deep learning is transformative for the student and indeed for the teacher as well: students and teacher engage in a process, where not only technical skills, but their use and interpretation in the context of the society is discussed. It is the job of the teacher to prepare students for the full picture of their future responsibility. It is seen as helpful for teaching engineering to use interesting, relevant, actual problems that reach out beyond engineering and into the lives of the future engineers, instead of teaching only by using simple theoretical tasks. By doing so other competences, like *interpersonal competence* can be developed as well.

In mechanical engineering there are some possible approaches to revise an existing programme and to include the identified competences:

(a) Create a bachelor in sustainable engineering science. Even though this is what is needed in the long run, this approach does not seem viable: Existing departments are blocked by competing orientations, industry is not interested in new degrees, most engineering students wouldn't be reached at all as they are still in the old-fashioned *business-as usual* programmes, it is not obvious if graduates from such a programme will experience a reasonable job market.

(b) Use and specify the mandatory ad-on interdisciplinary courses for the purpose. The impact on all students could be well increased through prescribing the intended learning outcomes of interdisciplinary courses and co-teaching them where one of the teachers is an engineer and both teachers have sufficient *interpersonal competence*.

(c) Introducing identified competences into specific courses of an existing bachelor. This can be done either gradually through the re-evaluation process as a joint undertaking in programme development by the department and its stakeholders or it can be done on an individual basis in the short-term. Possibilities reach from individuals just adding content (see below for a meaningful way to do so), the addition of a single course (on climate change and technology; either mandatory or as an optional choice), implementing relevant aspects of *systems thinking*, *anticipatory competence* and others in the curriculum.

(d) Create a master in sustainable engineering science.

Each approach has its own merits. Most will need long term consensus-seeking in organisations and thus are slow or impossible to implement. What can be done directly is to include relevant aspects in one's own courses. The major obstacle is having a crammed curriculum and no sufficient time for new aspects, but this could be overcome.

Teaching in universities of applied sciences relies heavily on the use of good examples. In depth lecturing of deep theory is reduced to the necessary; this creates the time that enables unique teaching/learning experience connected with the application of theory to examples. Different courses and different teachers make use of different approaches to the creation of examples: Anything between very generic problems (where the result is yet another equation) and real world problems using data from real machines is observed. Climate change as one of the great engineering challenges could become a relevant part of examples used for teaching and thus being placed in existing courses. It makes sense to organise and present the material as developing from the technical core content of the course: this to overcome the scepticism to any interdisciplinary or non-technical material that is quite common in engineering students. Up front, the example and material used needs to be part of technology: good examples will lead by themselves to the wickedness of the problem and interdisciplinary aspects can be discussed later. Using good examples is still only a first step, the generic competencies summarised above are not part of any intended learning outcome of existing courses. Students will not acquire them, they will only hear about them, and may consider them for their own. But doing so as a teacher and being visible about this can create more activity in the organisation and may lead to a revision of the programme.

Two examples are provided in the following that have been developed as teaching/learning examples in thermodynamics. Both make use of real numbers where possible; all assumptions and simplifications are obvious. The results are thus not exact (in the sense of right to the 5th digit), but meaningful. In using such examples another basic engineering competence is included, this is the ability to make a quick but meaningful assessment of a problem in short time. Some of the assumptions need to be explained by the teacher or be motivated by reference into literature (example is the 500 m thickness of the upper layer of the oceans used in Example 1). The thermodynamics behind these examples may seem simple, but as seen from a student's perspective is quite demanding. The aim of these examples is first training the application of the great thermodynamics toolbox and second to create unease with the results. Both examples use very large numbers; grasping the meaning of such large numbers and being able to set them in perspective is another competence included.

Example 1—Where Does the Heat Come From in Global Warming?
This task is focusing on early misconceptions about climate change, i.e. the question what heats up the system earth. Taxonomy level is application of basic concepts of thermodynamics such as; heat, lower heating value of fuels, ideal gas. The example is condensing many of the basic concepts learned in any engineering thermodynamics course into the steps needed for the solution, thus it does provide a clear opportunity for learning the usual skills. In addition it allows a discussion of some climate change basics and thus includes climate change literacy as a part of engineering thermodynamics.

Heating up the atmosphere? In this assignment you will calculate the heat needed to heat up earth's atmosphere and upper layer of the oceans by 0.5 K (this is the difference between the targets for global warming of 1.5 and 2.0 °C as is actually discussed).

Q: What is the mass of earth's atmosphere?
A: In a first step one calculates the surface area of the earth based on the average diameter of the planet $d = 12{,}730$ km, which is

$$O_{earth} = \pi \cdot d^2 = 509.1 \cdot 10^{12} \, \text{m}^2.$$

Using the weight of the atmosphere at ground level and the resulting force $F_{Atm} = m_{Atm} \cdot g$ causing a pressure p, one gets approximately

$$m_{Atm} = \frac{F_{Atm}}{g} = \frac{p \cdot O_{earth}}{g} = 5257 \cdot 10^{18} \, \text{kg}.$$

Q: How much heat is needed to rise the temperature of the atmosphere by 0.5 K?
A: Assuming isobaric conditions, one gets

$$Q_{Atm} = m_{Atm} \cdot c_{p,air} \cdot \Delta T = 2640 \cdot 10^{21} \, \text{J}.$$

Q: How much heat is needed to raise the temperature of the uppermost 500 m of the oceans by 0.5 K (this simplifies the problem, it becomes necessary to explain some basics about how oceans interact with atmosphere, but the value corresponds well with in-depth analysis)?

A: 70% of earth's surface is covered by oceans; water is well described as an ideal fluid, i.e. pressure is not affecting density, thus one estimates

$$m_{uo} = \Delta z \cdot O_{earth} \cdot 0.7 \cdot \rho_{water} = 178.1 \cdot 10^{18} \, \text{kg}.$$

and

$$Q_{uo} = m_{uo} \cdot c_{p,water} \cdot \Delta T = 374.2 \cdot 10^{21} \, \text{J} = 374200 \, \text{EJ}.$$

Q: What is the amount of coal with a lower heating value of $H_{i,coal} = 25 \, \text{MJ} \, \text{kg}^{-1}$ to be burned to generate the combined heat for atmosphere and ocean heating?

A: Assuming complete combustion and that all heat from the combustion is kept in the atmosphere and the upper ocean, one gets

$$m_{coal} = \frac{Q_{Atm} + Q_{uo}}{H_{i,coal}} = 15.07 \cdot 10^{15} \, \text{kg}.$$

Q: After burning this amount of coal, what is the resulting increase in carbon dioxide volume-content of the atmosphere?

A: We assume that the coal is only made from carbon, then the flue gas is carbon dioxide only and we get as an estimate

$$m_{CO_2} = m_{coal} \cdot \frac{M_{CO_2}}{M_C} = 49.74 \cdot 10^{15} \, \text{kg}.$$

Exhaust gases are usually assessed at standard conditions (0 °C, 101,325 Pa) and this is what we do. First we calculate the standard volume (index N) of the atmosphere

$$V_{N,Atm} = \frac{m_{Atm} \cdot R_{air} \cdot T_N}{p_N} = 4070 \cdot 10^{15} \, \text{m}_N^3$$

and of the carbon dioxide generated from burning the coal

$$V_{N,CO_2} = \frac{m_{CO_2} \cdot R_{CO_2} \cdot T_N}{p_N} = 25.33 \cdot 10^{15} \, \text{m}_N^3.$$

From this would result an increase of the atmospheric CO_2 by a volumetric rate of

$$r_{CO_2} = \frac{V_{CO_2}}{V_{Atm} + V_{CO_2}} = 6184 \, \text{ppm}.$$

> Q: Discuss the assumption used: where does the heat come from and where does it not come from? Carbon Dioxide increased in the atmosphere from 280 ppm (1750) to about 410 ppm (2018).
> A: ...

The assignment leads to an obvious contradiction, once levels of carbon dioxide in the atmosphere have been established in the course. It can't be the dissipative heat from industry that is causing global warming. The ensuing discussion in the classroom showed that some students have no clear understanding of the causes of global warming and would have accepted dissipation as the source. A number to be used for illustrating the size of the results could be humanities energy demand, which is today at about 600 EJ per year. It would make sense from a didactical perspective and if the curriculum allows it, to directly go to radiation heat transfer and explain greenhouse gases in more depth.

Example 2—Carbon Capture and Storage

This example is based on learning to use thermo-physical data from tables or diagrams. Such data and diagrams are used for real-gas thermodynamic cycles as in steam turbines or refrigeration. The resulting numbers provide an impression of the size and energetic cost of the undertaking, if taken seriously.

> **Carbon capture and storage—what does it cost energetically?** The idea behind CCS is quite simple: take the carbon dioxide out of the atmosphere and store it underground. CCS is part of nearly all emission pathways discussed for reaching the Paris/COP 21 climate targets: your government has bought into the scheme.
>
> We simplify the problem by assuming that the average atmospheric condition is 12 °C at 1 bar. We look at a relevant contribution, i.e. reducing atmospheric CO_2 by 100 ppm by storing it 700 m below ground at average conditions of 200 bar and 32 °C. This assumes that sufficient long term storage capacity is available at this depth.
>
> Q: Calculate the mass equivalent to 100 ppm CO_2.
> A: One starts with the mass of the atmosphere from Example 1, calculates the amount of air (moles) from this (the molar mass of air comes from a table or from the specific gas constant of air)
>
> $$n_{CO_2} = 0.0001 \cdot n_{air} = 0.0001 \cdot \frac{m_{air}}{M_{air}} = 18.15 \cdot 10^{12}\,\text{kmol},$$
>
> and
>
> $$m_{CO_2} = n_{CO_2} \cdot M_{CO_2} = 798.9 \cdot 10^{12}\,\text{kg}.$$

Q: What is the minimum work required to capture this 100 ppm CO_2 directly from air?

A: The necessary calculation depends on data given or content learned: The common solution would be through minimisation of the Gibbs free energy, but this is not a usual part of all thermodynamic curricula. Therefore a simpler approach is the use of the Gouy-Stodola theorem: One starts from calculating the mixing entropy caused from adding 100 ppm of CO_2 into air

$$\Delta S_{Mi} = -R \cdot \sum_i n_i \cdot \ln \frac{n_i}{n_{Mi}} = 1.541 \cdot 10^{18} \, \text{J K}^{-1}.$$

The lost work through the mixing is

$$W_{ab} = W_{loss} = T_u \cdot \Delta S_{Mi} = 439.4 \, \text{EJ}.$$

This is the theoretical minimum work required to capture 100 ppm from air. In the literature one finds as theoretical minimum work about 22 MJ kmol^{-1}, which results in a nearly identical lost work of

$$W_{ab} = n_{CO_2} \cdot W_{M,ab} = 399.4 \, \text{EJ}.$$

Q: What is the minimum required work to store the captured CO_2? Assume ideal gas conditions.

A: For ideal, that is isentropic compression one gets as temperature at 200 bar

$$T_L = T_{Atm} \cdot \left(\frac{p_{store}}{p_{Atm}} \right)^{\frac{\kappa-1}{\kappa}} = 695 \, ^\circ\text{C}$$

and thus as required minimum work

$$W_p = m_{CO_2} \cdot c_{p,CO_2} \cdot \Delta T = 446.1 \, \text{EJ}.$$

This gas needs to be cooled down to 32 °C and isobaric removed heat is

$$Q = m_{CO_2} \cdot c_{p,CO_2} \cdot \Delta T = -427.4 \, \text{EJ}.$$

Q: What is the minimum required work to store the captured CO_2? Use the log T-s diagram.

A: This answer needs some preparation, as available T-s diagrams of CO_2 do not provide the data for 1 bar (this is below the triple point pressure of CO_2). Thus students need to calculate the conditions at atmosphere via ideal gas assumptions starting with a specific volume of

$$v_{CO_2, Atm} = \frac{R_{CO_2} \cdot T}{p} = 0.5386 \, \text{m}^3\text{kg}^{-1}$$

for the environmental condition. This provides a specific entropy for that state of about $s = 2.7 \, \text{kJ} \, \text{kg}^{-1} \, \text{K}^{-1}$.

Based on this one gets after isentropic compression as temperature 525 °C and a necessary work of about 395 EJ. The ideal gas assumption is thus quite OK for the purpose. Isobaric cooling assessed in the diagram provides a higher value of 607 EJ, as the real fluid undergoes a phase transition to a condensed liquid.

Q: What is the theoretical minimum required work for CCS of 100 ppm CO_2? What would be a realistic value assuming usual technical equipment?

A: The theoretical value is 885 EJ for ideal gas and 835 EJ for real gas assumptions.

At this point a discussion of technical progress could be initiated: how good are usual technical solutions compared with the ideal? Did we include all relevant energetic costs?

An educated guess (based on experience and arguments in Smil 2017) leads to: (i) capture schemes seem to need real work that is between a factor of 2 and 5 larger, than the thermodynamic minimum; (ii) work for compression to such high pressure and intercooling will be about a factor of 1.5–2 larger than isentropic work; (iii) we did not include manufacturing of the infrastructure, transport, maintenance, leakages, accidents; (iv) most of today's technologies are at a factor of 10 above thermodynamic minimum and observed increase in efficiency is slow even over a 50 year time-frame.

Summarising, a factor of 4 is on the optimistic side and will result in about 3500 EJ for clean up.

Q: How much electricity can be generated from natural gas causing an emission of 100 ppm CO_2?

A: The basic assumption here is, that the emitted CO_2 is not leaving the atmosphere—to keep the calculation consistent. The mass of natural gas is then

$$m_{CH_4} = n_{CO_2} \cdot M_{CH_4} = 291.2 \cdot 10^{12} \, \text{kg}$$

and with a lower heating value of $H_{l,CH_4} = 50 \, \text{MJ} \, \text{kg}^{-1}$

$$Q = m_C \cdot H_{l,C} = 14560 \, \text{EJ}.$$

Usual large scale gas-turbine power plants have about 40% conversion efficiency and we get

$$W_{el,CH_4} = Q_{CH_4} \cdot \eta = 5820 \, \text{EJ}.$$

> Q: discuss the results in view of energy cost, financial cost, intergenerational justice
>
> A: ...

This example needs some time and explanation: students will not be prepared to solve this without any help and it is not intended for an exam, it might be part of an assessed group activity. Quite some basic skills in thermodynamics are used throughout and give a good idea, what kind of problems could be assessed with them. But the major issue comes at the end: Note that your government has already accepted CCS as the ultimate method to clean up carbon dioxide. So it will be the generation of our children that has to implement CCS. CCS on a meaningful scale will bind impressive amounts of energy and resources that are then no longer available for them to meet the needs of their generation, but needed to clean up our mess. Here the Brundtland definition on sustainability is used on purpose. Their energy expense needed for CCS is about the same—or when being really optimistic, a factor of two smaller—than our benefit. Numbers become even worse, when coal instead of natural gas is used for calculations. This is an example that directly leads to wicked problems and to ethical implications of today's decisions, the definition of super wicked problem introduced by Levin et al. (2012) is well illustrated, especially the aspect of irresponsible discounting the future. If the course allows, one could evaluate the same problem using a cost-benefit analysis. Assuming usual depreciation rates one would show that CCS would make sense from a purely monetary perspective and would not create a burden on future generations under the usual assumptions of a long term stability of our financial system and that the actual owners of the wealth would actually be willing to spend that wealth for society—both assumptions provide a very low confidence. For a more realistic assessment of cost refer to Hansen and Kharecha (2018).

Summary and Outlook

Climate change is at the core of responsibilities of engineers and engineers will need to participate in any substantial activity aiming at governing the conditions of climate change or reducing its environmental impact. To do so engineers will need specific knowledge, skills and competences that are usually not learned in today's curricula. As climate change is happening now, there is a need to prepare students for the wicked problems and realities of the changing world they will live in. Some universities are in the process to change their curricula accordingly, others fully lack any sense of urgency with respect to any aspects of our changing world: this can be explained through deadlock caused by conflicting orientations. Organisations dominated by orientations where climate change is just another technical issue that will be handled once it becomes sufficiently urgent, something outside one's own discipline, or something that is to be ignored as it does not create any monetary value, will not consider implementing competences needed for addressing climate change.

One possibility to overcome such a stalemate and start change is to ignore this in one owns teaching. There is quite some freedom in teaching, especially when it comes to examples, tasks, and problems used. In mechanical engineering the mandatory courses in thermodynamics and fluid dynamics are well suited to provide the technical knowledge about climate and climate change that engineers will need for a meaningful understanding. Well-crafted examples can be used to explain the workings of climate change and to illustrate the wickedness of many problems connected to sustainable development. This approach does not enable us to teach and learn in detail the generic competences identified as the toolbox for sustainable engineering science. The teacher is only able, through the arguments given and his personal example, to let the students glimpse at these and become interested: their meaningful teaching and learning is a long process that can't be done as a side issue. But the approach shown here can be done right now, in many different settings and is a first step in accepting and living one's own responsibility.

The second argument for starting immediately is to show other members of the organisation that one treats climate change with the urgency and seriousness is demands. It may help others in the organisation to overcome their hesitations. This can lead to a coalition of the willing and may help to change the organisations orientation over time.

When looking into the details of the arguments presented, one sees that this paper is not a well-formed argument—to the contrary, many strands are open ended. This mirrors the wickedness of the larger problem: how to do meaningful *sustainable engineering science* and really start to govern climate change in a society that is not prepared to do so. The expectations to technology and its ability to eventually solve the task of climate change are better described as wishful thinking or a belief in magic. This faith is well fed by the neglect of any considerations of scale or feedbacks in the public discourse itself. Thus to expect that engineering departments or students will readily switch to an opposing paradigm in their problem solving competence would be over-optimistic as well. Still any attempt to change this does matter.

The examples given are part of a larger and growing set. Many of the examples are inspired by the public discourse and look at actual political issues or decision processes. Please contact the author for more information or collaboration.

References

Allenby B (2006) Macroethical systems and sustainability science. Sustain Sci 1:7–13
Biggs J, Tang C (2011) Teaching für quality learning at university. McGraw Hill, Maidenhead
Bridle J (2018) New dark age. Technology and the end of the future. London, Verso
Flyvbjerg B (2009) Survival of the unfittest: why the worst infrastructure gets built—and what we can do about it. Oxford Rev Econ Policy 25:344–367

Hansen J, Kharecha P (2018) Cost of carbon capture: can young people bear the burden? Joule 2:1405–1407

Jarrett LA, Ferry B, Takacs G (2012) Development and validation of a concept inventory for introductory-level climate change science. Int J Innov Sci Math Educ 20:25–41

Kleidon A (2017) Thermodynamik foundation of the earth system. Cambridge University Press

Landfester A, Linow S, van de Loo F (2019) Maschinenbaustudium im Spannungsfeld von Ingenieurskompetenzen, Digitalisierung und Nachhaltiger Entwicklung. In: Leal W (ed) Digitalisierung und Nachhaltigkeit: Chancen und Perspektiven für deutsche Hochschulen. Springer, Berlin

Levin L, Cashore L, Bernstein S, Auld G (2012) Overcoming the tragedy of super wicked problems: constraining our future selves to ameliorate global climate change. Policy Sci 45:123–152

Linow S, Führ M, Kleihauer S (2017) Aktivierende Ringvorlesung mit begleitender Konzept-Werkstatt Herausforderung: Nachhaltige Entwicklung—Klimaschutz in und um Darmstadt. In: Leal W (ed) Nachhaltigkeit in der Lehre. Eine Herausforderung für Hochschulen. Springer, Berlin

McCright AM, O'Shea BW, Sweeden RD, Urquhart GR, Zeleke A (2013) Promoting interdisciplinarity through climate change education. Nat Clim Change 3:713–716

Mezirov J (1997) Tranformative learning: theory to practice. New Dir Adult Cont Educ 74:5–12

Pahl G, Beitz W, Feldhusen J, Grote KH (2007) Engineering design. A systematic approach. Springer, Berlin

Rittel HWJ, Webber MM (1973) Dilemmas in a general theory of planning. Policy Sci 4:155–169

Seager T, Selinger E, Wiek A (2012) Sustainable engineering science for resolving wicked problems. J Agric Environ Ethics 25:467–484

Smil V (2017) Energy and civilization. A history. MIT Press, Cambridge

SuK. https://fbgw.h-da.de/begleitstudium/begleitstudium/

United Nations (2015) Transforming our world: the 2030 agenda for sustainable development. New York. https://sustainabledevelopment.un.org/sdgs

VDI (2002) Ethische Grundsätze des Ingenieurberufs. Düsseldorf. https://www.vdi.de/fileadmin/media/content/hg/16.pdf

Wiek A, Withycombe L, Redman CL (2011) Key competencies in sustainability: a reference framework for academic program development. Sustain Sci 6:203–218

Chapter 4
Climate Change Education Across the Curriculum

Rahul Chopra, Aparna Joshi, Anita Nagarajan, Nathalie Fomproix and L. S. Shashidhara

Abstract It is vital that both the current and future generations are better equipped to address the problem of climate change. This can be achieved by adopting appropriate pedagogical methods aimed at helping students to improve their understanding of the science of climate change and to acquire necessary skills to mitigate its impact. Learning is more effective when students are challenged to identify the cause and effect of a problem that they can relate to in their life. Thus, climate change is both a problem to be addressed and a problem that can be adopted for more effective teaching. Model teaching and learning modules have been developed as part of the Trans-disciplinary Research Oriented Pedagogy for Improving Climate Studies and Understanding (TROP ICSU) project (https://tropicsu.org/) as a proof of concept of integrating climate change-related topics across the curriculum. These innovative educational resources are locally rooted in their context, but globally relevant for their science. They are designed and packaged such that teachers across the world can use them to impart transdisciplinary training that is essential for addressing the

---

On behalf of all the partners of the project: Trans-disciplinary Research Oriented Pedagogy for Improving Climate Studies and Understanding (TROP ICSU).

The TROP ICSU project is a global project funded by the International Science Council (ISC). Its project partners are: the International Union of Biological Sciences (IUBS); the International Union of Quaternary Research (INQUA); the International Union of Geological Sciences (IUGS); the International Union of History and Philosophy of Science and Technology (IUHPST); IMAGINARY; the International Science Council—Regional Office for Africa (ISC-ROA); the Indian National Science Academy (INSA); the National Research Foundation of South Africa (NRF); the International Union of Soil Sciences (IUSS); the International Mathematical Union (IMU); the International Union of Geodesy and Geophysics (IUGG); the International Union of Forest Research Organizations (IUFRO); the Committee on Data of the International Council for Science (CODATA); the Australian Academy of Science (AAS); the Mongolian Academy of Sciences (MAS); Universidad Técnica Particular de Loja, Ecuador (UTPL); the African Union of Conservationists (AUC); the National Research Centre, Plant Protection Department, Egypt; the United Nations Educational, Scientific, and Cultural Organization (UNESCO); and the World Climate Research Programme (WCRP).

R. Chopra · A. Joshi · A. Nagarajan · L. S. Shashidhara
Indian Institute of Science Education and Research (IISER), Pune, India

N. Fomproix · L. S. Shashidhara (✉)
International Union of Biological Sciences (IUBS), Paris, France
e-mail: ls.shashidhara@iiserpune.ac.in

© Springer Nature Switzerland AG 2019
W. Leal Filho and S. L. Hemstock (eds.), *Climate Change and the Role of Education*,
Climate Change Management, https://doi.org/10.1007/978-3-030-32898-6_4

problems of climate change. All educational resources are reviewed and validated by subject and education experts before being made available for use.

Keywords Climate change education · Climate literacy · Education for sustainable development · Teaching resources · Lesson plans

Climate Change Education: Building Awareness and Skills for Adaptation, Mitigation, and Resilience

Climate change is a critical factor that affects sustainable and equitable development, increases conflicts, and causes extinction of species. Sea-level rise, habitat loss and changes in biodiversity, ocean acidification, and environmental migration are some of the serious effects of climate change that pose tremendous challenges and necessitate immediate climate action.

The impacts of climate change are not restricted to a single country or location. Current and future citizens—including policymakers, scientists, and researchers—are faced with the daunting task of determining actions that will help in adapting to and mitigating these impacts, while taking steps to prevent or reduce anthropogenic climate change.

Therefore, to address the challenges posed by climate change, current and future generations must be equipped with the relevant awareness, knowledge, and skills.

The United Nations Sustainable Development Goals (SDGs) (United Nations Sustainable Development 2019a) have been designed with the objective of achieving a sustainable future globally. To address each challenge in the achievement of this overall objective, fourteen SDGs have been defined. Goal 4 (Quality Education) emphasizes the importance and need of quality education in improving people's lives and achieving sustainable development (United Nations Sustainable Development 2019c). Goal 13 (Climate Action) highlights the necessity of urgent action in combating climate change and its impacts (United Nations Sustainable Development 2019b).

The objective of the project Trans-disciplinary Research Oriented Pedagogy for Improving Climate Studies and Understanding (TROP ICSU) is to equip students across the world to identify solutions for sustainable and equitable development in the context of climate change.

Recognizing the urgent need to educate forthcoming generations about the causes and effects of global climate change, the TROP ICSU project adopts the use of digital (or ICT-based) tools and teaching aids to impart this education. Further, the project proposes an integrated approach that seamlessly blends the teaching of a topic in a discipline with the teaching of a climate topic and provides sample teaching aids as a blueprint or template for this approach. It is envisioned that the educational resources on the TROP ICSU website will improve learning outcomes in various disciplines and will simultaneously increase climate awareness through the use of an engaging learning experience.

Thus, the project aligns with the UN SDGs 4 and 13 by using quality education as a means to equip current and future generations with the knowledge and skills required for climate change adaptation, mitigation, and resilience. Effective learning can be achieved by challenging students to identify the causes and effects of a problem that they face in everyday life—in this case, the problems resulting from climate change.

Existing Resources for Climate Change Education

In the context of climate change education and digital educational resources, a range of useful teaching resources are available online and are accessible to educators and students across the world. Some of these resources are listed in this section.

Portals such as Science Education Resource Center at Carleton College (SERC) (2019), Climate Literacy and Energy Awareness Network (CLEAN) (2019), University Corporation for Atmospheric Research (UCAR) (UCAR Center for Science Education 2019), and NASA's Global Climate Change (Climate Change: Vital Signs of the Planet 2019) contain climate-related teaching aids.

Examples of other pedagogical resources include large datasets/repositories of climate data (IRI at Columbia University 2019) and computer-based simulators/models for teaching climate-related topics (Archer 2019a).

Additional relevant resources include teaching material such as video lectures, models, and simulators that focus on climate and earth science (Archer 2019b), and ideas for teaching Earth-related topics, including climate change (Earth Learning Idea 2019).

For learning that is focused on climate science and meteorology, useful resources include the UN CC:e-Learn resource at The One UN Climate Change Learning Partnership (UN CC:e-Learn 2019); the COMET program developed by UCAR and the National Oceanic and Atmospheric Administration (NOAA) for better understanding of meteorology, and specifically, MetEd (2019), which contains training resources for the geoscience community and meteorologists; WMO Global Campus and its programs such as WMO Learn (World Meteorological Organization 2019); and climatology toolkits for technical practitioners, developed by WMO in partnership with UN CC:Learn.

For a basic introduction to climate change, aids such as a set of introductory videos (The National Academies of Sciences, Engineering, and Medicine 2012) and a guide for teachers (The Teacher-Friendly Guide™ to Climate Change 2017) can serve as a starting point.

Some portals provide examples of teaching specific disciplines such as Chemistry (KCVS 2019) or Mathematics (Pfaff 2019) through the context of climate change. Other websites provide visualization tools to learn science topics in an interactive, hands-on, and engaging manner (PhET 2019).

New ideas and methods to develop curriculum at the interface of a discipline such as Psychology and climate change (Maier et al. 2018) are also being proposed.

Challenges in Climate Change Education

Although several teaching resources for climate change education exist and are available for usage by teachers across the world, climate change-related topics are taught in few classrooms globally. Further, the field of climate change is underrepresented in the school and undergraduate syllabi in most countries. Typically, climate science courses are taught only in undergraduate Earth Sciences, Geography, or Environmental Sciences majors. Discussion and teaching of climate-related topics is minimal, if any, in other disciplines (Wise 2018).

Climate change education is vital for a sustainable future (Kagawa and Selby 2013). However, while several pedagogical resources for climate change are available (as described earlier in this section), their widespread usage and dissemination in classrooms across the world is impeded by barriers and challenges such as:

- curricular constraints (Dupigny-Giroux 2010)
- lack of confidence and other challenges faced by teachers owing to their limited knowledge of climate-related topics (and the corresponding lack of training) (Boon 2010, Plutzer et al. 2016)
- limited awareness about the existence, location, and usability of such teaching resources
- research-oriented nature/focus of some useful resources (thus, potentially restricting their direct usage in pedagogy)
- limited number of local and region-specific climate-related examples and activities that would be more relevant, and therefore, of more interest, to students and teachers.

TROP ICSU aims to address these limitations while increasing climate awareness and the understanding of the science behind climate change by

- integrating the teaching of climate topics with the teaching of topics in existing curriculum in the Sciences, Mathematics, Humanities, and Social Sciences
- curating and providing a set of basic, introductory climate-related resources for teachers of all disciplines
- providing a curated suite of teaching tools and lesson plans that are validated by scientific experts and teachers; creating lesson plans as a blueprint or template of how climate education can be integrated with different disciplines
- modifying research-oriented resources to create simpler teaching aids for classroom instruction
- curating and creating region-specific examples and activities to increase the relevance of resources in classrooms across the world.

TROP ICSU: An Integrated Approach to Climate Change Education

Trans-disciplinary Research Oriented Pedagogy for Improving Climate Studies and Understanding (TROP ICSU) is a global project funded by the International Science Council (ISC). The International Union of Biological Sciences (IUBS) is the lead partner and the International Union for Quaternary Research (INQUA) is the co-lead partner. The project partners include other international unions, national academies of various countries, national research centers, and United Nations agencies.

TROP ICSU aims to integrate relevant pedagogical resources in the education system to help future citizens across the globe in improving their understanding of the science of climate change and in developing the necessary skills for climate change adaptation, climate change mitigation, and climate resilience.

To achieve this aim, the project collates, curates, and creates digital teaching resources that can be used to teach topics in the Sciences, Mathematics, Social Sciences, and the Humanities with the help of climate-related examples, case studies, and activities. Thus, the project adopts an integrated approach and intends to serve as a proof of concept of integrating climate change education with core curriculum.

The goal of the TROP ICSU project is not to introduce climate education as a stand-alone topic, but to integrate it with the core Science, Mathematics, Humanities, and Social Science curriculum.

A suite of educational resources (teaching tools and lesson plans)—for the high school and undergraduate/university levels—is collated and created by adopting a meticulous methodology to select, review, and validate the resources. These resources typically include computer-based or digital tools and interactive, hands-on activities that provide an immersive and engaging learning experience. They introduce examples and case studies from climate science and climate change while enhancing the conceptual understanding of topics in various disciplines. These educational resources are locally rooted but globally relevant for their science and are designed to promote interdisciplinary thinking that is essential for research and action on climate change.

The TROP ICSU educational resources are free to download and can be used by teachers across the world to supplement their existing teaching methods.

In addition to the curation of existing relevant resources, another key aspect of the project is the creation of new resources and the modification of resources from complex scientific research to make them more suitable, relevant, and accessible for classroom instruction.

Thus, the project aims to provide a platform that can serve as a training and educational resource to equip current and future students with the knowledge and skills for effective policy-making and problem-solving in the context of climate change adaptation, mitigation, and resilience.

The TROP ICSU project is fully funded by the International Science Council (ISC) for the period 2017–2019. The first global Working Group meeting was held in April 2017 in Paris and was attended by representatives from all the partner organizations.

The implementation commenced in June 2017 and is ongoing till date. Workshops for teachers are being organized in countries across the world with the valuable support of the project partners.

TROP ICSU Educational Resources: Integrating Climate Topics Across the Curriculum

On the TROP ICSU website, the teaching resources are of two types:

- Teaching Tools
- Lesson Plans.

These teaching resources span disciplines such as the Biological Sciences, Chemistry, Physics, Mathematics, Statistics, Earth Sciences, Economics, Environmental Sciences, Geography, Humanities, and the Social Sciences. They cover a wide range of topics—such as Biodiversity, Species, Human Health, pH Scale, Isotopes, Planck's Law, Blackbody Radiation, Calculus, Trend Analysis, Confidence Intervals, Historical Climate Change, The Greenhouse Effect, Cost-Benefit Analysis, Tragedy of the Commons, Food Security, Glaciers, Environmental Migration, Climate Justice—that typically constitute an integral part of the curriculum of these disciplines. Table 4.1 shows a more detailed list of topics in different disciplines that can be taught using a climate-related example, activity, or case study.

The teaching tools are typically computer-based tools such as interactive visualizations, models and simulators, classroom/laboratory activities, online readings, games, mobile applications, and video micro-lectures. They are classified by Discipline, Climate Topic, Grade Level, Language, and Region.

The lesson plans include a set of teaching tools and detailed guidelines on using these tools to integrate the teaching of a topic in a specific discipline with a topic in climate science or climate change. Lesson plans are categorized by Discipline.

In the remainder of this section, lesson plans in the TROP ICSU project will be discussed in more detail.

Lesson Plan

The content in a lesson plan describes how a topic in a particular discipline (such as Blackbody Radiation in Physics, Carbon Compounds in Chemistry, Species and Speciation in Biology, or Cost-Benefit Analysis in Economics) can be taught by using examples, case studies, and exercises related to climate change.

The content of each lesson plan is organized into various sections and includes a step-by-step guide for improved ease-of-use and relevance in classrooms across the world. Educators can use these lesson plans as supplementary aids or can integrate

Table 4.1 Sample list of topics in different disciplines that can be taught using a climate-related example, activity, or case study

Discipline	Topics in discipline with connections to climate change
Chemistry	Biogeochemical Cycles, The Carbon Cycle, Atomic Number, Atomic Mass, Isotopes, Spectroscopy, Molecular Structure of Compounds, The Electromagnetic Spectrum, pH Scale, Acids and Bases, Ocean Acidification, Chemical Weathering, The Urey Reaction, Carbon Chemistry, Phase Diagrams, Aerosols, Fertilizers
Biological Sciences	Biodiversity, Species and Speciation, Species Distribution and Species Habitats, Range Shift, Survival Skills, Phenology, Phenophases, Leaf-out, Flowering and Bud Burst, Sex Determination, Human Health, Epidemiology, Vector-borne Diseases, Microbes, Microbial Growth, Insect Behavior
Earth Sciences	Planet Earth and its Climate, Earth's Climate System, Past Episodes of Climate Change, Historical Climate Change, Recent Climate Change, Future Projections, The Roles of the Atmosphere, Cryosphere, Hydrosphere, Lithosphere, Biosphere, Anthroposphere, and their interactions in determining the climate of planet Earth, Water Resources and Climate Change, Climate Change and Food, Adaptation and Mitigation, Biogeochemical Cycles, The Geologic Carbon Cycle, The Silicate Thermostat, Milankovitch Cycles, Isotopes and Climate Reconstruction, The Greenhouse Effect, El Niño Southern Oscillation, Thermohaline Circulation, Coriolis Force, Hadley Circulation
Geography	Glaciers, Glacial Retreat and Glacial Melting, Sea-Level Rise, Coastal Floods, Droughts, Forest Fires, Climate Refugees and Environmental Migration, Agro-ecosystems, Food Production and Crop Yields, Climate Resilient Agriculture, Water Resources and Water Security, Climate Change and Cities, Urban Heat Islands
Physics	Blackbody Radiation, Stefan-Boltzmann Law, Planck's Law, Heat Transport, Convection, Energy and Mass Conservation, The Electromagnetic Spectrum, Coriolis Force, Hadley Circulation, Planetary Energy Balance
Mathematics	Calculus, Integration, Differentiation, Trigonometry, Sine and Cosine Functions, Polynomial Differentiation, Tangent Line Problems, Area Under A Curve, Riemann Sums, Numerical Modeling, Mathematical Modeling, Computer Programming, Direct and Inverse Variations
Statistics	Trend Analysis, Linear Regression, Quadratic Regression, Time Series, Uncertainty, Correlation Coefficients, Confidence Intervals, Errors, Student's t-Distribution
Social Sciences	Climate Refugees, Environmental Migration, Climate Justice, Social Vulnerability, Urban Climate Resilience, Climate Change and Behavior, Climate Change and Children, Food Security, Human Health and Disease, Climate Change and Cities, Poverty and Climate Change, Tribes and Climate Vulnerability, Peace and Conflict, Global Security

(continued)

Table 4.1 (continued)

Discipline	Topics in discipline with connections to climate change
Economics	Economics of Climate Change, Cost-Benefit Analysis, Tragedy Of The Commons, Carbon Emissions Abatement, Marginal Abatement Cost, Global Economy and Energy, Economics and Geopolitics of Oil, Carbon Taxes, Trade Policies, Cap and Trade, Supply Demand Analyses, Prisoner's Dilemma, Principal-Agent Problems
Humanities	Climate Change Literature, Climate Fiction or Cli-fi, Climate and Culture, Gender and Climate, Climate Change and Human Rights, Climate Refugees, Environmental Migration, Climate Change and Children, Historical Climatology, The Decline of Civilizations, Climate Change and the Roman Empire, Postcolonial studies, Ethics, The Anthropocene, Climate Change Non-Fiction
Environmental Sciences	Water Security, Human Health and Disease, Climate Justice, Climate Refugees, Food Security, Agro-ecosystems, Crop Yields, Species Extinction, Energy—Traditional and Renewable, Carbon Emissions, Carbon Footprint, The Anthropocene, Migration and Migratory Patterns, Decline of Civilizations

them with their existing teaching modules. A lesson plan can also be used by students for independent learning.

A typical lesson plan may consist of an introductory video micro-lecture, a central hands-on classroom/laboratory activity, reading material, and additional resources for further exploration of the topic.

Components/Structure of a Lesson Plan

Each lesson plan on the TROP ICSU website consists of the following sections:

- An Introduction: This section contains information about the discipline for which this lesson plan can be used, the level (high school or undergraduate) for which it is appropriate, and the topic(s) in a discipline that can be taught using this resource. It also consists of a brief summary of the activities in the lesson plan and the corresponding learning for students.
- Key Information/Tags corresponding to the lesson plan: This table includes fields that would help teachers determine the usefulness of the lesson plan for their classroom instruction purposes.
- Contents: This table consists of a list of teaching tools in the lesson plan, with a brief description of each tool.
- A Step-by-Step User Guide: This section contains a detailed set of steps that serve as guidelines/recommendations for the usage of the lesson plan and its constituent tools. These guidelines provide a possible plan of action; they can be customized by a teacher for a specific target audience, based on individual preferences and the requirements of a particular classroom setting and students.

- Questions/Assignments: This list is a set of sample questions that can be used for discussion in class or for evaluation.
- Learning Outcomes: This list contains a set of learning outcomes corresponding to this lesson plan.
- Additional Resources: This section consists of a recommended list of additional teaching tools for further exploration of the topics in the lesson plan.
- Credits: This section mentions the credits for each teaching tool and additional resource used in the lesson plan.

Methodology for Creation of a Lesson Plan

A meticulous approach is followed for the creation of each lesson plan.

- First, a new idea for a lesson plan is discussed. A discipline topic and a climate topic that it can be linked with are determined.
- Teaching tools within a lesson plan are selected in the following manner:
  - One teaching tool is selected as the central or core resource for this lesson plan.
  - Other teaching tools that can be used for the introduction, the conclusion, and further investigation of the topic are selected.

- A detailed step-by-step guide for teachers is then created. The instructions specify how the teaching tools in this lesson plan can be used to explain the topic in the discipline and to link it with a climate-related topic. Clarity and simplicity are prioritized in the writing of these guidelines.
- Seamless connectivity between the climate topic and the corresponding topic in the curriculum of the discipline is ensured. "Forced" or "convoluted" connections are avoided.
- The lesson plan is internally reviewed for scientific accuracy and ease-of-use in the classroom.
- The lesson plan is published on the TROP ICSU website for its global review.
- The teaching tools used in a lesson plan are reviewed by climate science experts.
- The lesson plans are reviewed by high school and undergraduate teachers/educators at workshops across the world. They are reviewed for connectivity between the discipline topic(s) and climate topic(s), feasibility of use in regular teaching, and overall flow.
- The lesson plans are edited/modified by the TROP ICSU team, based on feedback and suggestions from the review by scientific experts and pedagogy experts.
- Peer-reviewed and validated teaching resources will be formally published on the TROP ICSU website.

The project invites teachers across the world to submit ideas and suggestions for lesson plans that link a discipline topic to a climate topic. This engagement is built and achieved through workshops with teachers and educators, discussions with pedagogy experts, and meetings with climate science experts. After a lesson plan idea or outline is received, the idea is further developed and refined collaboratively between the TROP ICSU team and the contributor.

Further, the lesson plan ideas and the overall project goals are being disseminated with the help of the global partners of the project.

New lesson plans are currently being developed in all the disciplines for topics that can be taught using climate-related educational resources (Table 4.1).

Thus, teachers and educators can choose a lesson plan from the TROP ICSU website to teach a topic in Science, Mathematics and Statistics, Social Science, or Humanities with the help of examples, case studies, and exercises related to climate change. These lesson plans are meant to serve as a blueprint or template and can be customized by teachers based on their preferences and requirements. These teaching resources are envisioned to be "by teachers, for teachers" across the world. Several teacher-submitted lesson plans have now been published on the TROP ICSU website.

Some examples of lesson plans available on the project website are listed in Table 4.2.

Benefits of Using the TROP ICSU Educational Resources

The TROP ICSU educational resources can be used as effective teaching/learning aids for the classroom and can also play an important role in continuing professional development (CPD) for teachers and educators.

Effective Teaching/Learning Aids for the Classroom

With increasing emphasis on education for sustainable development, the TROP ICSU resources harness the power and accessibility of digital pedagogical tools to increase climate change awareness and understanding while maintaining focus on the learning objectives of discipline-specific curricula. The suite of educational resources is carefully curated with the aim of building key skills such as problem-solving, critical thinking, and systems thinking.

These locally rooted, culturally appropriate, and globally relevant pedagogical resources for climate change can be highly effective teaching and learning aids for the classroom.

The TROP ICSU approach for curation, selection, modification, and creation of educational resources focuses on resources that:

1. Use local data, thus promoting the teaching and understanding of climate change in the local context
2. Include team-based activities, thus enabling the learning and exploration of climate change as a group

4 Climate Change Education Across the Curriculum

Table 4.2 Examples of lesson plans in various disciplines

Discipline	Lesson plan topic
Biological Sciences	Teaching Phenology in Plants (Leaf-out, Flowering) through Climate-related Examples; Teaching about Human Health and Climate Change; Impact of Climate Change on Photosynthesis; Teaching Microbial Life and Climate Change; Impact of Climate Change on Sex Determination of Sea Turtles; Teaching Biodiversity Through Climate-Related Examples
Chemistry	Teaching the pH Scale, and Acids and Bases through Climate-related Examples; Teaching the Chemistry of Carbon Compounds through Climate-related Examples; Buffers, Buffer Action, and Ocean Acidification; Phase Diagrams and Phase Equilibria; Infrared Spectroscopy and the Greenhouse Effect; Fertilizers and Climate; Hydrocarbons and Climate Change
Earth Sciences	Teaching Glaciology, Glaciers and Glacial Retreat, and the Cryosphere-Climate Relationship; Isotopes and Isotopic Ratios, and Isotopic Compositions as Climate Proxies; Permafrost and Climate Change; Heat Transport in the Atmosphere, Hadley Circulation and Climate
Economics	Teaching Cost-Benefit Analysis through Climate-related Examples; Teaching the Tragedy of the Commons through Climate Change
Environmental Sciences	Teaching about Hazards and Disasters through Climate-related Examples (Sea-level Rise and Flooding due to Melting of Polar Ice); Climate Refugees and Environmental Migration; Climate Change and Food Security
Humanities	Climate Refugees and Environmental Migration
Mathematics	Teaching Introductory Calculus (Integration) by using CO_2 Emissions Data; Teaching Introductory Calculus (Differentiation) by using Atmospheric CO_2 Data
Physics	Heat Transport in the Atmosphere, Hadley Circulation, and Climate; Coriolis Force, Coriolis Effect, and the Impact of Coriolis Effect on Climate
Social Sciences	Climate Change and Food Security, and Climate Change and Agriculture
Statistics	Analyzing Trends and Calculating Uncertainty by using Hurricane Data Records

3. Enable learners to explore various possible solutions, thus facilitating interactive learning
4. Reveal the scientific and mathematical concepts used and show the interaction between different components of the system, thus facilitating systems thinking and deep learning
5. Allow users to determine policies for mitigating and adapting to climate change, thus equipping our future citizens with leadership skills for sustainable and equitable development.

A connection between a classroom exercise and a real-life scenario such as a challenge resulting from climate change would help students apply scientific, analytical, and problem-solving skills to real-world problems.

Continuing Professional Development (CPD) for Teachers/Educators

The use of TROP ICSU teaching resources introduces teachers and educators to an innovative teaching practice that combines the use of digital pedagogy with an integrated, multidisciplinary approach to teaching climate science and climate change. These resources can also serve as learning material for teachers to understand some topics in climate science and climate change.

The usage of digital pedagogical tools and the creation of new lesson plans that integrate the teaching of topics in their discipline with an introduction to climate-related topics can assist teachers in CPD. Teachers and educators can improve their knowledge and skills by developing innovative pedagogy, ICT-based curriculum and courses, and e-content.

Workshops and Feedback (Validation)

The TROP ICSU team, with the support of its global partners, is conducting workshops across the world to provide teachers and educators with an engaging, hands-on introduction to the teaching resources.

The workshops introduce the participants to the TROP ICSU teaching resources for teaching topics in the Sciences, Mathematics, Social Sciences, and Humanities using climate-related examples, case studies, and activities. In addition, participants are invited to review the educational resources of the TROP ICSU project and to provide their feedback on the appropriateness and ease-of-use of the teaching tools and lesson plans.

The evaluation form includes questions on:

- the effectiveness of the teaching resource in explaining the topic in the discipline
- the effectiveness of the teaching resource in integrating the discipline topic with climate science
- the effectiveness of the teaching resource description (on the project website) in showing how to teach the topic in a discipline using a climate-related example
- the appropriateness of the different teaching tools used in a lesson plan
- the usefulness of each component (contents, step-by-step user guide, questions, learning outcomes, additional resources) of a lesson plan
- whether students will become more aware of climate change if this teaching resource was used in the classroom

- whether the teacher/educator would use a resource in their classroom
- suggestions for improving a teaching resource
- the likelihood of the teacher/educator developing their own teaching resources
- recommendations on other teaching tools that explain the topic in the discipline more effectively.

To date, workshops with teachers and educators from India and East Africa have been conducted in New Delhi, India (95 undergraduate teachers) and Kampala, Uganda (73 high school teachers and 88 undergraduate teachers).

A summary of responses from the workshop participants regarding the effectiveness of the lesson plans is provided below and is shown in greater detail in Fig. 4.1. The possible implications of these preliminary results are also discussed below.

- **Explaining the topic(s) in the discipline**

  As observed in Fig. 4.1a, **100%** of the responses from the undergraduate teachers from East Africa, **100%** of the responses from the high school teachers from East Africa, and approximately **72%** of the responses from the undergraduate teachers from India stated that the reviewed lesson plan was **very effective or moderately effective** in explaining the topic in the discipline.

  These responses indicate that, according to a majority of the participants, the choice of teaching tools and the flow of a lesson plan are appropriate for teaching a topic in a discipline.

- **Integrating the discipline topic(s) with climate science**

  As observed in Fig. 4.1b, approximately **93%** of the responses from the undergraduate teachers from East Africa, approximately **92%** of the responses from the high school teachers from East Africa, and approximately **82%** of the responses from the undergraduate teachers from India indicated that the reviewed lesson plan was **very effective or moderately effective** in integrating the discipline topic(s) with climate science.

  These responses indicate that, according to a majority of the participants, the primary purpose of the teaching resources (lesson plans)—to seamlessly blend or integrate the teaching of a topic in a discipline with the teaching of a climate-related topic—seems to be achieved by the reviewed lesson plan.

- **Increasing students' awareness of climate change**

  As observed in Fig. 4.1c, **100%** of the responses from the undergraduate teachers from East Africa, approximately **87%** of the responses from the high school teachers from East Africa, and **88%** of the responses from the undergraduate teachers from India stated that the use of the reviewed lesson plan will make their students more aware of climate change.

  These responses suggest that, according to a majority of the participants, the eventual goal of the project—to increase the awareness and knowledge of climate change in order to build necessary skills for climate adaptation, mitigation, and resilience—may be realized through the usage of these teaching resources in the classroom.

In your opinion, how effective is this lesson plan in explaining the topic(s) in discipline?

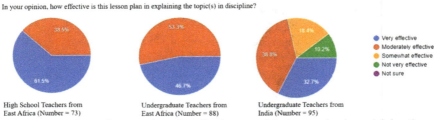

(a) Responses from educators about the effectiveness of the lesson plan in explaining the topic(s) in discipline

In your opinion, how effective is this lesson plan in integrating the discipline topic(s) with climate science?

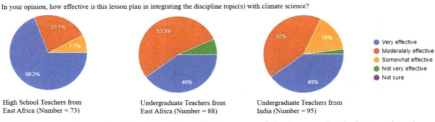

(b) Responses from educators about the effectiveness of the lesson plan in integrating the discipline topic(s) with climate science

Do you think that your students will become more aware of climate change if you use this lesson plan in your classroom?

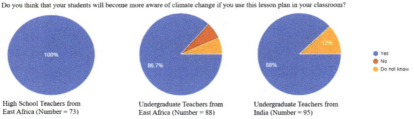

(c) Responses from educators about whether the usage of this lesson plan in the classroom would increase students' awareness of climate change

Would you use this lesson plan in your classroom for your students?

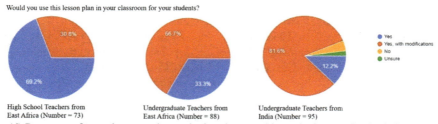

(d) Responses from educators about whether they would use the lesson plan in their classroom for their students

Fig. 4.1 Summary of feedback received from high school and undergraduate teachers at TROP ICSU workshops in Kampala, Uganda and New Delhi, India

- **Using the lesson plan in the classroom**
  As observed in Fig. 4.1d, **100%** of the responses from the undergraduate teachers from East Africa, **100%** of the responses from the high school teachers from East Africa, and approximately **94%** of the responses from the undergraduate teachers from India indicated that they **would use the lesson plan in their classroom as is or with some modifications**.
  These responses serve as preliminary indicators about the likelihood of the lesson plans being used by teachers across the world. It can be inferred that a majority of the participants find these lesson plans suitable for usage as teaching aids in the classroom. The lesson plans are designed to serve as templates, and therefore, it is expected that teachers would customize them according to their specific requirements.

  Some key suggestions received from teachers are: to add more details in the step-by-step guide, to include evaluation questions that range from a basic to an advanced level, to create separate lesson plans (on the same topic) for high school and undergraduate levels, and to increase the number of region-specific and location-specific activities, examples, and data sets.

  In overall feedback, the educators have expressed keen interest and enthusiasm in learning about digital pedagogy, the creation and usage of lesson plans, and the integration of climate change education with existing curriculum in various disciplines and subjects.

Conclusions and Future Work

To address the challenges of climate change, current and future generations require transdisciplinary training that will help them identify solutions for climate change adaptation, mitigation, and resilience. With a view to equipping students with the relevant awareness, scientific knowledge, and skills, the TROP ICSU project provides educational resources that integrate the teaching of a topic in a discipline with a climate-related example, case study, or activity. This integrated approach can help not only in enhancing the conceptual understanding of a topic in a particular discipline (Sciences, Mathematics, Social Sciences, or Humanities) but also in understanding the science of climate change. The educational resources include a carefully curated suite of appropriate and relevant teaching tools, and a collection of detailed lesson plans that are written in the form of step-by-step guides for teachers. These teaching aids for high school and college/university teachers span all disciplines and are reviewed by experts for scientific accuracy and ease-of-use in classroom instruction.

In preliminary evaluation, educators participating in the TROP ICSU workshops have provided positive feedback about the usefulness and potential effectiveness of the TROP ICSU educational resources in increasing climate awareness and building relevant knowledge and skills in the classroom (as shown in Fig. 4.1). A majority of the participants in these workshops have indicated that the teaching resources provided on the TROP ICSU website are effective in:

- explaining a topic in a specific discipline,
- integrating a discipline topic with climate science, and
- improving students' awareness of climate change.

Further, a majority of the teachers in these workshops have expressed their willingness and interest toward using the lesson plans in their classroom. Thus, preliminary results suggest that the integration of climate change education with existing curriculum in various disciplines is achieved by the sample lesson plans on the TROP ICSU website. The suggestions received from teachers and experts in pedagogy can be used to further refine these educational resources.

In future work, the TROP ICSU team will continue to conduct more workshops for teachers across the world, seek review and suggestions from educators on the appropriateness and ease-of-use of the teaching resources, collaborate with partners to further validate and enhance the content, and establish new partnerships for a wider dissemination of the project across the world.

The number and scope of the lesson plans and teaching tools will be increased to widen the coverage of disciplines and the topics within disciplines. To accomplish this goal, the team is engaging with educators and climate science experts worldwide for the creation of teacher-submitted lesson plans that are scientifically correct.

The TROP ICSU project and its resources are expected to grow extensively with the help of contributions by and for teachers across the world. The team will engage with partners and collaborators for widespread dissemination of the idea of integrating climate change topics with core curriculum in various disciplines through the adoption of these teaching tools and lesson plans at a global level.

It is expected that these innovative educational resources will equip future citizens with the knowledge and skills required to address the problems of climate change, build climate resilience, and take concrete steps toward sustainable development.

Acknowledgements We thank the International Science Council for financial support to the project; all the partners of the TROP ICSU project, climate change experts, and educators from across the world for their valuable suggestions and feedback on various aspects of the project; and the members of the executive council of IUBS and the staff of IISER Pune, India for their help and support.

References

Archer D (2019a) Global warming: understanding the forecast by David Archer. Forecast.uchicago.edu. Available at http://forecast.uchicago.edu/models.html. Accessed on 7 Feb 2019

Archer D (2019b) Global warming: understanding the forecast by David Archer. Forecast.uchicago.edu. Available at http://forecast.uchicago.edu/index.html. Accessed on 7 Feb 2019

Boon H (2010) Climate change? Who knows? A comparison of secondary students and pre-service teachers. Aust J Teach Educ 35(1):104–120

Climate Literacy and Energy Awareness Network CLEAN (2019) Available at https://cleanet.org/index.html. Accessed on 7 Feb 2019

Climate Change: Vital Signs of the Planet (2019) For educators | NASA global climate change. Available at https://climate.nasa.gov/resources/education/. Accessed on 7 Feb 2019

Dupigny-Giroux L (2010) Exploring the challenges of climate science literacy: lessons from students teachers and lifelong learners. Geogr Compass 4(9):1203–1217. https://doi.org/10.1111/j.1749-8198.2010.00368.x

Earth Learning Idea (2019) Available at https://www.earthlearningidea.com/home/Teaching_strategies.html#climatechange. Accessed on 7 Feb 2019

IRI at Columbia University (2019) IRI/LDEO climate data library. Available at http://iridl.ldeo.columbia.edu/index.html?Set-Language=en. Accessed on 7 Feb 2019

Kagawa F, Selby D (2013) Ready for the storm: education for disaster risk reduction and climate change adaptation and mitigation. J Educ Sustain Dev 6(2):207–217. https://doi.org/10.1177/0973408212475200

KCVS (2019) Visualizing the chemistry of climate change. Available at http://kcvs.ca/vc3/Lessons/homePage.htm. Accessed on 7 Feb 2019

Maier KJ, Whitehead GI, Walter MI (2018) Teaching psychology and climate change. Teach Psychol 45(3):226–234

MetEd (2019) MetEd: teaching and training resources for the geoscience community. Available at https://www.meted.ucar.edu/index.php. Accessed on 7 Feb 2019

Pfaff T (2019) Sustainability math—a quantitative literacy and mathematics resource for instructors. Sustainabilitymath.org. Available at http://sustainabilitymath.org/. Accessed on 7 Feb 2019

PhET (2019) New sims—PhET simulations. Available at https://phet.colorado.edu/en/simulations/category/new. Accessed on 7 Feb 2019

Plutzer E, McCaffrey M, Hannah A, Rosenau J, Berbeco M, Reid A (2016) Climate confusion among U.S. teachers. Science 351(6274):664–665

SERC (2019) Available at https://serc.carleton.edu/index.html. Accessed on 7 Feb 2019

The National Academies of Sciences, Engineering, and Medicine (2012) Climate change: lines of evidence. Available at https://www.youtube.com/playlist?reload=9&annotation_id=annotation_709415&feature=iv&list=PL38EB9C0BC54A9EE2&src_vid=qEPVyrSWfQE. Accessed on 7 Feb 2019

UCAR Center for Science Education (2019) Climate change activities| UCAR center for science education. Available at https://scied.ucar.edu/climate-change-activities. Accessed on 7 Feb 2019

UN CC:e-Learn (2019) One UN climate change learning partnership. Available at https://unccelearn.org/. Accessed on 7 Feb 2019

United Nations Sustainable Development (2019a) About the sustainable development goals—united nations sustainable development. Available at https://www.un.org/sustainabledevelopment/sustainable-development-goals/. Accessed on 7 Feb 2019

United Nations Sustainable Development (2019b) Climate change—united nations sustainable development. Available at https://www.un.org/sustainabledevelopment/climate-change-2/. Accessed on 7 Feb 2019

United Nations Sustainable Development (2019c) Education—united nations sustainable development. Available at https://www.un.org/sustainabledevelopment/education/. Accessed on 7 Feb 2019

Wise SB (2018) Climate change in the classroom: patterns, motivations, and barriers to instruction among colorado science teachers. J Geosci Educ 58(5):297–309

World Meteorological Organization (2019) WMOLearn: a global campus resource. Available at https://public.wmo.int/en/resources/training/wmolearn. Accessed on 7 Feb 2019

Zabel IH, Duggan-Haas DA, Ross R (2017) The teacher-friendly guide to climate change

Chapter 5
Researching Climate Change in Their Own Backyard—Inquiry-Based Learning as a Promising Approach for Senior Class Students

Sebastian Brumann, Ulrike Ohl and Carolin Schackert

Abstract Inquiry-based learning (IBL) enables students to personally experience climate change, which is often perceived as a rather abstract phenomenon. This article outlines characteristics of the IBL approach and sums up its most important benefits for climate change education. Recent relevant school projects, which address the topic of climate change, are presented. The article then describes main features of a prototypical IBL concept, which is being developed at the Chair of Geography Education at the University of Augsburg. In this project, Bavarian senior class high school students conduct research on the effects of climate change in their immediate surroundings. They work on individually generated study questions using anthropo-geographic and physiogeographic research methods. In order to do so, students are led towards a deeper understanding of the regional implications of climate change by working with online learning modules that additionally present adequate research methods. They also present their scientific findings in front of peers, a broader public and climatologists. Finally, first findings of the project portrayed above are discussed.

Keywords Climate change education · Inquiry-based learning · Fieldwork · Regional implications of climate change · Complexity

Teaching Climate Change—A Relevant and Challenging Matter

Current and pressing challenges of global climate change not only require corrective measures by governments and economy but also climate literate behaviour of the individual. Students, in particular, can make climate conscious decisions in areas

---

S. Brumann (✉) · U. Ohl · C. Schackert
Chair of Geography Education, University of Augsburg, Augsburg, Germany
e-mail: sebastian.brumann@geo.uni-augsburg.de

U. Ohl
e-mail: ulrike.ohl@geo.uni-augsburg.de

C. Schackert
e-mail: carolin.schackert@geo.uni-augsburg.de

© Springer Nature Switzerland AG 2019
W. Leal Filho and S. L. Hemstock (eds.), *Climate Change and the Role of Education*,
Climate Change Management, https://doi.org/10.1007/978-3-030-32898-6_5

such as personal mobility, holiday destinations, recreational activities, nutrition, consumerism, energy consumption and, partly, housing (Chiari et al. 2016).

Studies have shown that a certain climate literacy, i.e. basic knowledge about climate change as an issue as well as its causes and possible effects, does exist, at least in Central European societies (Chiari et al. 2016). At the same time, there is a frequently described discrepancy between a person's awareness or knowledge of climate change on the one hand and their actions on the other hand: although the risks of climate change are considered relevant, most people do little to counteract it (Renn 2018). A Europe-wide survey showed that especially people below the age of 24 contribute considerably less to climate-conscious behaviour than older groups (European Commission 2014). Personal factors that influence climate literate behaviour and help to explain this discrepancy are diverse. Chiari et al. (2016) illustrate that these factors can be both personal (as for example knowledge, values and attitudes, interests, perceived self-efficacy) and situational (as for example available infrastructure, economic circumstances or social context).

School education, especially geography education, is responsible for contributing to adequate climate change education by positively influencing the above-mentioned factors, as best as educational settings allow. In this context, teaching knowledge of climate change as well as the ability to act sustainably are crucial goals. As already mentioned, it is clear that the dissemination of knowledge alone does not automatically lead to climate literate behaviour. However, knowledge is an essential prerequisite for it. Results from a study by Ranney and Clark (2016) among American citizens are encouraging, as they show that the acceptance for anthropological climate change as well as the willingness to act grew with an understanding of the greenhouse effect. Additionally, these findings could be replicated in another study with German participants (Ranney and Clark 2016).

The didactic challenges that arise in connection with climate change education result in particular from the topic's complexity as well as a perceived temporal and spatial detachment from impacts of climate change.

The Complexity of Climate Change

Climate change is a complex issue on both a factual and an ethical level, as it is characterised by a tension between numerous interconnected aspects such as economic interests, ecological goals, cultural orientations, social norms and political decisions (Meyer et al. 2018). Approaches from both sciences and humanities are necessary to cover the subject adequately, assess its risks and find viable solutions (Meyer et al. 2018). As another facet of its factual complexity, it is local and global factors that play a role as well. With this in mind, factual complexity is then complemented by an ethical dimension. This becomes visible in the ongoing discussion on climate justice, for example.

In light of these complexities, it is not a surprise that international research in the field of Conceptual Change has uncovered challenges in teaching about climate change. Much of the research conducted shows that students often have wrong, highly persistent conceptions of climate change, especially with regard to the causes of climate change and possible solutions and, to a lesser degree, with respect to the impact of climate change (see Felzmann 2018 for a summary). Many children and adolescents mistakenly assume that the ozone hole is the cause of the anthropological greenhouse effect. Furthermore, without further differentiation, students often attribute the greenhouse effect to "emissions" that supposedly destroy the ozone layer. Another common misconception is that of a confined CO_2 or greenhouse gas layer in high altitudes (Schuler 2011; Felzmann 2018). Reinfried and Tempelmann (2014) show that scientific misconceptions formed prior to classes on the topic of climate change play a significant role in it. Depending on the characteristics of these scientific misconceptions, they may or may not be modified by learning environments, which are specifically designed to target them. Overall, research has shown that actively formulating and working with misconceptions is more effective than ignoring them (Felzmann 2018).

Attempts to explain the above-mentioned difficulties go beyond the topic's complexity as they also include the fact that, especially to people living in Central Europe, the mechanics of climate change are only perceivable to a small degree. Correspondingly, most people do not have personal experiences with climate change (Renn 2018). According to Reinfried (2007), in this case, many people might fall back on unscientific categories, which might lead to a causal connection of aspects such as the metaphoric term ozone hole, personally perceived temperature fluctuations, conceptions of a greenhouse glass ceiling and self-experienced higher concentrations of near-ground ozone.

Perceived Spatial and Temporal Detachment

In Central Europe and other regions of the world, the effects of climate change can only be perceived to a small degree, which causes a so-called "psychological distancing", i.e. a perceived spatial and temporal detachment from impacts of climate change. Consequently, people tend to ascribe them to other parts of the world (e.g. islands in the South Pacific Ocean) and a distant future (Chiari et al. 2016). In a study conducted with students of German high schools, Fiene (2014) shows that adolescents predominantly believe that climate change only happens on a global scale and that they consider climate change irrelevant to their own home. Consequently, they pay scant attention to the gradual nature of climate change, whose effects are barely perceivable (Renn 2018). Furthermore, they do not feel affected, have a low sense of responsibility and little willingness to act against the impacts of climate change. Due to psychological distancing and the misguided belief that climate change is merely a global phenomenon, many people share the conviction that their opportunities for actions are very limited and that they, as individuals, cannot make a difference (Renn 2018).

Inquiry-Based Learning as a Promising Approach

Researching climate change in their own backyard through inquiry-based learning (IBL) can be an opportunity for students that may help to overcome psychological distancing as it is based on individual interests, takes place close to home and in social interaction with peers.

In the following chapters, we will discuss potentials of inquiry-based learning for the topic of climate change in more detail. In addition, we will introduce our research project, in which students apply this very approach and summarise our experiences in a résumé.

IBL as a Promising Approach for Climate Change Education

It is obvious that there is no universal solution for the challenges of climate change education, but certain approaches seem to be particularly suitable for such settings. One exemplary approach is inquiry-based learning, which facilitates and optimises climate change education, especially with regard to the difficulties outlined above.

What Is IBL?

Inquiry-based learning is not a uniformly defined concept, but rather an array of didactic approaches that cover a wide range of educational areas and disciplines and can take different forms as well as names (Reitinger 2013; Brumann and Ohl 2019). However, all of them share a core of principal characteristics that, in light of the school context, can be described as a form of learning that aims for searching and finding insights, which are new at least to the learner, but can also be of interest to third parties. Attitudes and methods of inquiry-based learning are analogous to the fundamental mindset and approaches of scientific work (Messner 2009; Huber 2009). This means that, during their learning process, students partly or fully complete a typical scientific research cycle by developing questions and hypotheses, choosing and applying appropriate methods and then analysing, interpreting and presenting results (Huber 2009; Reitinger 2013; Pedaste et al. 2015). The goal is to apply and consequently acquire scientific core competences. These include both receptive research skills, such as information literacy, statistical literacy and critical thinking, as well as productive research skills. The latter include cognitive competences such as knowledge about research processes and methods, the generation of hypotheses or data analysis, affective-motivational competences such as research-related self-efficacy or a tolerance of uncertainty and ambiguity, and the social competence of cooperation in a learning community (Gess et al. 2017). Furthermore, IBL should at best require and encourage a so-called research attitude. According to Gess et al. (2017), this includes a reflexive distance—i.e. a reflective, questioning attitude—epistemic curiosity, in the sense of an intrinsically motivated tendency to find out new things (Kidd and Hayden 2015), as well as differentiated epistemological beliefs, i.e. the assumptions of a person about the nature of knowledge and the process of knowledge acquisition (Klopp and Stark 2016).

The range of didactic approaches, which correspond, at least in essence, to this basic character of IBL, is broad and varies in a number of dimensions. On the one hand, there are forms of teaching, which (for didactic and/or organisational reasons) are greatly simplified; they are pre-structured, narrowly focused and, hence, rather less scientific. On the other hand, there are approaches that can be very close to the model of scientific research, especially with an increasing age of the learners and in higher and adult education. As a rule, such concepts are only feasible as long-term projects. Their characteristics are that students work successively, on the basis of their own curiosity and questions and with a high level of self-control on complex topics that are closely connected to their everyday lives. With respect to content and methodology, the concepts align with (at times integrative) approaches from the humanities and natural science and/or humanities. Findings are then often of importance to third parties, i.e. their relevance goes beyond the individual learning processes (Brumann and Ohl 2019).

Potentials: How Can IBL Face the Above Mentioned Challenges?

There are many good reasons that argue for inquiry-based learning from an educational and scientific theoretical point of view. Also with regard to the challenges that successful education of topics relevant to sustainability, such as climate change, brings along, inquiry-based learning has considerable potential. For example, since the students are constantly asked to autonomously find, define and structure problems as well as plan, conduct and analyse investigations in communication and cooperation with other learners, IBL requires and fosters specific competences, which are highly important with regard to student's everyday lives. Moreover, the management of resources, the making of decisions and the fact that students have to be able to endure ambiguity enhances this effect further. Kuisma (2017) showed that a corresponding set of so-called "twenty-first century skills" can be promoted through inquiry-based learning. Many of these "design competences" can also be considered as important elements of education for sustainable development (ESD) (BLK Program Transfer-21 2007).

Among other things, Tilbury (2011) highlights the critical questioning of information as central to ESD. Especially in times of the internet and the concomitant complexity and controversy of available information, this is generally considered an important key skill. Relevant scientific studies have shown that inquiry-based learning has a positive effect on the ability to think critically (for example Apedoe et al. (2006) in the field of geology, or Uzunöz et al. (2018) in the sports sciences). Al-Maktoumi et al. (2016) observed that students participating in a hydropedological research project in Oman adopted an increasingly critical questioning attitude. In some cases, on the basis of own experiences, this attitude even led to the questioning of common textbook knowledge. Especially in regard to science education, Duran and Dökme (2016) were able to show that IBL fosters a better development of critical thinking compared to conventional approaches.

In addition, IBL also offers a great potential for dealing with the above-mentioned students' scientific (mis-)conceptions. In order to make use of these conceptions and to enable a Conceptual Change in the sense of constructivist learning processes (see e.g. Reinfried 2007), Schuler (2011) postulates the need for interdisciplinary and systematic learning as well as the requirement to teach students the ability to acquire knowledge by themselves and to critically question this information. In addition, education about global climate change should aim at the development of differentiated knowledge about problematic consequences, perpetrator roles and opportunities for action (Schuler 2011). Since these knowledge structures are often missing, it seems to be very important to make concrete, exemplary experiences. By the application of IBL, these experiences can possibly be produced to some degree. Because of a certain analogy between student's mental misconceptions and scientific hypotheses, which both have to be adapted or replaced when they fail through experience, this seems to be a promising approach. With the help of inquiry-based learning, students can succeed in empirically modifying the existing ideas by verifying concrete facts and scientific observations. According to this potential, Kukkonen et al. (2013) show that IBL has a positive effect on the students' perception of the greenhouse effect and the understanding of this phenomenon. More general, Chinn et al. (2013) show the potential of IBL for the promotion of Conceptual Change on the basis of selected studies and own investigations. The authors suspect that this is mainly due to the active negotiation of "evidence-theory linkages" in doing research.

Another advantage of researching a concrete—at best local—phenomenon is a potentially strong relevance to one's everyday life, which has been shown to be particularly favorable for the communication of climate issues among adolescents (Chiari et al. 2016). In doing so, IBL establishes a temporal and spatial proximity between the learner and the topic, because usually research projects are carried out in immediate vicinity to the student's own backyard. Accordingly, a special challenge of IBL is to put concrete experiences in a meaningful context with the rather abstract concept of climate change. However, if this is achieved through targeted didactic action, there is a tremendous opportunity to correspond to the general principle of meaningful global thinking. Pretorius et al. (2016) refer, for example, to the potential of IBL to apply relevant knowledge and competences on site and to reflect local challenges and opportunities. In other studies (e.g. Klein 1995), this could be identified as a necessity especially for geographic inquiry even more explicitly. Thus, access to a topic will be extended to an active and practical learning dimension, which is an important attribute of ESD (Tilbury 2011; Pretorius et al. 2016). However, this does not necessarily result in the reduction of psychological distancing, which in turn could favor an increased willingness to act, because environmental psychology has shown that people do not automatically care about proximal places and the elements, properties and qualities that constitute them (Brügger et al. 2015). Nevertheless, it is still very likely to create proximity to the object of learning through inquiry-based learning, because individual curiosity, interests and questions can be seen as the starting point of the learning processes. As a consequence, students devote themselves to these very aspects of their own immediate surroundings, which feature an exceptionally high potential of intrinsic motivation.

According to many research findings, own observation and research foster a better understanding of scientific concepts in particular (Markaki 2014). Especially in the field of Earth Sciences, two studies by Mao et al. (1998) and Chang and Mao (1999) should be mentioned, which showed that IBL promotes a generally greater learning success in researching geosciences compared to traditional approaches. In addition, the study by Klein (1995) also showed a notable increase in learning with regard to different geographic concepts and competences—including one topic on global climate change. Moreover, Namdar (2018) designed inquiry-based activities specifically on the topic of global climate change to provide future teachers with a tool to teach these relationships. The accompanying study showed a significant improvement in their understanding of global climate change.

An elaborate expertise, however, only represents one part of the felicity conditions for climate change education. The reduction of distance and abstractness can certainly facilitate the generally difficult step from knowledge to action. However, as explained above, a number of other important factors have to be added. In the research literature evidence can be found that these factors can possibly be supported by means of inquiry-based learning. It is a great challenge for teaching to overcome fear and feelings of powerlessness in relation to climate change and to replace it with the sense of self-efficacy. Gray (2018) for example shows that research in the field of earth sciences has a positive effect on the feeling of self-efficacy expectations of future teachers. Also, Sjödahl Hammarlund et al. (2013) concluded on the basis of their research that IBL promotes participants' motivation and self-efficacy expectations. This connection may be explained by the fact that a research cycle requires the learners to deal with unexpected events, solve problems independently and overcome setbacks. Learners can gain a strong sense of efficacy in situations of that sort, if the design of the IBL allows positive experiences. Additionally, these experiences may be a potential antidote to the mass media news broadcast that overwhelmingly has a threatening character.

Furthermore, tying in with the specific interests of each target group and the often heterogeneous interests within it, contributes to the success of climate communication. Various studies have shown that IBL leads to increased motivation among learners (e.g. Tuan et al. 2005; Bayram et al. 2013). The authors explain this with the comparatively great freedom of choice learners have in IBL, in the sense that learners have the possibility to control their own learning processes and choose the topics and methods based on their interests.

Finally, the willingness to adopt climate-friendly behaviour depends on certain values: while helpfulness or benevolence seem to behave a positive effect, values that focus on individual benefits tend to have negative effects on climate-friendly behaviour (Chiari et al. 2016). As values and attitudes are generally considered to be rather difficult to change, it is also difficult to make statements regarding the effect of inquiry-based learning. However, it can at least be assumed that one essential characteristic of scientific activity and, consequently, also IBL can make a contribution: research requires cooperation and communication as well as negotiating and bridging different perspectives, while "single-fighter mentalities" often make success more difficult.

Obviously, like any other form of learning, IBL does not per se achieve the afore-mentioned effects. The success of the method depends on the design of the respective concept. This includes a large number of didactic decisions, which can vary depending on the context. All in all, forms of inquiry-based learning that are very similar to science, which themselves imply a series of corresponding conceptual considerations and conditions of success, are more likely to produce the intended effects.

Recent Didactic Approaches that More or Less Meet the IBL Concepts

In light of what has been described above, it is not surprising that current educational projects on climate change include, more or less extensively, elements of inquiry-based learning.

A selection of the many examples of concepts that have been published in German recently are presented here:

ReKli:B

The interdisciplinary project *ReKli:B*, which took place between 2012 and 2016 at the Heidelberg University of Education, aims at promoting children's and adolescents' ability to evaluate regional climate change and their willingness to act upon it. Methodically, the concept is conducted with partner schools in the form of field studies, in-depth analyses in a geo-ecological laboratory and model- and experiment based approaches. Using local examples, it aims at making the ecological interrelations of climate change as comprehensible as possible. Based on several learning modules, participating students can conduct these research activities for different topics over a course of several lessons (Siegmund et al. 2017).

Climate ChangeS Cities

With *Climate ChangeS Cities* by the Ruhr-University in Bochum, the Trier University and the Heidelberg University of Education a new German educational project on climate change is under way. Inspired by *ReKli:B*, the project's emphasis is, again, on the enhancement of crucial competences such as evaluation and judgment skills and the capacity to act, and, in terms of content, the assessment of climate-related phenomena in urban spaces. *Climate ChangeS Cities* is based on the IBL approach described above, as students are enabled to self-directedly conduct investigations in their living environments, such as field observations in the city and laboratory work. Additionally, here based on the "experimental algorithm", students run through different stages of inquiry. According to first observation results, *Climate ChangeS Cities* seems to be a promising approach (Feja et al. 2019).

k.i.d.Z.21

The University of Innsbruck developed *k.i.d.Z.21* as a bi-national project in cooperation with a German school. It allows students to engage in individual subprojects, which can include questions on climate change from both the humanities and the sciences. Particularly noteworthy is the included five-day research stay in the Ötztal Alps, which ends the project of several months. As the effects of climate change are especially visible in mountain regions, the trip gives students the opportunity to experience the effects of climate change first hand and to investigate them independently by using geographical methods (Oberrauch et al. 2015).

The Augsburg Concept as an Exemplary IBL Approach in Climate Change Education

The examples presented above give an insight into possible designs of inquiry-based learning in connection with the topic of climate change. Almost always, the details of the concept depend on the organisational and thematic requirements of the respective educational context. Very often, time is a limiting factor, especially in more scientific forms of IBL. Consequently, finding ways to implement IBL as prototypically as possible into the given educational contexts, is particularly challenging. At Bavarian secondary schools, the so called "W-Seminare" in years 11 and 12 are a framework that offers the chance to teach propaedeutics and therefore prepare students for scientific work at universities. Over the course of one and a half years, students are enabled to develop science-related competences and a general ability to study. From the perspective of one school subject, the W-Seminars are also meant to make inquiry-based learning possible (ISB 2011). Although there are already numerous best practice examples of W-Seminar concepts, the full potential, especially with regard to the intention to give students' learning processes a clear scientific character, has not yet been exhaustively realised. Aiming at meeting both the requirements of propaedeutics and climate change education, the Chair of Geography Education at the University of Augsburg is currently developing a subproject of the Bavarian *BaySICS* climate research project initiated in 2018. The basic idea of the concept is to involve students, based on the concept of Citizen Science, in researching regional and local implications of global climate change in Bavaria. They use the one-and-a-half-year period to complete a full scientific research process, which includes acquiring basic competences, finding and developing individual research questions, planning and conducting data collections, analysing, interpreting and communicating results and, eventually, writing a scientific paper (see Fig. 5.1). In the following the didactic conception of the individual phases is briefly described:

Development of the Factual Basis for the Topic of Climate Change
The introductory lessons serve to give the students an organisational overview, make a first connection between the topic and their everyday lives and arouse curiosity. Following that, students shall delve into the topic for several weeks, paying special attention to regional aspects. A modular structure, which has already proven beneficial in other projects, is used to introduce the factual basics prior to the research

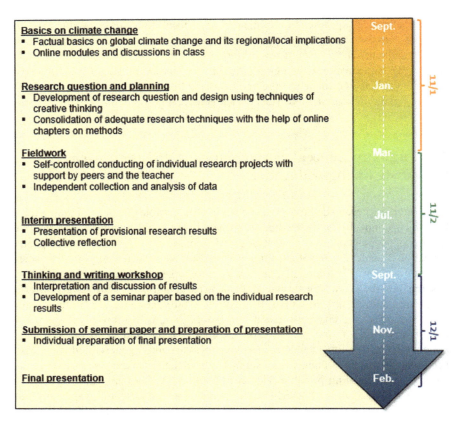

Fig. 5.1 Overview of sequences of the W-Seminar on regional/local implications of climate change in Bavaria

processes. There are several good reasons to make use of an online-based implementation: one main argument is the close relation to students' everyday lives, as most of them are confronted with online media contents on a daily basis. Therefore, it can be assumed that students not only show an affinity to such contents, but also basic technical competences in intuitively handling them.

At the same time, in contrast to analogue forms, online modules offer the chance to combine a wide range of learning materials (e.g. YouTube videos, audios, documents, hyperlinks to external databases, mapping services etc.) and thus different approaches. The *Onlinekurslabor* of the University of Augsburg, within which the modules have been realised, also offers the possibility of integrating a number of interactive tasks (e.g. multiple choice questions, drag and drop tasks, cloze tests) that already allow a certain amount of self-control in this first phase. As the concept is intended to be used at many different Bavarian schools, the online platform allows easy multiplication and the creation of a network for the participating schools. Additionally, a didactic system that includes current scientific findings—which are indispensable to a topic such as climate change—should be constantly modifiable. An online platform makes it possible to revise the content at a central point for all users.

At this point, the online course laboratory includes modules on "General basics of climate change", "Climate change perception", "Environmental protection", "Climate change and forest ecosystems", "Climate change related pollen pollution", "Phenology as an indicator of climate change" and "Climate change and tree line changes". Other modules on important anthropogeographic and physiogeographic topics are currently being developed.

Development of Individual Research Questions and Planning of Data Collection

The development of individual research questions based on personal interests plays an important role in inquiry-based learning (see Chap. 2.1). Accordingly, in a second phase of the W-Seminar, students elaborate regionally embedded research questions as well as suitable methodical designs based on their own ideas and interests that they developed earlier on closer examination of the topic. On the one hand, creative techniques can be used to help students further develop and/or narrow down their topic. On the other hand, measures of structuring, orientation and support ensure the emergence of a scientific research question. In addition, short application-oriented "how-to" chapters on climate-related research methods in the online course laboratory help students to create an optimal matching between their question and survey design.

Fieldwork

Based on the now established individual research questions, they start the actual fieldwork in their own living environments comprising the processes of searching and discovering, observing and measuring, inquiring and documenting, as well as reflecting and analysing. Regular reflective and metacognitive approaches should empower students to expand their thinking through their own inquiry as well as to address the nature of science and their own epistemic beliefs. However, in such a significantly self-controlled phase, constant individual support as well as the exchange and cooperation in the peer group play a pivotal role. This is realised in the form of regular meetings, in which central findings can be discussed. In order to satisfy these aspects in the best possible way, it is particularly important to allow sufficient time. Due to the generous time frame of the W-Seminar concept, it is possible to allocate around three months for the fieldwork.

Interim Presentations

An intermediary milestone are the presentations of the preliminary research results that take place before the summer holidays. They fulfil several functions: On the one hand, they mark a first success for the students, because by that time some data has been collected, first findings are noted and preliminary interpretations are made, which means that every student has already produced knowledge in a "raw form". On the other hand, the preparation and structuring of this knowledge also creates the potential for further mental differentiations. Furthermore, drawing up a presentation is, at the same time, a precursory product, which sets the ground for the subsequent learning phases. This not only serves as an orientation for the individual student, but also gives the teacher supervising information about the progress of the respective research projects. Lastly, it should also be mentioned that these presentations strengthen the students' communication skills.

Scientific Writing and Thinking Workshops and Conclusion of the Seminar

The last third of the W-Seminar focuses on the written elaboration of the individual research results in form of a term paper. Similar to the second phase of the seminar, the students should be provided with different measures of support to work out a written scientific paper comprising the necessary quality criteria. The central idea is to bring the social dimension of the acquisition of scientific knowledge into the classroom: Corresponding to symposiums, peer review procedures, interdisciplinary research projects, postgraduate programmes and the like, which characterise the scientific business as such, the students should perceive their own work as part of a knowledge-building community (see Scardamalia and Bereiter 2006). Accordingly, in this phase the individual inquiry is supplemented by regular seminar sessions with a workshop character, which primarily serve joint reflections and discussions, constructive criticism and mutual support. Besides the completion of the written term paper, the conclusive stage of the research process will be a final presentation, which gives the students the opportunity to present their results to an extended school public and to exchange ideas with interested specialists.

First Insights

The pilot stage of the seminar concept started in September 2018 with 14 students at a school in Augsburg. First experiences show that the following factors of the learning environment are relevant to successful learning:

Social Interaction

In the context of the previous activities, the role of sufficiently high proportions of social interaction became particularly obvious. This was especially evident in the first phase while working with the online learning modules. Whenever students discussed contents of the learning modules in class, a respectable learning progression and high level of motivation were apparent. Thus, several factors are potentially important: Firstly, the verbalisation of thoughts in the classroom requires a mental pre-structuring. Secondly, it could also be observed that the interrelated utterances of students and the teacher had a catalysing effect on the development of a common understanding. Thirdly, a classroom discussion establishes a particularly strong reference to everyday life. Although the learning modules were already designed with regard to the proximity of the contents to students' everyday life—e.g. by working with regional examples or tasks that aimed at the student's own experiences—in the vocal debriefings, students often resorted to self-selected current examples of specific local events or viral media issues. Moreover, in the context of oral interaction, it is also possible to respond specifically to aspects that arise situationally. The desire for opportunities for exchange and a compulsory backup of (intermediate) results explicitly mentioned by the students stresses the considerable potential of classroom discussions further. In contrast, stages in which students had to work autonomously with the online modules for longer periods turned out to be unfavourable in respect of progression and motivation.

Interaction of the Factual and Methodical Discussion

Especially in terms of motivation, but also in terms of the concrete comprehensibility of contextual relationships, the close interconnection of technical and methodological parts became obvious. With regard to opportunities in which students can learn about a scientific method, or in which they can exemplarily apply it, indicators for an increased curiosity and/or increased interest were observed. In addition, in such situations, the students were particularly well equipped to grasp the thematic aspects on the level of content that potentially could be explored with the respective method. On the one hand, this can be justified by the combination of procedural and declarative knowledge. On the other hand, the testing of a specialised method, such as the microscopy of pollen of nearby plants, the measurement of climate parameters in the schoolyard, or the interviewing of passers-by on campus, again establishes a relationship between scientific contents and genuine life-world situations. Accordingly, it is not surprising that those phases, which are solely based on the processing of the online modules, do not show comparable effects.

Measures of Guidance

The role of adequate measures of guidance, which has been emphasized for the success of inquiry-based learning processes by earlier studies (e.g. Lazonder and Harmsen 2016), could also be observed during the W-Seminar. Prompts and heuristics proved very useful for the learner group who had not worked in this way before and to foster students' awareness of when and how to perform certain tasks and actions. Especially in the case of specialised methods, detailed explanations were used to guide the learners step by step towards the correct application of the method. In order to allow students to focus on the contents only, the first interim presentations were scaffolded, i.e. the corresponding Power Point presentations were pre-structured on the basis of individual building blocks. Situations that lacked constraints and scaffolding turned out to be rather hindering for the learning process (for a detailed typology of IBL guidance see Lazonder and Harmsen (2016)). In general, different forms of support can be applied dynamically over the course of the seminar, as every phase of the research cycle may require specific strong control measures. However, for certain recurring inquiry skills such as communicating research results, the intensity and specificity of guidance measures can fade out successively, as students improve their competences during the course of the project.

Additional Influencing Factors
Even though a corporate design for the materials may seem trivial, it proved to be very favourable for the perception of identity and professionalism. Additionally, a learning environment, which is tailored to the perspectives and needs of the students and ascribes the teacher the role of a mentor and learning companion, turned out to be profitable, as it entails a strong learner-orientation. Conversations with the participating students also revealed that framing the concept with the Citizen Science Approach as well as the close exchange with the university, promote the feeling of participating in a superordinate societal task and thereby a sense of responsibility.

Conclusion

So far, experience with climate change education has shown that especially the factual and ethical complexity of the field and the perceived spatial and temporal detachment are key concerns. According to research findings, inquiry-based learning seems to be a promising approach to deal with these difficulties, as it fosters (among others) self-directed learning, critical thinking skills, the redesign of scientific misconceptions, local action on global issues, as well as the development of a feeling of proximity and perceived self-efficacy. The Augsburg IBL concept was designed to specifically fit the needs of senior class climate change education. Based on first insights, it seems suitable to achieve its intended goals. Nevertheless, important design principles that should be followed to obtain the desired results could also be identified.

References

Al-Maktoumi A, Al-Ismaily S, Kacimov A (2016) Research-based learning for undergraduate students in soil and water sciences. A case study of hydropedology in an arid-zone environment. J Geogr High Educ 40(3):321–339. https://doi.org/10.1080/03098265.2016.1140130

Apedoe XS, Walker SE, Reeves TC (2006) Integrating Inquiry-based Learning into undergraduate geology. J Geosci Educ 54(3):414–421. https://doi.org/10.5408/1089-9995-54.3.414

Bayram Z, Oskay ÖÖ, Erdem E, Özgür SD, Şen Ş (2013) Effect of inquiry based learning method on students' motivation. Procedia—Soc Behav Sci 106:988–996. https://doi.org/10.1016/j.sbspro.2013.12.112

BLK-Programm Transfer 21 (ed) (2007) Orientierungshilfe Bildung für nachhaltige Entwicklung in der Sekundarstufe I. Begründungen, Kompetenzen, Lernangebote, Berlin. http://www.transfer-21.de/daten/materialien/orientierungshilfe/orientierungshilfe_kompetenzen.pdf

Brügger A, Dessai S, Devine-Wright P, Morton TA, Pidgeon NF (2015) Psychological responses to the proximity of climate change. Nat Clim Chang 5(12):1031–1037. https://doi.org/10.1038/nclimate2760

Brumann S, Ohl U (2019) Forschendes Lernen im Geographieunterricht. In: Obermaier G (ed) Bayerischer Schulgeographentag 2018 Bayreuther Kontaktstudium Geographie, vol 10. Verlag Naturwissenschaftliche Gesellschaft Bayreuth e.V, Bayreuth, pp 25–41

Chang C-Y, Mao S-L (1999) Comparison of Taiwan science students' outcomes with inquiry-group versus traditional instruction. J Educ Res 92(6):340–346. https://doi.org/10.1080/00220679909597617

Chiari S, Völler S, Mandl S (2016) Wie lassen sich Jugendliche für Klimathemen begeistern? Chancen und Hürden in der Klimakommunikation. GW-Unterricht 141(1):5–18. https://doi.org/10.1553/gw-unterricht141s5

Chinn CA, Duncan RG, Dianovsky M, Rinehart R (2013) Promoting conceptual change through inquiry. In: Vosniadou S (ed) International handbook of research on conceptual change, 2nd edn. Routledge, New York, pp 539–559

Duran M, Dökme I (2016) The effect of the inquiry-based learning approach on student's critical-thinking skills. EURASIA J Math Sci Technol Educ 12(12):2887–2908. https://doi.org/10.12973/eurasia.2016.02311a

European Commission (ed) (2014) Climate change. Special eurobarometer 409. http://www.ec.europa.eu/commfrontoffice/publicopinion/archives/ebs/ebs_409_en.pdf

Feja K, Lütje S, Neumann L, Mönter L, Otto K-H, Siegmund A (2019) Climate changes cities—a project to enhance students' evaluation and action competencies concerning climate change impacts on cities. In: Leal Filho W, Lackner B, McGhie H (eds) Addressing the challenges in communicating climate change across various audiences. Springer International Publishing, Cham(CH), pp 159–174

Felzmann D (2018) Vorstellungen von Lernenden zu Ursachen und Folgen des Klimawandels und darauf aufbauende Unterrichtskonzepte. In: Meyer C, Eberth A, Warner B (eds) Klimawandel im Unterricht. Bewusstseinsbildung für eine nachhaltige Entwicklung, Diercke, Braunschweig, pp 53–63

Fiene C (2014) Wahrnehmung von Risiken aus dem globalen Klimawandel. Eine empirische Untersuchung in der Sekundarstufe I. Dissertation, Pädagogische Hochschule Heidelberg. https://www.opus.ph-heidelberg.de/files/47/Dissertation_Christina_Fiene.pdf

Gess C, Deicke W, Wessels I (2017) Kompetenzentwicklung durch Forschendes Lernen. In: Mieg HA, Lehmann J (eds) Forschendes Lernen. Wie die Lehre in Universität und Fachhochschule erneuert werden kann. Campus Verlag, Frankfurt, pp 80–90

Gray K (2018) Assessing gains in science teaching self-efficacy after completing an inquiry-based earth science course. J Geosci Educ 65(1):60–71. https://doi.org/10.5408/14-022.1

Huber L (2009) Warum Forschendes Lernen nötig und möglich ist. In: Huber L, Hellmer J, Schneider F (eds) Forschendes Lernen im Studium. Aktuelle Konzepte und Erfahrungen, (Motivierendes Lehren und Lernen in Hochschulen), vol 10. UVW Univ.-Verl. Webler, Bielefeld, pp 9–36

ISB/ Staatsinstitut für Schulqualität und Bildungsforschung (2011) Wissenschaftspropädeutisches Arbeiten im W-Seminar. Grundlagen—Chancen—Herausforderungen, 1st edn. München

Kidd C, Hayden BY (2015) The psychology and neuroscience of curiosity. Neuron 88(3):449–460. https://doi.org/10.1016/j.neuron.2015.09.010

Klein P (1995) Using inquiry to enhance the learning and appreciation of geography. J Geogr 94(2):358–367. https://doi.org/10.1080/00221349508979744

Klopp E, Stark R (2016) Persönliche Epistemologien—Elemente wissenschaftlicher Kompetenz. In: Mayer A-K, Rosman T (eds) Denken über Wissen und Wissenschaft. Epistemologische Überzeugungen, Pabst Science Publishers, Lengerich, pp 40–70

Kuisma M (2017) Narratives of inquiry learning in middle-school geographic inquiry class. Int Res Geogr Environ Educ 27(1):85–98. https://doi.org/10.1080/10382046.2017.1285137

Kukkonen JE, Kärkkäinen S, Dillon P, Keinonen T (2013) The effects of scaffolded simulation-based inquiry learning on fifth-graders' representations of the greenhouse effect. Int J Sci Educ 36(3):406–424. https://doi.org/10.1080/09500693.2013.782452

Lazonder AW, Harmsen R (2016) Meta-analysis of inquiry-based learning: effects of guidance. Rev Educ Res 86(3):681–718. https://doi.org/10.3102/0034654315627366

Mao S-L, Chang C-Y, Barufaldi JP (1998) Inquiry teaching and its effects on secondary-school students' learning of earth science concepts. J Geosci Educ 46(4):363–367. https://doi.org/10.5408/1089-9995-46.4.363

Markaki V (2014) Environmental education through inquiry and technology. Sci Educ Int 25(1):86–92

Messner R (2009) Forschendes Lernen aus pädagogischer Sicht. In: Messner R (ed) Schule forscht. Ansätze und Methoden zum forschenden Lernen. Körber-Stiftung, Hamburg, pp 15–30

Meyer C, Eberth A, Warner B (2018) Einführung. In: Meyer C, Eberth A, Warner B (eds) Klimawandel im Unterricht. Bewusstseinsbildung für eine nachhaltige Entwicklung, Diercke, Braunschweig, pp 4–5

Namdar B (2018) Teaching global climate change to pre-service middle school teachers through inquiry activities. Res Sci Technol Educ 12(2):1–23. https://doi.org/10.1080/02635143.2017.1420643

Oberrauch A, Keller L, Riede M, Mark S, Kuthe A, Körfgen A, Stötter J (2015) k.i.d.Z.21—kompetent in die Zukunft—Grundlagen und Konzept einer Forschungs-Bildungs-Kooperation zur Bewältigung der Herausforderungen des Klimawandels im 21. Jahrhundert. GW-Unterricht 139(3):19–31

Pedaste M, Mäeots M, Siiman LA, de Jong T, van Riesen SAN, Kamp ET, Manoli CC, Zacharia ZC, Tsourlidaki E (2015) Phases of inquiry-based learning: definitions and the inquiry cycle. Educ Res Rev 14:47–61. https://doi.org/10.1016/j.edurev.2015.02.003

Pretorius R, Lombard A, Khotoo A (2016) Adding value to education for sustainability in Africa with inquiry-based approaches in open and distance learning. Int J Sustain High Educ 17(2):167–187. https://doi.org/10.1108/IJSHE-07-2014-0110

Ranney MA, Clark D (2016) Climate change conceptual change: scientific information can transform attitudes. Top Cogn Sci 8(1):49–75. https://doi.org/10.1111/tops.12187

Reinfried S (2007) Alltagsvorstellungen und Lernen im Fach Geographie. Zur Bedeutung der konstruktivistischen Lehr-Lerntheorie am Beispiel des Conceptual Change. Geographie Schule 168:19–28

Reinfried S, Tempelmann S (2014) Wie Vorwissen die Lernenden beeinflusst—Eine Lernprozessstudie zur Wissenskonstruktion des Treibhauseffekt-Konzepts. Zeitschrift für Geographiedidaktik 42(1):31–56

Reitinger J (2013) Forschendes Lernen. Theorie, Evaluation und Praxis in naturwissenschaftlichen Lernarrangements (Theorie und Praxis der Schulpädagogik), vol 12, 2nd edn. Prolog-Verlag, Immenhausen

Renn O (2018) Klimaveränderungen als systemisches Risiko erkennen—Wege zur Handlungsbereitschaft. In: Meyer C, Eberth A, Warner B (eds) Klimawandel im Unterricht. Bewusstseinsbildung für eine nachhaltige Entwicklung, Diercke, Braunschweig, pp 77–85

Scardamalia M, Bereiter C (2006) Knowledge building: theory, pedagogy, and technology. In: Sawyer RK (ed) The Cambridge handbook of the learning sciences (1. publ). Cambridge University Press, Cambridge, pp 97–118

Schuler S (2011) Alltagstheorien zu den Ursachen und Folgen des globalen Klimawandels. Erhebung und Analyse von Schülervorstellungen aus geographiedidaktischer Perspektive (Bochumer Geographische Arbeiten), vol 78. Europäischer Univ.-Verl, Bochum

Siegmund A, Brockmüller S, Schuler C, Volz D (2017) Regionalen Klimawandel beurteilen lernen—interdisziplinärer Ansatz schulischer und außerschulischer Umweltbildung am Beispiel des UNESCO Geo-Naturparks Bergstraße-Odenwald. Abschlussbericht des Umweltbildungsprojekts. Deutsche Bundesstiftung Umwelt, Heidelberg. https://www.dbu.de/OPAC/ab/DBU-Abschlussbericht-AZ-29231.pdf

Sjödahl Hammarlund C, Nordmark E, Gummesson C (2013) Integrating theory and practice by self-directed inquiry-based learning? A pilot study. Eur J Physiother 15(4):225–230. https://doi.org/10.3109/21679169.2013.836565

Tilbury D (2011) Education for a sustainable development: an expert review of processes and learning. UNESCO, Paris

Tuan H-L, Chin C-C, Tsai C-C, Cheng S-F (2005) Investigating the effectiveness of inquiry instruction on the motivation of different learning styles students. Int J Sci Math Educ 3(4):541–566. https://doi.org/10.1007/s10763-004-6827-8

Uzunöz S, Erturan Ilker G, Arslan Y, Demirhan G (2018) The effect of different teaching styles on critical thinking and achievement goals of prospective teachers. Sportmetre 17(2):80–95

Chapter 6
Energy Transitions: Linking Energy and Climate Change

John H. Perkins

Abstract Climate change and energy are clearly linked because of emissions of carbon dioxide (CO_2) and methane (CH_4) from production and combustion of fossil fuels. Educational programs on climate change, however, have not developed robust curricula to help students see both the challenges and requirements for mitigating climate change, which requires a transition away from fossil fuels to renewable energy used efficiently. A case review of disputes about exploratory hydraulic fracturing (fracking) for natural gas in Lancashire, UK, illustrates key points relevant to climate change, its mitigation, and energy transition. These points can be organized in a theoretical framework—political ecology—to construct chains of explanation that clarify issues underlying mitigation/transition pathways. Three lines of inquiry specify key issues: (a) the scale of energy challenges at national levels, (b) the relative strengths and weaknesses of competing primary energy sources, and (c) three budgets or constraints (Carbon, Energy, Investment). The framework and lines of inquiry identify key points for debates about mitigation and energy transitions. These findings derive from synthesis and analysis of scholarly literature, agency reports, and newspaper reports.

Keywords Climate change · Mitigation · Energy transition · Fracking · Natural gas · United Kingdom

Education, Climate Change, and Energy

The sciences of climate change and energy both date to the 19th century. In the 1820s, Joseph Fourier inferred the importance of greenhouse gases in warming the earth (Weart 2003), and Louis Agassiz declared in 1837 that glaciers had at one time extended much further south than they do now (Wright1898). The practical and commercial implications of climate science, particularly for agriculture, were

---

J. H. Perkins (✉)
Member of the Faculty Emeritus, The Evergreen State College, 236 Cambridge Avenue, 94708 Kensington, CA, USA
e-mail: perkinsj@evergreen.edu

© Springer Nature Switzerland AG 2019
W. Leal Filho and S. L. Hemstock (eds.), *Climate Change and the Role of Education*,
Climate Change Management, https://doi.org/10.1007/978-3-030-32898-6_6

intuitively obvious. Not until late in the 19th century, however, did scientific evidence suggest humans could affect climate (Arrhenius 1896).

Climate science blossomed after the 1950s, which after the 1980s led to worrisome projections about increasing concentrations of carbon dioxide (CO_2) and other greenhouse gases in the atmosphere (Weart 2003). The World Meteorological Organization of the United Nations organized the Intergovernmental Panel on Climate Change (IPCC) in 1988, and from the beginning IPCC reports included the human dimensions of climate science as well as the physical science (IPCC 1990).

Consequences of climate change also stimulated extensive educational efforts in secondary and higher education and the general public. Climate change science has long been a staple of environmental courses, exemplified by appearance of the 16th edition of a textbook, *Environmental Science* (Miller and Spoolman 2019). Moreover, in 2013, a curricular reform movement in the USA embedded climate change science as a necessary component of pre-university science education (Next Generation Science Standards 2019). The new standards have been widely embraced (National Science Teachers Association 2019).

The trajectory of energy science and education was somewhat different. Extensive inquiries in the 18th and 19th centuries—prompted by the invention of steam engines—generated the concept of energy. Human control was paramount, so both research and education focused on prediction and control of energy as heat, mechanical motion, and electricity. The laws of thermodynamics were the bedrock of modern physics and chemistry, and they quickly spread to other natural sciences. Energy as a physical concept expanded further with identification of energy in quanta and quantum mechanics (Smith 1998; Coopersmith 2010).

Similarly, energy science quickly spilled into what became the social sciences, based on commercial and political dimensions of energy (Carnot 1899; Jevons 1866). The drive to control energy for practical purposes, however, led to specialization in energy education. In the physical sciences and engineering, energy education divided into the study and mastery of fuel-technology pairs. For example, engineering schools trained petroleum engineers in the science and art of producing and using petroleum; nuclear engineers knew how to prepare and manage nuclear fission of uranium for making electricity, and, not incidentally, for making weapons. A nuclear engineer was not skilled in the science and arts of petroleum, and vice versa.

A similar specialization developed in the social sciences. Economists and business administration specialists could predict the likelihood of profits from developing a primary energy source, like petroleum or uranium. Policy and political specialists knew about the domestic and international dimensions of controlling and managing deposits of primary energy sources. Education about energy for the general public, schools, and universities tended not to exist, except as disciplinary degree programs. Energy seldom appeared in general education, courses for non-majors, or environmental studies.

The different trajectories of climate and energy education explain the differences found today. Climate change science has made many appearances in educational institutions. General education and courses for non-majors abound, and specialized courses of study exist in both the natural and social sciences. Students learned that uses of fossil fuels must decline, but they seldom learned about the means to accomplish the reduction.

In contrast, energy education remains mostly in specialized programs for majors, e.g. in physical science departments and engineering schools. Broad, non-traditional, interdisciplinary instruction in energy, such as might be relevant to environmental studies are limited in number. In the USA, for example, a survey in 2016 found that, at most, only 27% of 1690 institutions of higher education had such degree programs, minors, and certificates. This means that 73% or more did not (Luzadis et al. 2018). In short, non-science students had few opportunities to learn about energy.

This paper addresses the gap in energy education, particularly as it relates to climate change. It argues that non-science and general education students can learn about energy in ways different from the intense focus on physical science and fuel-technology pairs. They learn a concept of *energy transition*—the development of energy resources to replace fossil fuels—and about national and local perceptions of energy production and use. Study of the comparative strengths and weaknesses of different energy options will enable them to grasp the challenges of energy transition, which must occur to mitigate climate change (Matthews and Caldeira 2008; Hansen et al. 2013).

The first section examines a specific case: disputes about hydraulic fracturing (fracking) to extract natural gas in the United Kingdom. This case illustrates the challenges of energy transitions away from fossil fuels. The second section develops a framework for understanding the case, which environmental studies and sciences can utilize to communicate the challenges of energy transitions. These findings derive from synthesis and analysis of scholarly literature, agency reports, and newspaper reports.

Fracking Disputes in the UK

Fracking has stirred controversy in the UK since preliminary drilling in 2011, and national newspapers, including *The Guardian*, *The Times*, and *The Telegraph* have covered it. These disputes increased substantially in 2018. For example, on 15 October 2018, *The Guardian* reported that protesters blocked efforts by Cuadrilla Resources to explore the potential production of gas from the Bowland-Hodder shales near Blackpool on Preston New Road in Lancashire County. One protester locked his neck to the top of a scaffold, while another locked herself to the base. Two other activists locked themselves together with chains attached to a barrel in the middle of a road and blocked traffic. Protesters were angered by the national government's overruling the County's rejection of Cuadrilla's request for a permit and by jail sentences in September for a "public nuisance" that blocked trucks from bringing equipment to the site (The Guardian 2018).

When citizens resort to civil disobedience, it generally means that frustrations are high. It would be possible to dismiss the disputes as ordinary conflicts between strongly opposed factions, one an industry seeking a permit and the second frustrated citizens who oppose it: certainly at this site and possibly everywhere.

Does this dispute have relevance to a class on climate change or energy? After all, squabbles abound in life, but that doesn't necessarily make them important to university education. To the extent that the class aims to include mitigation of climate change, however, this dispute illustrates important issues. But to make sense of the events, instructors will have to delve deeper into the context and facts about the disruptions on Preston New Road. Without context, students may not understand positions held by different actors, and they will gravitate to confusion or to conclusions reflecting only the assumptions held by their families and friends (Henderson and Duggan-Haas 2014).

Begin with the gas, the natural resource Cuadrilla wants to sell. The Bowland-Hodder shales (hereafter "Bowland shales") extend across central Britain and offshore beneath the Irish Sea and North Sea. The British Geological Survey concluded in 2013 that the onshore shale formations may contain between 822 and 2281 trillion cubic feet of natural gas (23.3–64.6 trillion cubic meters). It's uncertain how much of this gas can be extracted (Andrews 2013). A Parliamentary study in 2013 estimated extraction rates of 8–20% (based on US data), which made the Bowland shales a significant source (Parliamentary Office of Science and Technology 2013). Other studies have been far more pessimistic (based on UK data) (McGlade et al. 2013; Ahmed and Rezaei-Gomari 2019).

Despite uncertainty, consider that in 2017, the UK had proven gas reserves[1] of 6.5 trillion cubic feet (0.18 trillion cubic meters), or at most about 1% of the lowest estimate of gas in the Bowland shales. In fact, the total estimated amount of gas in these shales equals as much as 33% of the entire world's proven reserves of gas (BP 2018), hence Cuadrilla's interest.

Commercial drive by Cuadrilla, however, does not fully explain the passion of protesters. After all, many firms operating in Britain earn high profits and generate disputes without prompting civil disobedience. To plumb the depths of anger at Preston New Road requires exploring at least six other topics:

- reliance of modern British society on energy in general and gas in particular;
- production methods;
- consequences of using gas;
- relations between local and national governments and their respective responsibilities;
- effects of fracking on local people; and
- motivations and constraints of actors.

---

[1] Amounts that with reasonable certainty can be commercially extracted, based on geologic data, current extraction technology, and current political and economic conditions.

Gas in the UK

The story *must* begin with the fact that Britain is a modern nation, which means that its national security, economy, and the lifestyles of its citizens require a perpetual flow of energy resources. In fact, Britain was the first fully modern state and the pioneer of the energy-intensive industrial revolution (Wrigley 2010). The country moved from agrarian, in which most people worked on farms and lived in rural areas, to modern, in which most people lived in cities and worked in industry and service jobs. No modern country has ever voluntarily abandoned energy-intensive lifestyles and reverted to agrarian life constrained by firewood as the energy resource (Perkins 2017).

This statement about Britain's modernity is so obvious that virtually no one bothers to think about it, let alone include it as a major premise of climate-change education. Significantly, however, Britain's status as a modern state—and the necessity its government leaders see to preserve it—underlies the context for both climate and energy education, from overarching theories down to the dust-up on Preston New Road. If climate-change education is to prepare students for tackling the challenges of reducing emissions of fossil fuels, it must also help them learn about energy and the absolute dependence of modern states on a perpetual flow of energy and energy services: hour-by-hour, day-by-day, and year-by-year. Gas plays a vital role in the UK.

In 2017, gas contributed about 38% of the UK's useful energy from all sources, approximately equal to the 38% contributed by oil and oil products. Overall, the UK produced about 63% of the total energy it used, but the country conducts extensive trade in fuels. Some of its production of fuels is exported, and many fuels consumed are imported. Imports came from Norway and other countries, and exports went to Belgium and Ireland. Total domestic production of gas in 2017 amounted to 53% of gas consumed (Department for Business, Energy and Industrial Strategy 2018).

Gas has many uses. In 2017, the two largest were in homes for space and water heating and cooking (34%) and in generating electricity (33%). Industries took 19%, and commercial plus government activities used 9%. The remaining 5% produced heat for sale, miscellaneous and non-energy uses (Department for Business, Energy and Industrial Strategy 2018).

Electricity is a vital secondary source of energy, used ubiquitously for heat, light, motion, mobility, communications, and data management. In 2017, gas was the single largest source of energy for making the UK's electricity (40%), followed by renewable energy (29%), uranium (nuclear power, 21%), coal (7%), and other fuels (3%) (Department for Business, Energy and Industrial Strategy 2018).

These simple numbers capture the point that the UK depends upon gas for powering the nation and maintaining itself as a modern state. The national government in London, the jurisdiction that carries most responsibility for ensuring stable availability of energy, clearly has an interest in assuring gas supplies. This duty does not mean that gas cannot be replaced by another source, but gas is key as things stand now.

One year's snapshot of the role of gas in the UK, however, doesn't capture the wider picture that undoubtedly drives energy decision-making on fracking at the national level. Historically, contributions of gas to the British economy have oscillated over the past three centuries. North Sea production launched the UK into a new relationship with gas in 1960, and gas consumption tripled by 1964 compared to 1920, although coal and oil dominated gas in the 1960s. Producers of gas manufactured from coal ceased operations in 1979 due to competition from natural gas from the North Sea. Unfortunately for the domestic gas industry, production of natural gas peaked in 2000 and has steadily declined since that time, due to depletion.

Imports of natural gas had begun in 1964 (Department of Business, Energy and Industrial Strategy 2019), and they surpassed domestic production in 2013 (Peng and Poudineh 2015). Dependence on imports of any vital commodity always raises concerns about security of supply, but most imports came from friendly European neighbors: Norway, Belgium, and The Netherlands. Significant supplies, however, came from Qatar as Liquified Natural Gas (LNG), perhaps more of a worry due to potential instability in the Middle East. Other countries, such as the USA and Russian Federation, supplied small amounts (Department for Business, Energy, and Industrial Strategy 2018).

Given the key role played by gas in the UK's economy and the country's dependence on imports of gas, the estimated reserves of gas in the Bowland shales takes on new significance. A policy-maker—thinking only about the stable availability of gas—might reasonably conclude that fracking the shales for gas could propel the UK once again into stronger energy security through gas produced domestically. Hence: support exploratory fracking on Preston New Road.

Next contextual question: where did fracking come from and why is it controversial? This technology for pulling oil and gas from underneath the soil is simple in concept and in some cases has been phenomenally successful on economic grounds. National governments may embrace fracking for energy security, but companies will pursue it seeking potential profits.

Fracking in its earliest form dates to using explosives to produce oil in the 1890s (Smith and Montgomery 2015), but the current technology changed quite a bit from the 1930s through the 1970s. Two practices, each introduced separately, came together in the 1970s, and in the early 2000s the techniques improved dramatically in extracting tightly bound gas and oil from shale formations. The major steps forward occurred in the USA.

The first step was "directional drilling." At first, wells were holes drilled straight down into the earth, but in the 1930s drillers mastered the practice of turning the drill-bit at up to a 90° angle (a right angle) to make the well hole go sideways or horizontally (Veatch et al. 2017). This practice expanded greatly in offshore oil drilling in California (Bourgoyne et al. 1986).

In the late 1940s, oil producers in Texas and Oklahoma introduced a second new practice: hydraulic fracturing (fracking), in which mixtures of fluids and sand were pumped at high pressure into vertical wellbores, and the shale formations they were seeking fractured just as they had with the explosives. The fluid fractured the rock, and the sand grains held the channels open to release the oil. This practice produced substantially more oil from wells and spread rapidly among oil drillers in the USA and around the world (Smith and Montgomery 2015).

Horizontal drilling and hydraulic fracturing joined hands in 1974 in a patented practice now usually just called fracking (Veatch et al. 2017). High pressure fluid reached far larger amounts of shale in a horizontal well than in a vertical one, and the fractured rock thus released far more oil and/or gas. Sand to hold open the fractured channels was joined by other kinds of materials (together called "proppants"), and after 2009 fracking technology expanded greatly in the USA. Fracking technology was so successful that current estimates attribute at least 30% of recoverable oil reserves and 90% of gas reserves in the USA to fracking (Smith and Montgomery 2015).

Fracking appealed to national governments interested in energy security and to companies seeking profit, but more than gas production was involved. Fracking increased trips by big lorries, noise, interference with normal traffic patterns, wear-and-tear on roadways, and fumes from the trucks. Add to that the visual impacts of derricks 125 ft. tall (38 m, building of 10–11 stories, 40% as tall as Big Ben tower) (Office of Oil and Gas 2019) to handle the drilling and injection processes, with noise and light pollution at night. And then add the smells of chemicals as they are injected and as gas and produced fluids come back to the surface. Then add the potential to contaminate drinking water and the necessity to store and/or dispose of large amounts of produced fluids, typically contaminated water. This toxic mix inevitably raised questions from local inhabitants near the fracking site and their representatives in local government. What a far-off national government saw as "energy security," a local government saw as a public nuisance. Recent studies suggest that proximity to fracking wells may be a threat to health and the need for continued research (Elliott 2018).

In addition, direct adverse consequences come from producing, refining, transporting, storing, and using gas, because natural gas is mostly methane (CH_4), a potent greenhouse gas. Some CH_4, "fugitive methane," inevitably escapes and adds to climate change. Even if the methane is burned as intended, combustion yields CO_2, and CO_2 is the greenhouse gas most responsible for temperature increases.

If gas replaces coal, as in the USA, then releases of CO_2 may diminish, because gas releases approximately 50% of CO_2 per unit of heat generated compared to coal (U.S. Energy Information Administration 2019). This reduction is highly important in the short-run, and some have argued that substituting gas for coal is a good or necessary way to mitigate climate change. This argument is weakened, however, because fugitive methane will exacerbate climate change. Also, even though gas produces less CO_2 than coal, over time the CO_2 will exacerbate climate change.

The UK has made strong commitments to the EU and the Paris Climate Agreement to curb emissions of greenhouse gases, especially CH_4 and CO_2, so at the very least an embrace of fracking to produce gas is contradictory, and potentially devastating, to political commitments. The Committee on Climate Change, a statutory advisory committee for the national government, concluded in its first report (2016) that fracking for gas would not be compatible with UK climate goals, unless three, stringent factors were also present: (a) effective enforcement of regulations about CH_4 emissions at the production, transport, and use sites; (b) use of gas from fracking strictly to replace imported gas; and (c) any increases in production from shale be offset by decreased production of greenhouse gases elsewhere (Committee on Climate Change 2016). Optimistically, the UK government argued it could meet these guidelines (Skea 2018), but protesting NGOs and citizens disagreed.

Jim Skea, a policy analyst at Imperial College and member of the Committee on Climate Change, summarized the dilemmas. Production would help with energy security and potentially with the balance of trade in energy products. If fracking is successful, however, it might interfere with the UK's ability to meet its self-determined climate goals. Finally, approval of exploratory fracking in Lancashire began only after the national government overruled the Lancashire County Council, on which both Labour and Conservative members were split, but a majority voted to deny permits to Cuadrilla (Skea 2018).

Given the above discussion, we can identify the key actors and their clashing positions:

- Cuadrilla Resources has been persistent in seeking permission for its project on Preston New Road, because it might be lucrative financially.
- Local government (Lancashire County Council) in 2015 denied a permit to Cuadrilla to frack on Preston New Road. The sentiment was not primarily partisan politics.
- The national government, in 2016, overruled Lancashire County Council based on the national government's concerns about energy balance and trade issues. In addition, successful production would result in tax revenues to support government programs.
- Environmental NGOs, such as Frack-Off, vehemently opposed the Preston New Road project, because of climate change and local threats to health and the environment.
- Activist members of the public vehemently opposed the project on Preston New Road. Probably a mix of concerns motivated them, including climate change and local threats and nuisances. Some committed civil disobedience and went to jail.

Integrating Climate Change Education and Energy Education

At a simple level, the links between energy and climate change leap out of the example of conflicts about fracking for natural gas on Preston New Road: natural gas plays a role in debates about climate change. Parties in favor of fracking (industry, national government) think that shale gas will not interfere with the UK's climate goals, but environmental NGOs and citizens disagree.

Instructors can include the Lancashire case as they explain the sources and effects of greenhouse gases, assess the consequences of commercial fracking, and explore conflicting opinions. Higher temperatures will result from fracking unless greenhouse gas emissions drop elsewhere. In addition, CO_2 contributes to ocean acidification, a serious consequence of increased atmospheric CO_2.

Climate change education that simply recommends reductions in the uses of fossil fuels, however—without delving into what such an action entails—will leave students less prepared for leadership and action. Moreover, they may despair that anything can be done. If all mitigation requires is reduction of use, why is it so difficult?

One response from climate change educators might be to say that these are issues for "someone else," maybe educators in engineering, economics, political science, or public administration. Or maybe from land management or philosophy, especially ethics. Mitigation clearly involves issues of engineering, policy, land use, and ethics, among others. On this line of thought, it is enough for climate change educators to point out the problems of producing and using more CH_4 and refer further study elsewhere.

More productively, climate change educators can seek models and frameworks that confront the challenges of mitigation. Climate change educators know that some students will learn nothing more about climate change than is taught in their classes. To turn them out without grappling with mitigation doesn't help them play a powerful role in mitigating climate change. Climate change educators can point the way to understanding mitigation challenges, even if they can't provide full insights. They have an obligation to their students and to society to teach about solutions, not just problems, and ultimately multi-disciplinary efforts are required (Blockstein 2019).

No single discipline is any better prepared to tackle all the challenges of mitigation, but those who teach climate change can contribute. Students will benefit from deeper inquiry into energy production and use, whether they go on to specialized work on climate change and energy or they go into other fields but play important roles as educated citizens and leaders.

I maintain that interdisciplinary educational programs on climate change can acquire more power by placing the challenges of mitigation within a larger theoretical framework that instructors who teach climate change can master. Such frameworks foster abilities to compare different case studies and help students see general principles that apply across many stories illustrating mitigation challenges.

Four elements comprise the approach needed for interdisciplinary instructors to link energy and climate change education:

- A framework to organize chains of explanations
- Recognizing scale of challenges
- Comparing strengths and weaknesses of nine primary energy sources
- Understanding three budgets or constraints.

Framework to organize chains of explanation

In other places, I have used the terms *energy economy* or *energy budget* to indicate the amounts of energy used and the different patterns of using primary energy sources

(Perkins 2017). These two terms are not in wide use, either in environmental studies or everyday life, but a contingent of scholars in environmental studies have proposed a framework, *political ecology*, that leads to them (Robbins 2012; Vayda and Walters 1999).

Political ecology is far less used than the term *political economy*, which appears frequently in parts of the social science literature to integrate the interweaving of human culture and politics with the creation and distribution of wealth. Like political economists, political ecologists seek chains of explanation to understand interweaving of human culture and politics with ecology, the distribution and abundance of organisms.

Figure 6.1 presents a visual schematic of the political ecological framework in general terms and shows the linkage between climate change and energy. Figure 6.2 uses the general scheme in Fig. 6.1 to show the links in the case of exploratory fracking underway by Cuadrilla Resources on Preston New Road.

These main points emerge from the figures:

- Technology (fracking) transforms raw natural resources (fossil fuels in the ground) into fuels and then heat for energy services. No technology means no fuel, heat, energy services, or benefits.
- Emissions of CO_2 and CH_4 cause climate change, an unintended consequence of extracting and using fossil fuels. Nobody wanted CO_2 or fugitive methane, but essentially everybody wanted the energy services.
- Some benefits occur immediately, e.g. to heat water or buildings or to generate electricity. Some heat goes not to transitory benefits but to new investments supporting technology (fracking) to produce more energy (gas). Without investments, no technology, no continuous supplies of fuel, no continuous energy services.

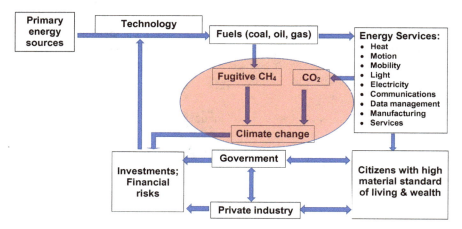

Fig. 6.1 Political ecology, general scheme. Investment is key to building technology to extract natural resources for energy services. Climate change, in shaded oval, results as an unintended consequence of using energy from coal, oil, and gas

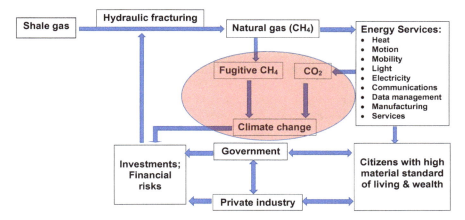

Fig. 6.2 Political ecology, specific for fracking. Investment in fracking technology extracts gas from shale to provide energy services. Climate change, in shaded oval, results as an unintended consequence of using energy from gas

- Investments in technology (fracking) carry financial risks. If these risks lead to economic failure, the technology will eventually go out of use. If the gamble succeeds, investors will continue production. Investors differ in their tolerance for losses and sometimes endure them for protracted periods, if potential rewards are large enough.
- Government plays a significant role in generation of investments and in tolerance of risks. They shape policies governing investments and risk. In some cases, government itself will be the investor, perhaps with high tolerance for risks. Successful investments inevitably produce constituencies (companies, investors, and employees) that advocate for favorable policies.
- Government also regulates technology (fracking) to reduce untended consequences: for example, rules on property rights, legal procedures, rights of citizens to protest, and taxes. Taxes discourage investments, but tax incentives promote them. Policies to protect against unintended consequences tilt toward higher risks or lower tolerance of them.
- If mitigation of climate change by curtailment of using fossil fuels is government's objective, then government should not promote investment in their extraction. Investors and others who benefit may argue against such policies.
- If citizens believe unintended consequences are intolerable and perceive government as favoring investment and unamenable to change, citizens may protest. If consequences of investment have some benefits and some negatives, citizens may be conflicted within themselves, or different groups of citizens may have conflicts.
- Different levels of government have responsibilities for different consequences of investment and may find themselves at loggerheads. Resolution of such conflicts will stimulate political actions that may be shifting and hard to predict.

Recognizing scale of challenges

Figures 6.1 and 6.2 depict the arena for chains of explanation among factors influencing investments and energy use. They also suggest an essential element about governing energy in modern states: scale of investment. Most students, and most citizens, know about energy use only at the individual or family scale. Energy measurements at national levels, however, are intrinsic to the challenges of mitigation.

Units of energy measurement can be a daunting subject, because the worlds of science and commerce use many different units for many purposes (Perkins 2017). Despite complexities, three simple points can lead students to effective learning. First, they most likely know a few common units to measure energy at the individual and family level. For example, they know about liters of petrol and diesel. Those who are living on their own may know electricity in terms of kilowatt-hours for electricity and therms for gas. They may have paid for these fuels, and the sophisticated ones will know they are purchasing *energy* or the ability to accomplish tasks like driving a car, lighting a room, or cooking a meal.

Second, they may have an intuitive knowledge that the more fuel or energy they purchase, the more they can do: drive more kilometers, light more bulbs, and cook more meals. In this way, energy is like money: if you have more, you can do more, and if you do more you must spend more money. Or they must use energy more efficiently so that less suffices: buy an efficient car or lightbulb.

Finally, even though students may know the units of liters of petrol or kilowatt-hours, they may not know that units measuring fuels or energy are all interconvertible and they can all be measured as heat. The most familiar measure of heat is likely to be calories to measure amount of energy in food.

Knowledge about fuels and energy at small scale helps transition to knowledge at national levels. The second point is key: human needs for energy services expand as one looks from individuals to whole populations. Part of expanding student familiarity with energy at national levels means recognizing that more people require larger amounts of energy. As a result, larger units of energy must be used.

The British and American governments use many different units to report national uses of energy, but this paper will use one from the metric system, the *Joule* (J). An individual's use of energy can usually be expressed in thousand Joules (kilojoules, kJ), million Joules (megajoules, MJ) or billion Joules (gigajoules, GJ). At the national level, other, less familiar units appear: the *petajoule* (PJ) or the *exajoule* (EJ). A PJ is 10^{15} J and an EJ is 10^{18} J. Big numbers to be sure, but conceptually these units are no more difficult to understand than a billion (10^9) or trillion (10^{12}) pounds or dollars, the amounts of national budgets.

The issue of scale for energy is easily demonstrated. Currently only nine primary sources of energy are available, and the USA (2017) draws 80% of its energy from fossil fuels, which release CO_2. To reduce these emissions by 80%, the country would lose 60% of its energy (Table 6.1). To minimize the losses of energy, coal and oil were cut to zero, because they release more CO_2 per unit of heat than gas (U.S. Energy Information Administration 2018a).

6 Energy Transitions: Linking Energy and Climate Change 99

Table 6.1 Reduction in emissions of CO_2 by reducing uses of fossil fuels (USA 2017)

Primary energy source	Energy provided (EJ)	CO_2 emissions (Million tons)	CO_2 emissions reduced 80% (Million tons)	Energy after reduction in CO_2 (EJ)	Energy remaining (%)
Coal	14.7	1318.0	0.0	0.0	0.0
Oil	38.2	2338.0	0.0	0.0	0.0
Gas	29.5	1472.0	1028.0	20.7	70.0
Uranium	8.9	0.0	0.0	8.9	100.0
Hydropower	3.0	0.0	0.0	3.0	100.0
Solar	0.8	0.0	0.0	0.8	100.0
Wind	2.4	0.0	0.0	2.4	100.0
Biomass	5.2	0.0	0.0	5.2	100.0
Geothermal	0.2	0.0	0.0	0.2	100.0
Total	103	5140.0	1028.0	41.1	40.0

Energy would drop by about 60% if CO_2 emissions from fossil fuels were to be reduced by 80%. Columns may not add exactly due to rounding errors

Economic and political catastrophe would result if such a drastic cut of emissions occurred rapidly without replacement with other energy sources and/or increased energy efficiency. The figures in Table 6.1 are thus imaginary in the sense that no government would ever voluntarily tolerate such a rapid reduction without replacement.

It may be easy to say, "Reduce the uses of fossil fuels," but the challenge of doing so is far from trivial. For the British situation on Preston New Road, national government leaders have understood the imperative for energy security and decided that the potential reserves of gas in the Bowland shales make fracking an important activity. In the government's view, decline of North Sea production amplifies the need.

Comparing strengths and weaknesses of primary energy sources

The third component needed to link climate change with energy education requires delving into strengths and weaknesses of each of the nine sources of energy currently available. More complete discussions of this topic emphasize the concept of sustainability, particularly the ability of energy sources to provide energy for the indefinite future without destroying common resources, such as the earth's climate (Perkins 2017). For the purposes of climate change education, three points require emphasis.

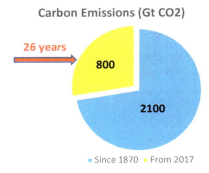

Fig. 6.3 The Carbon Budget, showing the amount of carbon emissions that can be emitted worldwide while maintaining a 66% chance of not driving global mean surface temperature above 2 °C. At current emission rates, the 800 Gt of CO_2-equivalent will be added to the atmosphere by fossil fuels alone in 26 years. Figure 3 adapted from Global Carbon Budget (2016)

First, promotion of energy efficiency is key. Modern civilization rests upon extensive uses of energy, but no source of energy is without harms to the earth's ecosystem. For this reason, the promotion of human well-being for the indefinite future should rest on parsimony: use energy, but always seek to use as little as necessary. Uses beyond this level add harms but no benefits.

Second, only nine choices of energy natural resources exist (coal, oil, gas, biomass, hydropower, uranium, wind, solar radiation, and geothermal heat). Energy services lost in mitigation must come from the other six sources or from improved energy efficiency.

Third, all fossil fuels, including gas, have fatal weaknesses, the most important of which is the release of CO_2; using gas also emits CH_4. If one accepts the argument that anthropogenic climate change poses potentially catastrophic risks for human well-being, then it follows that nations must transition to energy sources other than fossil fuels. The questions are which ones, how fast, and how much? The next section explores three constraints or budgets that must be coordinated to avoid unacceptable losses of energy from mitigation.

Understanding three budgets or constraints

Three budgets govern the speed and magnitudes of mitigation: (a) the Carbon Budget, (b) the Energy Budget, and (c) the Investment Budget.

The Carbon Budget (Fig. 6.3) comes from estimates of the amount of greenhouse gases that can be emitted without increasing the earth's temperature more than 2 °C. If humanity does not keep emissions below this estimated level, the potential for catastrophic risks from climate change will increase.

6 Energy Transitions: Linking Energy and Climate Change

In 2013, the IPCC estimated that the earth's temperature had better than a 66% chance of staying below 2 °C if the amount of greenhouse gases emitted after 1870 did not exceed the equivalent of 2900 giga-tonnes (Gt) of CO_2. By 2016, the equivalent of 2100 Gt of CO_2 had already been released, which left only a total of 800 Gt of CO_2 equivalent to have a reasonable chance staying below an increase of 2 °C, a level that will still cause severe consequences, some of which have already appeared (IPCC 2014; Le Quéré et al. 2016).

Emissions of CO_2 from fossil fuels alone will emit 800 Gt in 26 years at current rates of emission. Other sources of emission, such as land use changes, augment those from fossil fuels, so the time remaining before the maximum Carbon Budget is exceeded is less than 26 years. These estimates establish a timeline for mitigation to avoid higher probabilities of catastrophe from climate change. Success in the exploratory fracking on Preston New Road will result in more emissions of greenhouse gases, the exact opposite of the imperatives from the Carbon Budget.

The Energy Budget (Fig. 6.4) recognizes the need for energy and estimates the amounts needed to power a modern economy. The world used about 566 EJ of energy in 2017 (BP 2018), and this amount is projected to grow to about 686 EJ by about 2030 and about 765 EJ by 2040 (U.S. Energy Information Administration 2017). The

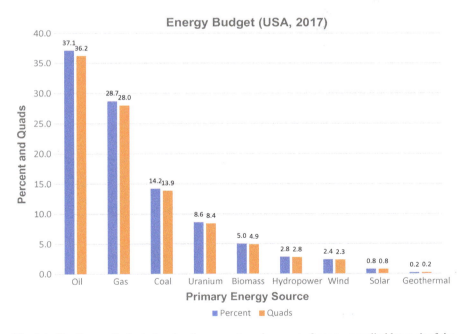

Fig. 6.4 The Energy Budget, showing the amounts and percent of energy supplied by each of the nine primary energy sources (USA 2017). Oil supplies most of the US energy, followed by gas, coal, uranium, biomass, hydropower, wind, solar, and geothermal. This is the amount of energy currently necessary to power the USA for one year

Energy Budget prompts the question, "Is it possible for primary energy resources other than fossil fuels to provide such amounts of energy to power the entire world's economy?"

Estimates indicate "yes, these amounts of energy can be procured." One study estimated, from wind and solar radiation, a minimum of 781 EJ and a maximum of 3805 EJ of electrical energy could be made available by 2070 (Deng et al. 2015). Such estimates always depend upon assumptions, and actual deployment of the technology needed may or may not match the assumptions. Nevertheless, these numbers suggest that—theoretically—energy from fossil fuels can be replaced by wind and solar radiation, which have far lower emissions of CO_2 and CH_4.

These estimates thus suggest that energy security in the future may not need gas that might be produced by fracking in the UK. In classes on climate change, the mitigation challenge thus becomes a matter of deploying alternative technology to provide energy from primary energy sources other than fossil fuels. Energy transition is the solution to the Carbon Budget that is compatible with the Energy Budget.

Third and finally, the Investment Budget looms (Fig. 6.5). An essential point to understand is that a decision to invest money entails first a selection of energy source. To decide on an investment in energy technology is also a decision to use that source's strengths and tolerate its weaknesses. That's why knowledge of comparative strengths and weaknesses is so important.

Then the next question arises. Based on current annual investments, do the amounts of money required to build an economy based on renewable resources look feasible? The answer to this question lies partly in the projected costs of energy procured, current estimates of which indicate a tentative "yes" but with considerable uncertainties.

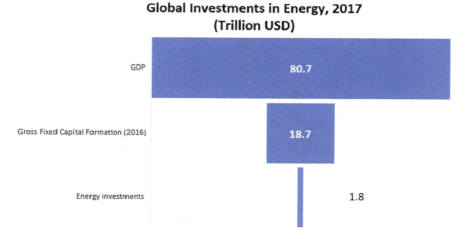

Fig. 6.5 The investment budget, showing the world's GDP, total gross fixed capital formation (all investments), and energy investments, 2017

For example, calculations of the Levelized Cost of Electricity (LCOE) estimate the cost of electricity delivered to the customer. Costs of electricity from wind and solar radiation are beginning to compete well with fossil fuels (U.S. Energy Information Administration 2018b). As LCOE of wind and solar begin to match those of fossil fuels, wind and solar gain strength as primary energy sources.

Another dimension of investments lies in their current magnitudes and the types of energy source selected. Investments are goods and services forming a part of the world's gross domestic product (GDP), which in 2017 was an estimated \$80.7 trillion. Of this, \$18.7 trillion or 23% of GDP went for investments in all sorts of things. Energy investments totaled \$1.8 trillion or 2% of GDP and 10% of total investments (Fig. 6.5) (World Bank 2019a; World Bank 2019b; International Energy Agency 2018).

The energy investments may look small, perhaps insignificant, compared to total GDP, but without these annual expenditures, the world's economy collapses. Most current investments (59%, 2017) support continued production of fossil fuels (International Energy Agency 2018), and thus in the long run they are not helpful to mitigate climate change.

Another uncertainty of wind and solar, however, focuses on the intermittency of wind and solar power, a real challenge. Expenditures for storage technology to solve intermittency may be large. Nevertheless, studies by the International Renewable Energy Agency in 2017 suggested that costs of storage technologies would continue falling in the next decade, just as they have in the last. Currently, installed storage capacity continues to increase (International Renewable Energy Agency 2017).

Conclusions

Clearly, education about climate change and energy share a common concern: emissions of CO_2 and CH_4 and how to reduce them to as near zero as possible, without losing the benefits of energy underpinning modern societies and nations. This paper advances the ability of educators in these two fields to join in meeting their common challenge. The two parts illustrate points, each of which must appear in learning programs about mitigation of climate change by energy transitions.

The case review of fracking disputes in Lancashire, UK, brought forth three features. First, energy resources occur in unrefined deposits on the earth. In this case, gas exists bound to shale formations, and fracking technology might extract it. Deploying fracking on a specific site immediately raises site specificity as a key concern, and site-specific issues will always arise, no matter whether it's gas, solar radiation, hydropower, uranium, or any other primary energy source.

Second, with few exceptions, most societies have levels of governance for large-scale issues and for local issues. These governments have different responsibilities, and their efforts to act for constituents may clash. In the Lancashire case, the local government saw nuisances, pollution, and health and environmental threats that local folks didn't want. The UK government, in contrast, saw issues of energy security, balances of trade, and other economic advantages to producing gas. Two governments disagreed, and the national government prevailed. Disagreements about governance can occur with all other energy sources.

Finally, all energy resources have strengths and weaknesses. Gas is plentiful, low cost, high energy density, and clean burning other than CO_2 and CH_4 emissions. Gas also can mitigate climate change, if it replaces coal and if CH_4 emissions are controlled. Alternative energy sources developed for energy transitions will each have their own strengths and weaknesses. Wind and solar, for example, are intermittent, but gas can generate electricity constantly.

In contrast to site-specific information, the political ecological framework and lines of inquiry support learning about mitigation-transitions in general and at scales larger than local. Students need to grasp that investment in technology must precede delivery of benefits from energy, and the investments must be continuous. Furthermore, an investment decision is the same thing as choosing an energy source.

Students must also learn the dimensions of energy at large scale. Climate change is a global problem, and it is meaningless for mitigation to have reductions of greenhouse gases at just a few locations. Mitigation of climate change must occur at local, national, and global levels, and so, therefore, must energy transitions.

Finally, students must learn to see the importance of congruent solutions to the Carbon, Energy, and Investment Budgets. Without congruence, mitigation will not succeed, nor will energy transition.

Acknowledgements I'm deeply indebted to comments and critiques by David Blockstein, Cathy French, and Andy Jorgensen. Comments from anonymous reviewers were very helpful for revisions. I remain responsible, however, for all errors and garbled communications.

Origins and Support for this Work

Portions of the material included in this paper originated from work leading to several publications since 2010, including the book, *Changing Energy: The Transition to a Sustainable Future* (Oakland: University of California Press, 2017). Almost without exception, the work has been self-financed by the author, who claims no conflicts of interest.

References

Arrhenius S (1896) On the influence of carbonic acid in the air upon the temperature of the ground. Philos Mag J Sci 41:237–276. Series 5

Ahmed M, Rezaei-Gomari S (2019) Economic feasibility analysis of shale gas extraction from UK's Carboniferous Bowland-Hodder Shale Unit. Resources 8(5). https://doi.org/10.3390/resources8010005

Andrews IJ (2013) The carboniferous Bowland shale gas study: geology and resource estimation. Br Geol Surv, London

Blockstein DE (2019) Personal communication. 26 Jan

Bourgoyne AT Jr, Millheim KK, Chenevert ME, Young FS Jr (1986) Applied drilling engineering. Society of Petroleum Engineers, Richardson

BP (2018) BP statistical review of world energy 2018. BP, London

Carnot S (1899) Reflections on the motive power of fire. In: Magie WF (ed) The second law of thermodynamics: memoirs by Carnot, Clausius, and Thomson. Harper and Brothers, New York

Committee on Climate Change (2016) Onshore petroleum: the compatibility of UK onshore petroleum with meeting the uk's carbon budgets. Committee on Climate Change, London. https://www.theccc.org.uk/publications/

Coopersmith J (2010) Energy, the subtle concept: the discovery of feynman's blocks from Leibniz to Einstein. Oxford University Press, New York

Deng YY, Haigh M, Pouwels W et al (2015) Quantifying a realistic, worldwide wind and solar electricity supply. Glob Environ Change 31:239–252

Department for Business, Energy and Industrial Strategy (2018) Digest of United Kingdom energy statistics 2018. Department for Business, Energy and Industrial Strategy, London

Department for Business, Energy and Industrial Strategy (2019) Historical gas data: production and consumption and fuel input 1920–2017. Found at https://www.gov.uk/government/statistical-data-sets/historical-gas-data-gas-production-and-consumption-and-fuel-input. 9 Jan

Elliott EG (2018) A community-based evaluation of proximity to unconventional oil and gas wells, drinking water contaminants, and health symptoms in Ohio. Environ Res 167:550–557

Global Carbon Budget (2016) http://www.globalcarbonproject.org/carbonbudget/archive.htm#CB2016. 13 Feb

Hansen J et al (2013) Assessing "dangerous climate change": required reduction of carbon emissions to protect young people, future generations and nature. PLoS ONE 8(12):e81648

Henderson JA, Duggan-Haas D (2014) Drilling into controversy: the educational complexity of shale gas development. J Environ Stud Sci. https://doi.org/10.1007/s13412-013-0161-9

International Energy Agency (2018) World energy investment, 2018. International Energy Agency, Paris

International Renewable Energy Agency (2017) Electricity storage and renewables: costs and markets to 2030. International Renewable Energy Agency, Abu Dhabi

IPCC. (1990). Climate change: the IPCC response strategies. IPCC, Geneva. https://www.ipcc.ch/reports/. Accessed 11 Feb 2019

IPCC (2014) Climate change 2014, synthesis report. IPCC, Geneva

Jevons WS (1866) The coal question: an inquiry concerning the progress of the nation and the probable exhaustion of our coal mines, 2nd edn. Macmillan and Co., London

Le Quéré C et al (2016) Global carbon budget 2016. Earth Syst Sci Data 8:605–649. https://doi.org/10.5194/essd-8-605-2016

Luzadis VA, Hirsch PD, Huang X (2018) Energy programs in higher education in the United States: assessing trends across two pathways to knowledge development. National Council for Science and the Environment, Washington

Matthews HD, Caldeira K (2008) Stabilizing climate requires near-zero emissions. Geophys Res Lett 35(2008):L04705. https://doi.org/10.1029/2007GL032388

McGlade C, Speirs J, Sorrell S (2013) Methods of estimating shale gas—comparisons, evaluation, and implications. Energy 59:116–125

Miller GT, Spoolman SE (2019) Environmental science, 16th edn. Cengage Learning, Boston

National Science Teachers Association (2019) About the next generation science standards. National Science Teachers Association, Arlington. https://ngss.nsta.org/About.aspx

Next Generation Science Standards (2019) Developing the standards. https://www.nextgenscience.org/developing-standards/developing-standards

Office of Oil and Gas (2019) What it looks like. Found at http://www.ukoog.org.uk/community/what-it-looks-like

Parliamentary Office of Science and Technology (2013) UK shale gas potential. Parliamentary Office of Science and Technology, London

Peng D, Poudineh R (2015) A holistic framework for the study of interdependence between electricity and gas sectors. Oxford Institute for Energy Studies, Oxford

Perkins JH (2017) Changing energy: the transition to a sustainable future. University of California Press, Oakland

Robbins P (2012) Political ecology: a critical introduction, 2nd edn. Wiley, Malden

Skea J (2018) The United Kingdom: to develop or not to develop? In: Gamper-Rabindran Shanti (ed) The shale dilemma. University of Pittsburgh Press, Pittsburgh

Smith C (1998) The science of energy: a cultural history of energy physics in Victorian Britain. University of Chicago Press, Chicago

Smith MB, Montgomery CT (2015) Hydraulic fracturing. CRC Press, Boca Raton

The Guardian (2018) Anger and blockades as fracking starts in UK for first time since 2011. Found at https://www.theguardian.com/environment/2018/oct/15/fracking-protest6ers-blockade-cuadrilla-site-where-uk-work-due-to-restart

U.S. Energy Information Administration (2017) International energy outlook 2017. US Energy Information Administration, Washington

U.S. Energy Information Administration (2018a) Monthly energy review May 2018. U.S. Energy Information Administration, Washington

U.S. Energy Information Administration (2018b) Levelized cost and levelized avoided cost of new generation resources in the annual energy outlook 2018. U.S. Energy Information Administration, Washington

U.S. Energy Information Administration (2019) Frequently asked questions: how much carbon dioxide is produced when different fuels are burned? Found at https://www.eia.gov/tools/faqs/faq.php?id=73&t=11. 28 Jan

Vayda AP, Walters BB (1999) Against political ecology. Hum Ecol 27(1):167–179

Veatch RW, King GE, Holditch SA (2017) Essentials of hydraulic fracturing: vertical and horizontal wellbores. PennWell Corp, Tulsa

Weart SR (2003) The discovery of global warming. Harvard University Press, Cambridge

World Bank (2019a) GDP (current US$). Found at https://data.worldbank.org/indicator/NY.GDP.MKTP.CD. 28 Jan

World Bank (2019b) Gross fixed capital formation (% of GDP). Found at https://data.worldbank.org/indicator/NE.GDI.FTOT.ZS?end=2017&start=1960&view=chart. 28 Jan

Wrigley EA (2010) Energy and the english industrial revolution. Cambridge University Press, Cambridge

Wright GF (1898) Agassiz and the ice age. Am Nat 32:165–171

Chapter 7
Delivery Mode and Learner Emissions: A Comparative Study from Botswana

Alexis Carr, Stanley Modesto, K. Balasubramanian, Kayla Ortlieb and John Lesperance

Abstract This paper analyses and compares the learning-related carbon emissions of primarily distance, CD-based and campus-based cohorts in Botswana. The study finds that the learning-related carbon footprint of campus-based learners is significantly larger than that of the CD-based learners, with travel being the greatest contributor to this disparity. Moreover, even when controlling for demographic variables, institution (as a proxy for delivery mode) is the most significant predictor of learner emissions. These findings suggest that blended and distance learning can contribute to the environmental sustainability of education by reducing learner emissions. The paper outlines implications for future research and discusses how universities can be encouraged to consider pedagogical design in their carbon reduction strategies.

Keywords Carbon emissions · Higher education · Learner emissions · Mode of delivery · Open and distance learning

Introduction

The education sector is responsible for a growing environmental footprint, attributable to both institutional and learner-generated emissions. In China, for example, the Ministry of Education (as cited in Li et al. 2015) estimates that the education

---

A. Carr (✉) · K. Balasubramanian · K. Ortlieb · J. Lesperance
The Commonwealth of Learning, Burnaby, Canada
e-mail: acarr@col.org

K. Balasubramanian
e-mail: Balakod@outlook.com

K. Ortlieb
e-mail: kmortlieb@gmail.com

J. Lesperance
e-mail: johntlesp@gmail.com

S. Modesto
Botswana Open University, Gaborone, Botswana
e-mail: stmodesto2006@gmail.com

© Springer Nature Switzerland AG 2019
W. Leal Filho and S. L. Hemstock (eds.), *Climate Change and the Role of Education*,
Climate Change Management, https://doi.org/10.1007/978-3-030-32898-6_7

sector is responsible for 40% of the energy consumption in the public sector. As global tertiary enrolments continue to rise, it is necessary to critically assess the environmental sustainability of higher education (HE) in its current form. Historically, the HE sector has not engaged intensively with sustainability. In fact, Tilbury (2011) points to the incompatibility of conventional HE and sustainability, asserting that "sustainability challenges the current paradigms and structures as well as the predominant practices in HE" (p. 1). However, with the introduction of the global Sustainable Development Goals (SDGs) and a growing focus on sustainability across productive sectors, modern higher education institutions (HEIs) are beginning to engage more meaningfully with sustainability through frameworks that seek to track, incentivise and improve sustainable practices. Within these emerging sustainability frameworks, much of the focus is on institutional infrastructure, consumption and waste, while little attention is given to how mode of delivery, particularly through its impact on learner behaviours, can affect emissions. While there is a general assumption that open and distance learning (ODL) is inherently 'green' or sustainable (Bourke and Simpson 2009), there has been "little research on the environmental sustainability of complex HE teaching models" or learner behaviours within these models (Caird et al. 2013, p. 1). Findings from the few studies that do exist suggest that this assumption has merit. In a 2011 study by Campbell and Campbell, the travel of 100 distance education students was tracked over the course of one semester. They estimated that by students not driving to campus for classes, the distance education mode resulted in a five to ten-ton reduction of greenhouse gas emissions. While this study showed a clear environmental benefit of reduced travel, it did not measure impacts of other learner behaviours, such as information and communications technology (ICT) use or paper consumption. Similarly, a study conducted by Güereca et al. (2013) of the Institute of Engineering at Universidad Nacional Autónoma de México estimated that the institution's emissions could be reduced by 22% if students and staff worked from home for two days a week and carpooled. However, the mode of delivery was not discussed in the study. Perhaps the most comprehensive project assessing the environmental impacts of different modes of delivery in higher education is the SusTEACH project, supported by the Open University, UK. The project compared the carbon emissions of ICT-enhanced courses with online courses and mainly face-to-face taught courses. The findings showed that online and blended ICT-enhanced distance teaching models had significantly lower environmental impacts than face-to-face teaching modes (Caird et al. 2013; Caird et al. 2015). This study used a comprehensive methodology that considered multiple contributors to emissions for both students and the institutions.

Recognizing the potential of ODL, the United Nations Environment Programme (2014) *Greening Universities Toolkit V2.0* mentions distance learning as a means to reduce emissions, which merits further study; however, it is not discussed in detail. Moreover, ODL is not duly considered or highlighted in many international policies and recommendations related to carbon emissions and higher education. Indeed, even though staff and student travel has been cited in several studies as a main source of an HEI's carbon footprint, recommendations rarely consider how mode of delivery could reduce emissions (Appleyard et al. 2018; Barros et al. 2018; Larsen et al. 2013; Li et al. 2015).

Although there is some evidence to suggest that distance teaching models can reduce learners' carbon emissions, the bulk of the current literature is based on analysis of institutions in the UK, and the assessment tools used in previous studies have not been applied widely in the developing world (see Caird et al. 2013, 2015; Davies 2015; Roy et al. 2008). Inferring similar results for developing countries is problematic for several reasons. Firstly, the electricity emissions factors (EFs) used in the carbon footprint calculation vary from country to country and depend on the energy source used for power generation. For example, in Canada hydroelectricity accounts for 59.3% of the country's electricity supply (Natural Resources Canada 2019), while in Botswana 100% of total installed capacity comes from fossil fuels (Central Intelligence Agency 2019). This means that the electricity EF of Botswana will be substantially higher than that of Canada. Additionally, developing countries may not have financial and regulatory frameworks in place to support or incentivise sustainable practices in HEIs; as a result, the emissions from developed countries may be lower than developing countries. On the other hand, learners in developing countries may have lower rates of consumption because of economic constraints, which could impact their learning-related carbon emissions. With limited financial resources, students in developing countries may be less likely to use vehicles to get to school and may purchase fewer physical materials or limit their resource use, resulting in lower emissions. Because the development context has the potential to greatly influence the carbon footprint of learners, further research into learner carbon emissions in the developing country context is needed.

Another gap in the literature on learner carbon emissions relates to the demographic factors that may influence consumption and energy use. While the SusTEACH methodology allows one to easily determine the components of the calculation that contribute the most to overall emissions, there is a paucity of information about the relationship between demographic factors, such as age, sex, or income and learner carbon emissions. Various studies on the relationship between demographics and personal carbon footprints (although not specifically learning-related carbon footprints) suggest that consumption patterns and thus personal emissions are indeed affected by individual demographic characteristics (Zhang et al. 2015). A 2009 European study by Räty and Carlsson-Kanyama demonstrated that men consume more energy than women do, with differences ranging from six percent in Norway and eight percent in Germany to 22% in Sweden and as high as 39% in Greece. They also concluded that energy consumption increased as household income rose. In addition to sex and income, age has been shown to affect personal carbon emissions (Räty and Carlsson-Kanyama 2009). According to Zagheni (2011) personal emissions in the US increase steadily from age 10 to around 60 and then drop. The study cites multiple behavioural changes associated with aging that account for this trend, such as a larger disposable income, travel for work, parenting responsibilities, and retirement. There is clear evidence pointing to the impact of demographic traits on overall personal carbon emissions; however, these demographic traits have not been thoroughly addressed in the existing research on learning-related carbon emissions vis-a-vis mode of delivery. Without controlling for the influence of demographic traits on learner carbon emissions, it could be suggested that differences between the carbon emissions of ODL and face-to-face learners are not the result of mode of delivery, but rather that they are determined by demographic traits that have not been considered. An exploration of demographic factors that may affect learner carbon footprint can strengthen the discourse on the environmental impact of ODL, with a view to better understand the relationships amongst demographics, mode of delivery and learner emissions.

Using self-reported data on a variety of learning-related activities in Botswana, the present study sought to determine whether there was a difference between the carbon emissions of CD-based ODL and face-to-face learners, as well as the main sources of learners' emissions. A regression analysis was run to ascertain whether mode of delivery was a determinant of learner carbon emissions, when controlling for demographic factors.

Methods

Sampling and Data Collection

The two institutions selected for this study were the Botswana College of Distance and Open Learning (BOCODOL) (now called Botswana Open University) and the Botswana Accountancy College (BAC). BOCODOL was a semi-autonomous distance teaching college. In addition to its headquarters in the capital city of Gaborone, Botswana, the college had five regional offices in Gaborone, Francistown, Palapye, Maun and Kang that operate through study centres country-wide. According to the institution's website its "mandate is to make education accessible to all Batswana especially for the out of school young and adults using Open and Distance Learning (ODL) methodology. Open learning seeks to break down the barriers to personal development by providing flexible learning environments, enabling people to study what is relevant to their needs, at a time and place convenient to them. The University aims to extend education and training using open and distance learning methods on a national scale," (Botswana Open University, n.d.). The college offered a school-based equivalency programme, short courses, certificates, diplomas, bachelor's and master's degrees, and has recently transitioned to a full-fledged open university. Botswana Accountancy College (BAC) is a conventional face-to-face institution with its main campus in Gaborone and another location in Francistown. BAC has programmes in accounting, finance, business, management, hospitality, leisure, taxation and ICT, with credentials ranging from short courses to master's degrees. It is renowned for its international partnerships with universities in the UK, as well as its strong research focus (Botswana Accountancy College, n.d.). From BAC, the face-to-face Bachelor's in Entrepreneurship and Business Leadership was selected, and from BOCODOL the comparable CD-based blended Bachelor's in Business and

Entrepreneurship was chosen. These programmes were purposively selected because they were the most similar units available for comparison. Within these programmes the graduating cohorts from the main campuses in Gaborone were sampled. Semi-structured questionnaires were developed based on the SusTEACH methodology of Caird et al. (2013). The surveys were reviewed by a local consultant, who made contextual adaptations (outlined in the next section 'Adapting the Susteach Methodology'). As survey methods may be prone to question misinterpretation (which can compromise validity), the survey was piloted with two students from each institution to identify potential interpretation difficulties and was revised accordingly. The revisions included the addition of more specific instructions about answering questions (i.e. instructing respondents to circle an answer) as well as changes or additions to word choice: For example, 'residence' was changed to 'home/residence'. The revised surveys were canvassed face-to-face at both the BOCODOL and BAC campuses. At the BOCODOL campus, the surveys were distributed at the end of a regularly scheduled contact class and completed by students independently. At BAC, surveys were distributed at the end of a regularly scheduled tutorial and were also completed independently. The survey canvassers were present while respondents completed the questionnaires and were available to clarify instructions and question items. Respondents were given the option of anonymity, and all respondents provided voluntary informed consent. Responses were recorded by respondents themselves, and then surveys were collected, and responses entered by the consultant and his assistants into a spreadsheet. The data analysis was conducted using IBM's Statistical Package for the Social Sciences (SPSS). All differences between CO_2 emissions reported were tested for significance using independent samples T-test and were reported only if significant ($p < 0.05$).

Population and Study Sample

Thirty-eight students were registered in the final year of the Bachelor's in Business and Entrepreneurship from BOCODOL and 52 students were registered in the final year of the BAC Bachelor's in Business and Entrepreneurship. Both programmes were based out of their respective institution's main campus in Gaborone. Of the 38 BOCODOL students, 26 were present at the contact class when the survey was canvassed, and all 26 responded, representing 68% of students in the programme. Of the 52 BAC students, 34 were present for the tutorial when the survey was conducted, and all responded, representing 65% of students. The sample size ensures a greater than 90% confidence level and a less than 10% margin of error to extrapolate to the all final year students in each programme. To collect data on institutional carbon footprint, various staff members from the respective institutions were interviewed and institutional records were analsyed. Table 7.1 outlines the breakdown of the research participants.

The generalisability of the findings from this micro-level study should be considered. While there are no specific indications that the students surveyed would be

Table 7.1 Research participants

Institution	CEO	Lecturers and heads of departments	Registrars	Student population	Students surveyed (the sample)
BOCODOL	1	3	1	38	26
BAC	1	3	1	52	34
Total	2	4	2	90	60

not be representative of other HE students in Botswana, the institutional, programme and learner characteristics of those surveyed may influence the results and should be considered when generalising the findings. Given that the surveys were administered after class, it is also possible that those who were absent from class during the survey are slightly different from those who were present, which may affect the representativeness of the respondents. As the surveys rely on participants' interpretations and memories there may be issues of reliability. Responses require students to make estimations based on past behaviours, so the data may be affected by recall bias. To mitigate this bias and facilitate recall, students were asked to estimate based on an average week rather than the entire length of the semester.

Adapting the SusTEACH Methodology

The SusTEACH methodology (Caird et al. 2013) is a tested approach to assessing the carbon-based environmental impact of higher education models. The comprehensive methodology involves the calculation of learner and institutional emissions, while the present study focuses in on only the learner emissions. While the SusTEACH methodology has been implemented in Europe, most extensively in the UK, the present study adapted the SusTEACH learner questionnaire and methodology for the developing country context of Botswana. One of the main adaptations is the calculation of the learner carbon footprint based on a semester, rather than 100 study hours (10 CAT hours), as the institutions surveyed do not calculate their programmes in terms of study hours. To resolve the collection of data based on weekly estimates, the number of weeks per semester (12) was used to multiply responses. Similarly, when data was solicited in terms of monthly estimates, it was multiplied by the number of months in the semester (three).

Certain questions and response options from the SusTEACH survey were modified based on feedback from local consultants who pilot-tested the questionnaire with learners from both institutions. Options which were deemed to be contextually inappropriate or irrelevant were excluded. For example, as there is no passenger train available for transport in Botswana, this option was removed from the survey. Similarly, the consultant reported that there were no recycling services, thus this was also removed as a response option and excluded as a factor in the analysis. Regarding the emissions calculations, in some cases values different from those used in the SusTEACH methodology were employed. Alternative values were considered due to the availability of other data sources and were selected at the discretion of the researchers. The same values were applied consistently to all responses, ensuring comparability. In cases where alternative values were used, the sources have been noted. Any differences between values from the SusTEACH methodology and those used in the present study are minimal, and it is expected that the findings are robust to these differences.

Travel Emissions

Travel emissions were calculated by multiplying the total programme-related distance travelled by the learner each semester (the number of trips multiplied by the round-trip km estimate) by the vehicle emissions factors (EF) (as cited in United Kingdom Department of Business, Energy and Industrial Strategy 2016), as no specific estimates for Botswana were found. All makes of personal vehicle reported used petrol and were classified as upper medium.[1] For car travel, the distance multiplied by the EF was divided by the number of people in the vehicle to account for reduced personal emissions because of carpooling.

Electricity Emissions

Electricity consumption was calculated both for home usage and classroom usage. For classroom usage, representatives of the institution provided the monthly Kwh used in the classroom for the specific student population of interest. This was divided by the number of students in the classroom (the total population) to get an estimate of the classroom consumption that could be attributed to each individual learner, and then multiplied by three, which is the number of months in each semester. The Kwh per student for one semester was then multiplied by the metric tons EF for electricity in Botswana[2] and multiplied by 1000 to convert metric tons to kilograms. On-campus device use was also calculated. An estimate of device wattage per hour for each type of device[3] was multiplied by the number of hours of use per week reported by students, and then divided by 1000 to get the Kwh. The Kwh was then multiplied

---

[1] Upper Medium Petrol Passenger Vehicle kg CO_2e emission factor of 0.21733 (rounded to 0.22).

[2] 0.000686166 Metric tons per KwH (as cited in Energy Research Centre, University of Cape Town, 2012).

[3] PC = 150 W; Laptop = 35 W; Tablet = 4 W; Mobile = 0.25Wh. Multiple sources were consulted, and these rounded averages were taken for ease of reference (see; Caird et al, 2013; Energy Saver, Office of Energy Efficiency and Renewable Energy, U.S. Department of Energy, n.d.). The SusTEACH methodology uses the following values: PC = 65−170 Wh; Laptop = 35 Wh; Tablet = 2.5–5.4 Wh; Mobile (other devices) = 0.17–0.3 Wh.

by the metric tons EF for electricity in Botswana, and this number was multiplied by 1000 to arrive at the emissions in kilograms. This number was multiplied by 12, the number of weeks in the semester. Internet usage was not calculated as neither programme was online-based; however, in future this figure could be considered to capture supplementary use of internet for course-related purposes.

The total at home electricity usage calculation comprised of lighting, heat and at-home ICT usage. Seasonal factors for light and heat were not included based on suggestions from the local consultants who reported minimal variations in daylight hours and temperatures. All respondents were living off campus at the time of the study, so adding term-time residential energy consumption on top of regular residence energy consumption was not necessary. Respondents were asked to report on *additional* number of hours of light and heat used for learning-related activities.

Lighting and heating emissions were calculated by taking the reported number of hours of additional light and heat used for coursework/study each week and multiplying it by the Botswana electricity EF. This number was then multiplied by 1000 to convert it to kgs, before multiplying it by the 12 weeks in the semester. Emissions from at home device use was calculated using an estimate of device wattage per hour multiplied by the number of hours of use per week (for school related activities) reported by students, and then divided by 1000 to get the Kwh. The Kwh was then multiplied by the metric tons emission factor for electricity in Botswana, and this number was multiplied by 1000 to arrive at the emissions in kilograms. The total was multiplied by the 12 weeks in the semester to arrive at the semester total.

ICT Production Emissions

The production emissions of ICTs purchased specifically for the programme of study were also calculated. The total production emissions presented by Berners-Lee (2011) for desktops and laptops were used: 800 kg Co_2 for an average desktop computer and 200 kg CO_2 for an average laptop computer. 150 kg CO_2 was estimated for an average tablet.[4] Mobile phones were not reported as purchased specifically for the programme, thus this figure was not required.

Paper Emissions

Paper-related emissions were calculated in terms of three categories: paper provided by the institution for course related work (i.e. worksheets, readings etc.); paper used personally for course related work (i.e. related printed materials, taking notes on blank pages, printed essays, etc.); and, supplementary paper for programme related logistics (including exams, registration etc.). The estimated number of sheets for

---

[4]The SusTEACH Methodology estimates 150 kg for an average tablet.

each of these three categories was multiplied by 0.01115, which is the approximate kgs of CO_2 produced by 1 sheet of paper.[5]

Learning Materials (Textbooks or CD) Emissions

Assigned learning material CO_2 emissions were calculated for each group. Required learning materials consisted of paper-based textbooks for the BAC students and CDs for the BOCODOL students. For the BAC students, the average kgs of CO_2 produced by the full lifecycle of a textbook was estimated at 15kgs.[6] This number was multiplied by the number of textbooks required in each semester (4) for a total of 60kgs of CO_2 emissions from textbooks per semester per student. As the BOCODOL students received learning materials as a CD rather than printed textbooks, an estimate of average kgs of CO_2 produced by a CD and plastic case, including transport[7] was multiplied by the number of CDs each semester (1) for a total of 0.5 kg of CO_2 per semester for the BOCODOL students' assigned learning materials emissions.

Total Semester CO_2 Emissions

The total semester CO_2 emissions calculation was the sum of the travel, electricity, ICT production, paper and learning material (textbook or CD) emissions.

Results

Demographics

The BOCODOL students range from age 26 to 51, with an average age of 35 compared to the BAC students who range from 19 to 26, with an average age of 24. As evident in Table 7.2, there were more females than males in both the BOCODOL and BAC groups, with similar proportions across the two groups. In terms of level of education, a larger proportion of BOCODOL students as compared to BAC students had completed college or university prior to enrolling in their current programme (Table 7.3).

---

[5]This is the same figure used in the SusTEACH methodology.

[6]Gattiker et al. (2012) report 9 kg CO_2e; Solanas et al. (2009) report 34 kg. The SusTEACH Methodology works on the assumption that a single book generates about 15 kg of CO_2e.That includes production, transport, and either recycling or disposal.

[7]0.5 kgs (as cited in Julie's Bicycle: Taking the Heat Out of Music 2009).

Table 7.2 Participants by gender and institution

BOCODOL participants		BAC participants		
Male	Female	Male	Female	No response
8 (31%)	18 (69%)	10 (30%)	21 (64%)	2 (6%)

Table 7.3 Participants' education level by institution

Level of education		Total
BOCODOL	BAC	
2	23	25
7	4	11
17	4	21
26	31	57

Travel Emissions

While BOCODOL students were more likely than their BAC counterparts to use a car as their main mode of transportation (see Table 7.4), they have significantly less programme related travel each semester, as seen in Table 7.5. The greater overall distance travelled by the BAC students contributes to their higher CO_2 emissions for travel each semester, as compared to the BOCODOL students (Table 7.6 and Fig. 7.1).

Table 7.4 Travel mode by institution

		BOCODOL	BAC	Total
Travel mode	Car	22	7	29
	Walking	0	2	2
	Bus	0	3	3
	Other public transport	4	20	24
Total		26	32	58

Table 7.5 km Travelled

		N	Mean	Std. deviation
KMs travelled per semester	BOCODOL	26	179.8077	64.83580
	BAC	33	3070.4697	3724.97366

7 Delivery Mode and Learner Emissions: A Comparative Study ...

Table 7.6 Total CO_2 emissions from travel

		N	Mean	Std. deviation
Total CO_2 from travel	BOCODOL	26	31.1748	19.83162
	BAC	33	194.8556	226.82267

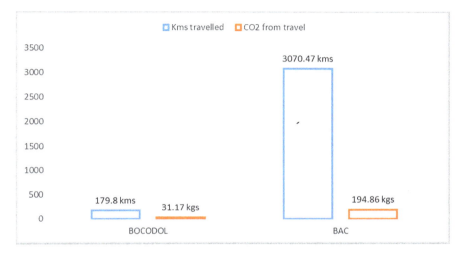

Fig. 7.1 Total CO_2 emissions from travel

Energy Consumption

On Campus

The on-campus electricity emissions were calculated as the sum of the classroom emissions and the emissions from on-campus ICT use. Overall, the BAC students have significantly higher emissions from on-campus energy use, as compared to the BOCODOL students. This result is expected, as the BOCODOL students are only required to be on campus for contact classes four times during the semester (Table 7.7 and Fig. 7.2).

Table 7.7 On campus CO_2 emissions

		N	Mean	Std. deviation
Total campus CO_2	BOCODOL	26	49.0928	3.11698
	BAC	33	137.2370	5.57144

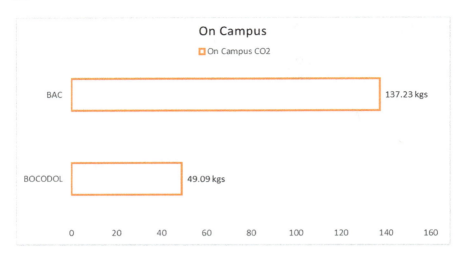

Fig. 7.2 On campus CO_2 emissions

Table 7.8 Additional heat and lighting

		N	Mean	Std. deviation
Additional home lighting and heat CO_2	BOCODOL	26	9.4985	18.28799
	BAC	33	24.6960	23.64462

At home/residence

The at-home energy usage consists of lighting, heat and ICT consumption of electricity at home for programme related work and study. While BOCODOL learners have higher emissions for at home ICT usage than the BAC students (Table 7.9), they have lower usage of additional heat and lighting at home than their counterparts at BAC (Table 7.8). Overall the BAC group has higher average home emissions than that BOCODOL group (Table 7.10 and Fig. 7.3).

Table 7.9 Home ICT usage

		N	Mean	Std. deviation
Home ICT CO_2	BOCODOL	26	14.7328	34.10466
	BAC	33	6.6220	10.90689

7 Delivery Mode and Learner Emissions: A Comparative Study …

Table 7.10 Total home emissions

		N	Mean	Std. deviation
Total home CO_2	BOCODOL	26	24.2313	48.45349
	BAC	33	31.3180	24.84029

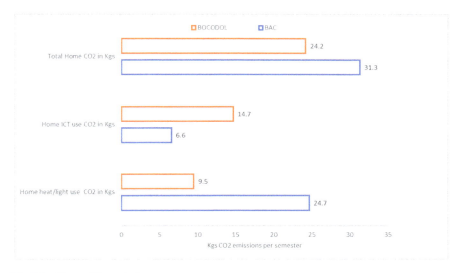

Fig. 7.3 Home CO_2 emissions

Production Emissions from ICT Purchased

Respondents reported whether they had purchased any ICT specifically for their programme or not. The BAC students were more likely to have purchased a laptop specifically for the programme, as evident in Table 7.11, while BOCODOL students were more likely to have purchased a tablet for their programme (Table 7.12). Only laptops and tablets were reported as being purchased specifically for the programme. Approximately 46% of BOCODOL students use existing ICTs for their programme while only 18% of the BAC students do. The total production CO_2 values used for ICT purchased was 200 kg for laptops and 150 kg for tablets. Due to the significantly

Table 7.11 Percentage of respondents that purchased a laptop by institution

			BOCODOL	BAC	Total
Purchased laptop?	No	Count	16	8	24
		%	61.5%	24.2%	40.7%
	Yes	Count	10	25	35
		%	38.5%	75.8%	59.3%

Table 7.12 Percentage of respondents that purchased a tablet by institution

			BOCODOL	BAC	Total
Purchased tablet?	No	Count	22	31	53
		%	84.6%	93.9%	89.8%
	Yes	Count	4	2	6
		%	15.4%	6.1%	10.2%

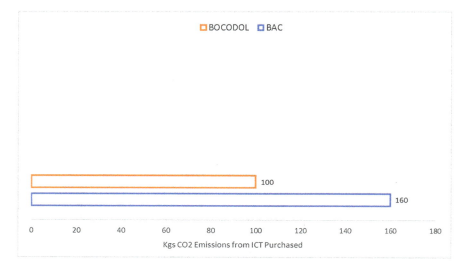

Fig. 7.4 ICT production emissions for devices purchased for the programme

higher proportion of BAC students who had purchased a laptop specifically for their programme as compared to the BOCODOL group, the average CO_2 emissions from ICT purchased was significantly higher for the BAC students (Fig. 7.4).

Paper Use

Respondents were asked to estimate the amount of paper they used in three different categories: Paper distributed by the institution; paper for their own personal use; and, miscellaneous programme-related paper such as exams, registration documents etc. The BAC students reported using significantly more paper than the BOCODOL students, resulting in much higher average emissions from paper (Table 7.13).

Table 7.13 Total paper CO_2 emissions

		N	Mean	Std. deviation
Total paper CO_2	BOCODOL	26	5.6686	2.02326
	BAC	33	30.0547	28.90344

Learning Materials

While the BAC students each received four assigned hardcopy textbooks for the semester, the BAC students received their learning materials on a single CD. The average total emissions from the BAC textbooks was estimated at 60 kgs per student while the emissions from the CD for the BOCODOL group was 0.5 kgs per student.

Total Carbon Footprint

Overall, the average carbon footprint of the face-to-face mode BAC group was nearly three times greater than that of the distance mode BOCODOL group (Fig. 5 and Table 7.14).

Fig. 7.5 Average learner CO_2 emissions (Kgs)

Table 7.14 Total CO_2 emissions (carbon footprint)

		N	Mean	Std. deviation
Total carbon footprint	BOCODOL	26	210.7146	110.68349
	BAC	33	614.1240	235.53545

Discussion

Order of Contributing Components in the Carbon Footprint Calculation

Table 7.15 lists the components of the overall carbon footprint calculation in order from the greatest difference in emissions between the two groups to the least. BAC students on average have higher emissions in every component.

The largest discrepancy between the emissions can be seen in the travel category: The travel related emissions generated on average by the face-to-face BAC students is more than six times higher than the average from BOCODOL students. The second largest difference is seen in the on-campus energy use, which is expected, as the BOCODOL students do not attend regular on-campus classes like their face-to-face counterparts. What is somewhat unexpected is that the BOCODOL students also have lower (although minimally) average emissions from at home energy consumption. While the BOCODOL students have higher energy consumption from ICT use for course related work at home, the use of additional lighting and heating is higher for the face-to-face students. One possible explanation is that the use of ICT by the BOCODOL group (as all materials are on CD) does not require as much additional lighting as reading from a textbook does, given that the devices have lighted screens. Another potential explanation is that the BOCODOL students do not consider the lighting and heating used for study time as 'additional' because they would expect to be using the electricity, for example, for work-related purposes regardless. This is plausible as 77% of BOCODOL students were employed, compared to only 9%

Table 7.15 Components of total carbon footprint

Factor	BOCODOL emissions (Kgs CO_2)	BAC emissions (Kgs CO_2)	Difference
Travel CO_2	31.17 (14.8%)	194.85 (31.7%)	163.68
On campus energy use	49.09 (23.3%)	137.24 (22.3%)	88.14
ICT purchased	100.00 (47.5%)	160.60 (26.2%)	60.60
Learning materials	0.5 (0.2%)	60 (9.8%)	59.5
Paper	5.67 (2.7%)	30.05 (4.9%)	24.38
Home energy use	24.23 (11.5%)	31.31 (5.1%)	7.08
Total	210.71	614.12	403.41

of BAC students. An alternative explanation could be that the workload of the BAC programme is heavier and requires more time than the BOCODOL programme. Further investigation into this finding, through interviews or focus group discussions, could help elucidate the reasons behind this unexpected difference.

Determinants of Carbon Footprint

The large and significant difference between the average carbon footprint of the BOCODOL group and the BAC group suggests that the institution, as a proxy for mode of delivery, may be a determining factor of the learner carbon footprint. To test this hypothesis and control for other possible determining demographic factors, a regression analysis on the dependent variable of overall learner carbon footprint was run. The independent variables included in the model were: age, sex, institution and parental university education. Age was a continuous variable, whereas sex (man = 1; woman = 0), institution (BOCODOL = 1; BAC = 0), mother university educated (Yes = 1; No = 0), and father university educated (Yes = 1; No = 0) were all dummy variables. In this model institution is assumed to be a proxy for mode of delivery, and father and mother's completion of university as a proxy for socio-economic status. The selection of these variables was grounded in the literature on demographic determinants of carbon footprint (see Räty and Carlsson-Kanyama 2009; Zagheni 2011). The results of this regression are presented in Table 7.16.

In this model, which accounts for approximately 52% of the variation in carbon footprint, institution emerges as the only significant, independent determinant of learner carbon footprint, even when controlling for age, sex, and parental university education. Being a BOCODOL learner results in a predicted decrease in emissions of approximately 435 kgs a semester, further supporting the contention that mode

Table 7.16 Regression model

R	R square	Adjusted R square		Std. error of the estimate	
0.748[a]	0.559	0.515		185.52358	
Model		Unstandardized coefficients		Standardized coefficients	Sig.
		B	Std. Error	Beta	
1	(Constant)	347.221	150.177		0.025
	Mother university educated	34.799	72.940	0.062	0.635
	Father university educated	65.731	72.452	0.116	0.369
	Sex	32.824	55.248	0.057	0.555
	Institution	−435.940	116.828	−0.824	0.000
	Age	8.240	6.440	0.245	0.207

[a]$p < 0.001$

of delivery is a major factor in students' learning related carbon emissions. While a more robust analysis, including other possible predictors, would help to strengthen this hypothesis, the present model suggests that mode of delivery has an impact on learning-related carbon footprint, even when considering demographic factors.

Conclusion

The comparison of learner carbon emissions data from both face-to-face and CD-based students in Botswana reveals that the average learning-related carbon footprint of the face-to-face group is nearly three times greater than that of the ODL group, and the difference is statistically significant. Within the overall carbon footprint, emissions from travel are by far the greatest contributor to this disparity. This suggests that regardless of the exact mechanism of delivery, ODL or blended modes can decrease emissions by reducing face-to-face contact hours. Nevertheless, an analysis of differences in carbon emissions by delivery mode within the ODL stream would help to better understand which ODL modes are more environmentally friendly. For example, a CD-based ODL programme, like the one examined in the present study, may differ from correspondence-based or online modes. These differences should be further investigated to inform programme design, with a view to developing greener ODL programmes.

While the findings of this study support the conclusions of similar studies comparing the carbon footprint of learners under different modes of delivery in the UK and North America, additional comparative studies of ODL and face-to-face programmes in developing country contexts can further strengthen these findings. Such studies could be replicated in diverse development contexts and locations to test the generalisability of the findings.

While such studies will contribute to the growing body of knowledge on ODL and environmental impact, the results can also be strategically leveraged for the promotion of ODL. Campbell and Campbell (2011) found that "feedback regarding the CO_2 savings associated with the online class format appears to have reinforced positive attitudes regarding distance learning and even helped to mitigate dissatisfaction with the online format." While ODL is often marketed to prospective students on the basis of its flexibility and ease-of-access, it is rarely positioned as an environmentally sustainable option. HEIs should consider marketing ODL courses from a sustainability angle to attract students and increase their satisfaction. Moreover, these findings can be promoted by institutions to improve faculty acceptance of ODL, which remains a major challenge for many institutions looking to mainstream ODL delivery modes. While a 2017 study by Versteijlen, Salgado, Groesbeek, and Counotte showed that few HEI professionals were aware of the potential for online education to reduce carbon emissions, Campbell and Campbell (2011) found that data on the positive environmental impact of distance learning can positively influence faculty opinions and acceptance of ODL. However, attention needs to be given to improving perceptions of ODL in terms of quality, as Versteijlen et al. (2017) also found that

"Professionals do not consider online education the most obvious measure to reduce travel carbon emissions, because they expect to meet resistance in their organisation, and they suspect it might deteriorate the quality of education," (p. 88). Highlighting the environmental benefits of ODL could go a long way in promoting its expansion and uptake, particularly in conjunction with advocacy efforts to address perceptions of quality.

HEIs should also consider other ways to reduce the impact of transportation-related emissions, by providing and incentivising sustainable transportation practices. For example, subsidies for public transit, and dedicated carpooling or ridesharing services can help reduce emissions generated by student travel (Appleyard et al. 2018; Barros et al. 2018). Additionally, investments in on-campus or near campus housing options can cut down student commutes, thereby reducing overall student generated carbon emissions (Appleyard et al. 2018; United Nations Environment Programme 2014).

Perhaps most importantly, the findings of this study suggest that the carbon reduction efforts of HEIs should have a greater focus on pedagogical design, not just infrastructural improvements or campus greening. As sustainability becomes a more important goal for HEIs around the world, institutions should consider how ODL modes of delivery can be upscaled, not only to increase access, but also to reduce the environmental impact of education.

References

Appleyard B, Frost AR, Cordova E, McKinstry J (2018) Pathways toward zero-carbon campus commuting: innovative approaches in measuring, understanding and reducing greenhouse gas emissions. Transp Res Rec J Transp Res Board 2672(24):87–97. Retrieved from https://doi.org/10.1177/0361198118798238

Barros MV, da Silva BPA, Piekarski CM, da Luz LM, Yoshino RT, Tesser DP (2018) Carbon footprint of transportation habits in a Brazilian university. Environ Qual Manage 28(10):139–148. Retrieved from https://doi.org/10.1002/tqem.21578

Berners-Lee M (2011) How bad are bananas? The carbon footprint of everything. Greystone Books, Vancouver, BC

Botswana Accountancy College (n.d.) About us. Retrieved from http://www.bac.ac.bw/about-us. 25 Jan 2019

Botswana Open University (n.d.) About us. Retrieved from http://www.bou.ac.bw/index.php/home/about-us. 24 Jan 2019

Bourke J, Simpson O (2009) Sustainability in education: Is distance learning an answer? Lower Hutt, New Zealand: the open polytechnic of New Zealand. https://repository.openpolytechnic.ac.nz/bitstream/handle/11072/1434/Bourke_Simpson_2009%20-%20Working%20Papers%20-%20WP_09_2.pdf?sequence=1

Caird S, Lane A, Swithenby E, Roy R, Potter S (2015) Design of higher education teaching models and carbon impacts. Int J Sustain High Edu 16(1):96–111. Retrieved from http://www.emeraldinsight.com/doi/pdfplus/10.1108/IJSHE-06-2013-0065

Caird S, Lane A, Swithenby E (2013) ICTs and the design of sustainable higher education teaching models: an environmental assessment of UK courses. In: Caeiro S, Leal Filho W, Jabbour C, Azeiteiro UM (eds) Sustainability assessment tools in higher education institutions. Springer, Switzerland, pp 375–385. Retrieved from https://doi.org/10.1007/978-3-319-02375-5

Campbell JE, Campbell DE (2011) Distance learning is good for the environment: savings in greenhouse gas emissions. Online J Distance Learn Admin 14(1). Retrieved from http://www. westga.edu/~distance/ojdla/winter144/campbell_campel.pdf

Central Intelligence Agency (2019 January 23). The world factbook 2019: Botswana. Retrieved from https://www.cia.gov/library/publications/the-world-factbook/geos/bc.html

Davies J (2015) An analysis of the sustainability of different methods of delivering higher education. In: Leal Filho W, Brandli LL, Kuznetsova O, Paço AMFD (eds) Integrative approaches to sustainable development at university level, world sustainability series. Springer, Switzerland, pp. 67–79. Retrieved from https://doi.org/10.1007/978-3-319-10690-8_5

Energy Research Centre, University of Cape Town (2012) Personal carbon footprint calculator for developing countries: working document. Cape Town: Energy Research Centre, University of Cape Town. Retrieved from http://www.c3d-unitar.org/docs/personal_carbon_footprint_calculator_documentation.pdf

Energy Saver, Office of Energy Efficiency and Renewable Energy, U.S. Department of Energy (n.d.) Estimating appliance and home electronic energy use. Retrieved from https://www.energy.gov/energysaver/save-electricity-and-fuel/appliances-and-electronics/estimating-appliance-and-home. 29 Jan 2019

Gattiker TF, Lowe SE, Terpend R (2012) Online texts and conventional texts: estimating, comparing, and reducing the greenhouse gas footprint of two tools of the trade. Decis Sci J Innovative Edu 10(4):589–613. Retrieved from https://doi.org/10.1111/j.1540-4609.2012.00357.x

Güereca LP, Torres N, Noyola A (2013) Carbon footprint as a basis for a cleaner research institute in Mexico. J Clean Prod 47:397–403. Retrieved from https://doi.org/10.1016/j.jclepro.2013.01.030

Julie's Bicycle: Taking the Heat Out of Music (2009) Executive summary: impacts and opportunities: reducing the carbon emissions of CD packaging. Julie's Bicycle: Taking the Heat Out of Music, London. Retrieved from http://www.musictank.co.uk/wp-content/uploads/2016/04/JB-Impacts-Opportunities-CD-Packaging.pdf

Larsen HN, Pettersen J, Solli C, Hertwich EG (2013) Investigating the carbon footprint of a university: The case of NTNU. J Clean Prod 48:39–47. Retrieved from https://doi.org/10.1016/j.jclepro.2011.10.007

Li X, Tan H, Rackes A (2015) Carbon footprint analysis of student behavior for a sustainable university campus in China. J Clean Prod 106: 97–108. Retrieved from https://www.sciencedirect.com/science/article/pii/S0959652614013857

Natural Resources Canada (2019) About electricity. Retrieved from http://www.nrcan.gc.ca/energy/electricity-infrastructure/about-electricity/7359

Räty R, Carlsson-Kanyama A (2009) Comparing energy use by gender, age, and income in some European countries. FOI, Swedish Defence Research Agency, Stockholm. Retrieved from http://www.compromisorse.com/upload/estudios/000/101/foir2800.pdf

Roy R, Potter S, Yarrow K (2008) Designing low carbon higher education systems: environmental impacts of campus and distance learning systems. Int J Sustain High Edu 9(2):116–130. Retrieved from https://doi.org/10.1108/14676370810856279

Solanas T, Calatayud D, Claret C (2009) 34 kg de CO_2. Government of Catalonia, Ministry of the Environment and Housing, Barcelona. Retrieved from http://www20.gencat.cat/docs/habitatge/Home/Secretariadehabitatge/Publicacions/34KgdeCO2/doc/34_Kg_CO2.pdf

Tilbury D (2011) Higher education for sustainability: a global overview of commitment and progress. In: Global University Network for Innovation (ed) Higher Education in the World 4. Global University Network for Innovation, Barcelona, pp 18–28. Retrieved from http://www.guninetwork.org/files/8_i.2_he_for_sustainability_-_tilbury.pdf

United Kingdom Department of Business, Energy and Industrial Strategy (2016) Conversion factors 2016-Condensed set (for most users) (Data file). Retrieved from https://www.gov.uk/government/publications/greenhouse-gas-reporting-conversion-factors-2016

United Nations Environment Programme (2014) Greening universities toolkit V2.0. Transforming Universities into green and sustainable practices: a toolkit for implementers. UNEP, Nairobi. Retrieved from https://www.unenvironment.org/resources/toolkit/greening-universities-toolkit-v20

Versteijlen M, Salgado FP, Groesbeek MJ, Counotte A (2017) Pros and cons of online education as a measure to reduce carbon emissions in higher education in the Netherlands. Curr Opin Environ Sustain 28:80–89. Retrieved from http://dx.doi.org/10.1016/j.cosust.2017.09.004

Zagheni E (2011) The leverage of demographic dynamics on carbon dioxide emissions: does age structure matter? Demography 48(1):371–399. Retrieved from https://doi.org/10.1007/s13524-010-0004-1

Zhang X, Luo L, Skitmore M (2015) Household carbon emission research: an analytical review of measurement, influencing factors and mitigation prospects. J Clean Prod 103:873–883. Retrieved from https://doi.org/10.1016/j.jclepro.2015.04.024

Chapter 8
Adolescents' Perceptions of the Psychological Distance to Climate Change, Its Relevance for Building Concern About It, and the Potential for Education

Moritz Gubler, Adrian Brügger and Marc Eyer

Abstract One of the greatest challenges of this century is climate change. Unfortunately, it is still unclear how to motivate people to engage in environmentally friendly behaviour. To be effective, education and communication strategies must take into account people's perceptions and beliefs. A root difficulty is that the general public tends to perceive climate change as a psychologically distant phenomenon—something that, if at all, happens not here, not now, and not to oneself. In this study, we explored perceptions of psychological distance to climate change with a highly relevant but so far overlooked population—adolescents. Swiss adolescents ($N = 587$) perceived climate change to be a certain and present risk. However, they perceived climate change to affect other places and other people more than themselves. Regression analysis revealed a significant inverse relationship between distance and concern: respondents who felt psychologically closer to the phenomenon expressed greater concern. The findings contribute to the understanding of how young people perceive climate change, which should assist in designing education strategies to make it more salient for individual behaviour.

Keywords Climate change education · Young people · Psychological distance · Climate change concern · Sustainable behaviour

M. Gubler (✉)
Institute for Research, Development and Evaluation, University of Teacher Education Bern, Fabrikstrasse 8, 3012 Bern, Switzerland
e-mail: moritz.gubler@phbern.ch

Climatology Group, Institute of Geography, University of Bern, Hallerstrasse 12, CH-3012 Bern, Switzerland

A. Brügger
Department of Consumer Behaviour, Faculty of Business, Economics and Social Sciences, University of Bern, Engehaldestrasse 4, 3012 Bern, Switzerland
e-mail: adrian.bruegger@imu.unibe.ch

M. Eyer
Institute for Upper Secondary Teacher Education, University of Teacher Education Bern, Fabrikstrasse 8, 3012 Bern, Switzerland
e-mail: marc.eyer@phbern.ch

© Springer Nature Switzerland AG 2019
W. Leal Filho and S. L. Hemstock (eds.), *Climate Change and the Role of Education*, Climate Change Management, https://doi.org/10.1007/978-3-030-32898-6_8

Introduction

The risks of anthropogenic climate change for human and natural systems are seen as the major societal challenge of the twenty-first century. In order to keep global warming at or below 2 °C, deep and rapid changes in individual and collective behaviours in favour of reducing energy consumption and adopting policies for mitigation and adaption are required (IPCC 2018). Thus, understanding the complex psychological processes and barriers involved in public engagement for climate change has been subject to extensive research in the psychological and social sciences (Gifford 2011; Hornsey et al. 2016; Stoknes 2014). A key barrier, identified in several studies, is that people in Western countries often perceive climate change as a future risk for far-away places and other people, something that happens—if at all—in remote times and spaces rather than in the here and now (Leiserowitz 2005; Lorenzoni and Pidgeon 2006; Milfont 2012). This perception is succinctly captured by the concept of "psychological distance": the extent to which an object is removed from an individual on the dimensions space, time, social similarity, and hypotheticality (uncertainty) (Liberman and Trope 2008).

According to construal level theory (Trope and Liberman 2010), an object perceived as psychologically distant is represented in abstract terms. This may make it difficult to imagine its impact and seriousness and consequently, the perceived personal relevance (Scannell and Gifford 2013) and concern (Jones et al. 2017; Lorenzoni et al. 2007; Singh et al. 2017) about the issue may be low. Lack of concern is problematic because concern is a key factor for triggering climate action (Semenza et al. 2008; Spence et al. 2011; Tobler et al. 2012).

Identifying factors that may contribute to increasing individual concern about climate change has therefore been the subject of numerous studies. In addition to cultural and sociodemographic variables such as knowledge (Milfont 2012; Shi et al. 2015), political ideology (Hornsey et al. 2016), worldviews (Kahan et al. 2012), and gender (McCright 2010), the perception of climate change as a distant and uncertain phenomenon was indeed found to be a critical barrier to concern (Jones et al. 2017; Lorenzoni et al. 2007; Singh et al. 2017). However, previous studies have almost exclusively focussed on samples of the general public made up of adults. Thus, little is known about how psychologically distant or close climate change is perceived by an important population segment: adolescents.

There are several reasons that adolescents—generally referred to as those between 13 and 19 years of age—are of special interest with respect to perceptions of climate change. First, they are the group that will witness the strongest impacts of climate change during their lifetimes (IPCC 2014). Second, they are being raised in an era of relatively high media coverage and public discussion of the issue. Third, over the last two decades, information about climate change has been increasingly integrated into formal education in order to empower this generation to take responsibility and shape their future (Monroe et al. 2017). More than ever before, contemporary adolescents may be very well informed about the issue, as they have been repeatedly confronted with it in classrooms, the media, and in discussions with their families and peers

(Corner et al. 2015; Ojala and Lakew 2017). Compared to adults, adolescents may perceive climate change as something that is already part of their lives, something that will impose strong impact in their lifetimes, and as a certain phenomenon that is really happening.

Logically, one would think that adolescents' potentially closer perception of climate change in terms of time and uncertainty would lead to enhanced personal relevance and increased levels of concern. Evidence against this supposition, however, comes from several studies that have found them to have spatially and temporally distant perceptions of climate change and correspondingly low personal relevance, which, it is argued, might act as a barrier to action (Ballantyne et al. 2016, Yang et al. 2014). However, taken as a whole, such research indicates that perceptions of psychological distance to climate change are probably complex, sometimes counter-intuitive, and may differ with respect to different dimensions.

The present research examines from which perspectives adolescents perceive climate change to be close or distant, and how such perceptions may explain their level of concern relative to other factors often used to account for it. With respect to alternative predictors, Stevenson et al. (2018) found that knowledge about climate change positively relates to concern; perceptions may also be influenced by gender (Stevenson et al. 2014), socioeconomic status (Stevenson et al. 2018), and value orientations: self-transcending values (e.g. altruism) are assumed to have a positive effect, whereas self-enhancing values (e.g. comfort, ease, luxury) have a negative effect (Corner et al. 2015).

Despite considerable previous research, the effects of psychological distance on concern about climate change among adolescents have not yet been directly addressed. The current study therefore explores the perceptions of different dimensions of psychological distance to climate change among ninth-graders from Switzerland. We are especially interested in learning if perceived psychological distance to climate change can account for concern about it. Therefore, we examine the predictive power of psychological distance relative to other predictors that have been identified as having an effect.

Methodology

Participants and Design

In October and November 2018, we conducted a written survey in 32 ninth-grade classes of the Canton ("state") of Bern, Switzerland. The participating schools were recruited through the network of internship teachers of the University of Teacher Education Bern and personal contacts. The data was collected at all levels of secondary school: basic (25.9%), general (31.6%), and upper (42.4%). Performance standards are lowest for the basic and highest for the upper secondary school level. To obtain a sample that is broad in terms of its socio-demographic characteristics and environmental surroundings, we chose schools that varied in terms of locality (urban, rural, and alpine). Compared to the canton-based statistics (ERZ 2018), the basic level is slightly underrepresented (by about 10%) whereas the general and upper secondary levels are correspondingly overrepresented. Students' average age was 14.9 years ($SD = 0.59$; range: 14 to 17 years); 51.7% of the participants were girls.

At that the time of the study, neither of the authors' faculties had an internal review board to grant ethical approval for the study. However, we certify that the research adhered to the American Psychological Association's ethical principles (APA 2017). Participation was voluntary and consent was obtained by asking participants to continue only if they had read and understood the instructions and information provided.

A paper-and-pencil questionnaire with open- and closed-ended questions was used. The survey began with an open-ended question about thoughts, images, or feelings when thinking about climate change. The students were then asked about their knowledge of climate change and perceptions of their psychological distance to it. The second part of the survey assessed students' concern about climate change. Next, they were queried about attitudes towards nature and value orientations. Socio-demographic questions concluded the survey. Some items were adopted from the literature and had to be translated into German as well as rephrased. We pretested the questionnaire quantitatively and qualitatively.

The survey was carried out during regular lessons. Participants were informed about the topic, purpose and procedure by the same instructor. Completion time was 21–47 min. Of 603 completed questionnaires, 16 were excluded because they had too many missing or suspiciously similar answers, which resulted in a total of $N = 587$ questionnaires used for analysis.

Measures

Perceptions of climate change focussed on the four theorized dimensions of psychological distance: space, time, social similarity and hypotheticality (Bar-Anan et al. 2006). To evaluate dimensions of psychological distance, we used questions partly retrieved from previous studies (see Table 8.1). As these were designed for older age groups or the general public, the wording of some questions was changed slightly. For instance, we used the students' levels of worry as a measure of concern about climate change. Self-reported knowledge about climate change was assessed by asking "How much do you think you know about climate change" (5-point Likert scale; 1 = "very little," 5 = "a lot"). The Inclusion of Nature in Self scale (Schultz 2002) served as a measure of attitude towards nature, where higher values (on a 7-point scale) indicate a more positive attitude towards nature. Value orientations were assessed by a 25-item scale (Speyer Value Inventory; Gensicke 2000) used in current youth surveys of Germany and Switzerland (Huber and Lussi 2016). Students rated the importance of different value orientations for their lives (1 = "not at all important," 7 = "extremely important"). Value orientations were aggregated as in the Young

8 Adolescents' Perceptions of the Psychological Distance ...

Table 8.1 Items assessing perceptions of psychological distance and climate change concern

Construct	Statement	Cronbach's α	Inter-item correlation
Spatial distance	Climate change is mostly affecting the area where I live[1][*] First thoughts about climate change are about how it will impact the area where I live[2][*] Climate change is mostly affecting areas that are far away[1]	0.5	0.25
Temporal distance	The impacts of climate change will be mostly felt far in the future The impacts of climate change can already be felt now[*] As long as I live, the impacts of climate change will be only felt very weakly	0.58	0.31
Social distance	Climate change impacts will be particularly strong for people that are like me and think like me[1][*] Climate change will particularly affect me, my family and my friends[2][*] Climate change will particularly affect people who have similar desires and goals in life as I do[*]	0.68	0.42
Hypothetical distance	I wonder if climate change is a serious threat at all Science agrees that humans are responsible for current climate change[1][*] I am certain that climate change is happening[1][*]	0.57	0.31
Concern about climate change	When I think about the effects of climate change, I worry a lot The more I know about the effects of climate change, the more I'm worried The thought of the effects of climate change worries me every day I classify the effects of climate change as harmless[*]	0.84	0.56

Note Response options for all items consisted of a 5-point agreement scale (strongly disagree–strongly agree)

[1] Spence et al. (2012)

[2] Jones et al. (2017)

[*] Reverse-scaled

Adult Survey Switzerland (Huber and Lussi 2016) into five value dimensions: private harmony and independence, duty and convention, materialism, idealism, and tradition.

Statistical Analysis

Psychological distance dimensions were reverse-scored such that higher values indicated greater amounts of distance (1: close perception; 5: distant perception). Similarly, the concern scale was scored so that higher values indicated higher levels of concern. We performed hierarchical regression models to predict concern about climate change using different sets of predictor variables. All statistical analyses were performed using the software JASP (2018).

Results

Perceived Psychological Distance to Climate Change

In terms of *spatial* distance, students generally endorsed statements that located climate change impacts farther away as compared to statements where impacts are situated close to the respondent (Fig. 8.1). Whereas 42.9% of respondents strongly or tended to agree with the statement "Climate change is mostly affecting areas that are far away," disagreement (strongly or tended) was rather low (20.9%). Conversely, disagreement was relatively higher (45.8%) than agreement (14.7%) with the statement that "Climate change is mostly affecting the area where I live." Disagreement

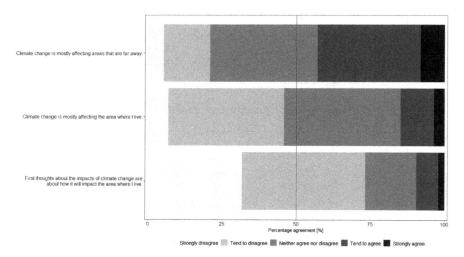

Fig. 8.1 Perceived spatial distance of climate change impacts. Respondents' agreement with each statement is provided in percent

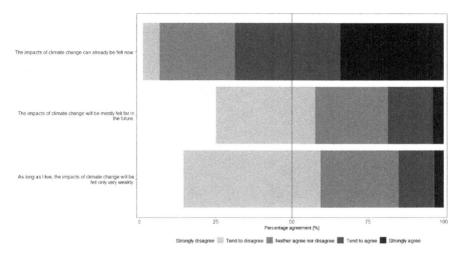

Fig. 8.2 Perceived temporal distance of climate change impacts. Percentage refers to the respondents' agreement

was highest (73.3%) for the statement that "First thoughts about the impacts of climate change are about how it will impact the area where I live," compared to only 9.4% of respondents who agreed.

With respect to *temporal* distance, students seemed to perceive climate change as something that happens now more than in the future (Fig. 8.2). Compared to only 6.7% who disagreed (strongly or tended), more than two-thirds (68.6%) agreed that "The impacts of climate change can already be felt now." Moreover, disagreement was stronger than agreement for the statements "The impacts of climate change will be mostly felt far in the future" (57.7 vs. 18.4%) and "As long as I live, the impacts of climate change will be felt only very weakly" (59.5 vs. 14.9%).

With respect to *social* distance, respondents clearly disagreed that climate change was a risk for people socially similar to themselves (Fig. 8.3). Some 60.5% disagreed (strongly or tended) with "Climate change impacts will be particularly strong for people that are like me and think like me." Disagreement was even higher for "Climate change will particularly affect people with similar desires and goals in life as I do" (64.9%) and "Climate change will particularly affect me, my family and my friends" (76.0%). Fewer than 10% agreed with these statements.

With respect to *hypotheticality* (uncertainty), climate change was perceived as being a rather certain phenomenon that is caused by humans (Fig. 8.4). A large majority (82.4%) agreed that "I am certain that climate change is happening" compared to only 5.6% disagreement. Agreement was also very high (74.4%) for "Science agrees that human beings are responsible for current climate change" (10.2% disagreed). Conversely, students overwhelmingly (79.3%) disagreed with "I wonder if climate change is a serious threat at all" (10.9% agreed).

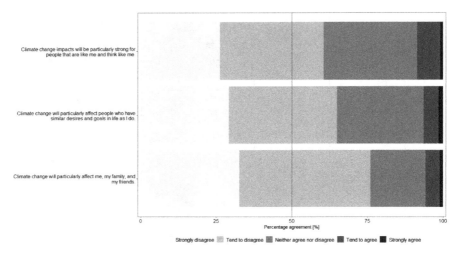

Fig. 8.3 Perceived social distance of climate change impacts. Respondents' agreement with each statement is provided in percent

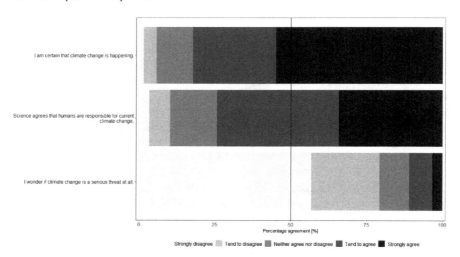

Fig. 8.4 Perceived hypothetical distance of climate change impacts. Percentage refers to the respondents' agreement

Predictors of Climate Change Concern

Next, we examined the relations among different variables by building hierarchical regression models with four different blocks to predict climate change concern. The first block was demographic information—gender, age, and school type. The second block was self-reported knowledge about climate change. The third block was different dimensions of psychological distance. The fourth block was attitude towards nature and value orientations. These four blocks (models) allowed us to evaluate the predictive power of each group of variables. We also ran the full model (Model 5) that included all predictors simultaneously in order to assess their relative contributions and to account for potential overlap among predictors.

Overall, concern about climate change was rather high: $M = 3.29$ (on a 1–5 scale, where higher values indicate greater levels of concern). Psychological distance (Model 3) was the best predictor of climate change concern ($R^2_{adjusted} = 0.44$, Table 8.2). Except for spatial distance, all distance dimensions showed significant negative relationships with concern, meaning that the greater the perceived distance, the lower the respondents' level of concern about climate change. Standardized regression coefficients show that hypotheticality explained more variance than the other dimensions, highlighting the importance of perceived uncertainty (or the conceptually similar construct of scepticism) with respect to climate change.

Attitudes towards nature and value orientations formed the second most powerful group of predictors, explaining about 26% of the variance in concern change (Model 4). Students with positive attitudes towards nature and idealistic (self-transcendent) value orientations had greater concern for climate change.

Notably, self-reported knowledge accounted for about 18% of the total variance (Model 2) and was positively correlated with concern (Model 5). This suggests that gains in knowledge about climate change could play an important role in increasing concern about it.

With respect to socio-demographic variables, girls were generally more concerned than boys. Students at upper secondary school level showed greater concern than those from basic/general secondary schools. Even so, demographic characteristics were overall the weakest block of predictors for concern ($R^2_{adjusted}$ Model 1 = 0.13).

As mentioned, in the full model (Model 5), the most powerful predictor was hypothetical distance, followed by temporal distance, self-reported knowledge, social distance, gender, idealistic values, attitude towards nature, and school type (standardized β-coefficients). The full model with all predictors combined explained about 54% of the variance of concern about climate change.

Table 8.2 Results of hierarchical regression analyses predicting climate change concern

Predictors	Mean/ SD	Model 1				Model 2				Model 3				Model 4				Model 5			
		B	SE	β	Sig.	B	SE	β	Sig.	B	SE	β	Sig.	B	SE	β	Sig.	B	SE	β	Sig.
Age	14.9/0.9	−0.069	0.06	−0.046														−0.061	0.04	−0.041	
Gender[1]	0.5/0.5	0.265	0.07	0.147	***													0.238	0.06	0.133	***
School type[2]	1.4/0.5	0.545	0.07	0.3	***													0.139	0.06	0.077	*
Self-reported knowledge[3]	2.9/0.9					0.415	0.04	0.43	***									0.167	0.03	0.174	***
Spatial distance[4]	3.5/0.7									−0.006	0.04	−0.005						−0.02	0.04	−0.016	
Temporal distance[4]	2.3/0.7									−0.33	0.04	−0.276	***					−0.212	0.04	−0.178	***
Social distance[4]	3.9/0.7									−0.236	0.04	−0.191	***					−0.184	0.04	−0.151	***
Hypothetical distance[4]	1.8/0.8									−0.531	0.04	−0.446	***					−0.412	0.04	−0.346	***
Values: private harmony and independence[5]	5.9/0.6													0.041	0.06	0.027		−0.035	0.05	−0.023	

(continued)

Table 8.2 (continued)

Predictors	Mean/SD	Model 1				Model 2				Model 3				Model 4				Model 5			
		B	SE	β	Sig.	B	SE	β	Sig.	B	SE	β	Sig.	B	SE	β	Sig.	B	SE	β	Sig.
Values: duty and convention[5]	5.3/0.7													0.123	0.05	0.106	*	0.024	0.04	0.02	
Values: materialism[5]	4.1/1													−0.007	0.035	−0.008		0.044	0.03	0.05	
Values: idealism[5]	5/0.9													0.293	0.04	0.302	***	0.122	0.03	0.125	***
Values: tradition[5]	3.5/1.1													−0.117	0.032	−0.143	***	−0.025	0.03	−0.03	
Attitude towards nature[6]	4/1.2													0.187	0.03	0.27	***	0.084	0.022	0.119	***
Adjusted R²/N				*0.13*	*581*			*0.18*	*579*			*0.44*	*585*			*0.26*	*574*			*0.54*	*562*

Note B = Unstandardized regression coefficient, *SE* = Standard Error, β = Standardized regression coefficient, *Sig.* = level of statistical significance

*** stands for $p < .001$

** stands for $p < .01$

* stands for $p < .05$

[1] male = 0, female = 1

[2] Basic/general secondary school = 1, upper secondary school = 2

[3] 5-point scale; very little–a lot

[4] 5-point scale; very close–very distant

[5] 7-point scale; not at all important–extremely important

[6] 7-point scale; Inclusion of Nature in Self scale Schultz (2002)

Discussion

Adolescents' Perceptions of Psychological Distance to Climate Change

The students in our sample perceived climate change to be psychologically rather distant in terms of space (Fig. 8.1) and social similarity (Fig. 8.3). In other words, they think climate change is a greater problem in areas remote from where they live and on people who are unlike themselves. Similar findings among high-school students from Sweden have been reported by Ballantyne et al. (2016). Young people from Australia (18–25 years) also perceived climate change to primarily affect distant places (Perera and Hewege 2013). Social similarity is an interesting concept in the context of young people, as adolescence is a time of potentially profound changes in identity (Crocetti and Rubini 2017; Tanti et al. 2011). This raises the question about which individuals or social groups are seen to be socially similar by adolescents. In this era of social media, social identification could be with role models in various places and socio-economic contexts. However, family members and peers are still the most important group for young people's social identification (Maccoby 2015). Our findings are consistent with this conclusion, as there was a strong relationship between perceived spatial and social distance, and both dimensions were relatively pronounced in our sample.

With respect to temporal distance (Fig. 8.2), students perceived climate change to be already upon us. This is consistent with findings among young people from the UK (Corner et al. 2015), Austria, and Denmark (Harker-Schuch and Bugge-Henriksen 2013). Another factor that could have contributed to the perception of climate change happening now is the high media coverage in Switzerland (and Europe) during the record-breaking dry and hot summer of 2018: the drought and heat were often directly linked to anthropogenic climate change taking place now. Contextual factors like these may be crucial for the development of climate change knowledge and attitudes (Tobler et al. 2012); their influence over time could be addressed in longitudinal studies.

With respect to hypothetical distance, adolescents perceived climate change as being real and caused by humans (Fig. 8.4). This is consistent with results of a literature review (Corner et al. 2015) and with studies among the general public (Poortinga et al. 2011), both showing that young people are generally not sceptical about the reality of climate change and accept the scientific consensus about the human contribution to it. This could be explained by the increasing scientific consensus about human causes during the last two decades (Kerr and Wilson 2018) when the generation of today's adolescents was raised and came into contact with information about climate change. However, we did not take into account the climate-related media attention of the students or the relative importance of different media channels for their scepticism. Exploring the effects of perceived consensus or public denial via different media channels (e.g. social media) is a topic for future research.

It has been hypothesized that perceived psychological distance to climate change might be particularly pronounced among young people due to competing priorities (e.g. career choices) or little personal experience with its impacts (Corner et al. 2015). We found evidence for different perceptions of psychological distance depending on the dimension involved. Although our data does not allow for direct comparisons with other segments of population, we suggest that adolescents do not necessarily perceive climate change as a more distant risk than adults. Adolescents perceived climate change impacts to be most pronounced for places farther away and for people in different social contexts; at the same time, however, they perceived it as quite certain and already happening. To explore potential differences between adults and adolescents, comparative studies would be an interesting avenue for future research.

Finally, psychological distance to climate change may refer not only to a perceptual process—it could also be part of an active, constructive process. Ojala (2013) found that psychological distancing is a strategy children and adolescents use to cope with negative emotions (e.g. worry) stemming from climate change. People can distance themselves from worrying about climate change by avoiding the topic or by cognitive and/or behavioural distraction (Ojala 2016). Either way, the result could be a perception that climate change happens to other people, in other places. In order to get a more nuanced picture, a combination of qualitative and quantitative methods is recommended.

Educational Potential of Reducing Psychological Distance to Climate Change

Concern for climate change is a key determinant of willingness to act (Hornsey et al. 2016; Semenza et al. 2008; Spence et al. 2011). Our results show that climate change concern among adolescents is relatively high, which is in line with findings of an inverse relationship between age and concern (Fløttum al. 2016; Milfont 2012). Furthermore, the regression analysis here revealed that concern is associated with the perception that climate change is psychologically close (Table 8.2), which is consistent with findings from the general public (Jones et al. 2017; Singh et al. 2017; Spence et al. 2012).

In addition to psychological distance, idealistic value orientations and positive attitudes towards nature were correlated with climate change concern. Similar findings have been reported previously (Corner et al. 2014; Kollmuss and Agyeman 2002). Self-reported knowledge was also associated with climate change concern. This adds to findings from the US showing positive relationships between educational interventions, knowledge, and concern (Flora et al. 2014; Stevenson et al. 2014, 2018). Stevenson et al. (2014) found that adolescents' knowledge reduced the effect of ideology-based worldviews on the acceptance of climate change (but see Kahan et al. 2012 for different findings with adults). Similarly to their proposition, we argue that during adolescence, beliefs and worldviews are still in a plastic state, which makes knowledge a likely trigger for climate change concern. Socio-demographic factors, however, played only a minor role in predicting climate change concern. We found girls more concerned about climate change than boys, although the difference was relatively small. This replicates findings among US adults (e.g. McCright 2010) and adolescents (see Stevenson et al. 2016 for detailed discussion), although Ojala (2015) reported no relationship between gender and climate change scepticism or engagement in Sweden.

Our finding that perceived psychological distance explains by far most of the variance in climate change concern, may have implications for communication and education efforts. When adolescents perceive climate change as a risk to people similar to themselves, they may be more likely to be concerned about it. Based on studies that found a direct relationship between concern and preparedness to take action (Stevenson and Peterson 2016; Stevenson et al. 2018; Valdez et al. 2018), we hypothesize that decreasing psychological distance could act as a trigger for behavioural intentions to reduce the effects of climate change through increased concern. Similarly, several scholars have proposed that reducing psychological distance might be an effective communication strategy to provoke concern about and subsequent public engagement with climate change (van der Linden et al. 2015).

Educational interventions that highlight anthropogenic climate change as a certain and immediate risk with potentially harmful consequences for close friends and family members could make the relevance of the phenomenon more salient, and thereby increase concern among adolescents. Although the effects of differently framed messages have been poorly studied to date (Corner et al. 2015), promising indications come from Yang et al. (2014). They found that American and Chinese youth who perceive climate change as personally relevant were more willing to respond to information and to engage in conversations about it. Other researchers have highlighted the importance of tailoring climate-related information to social and geographical contexts that students could relate to (Adams and Gynnild 2013).

However, educational interventions aiming to increase concern about climate change should not only take into account manipulations of dimensions of psychological distance to climate change. Previous studies among the general public have shown that making climate change closer may lead to a range of effects (see McDonald et al. 2015 for a detailed review). Although Jones et al. (2017) and Singh et al. (2017) found that reduced psychological distance led to greater concern and willingness to act, others failed to replicate the finding (e.g. Brügger et al. 2015). Other variables may also significantly contribute to the development of climate change concern, such as perceived self-efficacy (Corner et al. 2015); emotions such as hope, worry, and despair (Ojala 2016; Stevenson et al. 2018); and interaction with peers and family (Ojala 2013; Stevenson et al. 2016).

The upshot is that educational interventions aiming to reduce psychological distance among adolescents need to carefully control for these variables, along with climate change knowledge, values, and socio-demographic aspects. Educational interventions are likely to be most effective when they are designed from a holistic perspective and address different sorts of knowledge (about causes, impacts, and actions), use systematic approaches (e.g. interdisciplinary), allow for direct experiences, and promote discussions with friends and family (see Ojala and Lakew 2017 for a detailed review). Approaches focusing on perceived psychological distance may have great potential for climate change education and could promote understanding how this important segment of the population perceives and acts upon climate change.

Limitations

Our study has several limitations that need to be addressed. First, the geographically restricted sample does not allow for extrapolation of the results, as it is not representative on a cantonal or national level. However, as it is the first study directly examining perceptions of psychological distance among adolescents, it is a starting point for research in other spatial, socio-cultural, and educational contexts. Provided that the same items are used, comparative studies at similar points in time could give valuable insights regarding variations in perceptions of climate change among different countries or social milieus.

Second, we asked students to report their knowledge instead of assessing it with objective questions (e.g. from Stevenson et al. 2014). Students unconcerned or sceptical about climate change might self-report higher levels of knowledge than they actually have (Hamilton 2018). On the other hand, previous studies have used a similar approach (e.g. Milfont 2012), and we do not claim direct translation of knowledge into concern about climate change. Third, the reliability of some scales was less than ideal. Given the exploratory nature of the study, lower alpha values are perhaps acceptable (Hair et al. 2010). Moreover, inter-item correlations—an alternative measure of reliability—were mostly within the desirable range of 0.2–0.4 (Piedmont 2014).

Conclusions

The identification of effective communication and education strategies to foster engagement among young people requires an in-depth understanding of how they perceive climate change. The adolescents studied here tend to perceive climate change as something that affects other places and other people. On the other hand, to a greater extent than the general population, they perceive climate change to be real, human-made, and happening right now. The closer participants perceived climate change, the more concerned they were about its effects. We conclude that reducing the psychological distance of climate change could be an efficient educational approach to increase concern, a key variable for the development of behavioural intentions. However, a range of other potentially influential variables need to be carefully addressed when using distance-reducing approaches.

Acknowledgements The authors wish to thank the students, schools, and teachers for their participation and for administrative help.

References

Adams P, Gynnild A (2013) Communicating environmental messages in online media: the role of place. Environ Commun 7(1):113–130

APA, American Psychological Association (2017) Ethical principles of psychologists and code of conduct. Author, Washington

Ballantyne AG, Wibeck V, Neset T-S (2016) Images of climate change—a pilot study of young people's perceptions of ICT-based climate visualization. Clim Change 134(1):73–85. https://doi.org/10.1007/s10584-015-1533-9

Bar-Anan Y, Liberman N, Trope Y (2006) The association between psychological distance and construal level: evidence from an implicit association test. J Exp Psychol 135(4):609–622

Brügger A, Dessai S, Devine-Wright P, Morton TA, Pidgeon NF (2015) Psychological responses to the proximity of climate change. Nat Clim Change 5(12):1031–1037. https://doi.org/10.1038/nclimate2760

Corner A, Markowitz E, Pidgeon N (2014) Public engagement with climate change: the role of human values. Wiley Interdisc Rev Clim Change 5(3):411–422. https://doi.org/10.1002/wcc.269

Corner A, Roberts O, Chiari S, Völler S, Mayrhuber ES, Mandl S, Monson K (2015) How do young people engage with climate change? The role of knowledge, values, message framing, and trusted communicators. Wiley Interdisc Rev Clim Change 6(5):523–534. https://doi.org/10.1002/wcc.353

Crocetti E, Rubini M (2017) Communicating personal and social identity in adolescence. In: Oxford research encyclopedia of communication. Oxford University Press, USA. Retrieved from (https://oxfordre.com/communication/view/10.1093/acrefore/9780190228613.001.0001/acrefore-9780190228613-e-482). 29 Jan 2019

ERZ, Department of Education of the Canton of Berne (2018) Education statistics of the Canton of Berne 2017. ERZ, Berne. Retrieved from https://www.erz.be.ch/erz/de/index/direktion/organisation/generalsekretariat/statistik.assetref/dam/documents/ERZ/GS/de/GS-biev-statistik/ERZ_INS_2018_Bildungsstatistik_Kt_BE_Basisdaten_2017.pdf. 29 Jan 2019

Flora JA, Saphir M, Lappé M, Roser-Renouf C, Maibach EW, Leiserowitz AA (2014) Evaluation of a national high school entertainment education program: the alliance for climate education. Clim Change 127(3):419–434. https://doi.org/10.1007/s10584-014-1274-1

Fløttum K, Dahl T, Rivenes V (2016) Young Norwegians and their views on climate change and the future: findings from a climate concerned and oil-rich nation. J Youth Stud 19(8):1128–1143. https://doi.org/10.1080/13676261.2016.1145633

Gensicke T (2000) Germany in transition. Attitude to life, value orientations, civic commitment. In: Speyer research reports 204. Forschungsinstitut für öffentliche Verwaltung bei der Hochschule für Verwaltungswissenschaften, Speyer

Gifford R (2011) The dragons of inaction: psychological barriers that limit climate change mitigation and adaptation. Am Psychol 66(4):290–302. https://doi.org/10.1037/a0023566

Hair JH, Black WC, Babin BJ, Anderson RE (eds) (2010) Multivariate data analysis, 7th edn. Prentice Hall, Upper Saddle River

Hamilton L (2018) Self-assessed understanding of climate change. Clim Change 151(2):349–362. https://doi.org/10.1007/s10584-018-2305-0

Harker-Schuch I, Bugge-Henriksen C (2013) Opinions and knowledge about climate change science in high school students. Ambio 42(6):755–766. https://doi.org/10.1007/s13280-013-0388-4

Hornsey MJ, Harris EA, Bain PG, Fielding KS (2016) Meta-analyses of the determinants and outcomes of belief in climate change. Nat Clim Change 6:622–626. https://doi.org/10.1038/nclimate2943

Huber SG, Lussi I (2016) Value orientations of young adults in Switzerland. In: Huber SG (ed) Young adult survey switzerland. BBL/OFCL/UFCL, Berne, pp. 98–101. Retrieved from https://chx.mazzehosting.ch/sites/default/files/ch-x_yass_huber-et-al_band-1_2016-2017.pdf. 29 Jan 2019

IPCC, Intergovernmental Panel on Climate Change (2014) Climate change 2014: synthesis report. Contribution of working groups I, II and III to the fifth assessment report of the intergovernmental panel on climate change. In: Pachauri RK, Meyer LA (eds). IPCC, Geneva. (Core Writing Team)

IPCC, Intergovernmental Panel on Climate Change (2018) Summary for policymakers. In: Masson-Delmotte V, Zhai P, Pörtner HO, Roberts D, Skea J, Shukla PR, Pirani A, Moufouma-Okia W, Péan C, Pidcock R, Connors S, Matthews JBR, Chen Y, Zhou X, Gomis MI, Lonnoy E, Maycock T, Tignor M, Waterfield T (eds) Global warming of 1.5 °C. An IPCC special report on the impacts of global warming of 1.5 °C above pre-industrial levels and related global greenhouse gas emission pathways, in the context of strengthening the global response to the threat of climate change, sustainable development, and efforts to eradicate poverty. World Meteorological Organization, Geneva

JASP Team (2018) JASP version 0.9 (computer software). University of Amsterdam

Jones C, Hine DW, Marks ADG (2017) The future is now: reducing psychological distance to increase public engagement with climate change. Risk Anal 37(2):331–341. https://doi.org/10.1111/risa.12601

Kahan DM, Peters E, Wittlin M, Slovic P, Ouellette LL, Braman D, Mandel G (2012) The polarizing impact of science literacy and numeracy on perceived climate change risks. Nat Clim Change 2:732–735. https://doi.org/10.1038/nclimate1547

Kerr JR, Wilson MS (2018) Changes in perceived scientific consensus shift beliefs about climate change and GM food safety. PLOS ONE 13(7):e0200295, 1–17. https://doi.org/10.1371/journal.pone.0200295

Kollmuss A, Agyeman J (2002) Mind the gap: why do people act environmentally and what are the barriers to pro-environmental behavior? Environ Edu Res 8(3):239–260. https://doi.org/10.1080/13504620220145401

Leiserowitz AA (2005) American risk perceptions: is climate change dangerous? Risk Anal 25(6):1433–1442. https://doi.org/10.1111/j.1540-6261.2005.00690.x

Liberman N, Trope Y (2008) The psychology of transcending the here and now. Science 322(5905):1201–1205. https://doi.org/10.1126/science.1161958

Lorenzoni I, Nicholson-Cole S, Whitmarsh L (2007) Barriers perceived to engaging with climate change among the UK public and their policy implications. Glob Environ Change 17(3):445–459. https://doi.org/10.1016/j.gloenvcha.2007.01.004

Lorenzoni I, Pidgeon NF (2006) Public views on climate change: European and USA perspectives. Clim Change 77(1):73–95. https://doi.org/10.1007/s10584-006-9072-z

Maccoby EE (2015) Historical overview of socialization research and theory. In: DeLamater J, Ward A (eds) Handbook of socialization: theory and research, 2nd edn. Guilford Press, New York

McCright AM (2010) The effects of gender on climate change knowledge and concern in the American public. Popul Environ 32(1):66–87

McDonald RI, Chai HY, Newell BR (2015) Personal experience and the 'psychological distance' of climate change: an integrative review. J Environ Psychol 44:109–118. https://doi.org/10.1016/j.jenvp.2015.10.003

Milfont T (2012) The interplay between knowledge, perceived efficacy, and concern about global warming and climate change: a one-year longitudinal study. Risk Anal 32(6):1003–1020. https://doi.org/10.1111/j.1539-6924.2012.01800.x

Monroe MC, Plate RR, Oxarart A, Bowers A, Chaves WA (2017) Identifying effective climate change education strategies: a systematic review of the research. Environ Edu Res 1–22. https://doi.org/10.1080/13504622.2017.1360842

Ojala M (2013) Coping with climate change among adolescents: implications for subjective well-being and environmental engagement. Sustainability 5(5):2191–2209. https://doi.org/10.3390/su5052191

Ojala M (2015) Climate change skepticism among adolescents. J Youth Stud 18(9):1135–1153. https://doi.org/10.1080/13676261.2015.1020927

Ojala M (2016) Young people and global climate change: emotions, coping, and engagement in everyday life. In: Ansell N, Klocker N, Skelton T (eds) Geographies of global issues: change and threat. Springer, Singapore, pp 1–19

Ojala M, Lakew Y (2017) Young people and climate change communication. In: (ed) Oxford research encyclopedia of climate science. Oxford University Press, USA. Retrieved from http://oxfordre.com/climatescience/view/10.1093/acrefore/9780190228620.001.0001/acrefore-9780190228620-e-408. 29 Jan 2019

Perera L, Hewege C (2013) Climate change risk perceptions and environmentally conscious behaviour among young environmentalists in Australia. Young Consumers 14(2):139–154. https://doi.org/10.1108/17473611311325546

Piedmont RL (2014) Inter-item correlations. In: Michalos AC (ed) Encyclopedia of quality of life and well-being research. Springer, Dordrecht

Poortinga W, Spence A, Whitmarsh L, Capstick S, Pidgeon NF (2011) Uncertain climate: an investigation into public scepticism about anthropogenic climate change. Glob Environ Change 21(3):1015–1024. https://doi.org/10.1016/j.gloenvcha.2011.03.001

Scannell L, Gifford R (2013) Personally relevant climate change: the role of place attachment and local versus global message framing in engagement. Environ Behav 45(1):60–85. https://doi.org/10.1177/0013916511421196

Schultz P (2002) Inclusion with nature: the psychology of human-nature relations. In: Schmuck P, Schultz WP (eds) Psychology of sustainable development. Springer, Boston

Semenza JC, Hall DE, Wilson DJ, Bontempo BD, Sailor DJ, George LA (2008) Public perception of climate change: voluntary mitigation and barriers to behavior change. Am J Prev Med 35(5):479–487. https://doi.org/10.1016/j.amepre.2008.08.020

Shi J, Visschers VHM, Siegrist M (2015) Public perception of climate change: the importance of knowledge and cultural worldviews. Risk Anal 35(12):2183–2201. https://doi.org/10.1111/risa.12406

Singh AS, Zwickle A, Bruskotter JT, Wilson R (2017) The perceived psychological distance of climate change impacts and its influence on support for adaptation policy. Environ Sci Policy 73:93–99. https://doi.org/10.1016/j.envsci.2017.04.011

Spence A, Poortinga W, Butler C, Pidgeon NF (2011) Perceptions of climate change and willingness to save energy related to flood experience. Nat Clim Change 1:46–49. https://doi.org/10.1038/nclimate1059

Spence A, Poortinga W, Pidgeon N (2012) The psychological distance of climate change. Risk Anal 32(6):957–972. https://doi.org/10.1111/j.1539-6924.2011.01695.x

Stevenson K, Peterson N (2016) Motivating action through fostering climate change hope and concern and avoiding despair among adolescents. Sustainability 8(1):6. https://doi.org/10.3390/su8010006

Stevenson KT, Peterson MN, Bondell HD (2018) Developing a model of climate change behavior among adolescents. Clim Change 151(3):589–603. https://doi.org/10.1007/s10584-018-2313-0

Stevenson KT, Peterson MN, Bondell HD (2016) The influence of personal beliefs, friends, and family in building climate change concern among adolescents. Environ Edu Res 1–14. https://doi.org/10.1080/13504622.2016.1177712

Stevenson KT, Peterson MN, Bondell HD, Moore SE, Carrier SJ (2014) Overcoming skepticism with education: interacting influences of worldview and climate change knowledge on perceived climate change risk among adolescents. Clim Change 126(3):293–304. https://doi.org/10.1007/s10584-014-1228-7

Stoknes PE (2014) Rethinking climate communications and the "psychological climate paradox". Energy Res Soc Sci 1:161–170. https://doi.org/10.1016/j.erss.2014.03.007

Tanti C, Stukas AA, Halloran MJ, Foddy M (2011) Social identity change: shifts in social identity during adolescence. J Adolesc 34(3):555–567. https://doi.org/10.1016/j.adolescence.2010.05.012

Tobler C, Visschers VHM, Siegrist M (2012) Consumers' knowledge about climate change. Clim Change 114(2):189–209. https://doi.org/10.1007/s10584-011-0393-1

Trope Y, Liberman N (2010) Construal-level theory of psychological distance. Psychol Rev 117(2):440–463. https://doi.org/10.1037/a0018963

Valdez RX, Peterson MN, Stevenson KT (2018) How communication with teachers, family and friends contributes to predicting climate change behaviour among adolescents. Environ Conserv 45(2):183–191. https://doi.org/10.1017/S0376892917000443

van der Linden S, Maibach E, Leiserowitz A (2015) Improving public engagement with climate change: five "best practice" insights from psychological science. Perspect Psychol Sci 10(6):758–763. https://doi.org/10.1177/1745691615598516

Yang ZJ, Kahlor LA, Griffin DJ (2014) I share, therefore i am: a US—China comparison of college students' motivations to share information about climate change. Hum Commun Res 40(1):112–135. https://doi.org/10.1111/hcre.12018

Chapter 9
Addressing Climate Change at a Much Younger Age Than just at the Decision-Making Level: Perceptions from Primary School Teachers in Fiji

Peni Hausia Havea, Apenisa Tamani, Anuantaeka Takinana, Antoine De Ramon N' Yeurt, Sarah L. Hemstock and Hélène Jacot Des Combes

Abstract This study uses an explanatory design to investigate the role of primary education in addressing climate change in primary schools in Fiji. A self-administered questionnaire ($N = 30$) was conducted with primary school teachers from 14 primary schools in Fiji. Using frequency analysis, all teachers perceived that addressing climate change at a much younger age is more effective than just addressing it at the decision-making level. Furthermore, a Kendall tau-b was performed, and identified a significant correlation between the primary school teachers' location and recommendations for further training on climate change ($\tau b = .59, p < .001$) and work relevance and climate evaluation ($\tau b = .6, p < .001$). The same factors (e.g. work relevancy, helping primary education adapt to climate change, etc.) were explored qualitatively using desktop review, literature search and found addressing climate change at a

---

P. H. Havea (✉) · A. De Ramon N' Yeurt
PaCE-SD, USP, Lower Campus, Laucala, Suva, Fiji
e-mail: ilaisiaimoana@yahoo.com

A. De Ramon N' Yeurt
e-mail: antoine.nyeurt@usp.ac.fj

A. Tamani
Deutsche Gesellschaft Für Internationale Zusammenarbeit (GIZ) GmbH, Suva, Fiji
e-mail: tamani.apenisa@gmail.com

A. Takinana
Graduate School of Global Environmental Studies, Kyoto University, Kyoto, Japan
e-mail: nunu.takinana@gmail.com

S. L. Hemstock
School of Humanities, Bishop Grosseteste University, Longdales Road, LN1 3DY Lincoln, UK
e-mail: sarah.hemstock@bishopg.ac.uk

H. J. Des Combes
National Disaster Management Office, Government of Marshall Islands, Majuro, Republic of the Marshall Islands
e-mail: hjdc2000@yahoo.fr

© Springer Nature Switzerland AG 2019
W. Leal Filho and S. L. Hemstock (eds.), *Climate Change and the Role of Education*,
Climate Change Management, https://doi.org/10.1007/978-3-030-32898-6_9

much younger age to be significant. These results are expected to perfect not only the role of primary education but to contribute significantly to the achievements of a climate-resilient Fiji by 2030 and beyond.

Keywords Pacific · Climate change · Primary education · Policy · Teachers · Resilience

Introduction

According to the United Nations Educational, Scientific and Cultural Organization (UNESCO) (UNESCO 2017a), by 2030, all learners on primary education need to acquire knowledge and skills not only needed to know how to promote sustainable development but also averting climate change (UNDP 2015; UNESCO 2017b). For this state-of-the-art goal needs to be met by the Pacific region, then it is significant for Fiji and other Pacific Islanders to look at climate change, not just in terms of many people are being affected, but also how it influences political leaders and people positively so their lives become more resilient and sustainable in their own communities. (Dalelo 2011; Hammersley-Fletcher 2008; Karpudewan et al. 2015; Lenti Boero et al. 2009; Pacific Community et al. 2016).

In the literature, a post-positivist theory is what Bloom et al. (2010) called collegiality amongst primary school teachers. This positive spirit of collaboration can go beyond teachers to their learners and most importantly to where these children live leading to improving climate resilience effectively at the household level. This is expected to result in a top-down approach (Lata and Nunn 2012), where primary school teachers influence their students positively and their students influence the community, accordingly (e.g. improving community resilience [Climate Change Adaptation and Disaster Risk Reduction] to climate change) (Bibi Abdullah and Kassim 2012; Bynoe and Simmons 2014; Dalelo 2011; Hammersley-Fletcher 2008; Karpudewan et al. 2015; Özdem et al. 2014; Santos et al. 2008; Tsaliki 2017; Usman 2008; Van Dam et al. 2015).

However, this pathology of power (Farmer 2005) and leading-edge development to improve climate adaptation and disaster risk reduction has been overlooked by climate change educators and policy-makers who focused primarily on higher educators (e.g. secondary education and above) (Akrofi et al. 2019; Nagy et al. 2017; Verlie 2018; Walshe et al. 2018). As a result, the perceptions of primary school educators and praxis are missing out from the national adaptation plan, especially in countries that are severely and overly affected by climate change. Most importantly, Ledley et al. (2017) did not support this insurmountable types of climate solutions, especially when addressing climate change via education—preferably it needs to include all level of society and primary school teachers are no exception (Henderson et al. 2017).

This deficiency in research allows this paper to contribute to the literature in two principals ways. First, to provide empirical evidence that primary school educators may contribute to the national climate change convention and adaptation effectively. Second, to share experiences of how primary education in Fiji may impact climate adaptation positively for others in the Pacific. This state-of-the-art idea is such an important milestone in the formulation and the achievements of resilient Pacific Islanders for Fiji and other countries in the region at large (Buckland et al. 2018; Henderson and Mouza 2018). Simply, because apparently schools who have already integrated and/or planned for integrating climate change into their curricula have their children and peers becoming more ecocentric and less homocentric (Hestness et al. 2011; Lambert and Bleicher 2013; UN CC: Learn 2013) in their attempts to adaptation. To act on this problem, the Government of Fiji's response to this call could be by addressing the roles of primary education in climate change.

At the regional level, the SPC/GIZ programme 'Coping with Climate Change in the Pacific Island Region' (CCCPIR), has supported Pacific Island Countries (PICs) and regional organisations in building their capacities to cope with the anticipated effects of climate change that will affect communities across the region (SPC and GIZ 2016). Focusing on primary, secondary and TVET education, the CCCPIR programme supported the Ministry of Education and Training institutions in Fiji to integrate climate change into primary education aiming to equip children from aged 6–11 (Education Policy and Data Center 2014) with the knowledge and skills required to liven up coping with the effects of climate change. Significantly, from what is known from the literature, Fiji is amongst the first Pacific country in the world, along with Malawi from the African countries (UN CC: Learn 2013) and the UK in Europe (Oxfam 2015), to have climate change fully integrated in its primary schools curricula (S. Mesquita and Bursztyn 2016; SPC/GIZ 2018; UN CC: Learn 2013).

Integrating climate change into the school course of study for Fiji is critical if the country wishes to build a sustainable and resilient school campus (Özdem et al. 2014). From a climate change perspective, this is vital because if children would understand the causes and consequences of climate change, then, in reality, they and their families and/or their communities could do anything conceivable to protect themselves from these negative impacts of climate change and/or disasters caused by extreme weather events (Dalelo 2011; Karpudewan et al. 2015; Lenti Boero et al. 2009).

These children are tomorrow's business leaders, decision-makers and political leaders (e.g. which could be seen as islands of tomorrow, if high-quality education and proper nourishment will be given to them). Consequently, primary education plays a significant role in responding to climate change (Chang 2012; Hammersley-Fletcher 2008; Satchwell 2013; Thomalla and Djalante 2012; UN CC: Learn 2013; Vines et al. 2014; Vogel et al. 2015). Most importantly, that is why Fiji was chosen as a case study for this paper.

Fiji is a Melanesian republic state and archipelago comprised of 332 islands and 552 small islets (Bissessar 2017), of which 110 are permanently inhabited over a land area of 18,333 km^2, with a total ocean area of 194,000 km^2 (Department of Environment 1997; Pacific Community (SPC) 2012). The archipelago is located between the longitudes 17.4°E and 17.8°W and latitudes 12°S and 22°S (Pacific

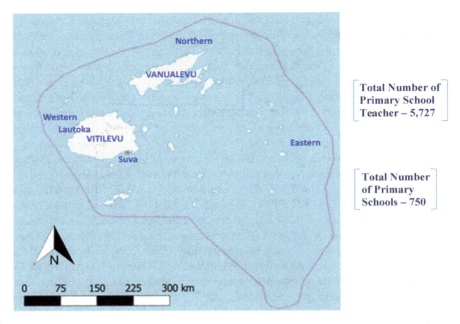

Fig. 9.1 Map of Fiji. Created by Peni Hausia Havea, 2019. Total number of primary schools and teachers were taken from the Ministry of Education (2016)

Community (SPC) 2012) (Fig. 9.1). According to the Ministry of Education (Ministry of Education 2016), there are a total of 750 primary schools in Fiji, including in special education. The total number of primary school teachers serving the nation in 2016 was 5727.

The nation is ranked number 15 most at risk in the world with a risk index value of 13.50% (Bündnis Entwicklung Hilft 2017). Resultantly, it is one of the highest risk countries to be affected by climate change and disasters caused by extreme weather events besides Vanuatu, Tonga, the Solomon Islands and Papua New Guinea. Because climate change is known to affect the livelihoods, health, well-being, national economies and education in Fiji and the Pacific (Chand and Walsh 2009; Government of Vanuatu 2015; Luetz and Havea 2018; Maeke 2013; Reardon and Oliver 1983; Woodroffe 1983), there is a need for critical evaluation on the role, primary education plays in addressing climate change challenges and opportunities for the people of Fiji. The reason is that the roles that primary education plays in addressing climate change are unclear (Ministry of Education 2016).

This study will be used to fill this gap in knowledge and needs for research on the perceptions of primary school teachers and integration of climate change education into the primary school curricula in the country. Hence, based on the perceptions of primary school teachers, this paper intends to provide a better understanding as to why addressing climate change at a much younger age is more effective and efficient in responding to the impacts of climate change than just at the decision-making level.

Methods

The study used a mixed method approach (Havea et al. 2017).

Methodology

A mixed-method design named explanatory (Creswell and Plano Clark 2011), was used to gather all the quantitative information for this study. These data were collected from workshops and training sessions of primary schools teachers in Fiji between 2016 and 2018. The teachers who were selected represented 14 different primary schools in the country. The quantitative data were mainly from surveys (N = 30) based on the evaluation of the teachers after their training on climate change based on the school curriculum. The qualitative aspect of the study focused on data collected from personal communications, desktop review, online searches and reviewed of project documents on CCCPIR and climate change in Fiji. The study approach was called explanatory (Creswell 2014) because this paper has relied heavily on the quantitative aspect of the study. Unfortunately, due to time constraint, there was no donor available at the time to fund this work.

Data Analysis

The data analysis used an explanatory design model (Creswell 2013). For the quantitative data, frequency plot analysis and Kendall tau-b were calculated. The frequency plots used r studio and SPSS for the Kendall tau-b. These results were then analysed using thematic analytical strategy (e.g. themes). In this strategy, the quantitative results were explored qualitatively and vice versa. Meaning, the same factors (e.g. result from the frequency plots and Kendall tau-b) were merged with the qualitative data by taking one piece of theme, information or idea and compared with the quantitative results, iteratively. This was done in order to provide a better understanding as to why addressing climate change at a much younger age, is more effective in responding to the impacts of climate change than just at the decision-making level. The data analysis was also performed using QGIS for mapping (Cronk 2017; Schwarts et al. 2015).

Limitations

Despite significant effort made to improve the reliability and validity of the study, there are always limitations with regard to the collection of data. In this instance, the researchers were unable to include findings from personal oral in-depth interviews. Therefore, individual teacher's in-depth perceptions and nuances are absent from the results. However, qualitative data has largely been drawn from the primary school teachers' written accounts and open-ended answers obtained during quantitative data collection. This information complements the quantitative analysis. Other information used to assist the qualitative analysis has been taken from reviewing project documents, desktop review, online stories, debating and vlogging (e.g. publish of opinion or polls online) of primary school teachers and others on the subject. This method allows this paper to reach its valid conclusion.

Results

Based on this sample and the data available for this study, the following results and implications are presented. Six reasons why addressing climate change at a much younger age is significant than just at the decision-making level were identified:

(1) national priority;
(2) teachers have optimal facilitation skills;
(3) sense of duty;
(4) climate awareness;
(5) combat climate skepticism and anxiety;
6) biblical message.

(A) **National Priority**

More Than 90% of Teachers in the Study Perceived that Integrating Climate Change on the Curriculum Aligned with the Ministry of Education Priorities and Implication for Integrating Climate Change into the School Curricula in Fiji

Of the 30 teachers selected to participate in the study, 96.7% (29) perceived that the training in integrating climate change into the school curriculum applied to their teaching careers and improving the knowledge of their primary schools' children to resilience (CCA and DRR) (Fig. 9.2).

During an online opinion polling for primary school teachers regarding the question "should climate change be taught in schools?", interestingly a participant expressed climate change education in the form of children right and as a national priority:

> Students have a right to know how to adapt to the impacts at a younger age. It is like any other subjects like English and Science. It is the duty of the Education System to teach climate change at a younger age in order to help them adapt better now and in the future. Students have a right to improve their knowledge regarding the solutions to climate change through education (Debate.org 2018).

9 Addressing Climate Change at a Much Younger Age Than …

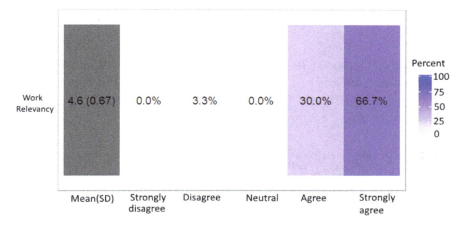

Fig. 9.2 Percentage of primary school teachers who perceived that the climate change training that they attended was relevant to their work

The same point emphasised during the workshop with the primary school's teachers in Fiji (Fig. 9.3) that such training of teachers on integration of climate change into the school curricula is not only relevant but is also aligned with national processes and activities. Prompting the Ministry of Education and line ministries for taking the lead in implementing this policy for Fiji and the region is vital (SPC/GIZ 2018).

The implication of this result on policy for the education in Fiji is that climate change will be integrated to all primary schools education in the country by 2030.

Fig. 9.3 Participants and trainers from workshop and training of primary school teachers in Fiji on the integration of climate change into the school curriculum

(B) **Facilitation Skills**

Training of Climate Change Education Teachers with Good Facilitation Skills at Fiji's Teachers Training: An Implication for Integrating Climate Change Education into the Fiji Teachers Training Programme

Adding to the above state of climate change relevancy, ensuring teachers facilitation skills to be professional and be able to cater to a younger age students degree of understanding is also significant. As indicated in this study, 96% of the primary school teachers who attended the training perceived that facilitation skills are essential (Fig. 9.4).

Importantly, to evaluate whether those facilitation skills and other attributes are relevant to integrating climate change into the school curriculum, a frequency analysis revealed that all primary school teachers in the study indicated that not only it is coherent but that the whole climate change integration training programme was excellent (Fig. 9.4). Outside the PICs, a Science Teacher from Algonquin Middle School Des Plaines, I11, USA goes beyond this state-of-the-art integration by stating: "I'd been considering making it a mandatory subject" (Harmon 2017) showed that some countries may already ahead on this curve.

The same point was made by Tameka Wallace on the online polling for primary school teachers. She stated explicitly:

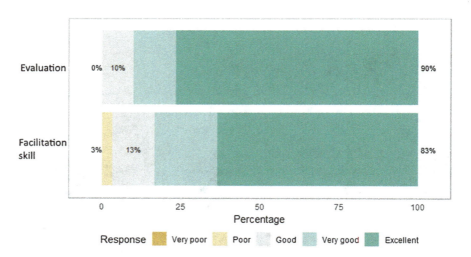

Fig. 9.4 Percentage of primary school teachers who rated the facilitator skills and evaluated the training as excellent

Yes, young people need to understand climate change. Climate change is a major issue that all young people need to learn about because it will affect their future. Students need to understand the factors that have led to climate change and the ways in which they can remedy this issue. In the future, the young people are the ones that will have to remedy the issue of climate change in the world (Debate.org 2018).

The implication of this result on the role of education in addressing climate change for Fiji are two folds. First, to have a policy that allows integrating of climate change education into the primary school education teachers training programme. Second, the training of teachers should focus and master the facilitation skills of the course.

(C) **Sense of Duty**

Teachers Felt that It Is Their Duty to Teach Climate Change to the Much Younger Age Student

Further to perfecting the facilitation skills, a Kendall tau-b was performed and concluded that there is a significant and positive correlation between the primary school teachers location (e.g. remoted areas) and recommendations for further training on climate change ($\tau b = .59, p < .001$) (Table 9.1).

Because there is a positive and significant relationship between climate change and primary school teachers who perceived to teach this in their schools, from a climate change perspective, this could mean that teachers felt it is their responsibility to educate these young generations about climate change. As pointed out in the workshop, although some schools have already taught climate change, the challenges are to have a strong partnership with the government and its partners to nationalise this in Fiji. Other challenges may include but are not limited to addressing issues such as need for staff training (e.g. logistics and venue) and prevent staff turn over effects on schools system (SPC/GIZ 2018).

There are two implications for this result. First, teachers show a high-level of compliance to integrate climate change into the school curricula. Because teachers seemed to agree with this cutting edge approach, the implication is that climate change will be integrated into school curricula at the national level. Second, because of their sense of duty to fulfil, teachers will do it no matter what. This indicates that most likely all primary school teachers in Fiji will accept this proposal and do it.

Table 9.1 The relationship between primary school teachers location and recommendations for integrating climate change into the school's curriculum

	Location	Recommendations
Location	–	0.588**
Recommendations	0.588**	–

*$P < 0.05$, **$P < 0.01$

Table 9.2 The relationship between work relevancy and climate evaluation

	Work relevancy	Climate evaluation
Work relevancy	–	0.600^{**}
Climate evaluation	0.600^{**}	–

$^{*}P < 0.05, \, ^{**}P < 0.01$

(D) Climate Awareness

Increase Level of Climate Change Awareness to Include All Primary Schools in Fiji

From what is known in this current study, teaching climate change at a much younger age is not only a sense of duty, but according to Sarah Murphy, a Science Teacher, at the Algonquin Middle School Des Plaines, Ill, USA: "Our students need to have an awareness of an issue/problem they can hopefully solve" (Harmon 2017) at a much younger age and throughout their lifetime. In supporting this point, a Kendall tau-b was performed and concluded that educating young children on climate change is relevant not only to their current living condition but also throughout their lifetime ($\tau b = .60, p < .001$) (Table 9.2).

Since climate change is integrated into the primary schools curricula for Fiji, the implication of this result will be on increasing the level of communication and awareness of students and peers about climate change. Be it climate change impacts or solutions. As a result of exposure to climate change at a much younger age, students tend to deal with it better as they grew old (Ahdoot 2015).

(E) Combat Climate Skepticism and Anxiety

To Combat Climate Skepticism and Ease Climate Anxiety While They Are at a Much Younger Age to Help Them Cope with Worriedness and Post-traumatic Stress Disorder When They Grow up Throughout Their Lifetime

More importantly, adding to the above state of affairs, Joshua Moses, an Assistant professor of Anthropology, Haverford College Haverford, Pa, USA, stated that climate change education "always include an empowering message. Some teachers felt their lessons might both persuade skeptics and ease the anxious" (Harmon 2017).

He then continued:

> Cooking creates an approachable platform to teach basic thermodynamic concepts. Everyone is interested in why a pizza stone results in a crispier pizza crust! Once they understand these concepts, the energy balance of the climate system is easy to understand (Harmon 2017).

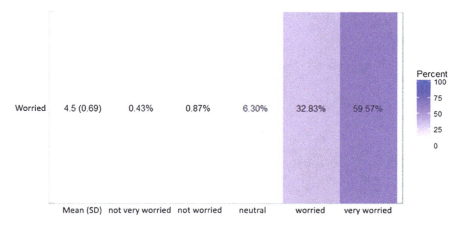

Fig. 9.5 How worried are you about the impacts of climate change and disasters on your well-being?

In one way or another:

> It is also a great way to promote climate optimism as there is much we can in our food system to have a significant impact, this is what students want to know. Always include an empowering message—we can do this! (Harmon 2017).

However, this is intriguing because according to Havea et al. (2018a), Tonga is one of the most worried nation affected by climate change and hazards because of the effects of climate change on their livelihoods, health and well-being (Havea et al. 2017, 2018a; Havea et al. 2018b), based on the response of 92.4% (425) of the 460 participants to a study on this issue (Fig. 9.5).

Nevertheless, albeit this study has conducted in Tonga, this paper believes that the result may yield a significant finding as well in Fiji due to its same system of belief as Tonga (Havea et al. 2017, 2018a). The implication of this result, is that learning about climate change at a young age helps in addressing climate scepticism and anxiety in schools and at the national level in Fiji and alike in the Pacific. Once this policy is implemented, people may know how to deal with the psychological impacts of climate change on their health through their children (UN CC: Learn 2013; WHO 2015). As a result, it is pivotal to integrate climate change into the school curriculum in Fiji and/or Tonga.

(F) **Biblical Message**

A Message from the Bible: To Teach or Educate Young Children About Climate Change and/or Hazards

In a final point, although this is controversial, due to Fiji's high proportion of Christians at 65% (Fiji Bureau of Statistics 2007) (Fig. 9.6), educating young children

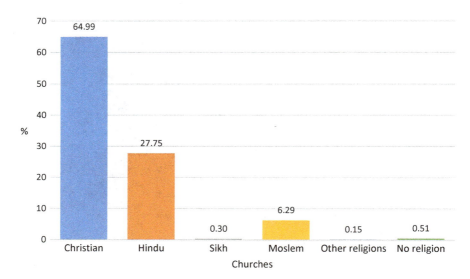

Fig. 9.6 Different types of religions in Fiji. The Christian churches include Anglican, Apostolic, Assembly of God (AOG), All Nation Christian Fellowship, Baptist, Roman Catholic (RC), Christian Mission Fellowship (CMF), Church of Christ, Gospel, Jehovah's Witness, Latter Day Saint (LDS), Methodist, Pentecostal, Presbyterian, Salvation Army, Seventh Day Adventist (SDA), United Pentecostal and Other Christians. *Source* Fiji Bureau of Statistics (2007)

about climate change impacts and adaptation while they are young enough to learn is also aligned with the Biblical teaching of King Solomon about being wise in Proverbs 19:18 (Howells 2010; The Bible Society in Australia 1988). As a result, teaching climate change at a much younger age is not only a human right issue and climate change concerned but also a God's will to educate children about this global phenomenon as a mean of making the right decision in their journey to be an adulthood, whether they are future leaders (e.g. political, business) or not.

Additionally, research suggests that Christian theological, hermeneutical and eschatological perspectives on environmental sustainability and the care of God's creation may also harness for climate change adaptation in the Pacific islands contexts (Luetz et al. 2018; Nunn et al. 2016). In the context of Fiji and Tonga, this is especially relevant.

As stated by the Permanent Secretary (PS) for Education, Heritage and Arts of Fiji, Mr Iowane Ponipate Tiko in 2016:

> The Ministry of Education, Heritage and Arts ensure that the children of this nation are provided with equitable and affordable education (Ministry of Education 2016), inclusive of climate change

indicated that the Ministry is also doing God's will.

In his final remarks, he stated that: "May GOD continue to bless our beloved Fiji" is also reflected his belief as Christian is significant to the state-of-the-art idea of teaching climate change at a much younger age in Fiji (Ministry of Education 2016).

The implication of this result in addressing climate change in Fiji is that people are pleasing their Christian God by doing His will (e.g. integrating climate change into the school's curriculum or doing the same thing) (Bible Society of the South Pacific 1966). This Biblical teaching, interpretation and Christian perceptions were influenced by churches as well. So, telling a climate change story to a kid (e.g. the story about Noah and others), are not only at the policy level but most importantly, it shows that climate solutions through primary education, art and God are work as well (Havea et al. 2018a; Ministry of Education 2016; The Bible Society in Australia 1988).

Discussion

Primary school teachers agreed on the importance to address climate change at a much younger age rather than just at the decision-making level have participate to this national adaptation development. Significantly, it is evidenced by this study also that they consider it is their duty ($\tau b = .59, p < .001$) and relevant to their roles as primary school teachers to teach climate change at their schools ($\tau b = .60, p < .001$). In addition, from a psychological perspective, for the children, the school system offers them a sense of security, better health, happiness and well-being (Erricker 2009)—a win-win situation. As a result, the benefits of implementing this approach for the people of Fiji and implication for policy are seen as a sine qua non condition for the building of a resilient Fiji by 2030 and beyond (Magee et al. 2016; Ministry of Education 2016). There are several reasons why this state-of-the-art development in primary education is favourable for the general population of Fiji.

From a legal perspective, according to Fiji's Constitution on Rights of Children, all children including those with disabilities have the right to education, good health care system and protection from the impacts of climate change and hazards (Government of Fiji 2013). Therefore, if this is the case, then every primary school student also has a right as a human being and/or individual to learn about climate change from a much younger age, and not when they are just at the decision-making level. This concern for human rights is also aligning with the Framework for Resilient Development in the Pacific (FRDP) (Pacific Community et al. 2016) as well as the United Nations Sustainable Development Goals (SDGs) (UNDP 2015). More importantly, if Fiji wishes to become more resilient, the integration of climate change into the schools' curriculum must also be nationalised because one way or another this will ensure that education and training programs are designed to allow and encourage individuals to understand and to take action on mitigation and adaptation, which is the hallmark of climate change education and training.

In addition, according to the Ministry of Education in Fiji (Ministry of Education 2016) and primary school teachers in this study, the integration of climate change into the school's curriculum is twofold. First, it is a national priority. Second, it is parallel with the Education for Sustainable Development (ESD) goals, objectives and targets by 2030 and beyond (UNESCO 2006, 2017a) and education for all. Based on this result, the government's decision to incorporate climate change issues at Fiji's primary, secondary and tertiary school curriculum is not only meant to create greater awareness but is also significant participation at the national level to mitigate and adapt to this global phenomenon.

As stated by Mr Hem Chand, the Director for Primary Education for the Ministry of Education in Fiji, while speaking to more than 300 Early Childhood Education (ECE) teachers conference in Lautoka, teachers have the power to develop solutions for climate change and participate in a building of a resilient Fiji. He said: "Imagine the social, cultural and environmental impact these practical classroom lessons would have on young minds of our ECE students who later become future leaders, policy makers and think tanks of our society" (Nasiko 2018).

He continued: "If we want to see the current generations of ECE students transform into champions of our environment and our future, ECE teachers have to play a part in their learning experience" (Nasiko 2018). In his final remarks, he then put teachers at the forefront of addressing these issues on climate change: "I earnestly appeal for all participants to consider your decisive roles and responsibilities in bringing into reality our common goals" (Nasiko 2018). As a result, Fiji could become a model so that other PICs can learn from it before adopting this approach to their education system. This proposition could be discussed with other PICs during the Pacific Leadership Forum and other high-level meetings with school leaders in the region.

More importantly, because climate change is a priority for Fiji political leaders (Government of Fiji 2018) and the general public are also aware of the importance of this issue to their children, there is an obligation of the school system and especially the teacher to educate them in accordance to their elders' wishes (Ministry of Education 1978). As indicated above, if all primary school teachers in the study agree to integrate climate change into the school curriculum in Fiji, then it is left now for the government to set their goals and planned to nationalise teaching of climate change to all primary schools, ideally between the next Conference of Party (COP24) meeting and the year 2030 or after.

This climate operation is fascinating because although Fiji is secular when it comes to climate change adaptation, both State and Church work together as one in building the capacity of the people of the nation to become more resilient by 2030 and beyond. From a religious viewpoint, this integration has been inline also with God's willing framework in the Bible (Howells 2010; The Bible Society in Australia 1988) and research on Christian theological, hermeneutical and eschatological adaptation, in the form of using Bible and Church to tackle climate change on the ground and/or to adapt to the effects of climate change negatively in Fiji (Luetz et al. 2018; Nunn et al. 2016).

For the primary school teachers, who have claimed themselves to be Christian in the study, they perceived that integrating climate change into the school curriculum and teaching children (e.g. discipline) about it is not only a duty attributed by law but also fulfils carrying out God's plan. A positive influences attributed by teaching of the Bible in the Churches that they belong to. In this paper, the concept of discipline in the Bible is broad. It could mean a branch of knowledge, to teach, to educate, or field of study (Bible Society of the South Pacific 1966; Lewis 2015; Oxford University Press 2019; The Bible Society in Australia 1988).

In the context of Fiji, this is highly relevant, since more than half of the total population (65%) belong to a Christian denomination (Fiji Bureau of Statistics 2007). Interestingly, although Fiji has already declared itself a secular state in 2013 (Government of Fiji 2013), the fact that Christianity is the dominant faith in the nation, reflect the belief that using Biblical principles to relate climate change to education and/or to adapt to the negative effects of climate change is significant. This application of Christian theological, hermeneutical and eschatological teachings is also relevant in the context of Tonga (Havea et al. 2017, 2018a).

Despite good intention by primary school teachers in Fiji to integrate climate change into the national school curriculum, there are still challenges. First, there is a need to develop relevant and locally relevant materials on climate change, secondly, train all primary school teachers in Fiji on climate change. Thirdly, the costs for logistics, training budget and reservations of venues must be met, accordingly. Fourthly, the school system in Fiji should have a plan to eliminate and account for the impacts of staff turnover due to relocation, new jobs, retirement inter alia (Iqbal 2010; Surji 2013). As a result, the findings of this study may lead to a policy for the Government of Fiji to integrate climate change first into Teacher Training Institution program before teachers are sent to their school location and employed as teachers.

And while doing that it is also critical to train teachers already in the school system who have not yet acquired necessary skills and knowledge to teach climate change. Realistically, this is the way forward for Fiji and for any other PICs who will be integrating climate change into their schools' curriculum.

Conclusion

The purpose of this article was to provide a better understanding of why addressing climate change at a much younger age is significant. This aim has achieved, and consequently, there were four lessons learnt from this paper.

First, the children have a right to have a good knowledge about the science of climate change so to help them mitigate and adapt to the effects of climate change including disasters caused by extreme weather events. As a result, they can become more resilient, now and in the future. Secondly, addressing climate change at a much younger age is also aligned with the Ministry of Education development goals. Thirdly, the primary school teachers felt that it is their obligations and duty to teach young children about climate change at a much younger age than just at the decision-making level. Fourthly, from the Christian Bible it is God's will to learn climate change impacts and adaptation at a much younger age.

Based on these lessons learnt, to ensure that the Government of Fiji is achieving a resilient Fiji by 2030 and beyond, this paper recommends the Ministry of Education to have primary teachers train on climate change and then integrate this topic to all primary schools in Fiji. The reason is simple: this top-down approach should start with the primary school teachers while they are still in training before entering the workforce. Once the teachers have mastered climate change, then it would be easy for them to convey these principles to the children they teach in order to save themselves from climate change and/or disasters caused by extreme weather events.

For the future, based on this study this paper recommended that researchers should explore the following areas of study:

(1) Cost-effectiveness analysis of integrating climate change into the primary school curricula in Fiji;
(2) Relationship between the teaching of climate change to children in the primary education in Fiji and community-based adaptation;
(3) Long-term effects (negative and positive) of educating young children about climate change impacts and adaptation: a longitudinal study.

If these realisations do not met, it is in the best interest of this paper to follow-through the Ministry of Education latest or the government updated annual plan. The main reason is that because this development originated and conducted by the Government of Fiji.

Acknowledgements The SPC/GIZ CCCPIR Programme in Suva supported this research, The University of the South Pacific, Suva, Fiji, the School of Humanities (Geography), at Bishop Grosseteste University, Lincoln, UK and the Graduate School of Global Environmental Studies, Kyoto University, Kyoto, Japan.

References

Ahdoot S (2015) Address causes of climate change to help alleviate effects on children: AAP. AAP News, USA

Akrofi MM, Antwi SH, Gumbo JR (2019) Students in climate action: a study of some influential factors and implications of knowledge gaps in Africa. Environments 6(12):1–15

Bibi Abdullah J, Kassim J (2012) Promoting learning environment among the Islamic school principals in the state of Pahang, Malaysia. Multicultural Edu Tech J 6(2):100–105

Bible Society of the South Pacific (1966) Koe Tohi Tapu Katoa (the Holy Bible in Tongan). The Bible Society in the South Pacific, Suva

Bissessar AM (2017) Ethnic conflict in developing societies: Trinidad and Tobago, Guyana, Fiji and Suriname. Springer, Switzerland

Bloom PJ, Hentschel A, Bella J (2010) A great place to work: creating a healthy organizational climate. New Horizons, LakeForest, IL

Buckland P, Goodstein E, Alexander R, Muchnick B, Mallia ME, Leary N, Andrejewski R, Barsom S (2018) The challenge of coordinated civic climate change education. J Environ Stud Sci 8(2):169–178

Bündnis Entwicklung Hilft (2017) World risk report analysis and prospects 2017. Bündnis Entwicklung Hilft, Berlin, Germany

Bynoe P, Simmons S (2014) An appraisal of climate change education at the primary level in Guyana. Caribb Geogr 19:89–103

Chand SS, Walsh KJE (2009) Tropical cyclone activity in the Fiji Region: spatial patterns and relationship to large-scale circulation. J Clim 22(14):3877–3893

Chang C (2012) The changing climate of teaching and learning school geography: the case of Singapore. Int Res Geogr Environ Edu 21(4):283–295

Creswell JW (2013) Qualitative inquiry and research design: choosing among five approaches. Sage, United State of America

Creswell JW (2014) Research design. Sage, Los Angeles

Creswell JW, Plano Clark VL (2011) Designing and conducting mixed methods research, 2nd edn. Sage, United State of America

Cronk B (2017) How to use SPSS—a step-by-step guide to analysis and interpretation, 9th edn. Routledge, New York

Dalelo A (2011) Global climate change in geography curricula for Ethiopian secondary and preparatory schools. Int Res Geogr Environ Edu 20(3):227–246

Debate.org (2018) Should climate change be taught in schools? Debate.org: USA. Retrieved from https://www.debate.org/opinions/should-climate-change-be-taught-in-schools. 28 Nov 2018

Department of Environment (1997) Convention on biological diversity: 1997 national report to the conference of the parties. Ministry of Local Government, Housing and Environment, Suva, Fiji

Education Policy and Data Center (2014) Fiji. EPDC, Washington

Erricker J (2009) The importance of happiness to children's education and wellbeing. In: de Souza M, Francis LJ, O'Higgins-Norman J, Scott D (eds) International handbook of education for spirituality, care and wellbeing. Springer, The Netherlands, Dordrecht, pp 739–752

Farmer P (2005) Pathologies of power: health, human rights, and the new war on the poor. University of California Press, USA

Fiji Bureau of Statistics (2007) Relationship ethnicity and religion by province of enumeration Fiji 2007. Fiji Bureau of Statistics, Suva, Fiji

Government of Fiji (2013) Constitution of the Republic of Fiji. Gov Fiji 14(80):2747–2840

Government of Fiji (2018) COP23. Government of Fiji: Fiji. Retrieved from https://cop23.com.fj/. 29 Nov 2018

Government of Vanuatu (2015) Vanuatu: post-disaster needs assessment (tropical cyclone Pam, March 2015). Government of Vanuatu, Vanuatu

Hammersley-Fletcher L (2008) The impact of workforce remodelling on change management and working practices in english primary schools. Sch Leadersh Manage 28(5):489–503

Harmon A (2017) A sense of duty to teach climate change. The New York Times, USA. Retrieved from https://www.nytimes.com/2017/06/28/us/teaching-students-about-climate-change.html?module=inline. 28 Nov 2018

Havea PH, Hemstock SL, Des Combes JH (2017) Preparing for better livelihoods, health and wellbeing—a key to climate change adaptation. In: Leal Filho W (ed) Climate change adaptation in pacific countries: fostering resilience and improving the quality of life. Springer, Cham, pp. 87–99

Havea PH, Hemstock SL, Jacot Des Combes, H (2018a) God and Tonga are my inheritance!—climate change impact on perceived spirituality, adaptation and lessons learnt from Kanokupolu, Ahau, Tukutonga, Popua and Manuka in Tongatapu, Tonga. In: Leal Filho W (ed) Climate change impacts and adaptation strategies for coastal communities—climate change management. Springer, Cham

Havea PH, Hemstock SL, Jacot Des Combes H (2018b) Improving health and well-being through climate change adaptation. In: Duncan LT (ed) Advances in health and disease, vol 4. Nova Science Publisher, Hauppauge, NY, pp. 215–230

Henderson J, Long D, Berger P, Russell C, Drewes A (2017) Expanding the foundation: climate change and opportunities for educational research. J Am Edu Stud Assoc 53(4):412–425

Henderson J, Mouza C (2018) Professional development design considerations in climate change education: teacher enactment and student learning. Int J Sci Edu 40(1):67–89

Hestness E, McGinnins JR, Riedinger K, Marbach-Ad G (2011) A study of teacher candidates' experiences investigating global climate change within an elementary science methods course. J Sci Teacher Educ 22(4):351–369

Howells K (2010) Making sense of bible prophecy. Lulu, European Union

Iqbal A (2010) Employee turnover: causes, consequences and retention strategies in Saudi Organizations. Bus Rev 16(2):275–282

Karpudewan M, Roth W-M, Abdullah MNSB (2015) Enhancing primary school students' knowledge about global warming and environmental attitude using climate change activities. Int J Sci Edu 37(1):31–54

Lambert JL, Bleicher RE (2013) Climate change in the preservice teacher's mind. J Sci Teacher Educ 24(6):999–1022

Lata S, Nunn P (2012) Misperceptions of climate-change risk as barriers to climate-change adaptation: a case study from the Rewa Delta. Fiji. Clim Change 110(1–2):169–186

Ledley TS, Rooney-Varga JN, Niepold F (2017). Addressing climate change through education. Oxford Res Encycl Environ Sci. (Oxford University Press, Oxford, UK)

Lenti Boero D, Clerici T, Perrucci V (2009) Children and marmots: a pilot study in mountain primary schools. Ethology Ecol Evol 21(3–4):415–427

Lewis A (2015) WordWeb Pro. WordWeb Software, Princeton University, USA

Luetz J, Havea PH (2018) We're not refugees, and we'll stay here until we die!—climate change adaptation and migration experiences gathered from the Tulun and Nissan Atolls of Bougainville, Papua New Guinea. In: Leal Filho W (ed) Climate change impacts and adaptation strategies to coastal communities. Springer, Switzerland

Luetz JM, Buxton G, Bangert K (2018) Christian theological, hermeneutical and eschatological perspectives on environmental sustainability and creation care—the role of holistic education. In: Luetz JM, Dowden T, Norsworthy B (eds) Reimagining Christian education—cultivating transformative approaches. Springer, Singapore

Maeke J (2013) Vulnerability and impacts of climate change on food crops in rainsed atoll communities: a case study of Bellona community in Solomon Islands. In: Master of science in climate change master thesis. The University of the South Pacific, Fiji

Magee AD, Verdon-Kidd DC, Kiem AS, Royle SA (2016) Tropical cyclone perceptions, impacts and adaptation in the Southwest Pacific: an urban perspective from Fiji, Vanuatu and Tonga. Nat. Hazards Earth Syst Sci (16):1091–1105

Ministry of Education (1978) Education act. Ministry of Education, Suva, Fiji

Ministry of Education, H. a. A (2016) Annual report. Ministry of Education, Suva, Fiji

Nagy GJ, Cabrera C, Coronel G, Aparicio-Effen M, Arana I, Lairet R, Villamizar A (2017) Addressing climate adaptation in education, research and practice: the CLiVIA-network. Int J Clim Change Strat Manage 9(4):469–487

Nasiko R (2018) Climate change in ECE plans. The Times of Fiji, Suva, Fiji

Nunn PD, Mulgrew K, Scott-Parker B, Hine DW, Marks ADG, Mahar D, Maebuta J (2016) Spirituality and attitudes towards nature in the Pacific islands: insights for enabling climate-change adaptation. Clim Change 136(3):477–493

Oxfam (2015) Climate challenge for 7–11 years. Oxfam, UK. Retrieved from https://www.oxfam.org.uk/education/resources/climate-challenge-7-11. 29 Oct 2018

Oxford University Press (2019) English Oxford living dictionary. Oxford University Press, UK. Retrieved from https://en.oxforddictionaries.com/definition/discipline. 16 Jan 2019

Özdem Y, Dal B, Öztürk N, Sönmez D, Alper U (2014) What is that thing called climate change? An investigation into the understanding of climate change by seventh-grade students. Int Res Geogr Environ Edu 23(4):294–313

Pacific Community, Secretariat of the Pacific Regional Environment Programme, Pacific Islands Forum Secretariat, United Nations Development Programme, United Nations Office for Disaster Risk Reduction, and University of the South Pacific (2016) Framework for resilience development in the Pacific: an integrated approach to address climate change and disaster risk management (FRDP) 2017–2030. The Pacific Community, Geoscience Division, Suva, Fiji

Pacific Community (SPC) (2012). Fiji country energy security indicator profile 2009. Secretariat of the Pacific Community, Suva, Fiji

Reardon GF, Oliver J (1983) The impact of cyclone Isaac on buildings of Tonga. J Wind Eng Ind Aerodyn 14(1–3):67–78

S. Mesquita P, Bursztyn M (2016) Integration of social protection and climate change adaptation in Brazil. Br Food J 118(12):3030–3043

Santos GMMC, Ferreira PJSG, Reis MJCS (2008) Promoting the educative use of the internet in Portuguese primary schools: a case study. Aslib Proc 60(2):111–129

Satchwell C (2013) Carbon literacy practices: textual footprints between school and home in children's construction of knowledge about climate change. Local Environ 18(3):289–304

Schwarts BM, Wilson JH, Goff DM (2015) An easy guide to research design and SPSS. Sage, USA

SPC and GIZ (2016) Coping with climate change in the Pacific Island region (CCCPIR) annual report—Vanuatu programme 2016. SPC-GIZ, Vanuatu

SPC/GIZ (2018) CCCPIR Component 5—climate change education. Suva, Fiji: SPC/GIZ

Surji K (2013) The negative effect and consequences of employee turnover and retention on the organization and its staff. Eur J Bus Manage 5(25):52–65

The Bible Society in Australia (1988) Good news bible, Australian edn. The Bible Society, Canberra in Australia

Thomalla F, Djalante R (2012) Disaster risk reduction and climate change adaptation in Indonesia: Institutional challenges and opportunities for integration. Int J Disaster Resilience Built Environ 3(2):166–180

Tsaliki E (2017) Teachers' views on implementing intercultural education in Greece: the case of 13 primary schools. Int J Comp Edu Dev 19(2/3):50–64

UN CC: Learn (2013) Resource guide for advanced learning on integrating climate change in education at primary and secondary level. United Nations Institute for Training and Research, Switzerland

UNDP (2015) Sustainable development goals (SDGs). UNDP, USA

UNESCO (2006) Pacific education for sustainable development framework. UNESCO, Fiji

UNESCO (2017a) Education for sustainable development goals: learning objectives. UNESCO, Paris, France

UNESCO (2017b) National commissions for UNESCO—annual report 2017. UNESCO, Paris, France

Usman LM (2008) Assessing the universal basic education primary and Koranic schools' synergy for Almajiri street boys in Nigeria. Int J Ed Manage 22(1):62–73

Van Dam K, Damen MAW, Sanders JMAF (2015) Are positive learning experiences levers for lifelong learning among low educated workers? Evid Based HRM Glob Forum Empirical Sch 3(3):244–257

Verlie B (2018) From action to intra-action? Agency, identity and 'goals' in a relational approach to climate change education. Environ Edu Res. (London, UK: Routledge)

Vines K, O'Toole D, Lee C, Jacobs B (2014) Integrated regional vulnerability assessment of government services to climate change. Int J Clim Change Strat Manage 6(3):272–295

Vogel C, Schwaibold U, Misser S (2015) Teaching and learning for climate change—the role of teacher materials and curriculum design in South Africa. South Afr J Environ Edu 31:78–97

Walshe RA, Chang Seng D, Bumpus A, Auffray J (2018) Perceptions of adaptation, resilience and climate knowledge in the Pacific: the cases of Samoa, Fiji and Vanuatu. Int J Climate Change Strat Manage 10(2):303–322

WHO (2015) Human health and climate change in Pacific Island countries. World Health Organisation, Geneva

Woodroffe CD (1983). The impact of Cyclone Isaac on the coast of Tonga, vol 37, no 3. University of Hawaii Press, pp 181–210

Chapter 10
The Benefits and Downsides of Multidisciplinary Education Relating to Climate Change

Lino Briguglio and Stefano Moncada

Abstract In this paper we present a literature review about the need for education to promote an understanding of climate change and its impacts, and the merits of teaching climate change in a multidisciplinary approach. We also refer to the external and internal multiplier effect of multidisciplinary education. We report on the results of a survey carried out by the Climate Change Platform (The Islands and Small States Institute of the University of Malta hosts the Climate Change Platform (CCP), with the objective of facilitating collaboration between University entities and individual academics in order to foster teaching and research initiatives relating to climate change, as well as strengthening cooperation with climate research centres outside Malta. During its three years of existence, the CCP, fully cognizant that the analysis of climate change involves various disciplines, has taken measures to encourage multidisciplinary teaching and research at the University of Malta, with a focus on small island states, which according the IPCC fifth assessment report (WGII, Chap. 29) are highly vulnerable to the harmful impacts of climate change. The paper will describe the approach adopted by the CCP in its endeavour to involve various Faculties, Institutes and Centres at the University of Malta to collaborate in teaching and research on climate change issues.) of the University of Malta, among lectures who teach subjects directly associated with climate change. It transpires from the literature and from the University of Malta survey that multidisciplinary climate change education is very important, given the complexity and the interlinkages of this field of study, but it also has a number of downsides, mostly related to the coordination work that will be needed when various disciplines are involved and the fear that the students could find it difficult to cope with many satellite subjects. The main message that emerges from the literature, as well as from the results of our survey, is that although a multidisciplinary approach to the teaching of climate

---

L. Briguglio (✉)
University of Malta, Msida, Malta
e-mail: Lino.briguglio@um.edu.mt

S. Moncada
Institute for European Studies and Focal Point of the Climate Change Platform, University of Malta, Msida, Malta
e-mail: stefano.moncada@um.edu.mt

© Springer Nature Switzerland AG 2019
W. Leal Filho and S. L. Hemstock (eds.), *Climate Change and the Role of Education*, Climate Change Management, https://doi.org/10.1007/978-3-030-32898-6_10

change education is highly desirable, its success depends on the extent to which it is well organised and suitably coordinated.

Keywords Education · Climate change · Education multiplier · Multidisciplinarity · University of Malta

Introduction

This paper argues that climate change education is of major importance for the dissemination of knowledge about the factors that lead to global warming and what needs to be done to reduce the harm caused by climate change. The paper also argues that climate change education is best imparted through a multidisciplinary approach, and that such an approach has an educational multiplier effect relating to climate change education, which in turn positively affects climate change awareness in society in general.

From the literature review it emerges that multidisciplinary education leads, amongst other things, to improved possibilities for lecturers to combine their expertise, enhancement of creative research, improvements in the participants' skills in communication, and a better understanding of how different disciplines complement each other.

The hypothesis to be tested in this paper is the following: Climate change education is best imparted through a multidisciplinary approach.

To test this hypothesis a literature review on the subject was carried out, and a survey was conducted among lecturers who teach climate-change related study units at the University of Malta.

The rest of the paper is organised as follows. The Section that follows presents a literature review on education for climate change and the benefits of a multidisciplinary approach in education. Section "Methodology of the Survey" describes the methodology used to conduct a survey on the benefits of multidisciplinary education in the case of climate change, with University lecturers participating in climate change study units at the University of Malta as respondents. The penultimate Section reports on the results of the survey. The final Section concludes the paper with a number of implications derived from the results of the survey.

Literature Review

The section is organised in three sub-sections respectively relating to the literature on the need for education on climate change, the pros and cons of multidisciplinary approaches and the multiplier effect of climate change education. The literature review was conducted by scanning the publications listed in Google scholar relating to these themes.

The Need for Education on Climate Change

Various authors contend that education is of paramount importance for understanding what climate change means, its likelihood, what harm it causes, what measures could reduce its intensity (mitigation) and its harm (adaptation) and why such measures need to be supported, provided that it is connected with its sustainable development. Bangay and Blum (2010) consider it to be a problem that often the main trust of education on climate change relates to technical solutions without seriously associating this matter with social, economic and environmental issues. They therefore see a need to associate climate change education with education for sustainable development. Such form of education could improve the capacity of the people to address environment and development issues. According to the same authors education of this kind has significant potential to both address climate change and to support wider goals relating to sustainable development.

This view is also expressed in Cordero et al. (2008) who maintain that climate change education should not only impart knowledge of science but also engage students and encourage positive responsiveness about the environment The authors carried out a survey with the aim of determining the effect of action-oriented learning on climate change literacy, while yielding additional insights on student misconceptions and the effectiveness of various teaching methods, connecting this matter this matter to a variety of social dimensions, including access to food, drinkable water, and sustainable energy. They found that students often fail to apply environmental knowledge to their own live, and that there were various misconceptions about climate change among students. This is also in line with the ideas expressed by Sterman and Sweeney (2007) and Uzzell (1999) that conventional environmental education focused exclusively on science without personal and social connections may not be the most effective educational model for moving toward social change.

Anderson (2012) defines climate change education as a comprehensive and multidisciplinary form of education and asserts that it should include relevant content knowledge on climate change, environmental and social issues, disaster risk reduction, and sustainable consumption and lifestyles. The author proposes a number of measures relating to climate change education, including that such education would have better results if it is based on integrated, cross-discipline curricula. grounded in the application and practicality of climate and environmental science. She emphasises the point that climate change education should not be introduced as a separate, stand-alone subject area, or solely within the science section, but instead integrated across existing subject areas such as science, citizenship education, geography, social studies, history, language, drama and the arts.

The same author also refers to the immediate institutional environment of students implying that schools and education systems themselves should be climate-proofed and resilient as well as sustainable and green. The author further argues that evidence shows that educational interventions are most successful when they focus on local, tangible, and actionable aspects of climate change, especially those that can be addressed by individual behaviour.

As is well known, through the Paris Agreement most countries agreed to keep green-house gasses under control.[1] According to Shapiro Ledley et al. (2017) the attainment of such pledges and actions of nations that are party to the Paris Agreement calls for an understanding of climate change which in turn requires appropriate education for this purpose. The authors maintain that addressing climate change will require action at all levels of society, including individuals, organizations, businesses, local, state, and national governments, and international bodies, and therefore it cannot be addressed by a few individuals with privileged access to information.

The authors further contend that education is needed because, in the case of climate change, learning from practical experience only is learning too late, as the impacts take a long time to be fully visible. In addition, climate change is a complex process and requires a systems thinking approach which can only be imparted through education, notably through multidisciplinary approaches. In their view, effective climate change education increases the number of informed and engaged citizens, building social will or pressure to shape policy, and building a workforce for a low-carbon economy.

The issue of risk and uncertainty often emerges in the literature on education for climate change. Stevenson et al. (2017), referring to IPCC (2014) argue that climate change education involves creatively preparing students for a rapidly changing, uncertain, risky and possibly dangerous future. Bangay and Blum (2010) contend that the reality of climate change calls for approaches to education that equip and empower people of all ages to deal with uncertain environmental, economic and political futures.

Some authors refer to a number of setbacks in climate-change education. Hamilton (2011) for example, refers to political prejudices as affecting this type of education. The author contends that persons with right-wing orientations tend to be less disposed to believe in climate change than those with left-wing orientations. The author notes also that in the USA, concern about climate change tended to increase with education among Democrats, but tended to decrease in the case of education among Republicans. The author suggests that certain biased media, including the many websites devoted to discrediting climate change concerns, counteract the effect of education, among deniers with pre-fabricated ideas relating to climate change.

Multidisciplinary Approaches in the Teaching on Climate Change

Many studies on climate change education refer to the need for a multidisciplinary approach in such education. Rhee et al. (2010) conducted a pilot implementation of an experimental interdisciplinary course on climate solutions undertaken at San

---

[1]The Paris Agreement's central aim is to strengthen the global response to the threat of climate change by keeping a global temperature rise this century well below 2 °C above pre-industrial levels and to pursue efforts to limit the temperature increase even further to 1.5 °C.

Jose' State University in 2008. The course, co-taught by seven faculty members from six colleges, was approved for a general education requirement and was open to students campus-wide. The lessons learned from the pilot effort were assessed from student, faculty, and administrative perspectives. The educational benefits to students from the interdisciplinary format were found to be substantial to students as well as to faculty development. The most compelling reason to implement such a form of education, involving team-taught courses, was found to be the enjoyment of the approach with most students rating the course as one of the best classes they had ever taken. Another advantage was that as a result of the opportunities to work in teams on solutions to climate change and by also seeing their colleagues' projects, most students appeared optimistic about what could be done and eager to continue this work. Furthermore, there were also benefits to the lecturers, resulting from the collaborative exercise. However, the authors admit that there were a number of difficulties such the those related to organization and coordination of different faculties as well as the administrative realities of resource allocation.

In another study on the benefits of a multidisciplinary approach to climate change education, Middleton (2011) refers to the environmental problems associated with global warming, which require complex syntheses from a vast set of disciplines including science, engineering, social science and the humanities. The author laments that most ecologists have narrow training, and are not equipped to utilise their environmental skills to help solve multidisciplinary problems.

In a textbook focussing on the need to apply a multidisciplinary approach to the teaching and understanding of climate change, Burroughs (2007) provides an explanation relating to the connection of many disciplines with the climate, including meteorology, oceanography, environmental science, earth science, geography, agriculture and social science. In his introduction to this book, Burroughs argues that there is nothing simple about climate change and that it pays to appreciate that the behaviour of the Earth's climate is governed by a wide range of factors all of which are interlinked in an intricate web of physical processes.

In the same vein, Todd and O'Brien (2016) contend that anthropogenic climate change is a complicated issue involving scientific data and analyses as well as political, economic, and ethical issues. They survey the reaction of students before and after engaging in a multidisciplinary study unit relating to climate change and found that students can learn to engage in political and moral challenges of climate change as well as science through a multidisciplinary format.

In an article in the *Nevada Today* website, titled "Understanding climate change through many disciplines" Trent (2010) argues that climate change is not a single science which deals with a single aspect, but is one that requires a cross-disciplinary approach. He gives the example of world-renowned theoretical physicist Geoffrey West, a mathematician—who teamed up with ecologists and biologists to tackle some of the world's most fundamental evolutionary questions. The author argues that by having scientists working in teams and forming multidisciplinary partnerships, they create opportunities to learn matters relating to climate change from different perspectives.

Nowotny et al. (2018), focussing on energy-related matters, propose a strategy of integration of different disciplines, in studying climate change mitigation. Energy is a vast field comprising many disciplines with different conceptual backgrounds and overlaps widely with climate change, pollution, water utilisation and management as well as population placement and growth. Therefore, the interdisciplinary approach in energy-related topics is needed in the development of climate change education programs.

De Tombe (2008), refers to the Compram methodology, associates climate change with complex societal problems which need to be handled. The Compram methodology, as proposed by De Tombe is often regarded as a framework with consecutive steps for analysing complex systems, such as climate change, which have major social implications.

However, there are downsides to multidisciplinary education. Pirrie et al. (1999) question the assumption that such approach is, by definition, a 'good thing'. They argue that the fact that this is so widely assumed to be the case is linked in part to the considerable conceptual confusion surrounding the use of terms. In a two-year evaluation of multidisciplinary education, the authors assessed the respondents' perceptions of multidisciplinary education; the factors which promote it; and those which inhibit its development. They found that there is a lack of consensus as to what the very term 'multidisciplinary' means.

The same authors also discuss the factors which inhibit the development of multidisciplinary education. These include the control over course development and innovation exercised by a particularly faculty, rendering difficult for course organizers to change the course descriptions.

The organization and implementation of multidisciplinary courses also make heavy demands on staff, in terms of personal commitment and effective teamwork. According to the authors, one of the biggest problem is getting people from different departments and faculties to work together, given that faculty members and faculty deans change over time, rendering it problematic to have constantly available lecturers. Another problem identified by the authors is that of finding suitable accommodation for large groups when multidisciplinary involves sharing a subject among many students.

The Multiplier Effect of Climate Change Education

As indicated above, several studies refer to advantages of climate change education and its positive effect on the students, on the faculty and on society in general. The spill-over effects of education (as in Riede et al. 2017), generate social, environmental and economic positive externalities. These effects, which may be termed the multiplier effects of climate change education, have been extensively discussed in the literature, but very few studies attempted to measure such an effect.

According to Mochizuki and Bryan (2015), the multiplier effect of education means that entire families and communities benefit when individuals share what they have learned. In the case of climate change education, sustainability is often given centre stage and the lessons learned regarding adaptation and mitigation can be passed on to others and to future generations. The authors argue that once informed, citizens are better able to participate in civil society and to influence decision-making in areas that affect them, particularly at the local level.

The multiplier effect of education in general was discussed in Kreuter et al. (1982) who devised a method for calculating the extent to which health education influenced new health promotion activities funded by other sources. This aspect of multiplier deals with monetary return. Such an approach was also used by Yen et al. (2015) and Elliott et al. (1988).

Santiago Fink (2018) writes about the negative physical multiplier effect of climate change per se, rather than education for climate change, describing this as a "threat multiplier". She argues that human health and the quality of life in general are dependent upon ecological processes, clean air, safe drinking water, and physical and mental well-being, which can be negatively affected by climate change. On the other hand, nature is a producer and multiplier of ecosystems services generating health and socio-economic co-benefits that improve the quality of life. If climate change education helps to identify the harm of climate change and the benefits of nature, the multiplier effect to which Santiago Fink refers. Mochizuki and Bryan (2015) also refer to the physical negative multiplier, describing climate change as a risk multiplier which disproportionately affects the most vulnerable groups in society.

There are also "internal" multiplier effects within the education establishments that impart climate change education. The multiplier in this case would measure how, with a give effort, better outcomes can be obtained from multidisciplinary approach compared to a monodisciplinary one. This "internal" multiplier effect can be expressed as a ratio $\Delta E/\Delta M$ where ΔE is the improvement in the overall educational outcome of a student or a group of students (say in terms of student satisfaction or student grades) as a result of ΔM representing the introduction of multidisciplinary education within a particular school or university.[2] This aspect of the educational multiplier has not been calculated in the literature of education in general and climate change education in particular, although, as shown above, identifying the positive spill-over effects and interlinkages of climate-change, through multidisciplinary tuition, are thought to improve outcomes within the institution itself.

Methodology of the Survey

This section explains the approach that researchers at the Climate Change Platform of the University of Malta adopted to conduct a survey among university lecturers

---

[2]Admittedly, such an experiment would be very difficult to carry out, especially because this would need to be based on the assumption that the two tuition approaches will be imparted to similarly gifted students and with the same teachers. In the case of climate change, as indicated above, this internal multiplier effect is likely to be relatively high, given that climate change lends itself very well to multidisciplinary approaches due to its interlinkages.

who taught study unit with "climate change" included in the title. There are 16 such study units offered by the University of Malta (see Annex 1). Some of these study units are taught by more than one lecturer and some lecturers contribute to more than one study unit.

The survey was drawn up so as to test the hypotheses set out in the introductory section of the present study. A qualitative survey was considered to be appropriate for such a test. Qualitative surveys, such as the one used for the purpose of this study, seek to describe a situation which cannot readily be measured quantitatively or which is better described in words rather than in numbers. The respondents to a qualitative survey are generally expected to express opinions, and views, thereby enabling the researcher to delve deeper into the issue being researched. However, given that certain answers, even if broadly conveying the same opinion, may be worded differently by different respondents, it may be more difficult to analyse the responses in a qualitative survey than is the case with a quantitative survey.

Another consideration relating to qualitative research, especially if the answers are open ended, and if the survey contains many questions, is that responding to such surveys is often very time consuming, and this may result in many of the persons contacted avoiding to respond. In order to avoid this pitfall, the questionnaire used in this study was intentionally kept as simple as possible to elicit maximum response, especially given the survey fatigue prevailing at the University of Malta.

The survey questionnaire is appended as Annex 2. It can be seen that the questionnaire consisted of just 4 questions with space for additional comments by the respondents.

The questionnaire was sent by email to the selected lecturers, with the objective of seeking their views regarding the pros and cons of multidisciplinary approach in climate change education. The views of the respondents as to which subjects, other than purely environmental studies, are associated with climate change were also sought. All the university lecturers who taught a study unit with "climate change" included in the title were contacted. These numbered 23, out of whom, 22 responded.

The respondents were not prompted with possible answers, and an open-ended format was utilised, so the responses, even when referring to the same issue, were often worded differently. For this reason, we adopted a coding method in order to group and summarise the responses under specific headings. This method is very often used in qualitative surveys with open ended questions or when the respondents are allowed to frame their responses in their own words. It is a form of assigning keywords or key phrases to responses which are broadly very similar, but could be worded differently. Coding the responses enables the researcher to effectively summarise the results of the survey.

Table 10.1 Institutional affiliation of respondents

Department/institute	Respondents
Earth Systems Institute	5
Environmental and Resources Law Department	4
Sustainable Development and climate change Institute	3
Centre for Environmental Education and Research	3
Sustainable Energy Institute	2
European Studies Institute	1
Islands and Small States Institute and Economics Department	1
Geography Department	1
Geosciences Department	1
Mechanical Engineering Department	1
Total	22

Results of Survey

This section summarises the responses by the University lectures to the survey questionnaire described in the previous section, followed by a discussion on same responses.

The Institutional Affiliation of the Respondents

The respondents were asked to mention the department to which they belonged, with the result shown in Table 10.1.

Subjects that Compliment Climate Change Taught in a Multidisciplinary Approach

The respondents were asked to list what in their view were the most important fields of study (other than purely environmental fields) that they associated with climate change, if this subject was to be taught in a multidisciplinary manner. The results are shown in Table 10.2.

The last column is a weighted average of the number of responses, with a 60% weight assigned to the first preference and 40% to the second preference.

Table 10.2 Two most important fields of study associated with climate change, if this subject is taught in a multidisciplinary manner

Field of study	First	Second	Simple average	Weighted average
Economics/development economics	6	2	4.0	4.4
Public policy/political studies/social studies	3	5	4.0	3.8
Law and diplomacy	1	6	3.5	3.0
Risk assessment studies	3	2	2.5	2.6
Energy studies	2	3	2.5	2.4
Geography	2		1.0	1.2
Health issues	1	1	1.0	1.0
Ethical behaviour		2	1.0	0.8
Consumer studies	1		0.5	0.6
Education	1		0.5	0.6
History	1		0.5	0.6
Transport studies	1		0.5	0.6
Did not indicate second preference		1	0.5	0.4
Total	22	22	22	22

Economics was the subject that was mentioned most often, even though there was only one respondent from the Economics Department. Interestingly, Public Policy was given a high score even if there was not even one respondent from the Public Policy Department.

Most Important Benefits

The respondents were asked to list up to two benefits associated with a multidisciplinary approach in the teaching of climate change. All respondents stated that multidisciplinary climate change education was beneficial. The benefits mentioned were coded and listed as shown in Table 10.3. Again here, the last column in the table is a weighted average of the number of responses, assigning 60% to the first preference and 40% to the second preference.

As expected, the most important two benefits mentioned related to the interrelationships and spill-over effects of the causes and impacts of climate change.

10 The Benefits and Downsides of Multidisciplinary … 179

Table 10.3 Benefits of climate change multidisciplinary education

Benefits	Benefit 1	Benefit 2	Simple average	Weighted average
Permits a better understanding of interrelationships and spill-over effects of different causes and impacts	8	8	8.0	8.0
Climate change itself is complex and requires a holistic and coordinated approach	11	3	7.0	7.8
Renders learning more enjoyable		4	2.0	1.6
Encourages cooperation between academic and generates sense of belonging among students with regard to many disciplines	1	2	1.5	1.4
Leads to increased awareness of risk and uncertainty	2		1.0	1.2
Encourages critical thinking		2	1.0	0.8
Leads to increased awareness that laws and regulation relating to climate change are multifaceted.		1	0.5	0.4
Did not indicate second preference		2	1.0	0.8
Total	22	22	22	22

Downsides

The respondents were asked to list up to two downsides associated with a multidisciplinary approach in the teaching of climate change. The respondents identified a number of downsides, which mostly relate to the coordination effort that is required by such an approach and the possibility that such an effort would be spread too thinly over many fields of study. The downsides mentioned were coded and listed as shown in Table 10.4. The weighted responses shown in the last column uses the same procedure as in the previous tables.

Table 10.4 Downsides of climate change multidisciplinary education

Downsides	Downside 1	Downside 2	Simple average	Weighted average
Highly demanding on the coordination effort of the school in terms of time-tabling and venues, and on the lecturers concerned, in terms of subject content, preparation and delivery of lecture	9	2	5.5	6.2
Limits the time available to cover the component disciplines properly (thin coverage of disciplines)	6	4	5.0	5.2
The students could be overwhelmed by the spread of the subjects, especially in those they are not familiar with	2	4	3.0	2.8
Some disciplines may send contradictory messages to the students	2	1	1.5	1.6
No significant downsides	3	3	3.0	3.0
No Answer		8	4.0	3.2
Total	22	22	22	22

Other Comments

Respondents were asked to put forward additional comments should they so wished. Most of the additional comments were elaborations on the points raised in response to the previous questions. The responses were again coded and listed as in Table 10.5.

Half of the respondents did not add additional comments, presumably because they considered their responses to the previous survey questions as being enough to explain their opinions.

Interestingly seven of the 11 respondents who wrote additional comments stated that multidisciplinary approach to teaching climate change is to be recommended in spite of downsides.

Table 10.5 Additional comments on of multidisciplinary climate change education

Additional comments	Topics
Multidisciplinary approach to teaching climate change is to be recommend in spite of its downsides	7
It is important to emphasise that the most important discipline to associate with climate change is energy studies	1
Multidisciplinary education requires a change in mentality in many institutes and departments at the University of Malta	1
The questionnaire restricted the choice of discipline to two … but in reality more than two disciplines are associated with climate change	1
In the responses to the questionnaire the choice of disciplines by different lecturers is likely to be biased towards the subjects they teach	1
No additional comments	11
Total	22

Discussion

The results of the survey as described above, generally indicate that multidisciplinary education relating to climate change has many benefits. Most of the benefits mentioned were also identified in the literature, as indicated in Section "Literature Review" of the present study.

When asked to mention subjects that compliment climate change taught in a multidisciplinary approach, respondents assigned major importance to Economics and Public Policy.

Economics as a field of study has often been associated with climate change (Stern 2007; Tol 2018), based mostly on the argument that carbon emissions are generally costless to the polluter, if they are not taxed or some other penalty is imposed on them, given that such emissions are externalities and therefore there is the need to put a price on such emissions in order to reduce them. In addition, climate change itself creates economic costs as a result of sea-level rise, especially on islands, extreme weather events, health problems and other negative effects which can and should be costed in order to make a case for climate-change mitigation and adaptation.

Climate change is caused by humans and affects human society in matters relating to health and safety, food security, agriculture production, transportation systems and others. It therefore calls for appropriate public policy measures such as moral suasion, adoption of economic instruments and command and control methods legally backed, to reduce carbon emissions and to promote adaptation and community resilience. These are some reasons why climate change is associated with public policy, at the national and international levels.

Law and Diplomacy, Risk Assessment and Energy Studies were also given major importance by the respondents in this regard. Other fields of study mentioned by the respondents were Geography, Health, Ethics, Consumer Studies, History and Transport Studies.

Interestingly, Energy Studies often associated with climate change, were assigned 5th place. The reason for this could have been that the respondents were asked not to include environmental fields of study, and some respondents may have associated energy studies with environmental studies.

When asked to identify the most important advantages of multidisciplinary climate change education, the item that received most mentions was that such an educational approach permits a better understanding of interrelationships and spill-over effects of different causes and impacts of climate change. A related advantage, also given many mentions was that climate change itself is complex and requires a holistic and coordinated approach. These advantages of multidisciplinary education, also identified in the literature, were mentioned in about 70% of the responses. Other advantages identified, but mentioned by fewer respondents, were that such a form if education renders learning more enjoyable, it encourages cooperation between academics and generates sense of belonging among students with regard to many disciplines, it encourages critical thinking, it leads to increased awareness of risk and uncertainty and it generates increased awareness of that laws and regulations relating to climate change is multifaceted.

The most important downsides of multidisciplinary education, identified by the respondents, related to the onerous coordination effort of the school in terms of time-tabling and venues, and on the lecturers concerned, in terms of subject content, preparation and delivery of lecture in the limited time available to cover the component disciplines properly, leading to the possibility of thin coverage of particular disciplines. These problems, also frequently mentioned in the literature, accounted for about half of the responses. Other downsides mentioned in this regard included that the students could be overwhelmed by the spread of the subjects and that some disciplines may send contradictory messages to the students.

The respondents were asked to add additional comments if they so wished. The majority of those who added such comments stated that multidisciplinary approach to teaching climate change is to be recommend in spite of downsides. This comment summarises the general thrust of the responses, as well as of the literature on the subject—a comment which while acknowledging that multidisciplinary education has its drawbacks, nevertheless, particularly in the case of climate change education, such an approach should be encouraged due to its various benefits.

Conclusion

In this paper we presented a literature review relating to the need for education to promote an understanding of climate change and its impacts, and the merits of teaching climate change in a multidisciplinary approach. We also commented on the external and internal multiplier effects of climate change education. We reported on the results of a survey carried out by the Climate Change Platform of the University of Malta, among lectures who teach subjects directly associated with climate change.

It transpires from the literature and from the University of Malta survey that multidisciplinary climate change education is of major importance, given the linkages and complexity of climate change, but it also has a number of downsides, mostly related to the coordination work that will be needed when various disciplines are involved and the fear that the students could find it difficult to cope with many satellite subjects.

The main message that emerges from the literature, as well as from the University of Malta survey, is that although a multidisciplinary approach to the teaching of climate change education is highly desirable, its success will depend on the extent to which it is well organised and suitably coordinated.

Annex 1

Study Units Offered at the University of Malta

Study unit code	Study unit title	Department/institute
EMP3007	Climate Change Science, Social Impacts and Regulation	Environmental Management and Planning
EMP3008	Climate Change: Science and Social Impacts	Environmental Management and Planning
ERL4000	Climate Change and International Law	Environmental and Resources Law
ERL5008	International Energy and Climate Change Law	Environmental and Resources Law

(continued)

(continued)

Study unit code	Study unit title	Department/institute
ERL5009	EU Energy, Environmental and Climate Change Law	Environmental and Resources Law
ERL5010	Comparative Energy, Environment and Climate Change Law	Environmental and Resources Law
EST5520	The European Union and Climate Change	European Studies
GEO1012	Climate and Biogeography	Geography
IEN5019	The Science and Management of Climate Change	Environmental Management and Panning
ISE5202	Climate Change: Myths, realities and Action	Sustainable Energy
ISS5011	Climate Change and Sea Level Rise	Islands and Small States
MSE2217	Understanding Local Environmental Issues: Climate Change	Environmental Education and Research
MSE3107	Sustainable Development and Climate Change Education	Environmental Education and Research
PHY3240	Fundamentals of Meteorology for Climate Studies	Geosciences
PHY3260	A Multidisciplinary Approach to Climate Change	Geosciences
SPI5701	Climate Change	Spatial Planning and Infrastructure

Annex 2

Questionnaire About a Multidisciplinary Approach in Teaching of Climate Change

1. Information about respondent:

Name of lecturer	Department	Study unit/s taught

2. What in your opinion are the two most important fields of study (other than purely environmental fields) that you associate with climate change, if this subject is taught in a multidisciplinary manner?

	Field of study 1	Field of study 2
Enter the two fields of study, in order of preference, here →		

3. Are there benefits associated with a multidisciplinary approach in the teaching of climate change? If yes, enter up to 2 such benefits below. If no significant benefit tick the box below.

	Benefit 1↓	Benefit 2↓	No benefits↓
Enter up to two benefits here →			

4. Are there downsides associated with a multidisciplinary approach in the teaching of climate change? If yes, enter up to 2 such downsides below. If no significant downsides tick the box below.

	Downside 1↓	Downside 2↓	No downsides↓
Enter up to two downsides here →			

Should you wish to put forward additional comments, please do so below.

Please return the filled in questionnaire to Lino Briguglio: lino.briguglio@um.edu.mt

References

Anderson A (2012) Climate change education for mitigation and adaptation. J Educ Sustain Dev 6(2):191–206

Bangay C, Blum N (2010) Education responses to climate change and quality: two parts of the same agenda? Int J Educ Dev 30(4):359–368

Burroughs WJ (2007) Climate change: a multidisciplinary approach. Cambridge University Press, Cambridge

Cordero EC, Todd AM, Abellera D (2008) Climate change education and the ecological footprint. Bull Am Meteor Soc 89(6):865–872

De Tombe D (2008) Climate change: a complex societal process; analyzing a problem according to the Compram methodology. J Organ Transform Soc Change 5(3):235–266

Elliott DS, Levin SL, Meisel JB (1988) Measuring the economic impact of institutions of higher education. Res High Educ 28(1):17–33

Hamilton LC (2011) Education, politics and opinions about climate change evidence for interaction effects. Clim Change 104(2):231–242

Intergovernmental Panel on Climate Change (IPCC) (2014) Climate change 2014: synthesis report. In: Contribution of working groups I, II and III to the fifth assessment report of the intergovernmental panel on climate change. IPCC, Geneva

Kreuter MW, Christensen GM, Divincenzo A (1982) The multiplier effect of the health education risk-reduction program in 28 states and 1 territory. Public Health Rep 97(6):510

Middleton BA (2011) Multidisciplinary approaches to climate change questions. In: Wetlands. Springer, Dordrecht, pp 129–136

Mochizuki Y, Bryan A (2015) Climate change education in the context of education for sustainable development: rationale and principles. J Educ Sustain Dev 9(1):4–26

Nowotny J, Dodson J, Fiechter S, Gür TM, Kennedy B, Macyk W, Baka T, Sigmundg W, Yamawakih M, Rahman KA (2018) Towards global sustainability: education on environmentally clean energy technologies. Renew Sustain Energy Rev 81:2541–2551

Pirrie A, Hamilton S, Wilson V (1999) Multidisciplinary education: some issues and concerns. Educ Res 41(3):301–314

Rhee J, Cordero EC, Quill LR (2010) Pilot implementation of an interdisciplinary course on climate solutions. Int J Eng Educ 26(2):391

Riede M, Keller L, Oberrauch A, Link S (2017) Climate change communication beyond the 'ivory tower': a case study about the development, application and evaluation of a science-education approach to communicate climate change to young people. J Sustain Educ 12. Online publication retrieved from http://www.susted.com/wordpress/wp-content/uploads/2017/02/Riede-et-al-JSE-Feb-2017-General-Issue-PDF1.pdf

Santiago Fink H (2018) The multiplier effect: climate change, health and nature. Online publications retrieved from https://www.urbanbreezes.com/the-multiplier-effect-climate-change-health-and-nature

Shapiro Ledley A, Rooney-Varga J, Niepold F (2017) Environmental issues and problems. Sustain Solutions. Online publication retrieved from: http://oxfordre.com/environmentalscience/view/10.1093/acrefore/9780199389414.001.0001/acrefore-9780199389414-e-56

Sterman JD, Sweeney LB (2007) Understanding public complacency about climate change: adults' mental models of climate change violate conservation of matter. Clim Change 80:213–238

Stern N (2007) The economics of climate change: the stern review. Cambridge University press, Cambridge

Stevenson RB, Nicholls J, Whitehouse H (2017) What is climate change education? Curriculum Perspect 37(1):67–71

Todd C, O'Brien KJ (2016) Teaching anthropogenic climate change through interdisciplinary collaboration: helping students think critically about science and ethics in dialogue. J Geosci Educ 64(1):52–59

Tol RS (2018) The economic impacts of climate change. Rev Environ Econ Policy 12(1):4–25

Trent J (2010) Understanding-climate-change-through-many-disciplines. Online publications retrieved from https://www.unr.edu/nevada-today/news/2010/understanding-climate-change-through-many-disciplines

Uzzell DL (1999) Education for environmental action in the community: new roles and relationships. Camb J Educ 29:397–413

Yen SH, Ong WL, Ooi KP (2015) Income and employment multiplier effects of the Malaysian higher education sector. Margin. J Appl Econ Res 9(1):61–91

Chapter 11
Climate Change, Disaster Risk Management and the Role of Education: Benefits and Challenges of Online Learning for Pacific Small Island Developing States

Diana Hinge Salili and Linda Flora Vaike

Abstract Climate change and disaster risk management education in the context of the Pacific Small Island Developing States is crucial, but even more so, is ensuring that access to this education is available to those who require it. The Pacific Small Island Developing States are extremely vulnerable to the impacts of climate change and disasters and have limited capacity to address this so the move by the University of the South Pacific to develop a post graduate diploma in climate change program and make this fully available online is an important one. This paper presents experiences and observations of the benefits and challenges of online learning in the context of the Pacific Small Island Developing States. Special focus is given to the documentation of staff and student experiences from across the Pacific Small Island Developing States region. Lessons learnt are shared and recommendations for actions to address the challenges are proposed. As the only institution providing post graduate climate change and disaster risk management education online in the Pacific Small Island Developing States region, this paper is aimed at informing individuals who wish to engage in climate change and disaster risk management related online learning and institutions who would like to venture into this mode of climate change and disaster risk management education.

Keywords Climate change · Disaster risk management · Online learning · Pacific Small Island Developing States · Experiences

---

D. H. Salili (✉) · L. F. Vaike
Pacific Center for Environment and Sustainable Development (PaCE-SD), University of the South Pacific, Suva, Fiji Islands
e-mail: diana.salili@usp.ac.fj

L. F. Vaike
e-mail: linda.vaike@usp.ac.fj

© Springer Nature Switzerland AG 2019
W. Leal Filho and S. L. Hemstock (eds.), *Climate Change and the Role of Education*, Climate Change Management, https://doi.org/10.1007/978-3-030-32898-6_11

Introduction

Vulnerable Pacific Small Island Developing States (PSIDS) currently face an unprecedented threat from the impacts of climate change and natural disasters (Carter 2015; Roy et al. 2018). The severity of the impacts of climate change and natural disasters on PSIDS are such that adaptation measures alone are no longer sufficient (Bataller 2010). Coupled with relevant education for Climate Change and Disaster Risk Management (DRM) however, is one way that the PSIDS can advance their adaptation and mitigation actions across all levels (Davidson and Lyth 2012). The role that education plays in the fight against climate change and the impacts of associated disasters is vital especially for PSIDS who are already pressed for technical expertise and the capacity to successfully implement national climate change and disaster risk reduction policies and projects (Barnett 2005; UNESCO 2019). Higher education in particular, plays an important role in educating students about climate change, however it has to include relevant and appropriate learning methods (Rees 2003; Cordero et al. 2008).

Online learning has significantly grown over the last decade and is emerging as the new paradigm of modern education (Pei-Chen et al. 2008; Li et al. 2016; Harasim 2000). The University of the South Pacific (USP) began offering online climate change and DRM courses in 2011 through its Pacific Center for Environment and Sustainable Development (PaCE-SD). The use of online and distance learning is critical to the development of the PSIDS where remoteness is a key challenge to a region which spans one-third of the earth's surface and consists of thousands of islands located far and wide across the Pacific (Bossu 2017). As a regional leader in education, USP may be an effective institution to lead efforts in designing corresponding online courses and resources. Online learning in the Pacific Islands has progressed over the last decade, largely owing to the move by USP to migrate courses from face to face modes to blended and fully online courses (ADB 2018; Bossu 2017).

An online course at USP is one where most or all of the content is delivered online and usually there are no conventional face-to-face classroom meetings (University of the South Pacific 2018). At USP's PaCE-SD, the Post-Graduate Diploma in Climate Change (PGDCC) program is offered fully online. This has been largely successful with steadily increasing enrollment numbers over the years, and benefits, challenges and lessons learnt from the delivery of the courses. While the benefits of providing climate change and DRM education online abound, the challenges are exacerbated by the geographical and socio economic features of different PSIDS creating an imbalance by default in adaptation and mitigation capacities of different PSIDS, one that may be reduced by education on climate change and DRM, hence the role of education in reducing PSIDS vulnerability cannot be stressed enough. Researches show that, the influence of education in developing countries is twofold due to its direct effect on aspects that reduce risk associated with climate change and disasters, and its mitigating effect on aspects that increase risks. (Wamsler et al. 2011; Armstrong et al. 2018).

This paper presents experiences and observations of the benefits and challenges of online climate change and DRM education in the context of the PSIDS. Special focus is given to the documentation of staff and student experiences from across the PSIDS in an effort to highlight the role that online learning plays in climate change and DRM education for PSIDS. Lessons learnt are shared and recommendations for actions to address the challenges are proposed.

Benefits of Online Learning in the Context of PSIDS—Experiences and Observations

Current literature reveals a myriad of reasons motivating institutions and students to engage in online learning (Yoo and Huang 2013; Johnson et al. 2015). The paradigm shift towards online learning is evident in PSIDS where higher education institutions are moving away from conventional teaching and learning methods to online modes of course and program delivery.

The benefits of online learning for developing countries like the PSIDS generally conform to online learning in other developing regions in the world. Information access, greater communication; synchronous and asynchronous learning; increased cooperation and collaboration, cost-effectiveness and pedagogical improvement are some of the pedagogical and socio-economic forces that influence the move towards online learning (Sife et al. 2007). More specific benefits to the PSIDS result from socio-economic, cultural and geographical circumstances as well as institutional organization and governing structures of higher education institutions. To this end, preservation of the "Pacific Way" and providing learners with a sense of community, while trying to address other forces such as globalization and advances in Information and Communication Technology (ICT), is paramount (Purcell and Toland 2004).

USP provides an ideal example of a higher education institution in the PSIDS Region that is moving from conventional face to face learning to online learning. The impact of online learning on student and overall institutional growth is still an area that is under researched although existing literature points to heterogeneous impacts (Hogan 2009; Purcell and Toland 2004). Despite the mixed impacts, important benefits that have stemmed from USP's online courses, particularly those under the PGDCC Program, cannot be overlooked.

Access to Online Tertiary Climate Change Education

An important feature of the PGDCC program is its online delivery. All courses in the program are offered online allowing students to study remotely from across the PSIDS. While the mode caters for access and cost-effectiveness (costs associated with airfare and daily stipend are minimized for some regional students), years of experience teaching the courses have shown that student interests in the program are mostly twofold: the first being the nature of the subject area and its significance in the PSIDS region and second is the quest for attainment of a graduate certificate and (or) degree after completion of an undergraduate degree. In the case of the former, a lot of students either enter the program for the purpose of upskilling or because of the desire to change their career path. Students who enroll for the purpose of upskilling are mostly those who are already practitioners in the field and therefore need to upskill their knowledge to become more competent and marketable. The change of career path can be attributed to a lot of reasons however it is common knowledge in the region that because climate change is the greatest development threat of our time, more and more projects and programs are being implemented allowing for greater job creation and security in the field of climate change and DRM. The increasing quest for attainment of a graduate certificate and (or) degree in this field comes about as more and more people compete for jobs in the market. The online nature of the PGDCC program is therefore ideal for all learners with the interests stated above.

Pacific Consciousness in Learning and Teaching

While student interests serve as a critical determining factor for increasing enrollment into the program, other factors are critical in terms of the overall learning journey and experiences of studying climate change and DRM online. These factors are important considerations when dealing with online courses in any PSIDS higher education institution as well as other developing countries. Experiences from the PGDCC Program have shown that teaching and learning climate change and DRM online involves a lot of planning from the instructors and ensuring constructive alignment from the individual course up to the program level and the overall university graduate attributes. A very important benefit for the region and more so for USP, lies in the preservation of culture and the greater need for Pacific consciousness. The preservation of culture is, in itself, threatened by climate change and having the PGDCC program contributes to information sharing and documentation of the diverse cultures across the PSIDS. Because courses are offered at the postgraduate level, all teaching and learning resources and assessment tasks have to support critical thinking skills and cognitive learning in students; while addressing the need to build capacity for industries as well as livelihoods in general. Critical thinking skills and cognitive learning is enhanced and measured through the use of online discussions administered through the Modular Object-Oriented Dynamic Learning Environment (MOODLE)—the online learning platform used by USP. While formative assessments of this nature measure student learning, they also foster greater collaboration and interaction among students. Experiences and observations have shown that such collaboration goes beyond the bounds of the virtual classrooms to industries, further strengthening the community of practice and professional network in the region.

Challenges of Online Learning in the Context of PSIDS—Experiences and Observations

Over the past years, USP has worked alongside its member governments, development partners, and other education stakeholders to increase its online learning presence (Asian Development Bank 2018; Bossu 2017). The delivery of climate change and DRM courses online however, has encountered challenges from the design to delivery and evaluation and improvement stages.

Course and Program Design

From the design stages of the climate change and DRM courses offered at USP's PaCE-SD, challenges with content and relevance to the PSIDS and inclusion of diversity and cultural inclusion have emerged. The program includes four (4) courses all of which, in design, are an attempt to meet the growing needs and the demand for quality climate change and DRM education for the PSIDS region. Challenges that emerged from this include, difficulty of teaching a vulnerability adaptation tool that is regionally recognized and used at the national level in all USP member countries, difficulty of incorporating Pacific consciousness and cultural inclusivity in the courses, requirements for continuous updates as the science continues to evolve and include more conclusive evidence of the impacts and projections of climate change and related disasters for the PSIDS, designing assessments that are constructively aligned to institutional, school and program learning outcomes as well as assessments that can be implemented online and in the region. While it is unclear whether a needs assessment was undertaken during the preparation and design stages of the courses, in theory and practice, a needs assessment identifying specific learning needs would have contributed to addressing the challenges associated with design (Faxon-Mills 2013; Sava 2012).

Course and Program Delivery

Delivery of the climate change and DRM courses online was met with challenges relating to access, time management and independent learning, availability and access to Open Educational Resources (OERs), online netiquette, feedback and student support. Access to fast and reliable internet connections in the region has been, and still is a major issue for online learning in the PSIDS and this is in part due to the varied but generally high costs of internet services. In many cases, this is partly due to national telecommunication monopolies and further exacerbated by different socio-economic strengths (Asian Development Bank and Asian Development Bank Institute 2015; Ogden 2013; Wellenius 2008). The main regulatory barrier that prevents a new entrance from starting up a business in a telecommunication sector in the Pacific Islands is the requirement to have an operating license granted by the government. In most PSIDS, the government owned monopoly provider has an exclusive license to provide telecommunication services for a fixed period, usually for a 10–15 year period. This form of license guarantees the monopoly status of the provider (McMaster 2006).

Time management and independent learning on the part of the learner is a challenge that exists for online learners and even more so, for learners in the PSIDS who have a full time job and need to travel frequently. Experiences from the PGDCC courses show that most of the students who are enrolled are working in a climate change and (or) DRM related field and given the nature of the job, need to travel frequently for work purposes. This creates pressure on the learner to make time for online learning simultaneously with their work demands. Women and girls have even more of a challenge as they have the additional responsibility of taking care of their households, working full time (if they are employed) and learning in an online environment (Kanai 2015). Education plays a more determinant role for women than for men in relation to their capacity to adapt to climate change so there is a need for women to have some understanding of climate change and associated risks (Wamsler et al. 2011; Armstrong et al. 2018). Creating a culture for independent learning has become a necessity for competence in online learning.

The availability of OERs with free licensing (such as the creative commons), and access to these OERs can be a barrier to enhanced online learning experiences (Weller 2014). As more and more OERs become available, and more higher education institutions utilise these, the need to create and deliver an enhanced online experience emerges, and this is not only in response to the 'competition' but also as a responsibility that the institution has to deliver quality Pacific-conscious education to learners from the PSIDS. OERs are a powerful tool for reducing inequalities in educational opportunities and promoting innovative strategies to improve the delivery of education across time zones, hence, their contribution to achieving Sustainable Development Goal 4 (SDG4) cannot be stressed enough. OERs make online learning more appealing owing to design, functions and access on different platforms and gadgets (Bliss and Smith 2017; Hegarty 2015).

In a region as geographically and culturally diverse as the PSIDS region, the delivery of online climate change and DRM courses needs to be inclusive of the different cultures and be wholly Pacific conscious to appeal to learners who are already experts in the climate change and (or) DRM field. The challenge of inclusivity is one that needs to be addressed at all levels of a higher education institution like USP. Accommodating learners with special needs is a challenge and specifically those who require special attention due to varying temperaments.

Student numbers in the PGDCC program have steadily increased over the years creating a higher demand for one on one feedback and assistance. Providing student analytics and monitoring progress has become more and more of a challenge, one that perhaps could be addressed with reducing the learner to educator ratio.

Evaluation and Improvement

Course evaluations over the years have confirmed an increase in the demand for one on one assistance. It is evident that while there are more and more available OERs, the added pressure on learners to be competent tech savvy learners has taken its toll on the learners. One reason for this is the fact that USP introduced MOODLE as a learning platform in 2008 and since then, more and more mature students are returning to study after at least ten (10) years of employment. The first challenge for them in a fully online program is mastering the use of the MOODLE platform. Including more OERs or components thereof creates stress on already heavily overloaded working students.

The need for educators who are competent using online open resources is a challenge and evaluations have shown the need to continue upgrading knowledge of OERs as they become available online. Completing the PGDCC program offered by USP has proven to be very useful for continuous provision of quality online climate change and DRM education.

Diverse cultural backgrounds are a challenge in an online learning environment if online learning netiquette is not adhered to. From experience, online netiquette is vital for creating an online learning environment where learners and educators are not bullied online. Online bullying has been known to discourage learners from continuing their course of study.

Lessons Learnt and Recommendations

Although unique circumstances in each USP member country lead to varying results and lessons learned, the following themes have emerged as recurring lessons, and should therefore be considered in parallel with the recommendations provided.

I. Delivering courses that accommodate constructive alignment ensures that intended learning outcomes are achieved and learners leave the program with the required skills. Constructive alignment needs to be enforced to ensure that assessments are contributing to meet course, program and institutional intended learning outcomes.

II. Pacific consciousness and inclusivity are required for a wholesome learning experience in the context of PSIDS. Design of course materials need to accommodate Pacific consciousness and inclusivity at all levels (including people with special needs).

III. Climate change science and DRM information is continuously evolving and curriculum and course updates need to be performed simultaneously and on a regular basis. A systematic and ongoing process of curriculum review and (re)design informed by up-to-date capacity needs of USP member countries is recommended. Course reforms should also be supported by needs assessments and high quality research.

IV. Online learning in the PSIDS region will only truly become fully accessible when telecommunication monopolies, political will and relevant institutional Memorandums of Understanding are aligned. It is highly recommended that higher education institutions like USP collaborate with governments of PSIDS to provide affordable internet services for online learners.

V. Online learners in PSIDS require additional funds to cover internet connectivity costs. Access and costs of online learning need to be factored into online course tuition fees—considering the available ICT resources at regional USP campuses and the number of students enrolled in online courses.

VI. E-learning resources play an important role in achieving intended learning outcomes and require collaboration to be fully utilized. Achieving intended outcomes requires that ICT for online learning initiatives be carefully planned, well-coordinated, based on empirical evidence, and that they adopt a long-term vision (Asian Development Bank 2018).

VII. Critical thinking skills are generally lacking for students who enter the program for the first time. Incorporating a diagnostic test before beginning the program is recommended to determine levels of understanding in the subject area.

VIII. Using a Research Skills Development Framework to guide the development and delivery of online rubrics for assessment tasks is encouraged as it fosters research skills. Academic staff need to build capacity in this area.

IX. Working students and students facing internet connectivity issues require some flexibility to meet deadlines. Flexibility needs to be built into course assessment requirements particularly submission deadlines for assessment tasks.

Conclusion

Long term investments by higher education institutions in formal, accredited and recognized climate change and DRM education could be the way forward as an adaptation and mitigation action implemented in parallel with national and regional adaptation and mitigation actions. The delivery of the PGDCC program by USP has contributed to the call by PSIDS leaders for capacity building and human resource development in the field of climate change and DRM. Having the program delivered fully online has resulted in a lot of benefits including access to tertiary education in climate change specifically tailored for the PSIDS.

While it is important to maximise on the benefits of online learning in a developing region like the PSIDS, challenges are still prevalent. Challenges with regards to course and program design, delivery and the necessary knowledge and skills required for effective online learning are areas that need to be addressed.

Recommendations and lessons learnt informed by years of observation and experience in the PGDCC program, although presented here, are not prescriptive for all online courses and programs in PSIDS, however they will only improve online learning in a highly diverse and developing region if considered and properly implemented.

References

Armstrong A, Krasny M, Schuldt J (2018) Climate change education outcomes. In: Communicating climate change: a guide for educators. Cornell University Press, Ithaca, London, pp 25–31. Retrieved from http://www.jstor.org/stable/10.7591/j.ctv941wjn.7

Asian Development Bank (2018) ICT for better education in the Pacific. ADB Avenue, Mandaluyong City, 1550 Metro Manila, Philippines. Retrieved from https://www.adb.org/sites/default/files/publication/428221/ict-education-pacific.pdf

Asian Development Bank, Asian Development Bank Institute (2015) Economic conditions and the recent performance of Pacific developing member countries. In: Pacific opportunities: leveraging Asia's growth. Brookings Institution Press, pp 8–35. Retrieved from http://www.jstor.org/stable/10.7864/j.ctt1gpccx2.9

Barnett J (2005) Titanic states? Impacts and responses to climate change in the Pacific Islands. J Int Aff 59(1):203–219. Retrieved from http://www.jstor.org/stable/24358240

Bataller M (2010) The challenges of adapting to climate change. In: Climate report: research on the economics of climate change, no 21. Retrieved from http://www.cdcclimat.com/IMG/pdf/21_Etude_Climat_EN_The_challenges_of_adapting_to_climate_change.pdf

Bliss TJ, Smith M (2017) A brief history of open educational resources. In: Jhangiani RS, Biswas DR (eds) Open: the philosophy and practices that are revolutionising education and science. Ubiquity Press, London, pp 9–27. https://doi.org/10.5334/bbc.b. Licence: CC-BY 4.0. Retrieved from https://www.google.com/search?q=Bliss+and+Smith+2017&rlz=1C1CHBF_enVU815VU815&oq=Bliss+and+Smith+2017&aqs=chrome..69i57.5796j0j1&sourceid=chrome&ie=UTF-8

Bossu C (2017) Pacific leaders in open, online and distance learning. J Learn Dev. Retrieved from http://www.jl4d.org/index.php/ejl4d/article/view/207/201

Carter G (2015) Establishing a Pacific voice in the climate change negotiations. In: Fry G, Tarte S (eds) The new Pacific diplomacy. ANU Press, pp 205–220. Retrieved from http://www.jstor.org/stable/j.ctt19w71mc.23

Cordero EC, Todd A, Abeller DM (2008) Climate change education and the ecological footprint. Am Meteorol Soc. https://doi.org/10.1175/2007bams2432. Retrieved from https://journals.ametsoc.org/doi/pdf/10.1175/2007BAMS2432.1

Davidson J, Lyth A (2012) Education for climate change adaptation—enhancing the contemporary relevance of planning education for a range of wicked problems. J Educ Built Environ 7(2):63–83, https://doi.org/10.11120/jebe.2012.07020063. Retrieved from https://www.tandfonline.com/doi/pdf/10.11120/jebe.2012.07020063

Faxon-Mills S, Hamilton L, Rudnick M, Stecher B (2013) Conditions that influence educators' responses to assessment. In: New assessments, better instruction? Designing assessment systems to promote instructional improvement. RAND Corporation, pp 21–33. Retrieved from http://www.jstor.org/stable/10.7249/j.ctt5hhtkh.12

Harasim L (2000) Shift happens: online education as a new paradigm in learning. Internet High Educ 3(1–2):41–61. Retrieved from https://www.sciencedirect.com/science/article/abs/pii/S1096751600000324#!

Hegarty B (2015) Attributes of open pedagogy: a model for using open educational resources. Educ Technol 55(4):3–13. Retrieved from http://www.jstor.org/stable/44430383

Hogan R (2009) Attitudes of indigenous peoples toward distance learning in the South Pacific: an empirical study. In: Siemens IG, Fulford C (eds) Proceedings of ED-MEDIA 2009–world conference on educational multimedia, hypermedia and telecommunications. Association for the Advancement of Computing in Education (AACE), Honolulu, pp 1064–1072. Retrieved from https://www.learntechlib.org/p/31621/

Johnson R, Stewart C, Bachman C (2015) What drives students to complete online courses? What drives faculty to teach online? Validating a measure of motivation orientation in university students and faculty. Interact Learn Environ 23(4):528–543. Retrieved from https://www.tandfonline.com/doi/full/10.1080/10494820.2013.788037?scroll=top&needAccess=true

Kanai A (2015) Thinking beyond the internet as a tool: girls' online spaces as postfeminist structures of surveillance. In: Bailey J, Steeves V (eds) Egirls, Ecitizens: putting technology, theory and policy into dialogue with girls' and young women's voices. University of Ottawa Press, pp 83–106. Retrieved from http://www.jstor.org/stable/j.ctt15nmj7f.7

Li X, Chen Q, Fang F, Zhang J (2016) Is online education more like the global public goods? Futures 81:176–190. https://doi.org/10.1016/j.futures.2015.10.001

McMaster J (2006) Toward a new Pacific regionalism: costs and benefits of deregulating telecommunication markets in the Pacific. Pacific studies series vol 3, working paper no. 15, University of the South Pacific, Suva, Fiji Islands. Retrieved from http://repository.usp.ac.fj/5258/1/Costs_and_Benefits_of_Deregulating_Telecommunication_Markets_in_the_Pacific.pdf

Ogden M (2013) Communications. In: Rapaport M (ed) The Pacific Islands: environment and society, revised edition. University of Hawai'i Press, pp 401–416. Retrieved from http://www.jstor.org/stable/j.ctt6wqh08.37

Pei-Chen S, Ray JT, Finger G, Chen Y-Y, Yeh D (2008) What drives a successful e-Learning? An empirical investigation of the critical factors influencing learner satisfaction. Comput Educ 50(4):1183–1202

Purcell F, Toland J (2004) Electronic commerce for the South Pacific: a review of E-readiness. Electron Commer Res 4(3):241–262. Retrieved from https://link.springer.com/article/10.1023/B:ELEC.0000027982.96505.c6

Rees WE (2003) Impeding sustainability? The ecological footprint of higher education. Plann High Educ 31:88–98

Roy J, Tschakert P, Waisman H, Abdul Halim S, Antwi-Agyei P, Dasgupta P, Hayward B, Kanninen M, Liverman D, Okereke C, Pinho PF, Riahi K, Suarez Rodriguez AG (2018) Sustainable development, poverty eradication and reducing inequalities. In: Masson-Delmotte V, Zhai P, Pörtner HO, Roberts D, Skea J, Shukla PR, Pirani A, Moufouma-Okia W, Péan C, Pidcock R, Connors S, Matthews RBR, Chen Y, Zhou X, Gomis MI, Lonnoy E, Maycock T, Tignor M, Waterfield T (eds) Global warming of 1.5°C. An IPCC special report on the impacts of global warming of 1.5°C above pre-industrial levels and related global greenhouse gas emission pathways, in the context of strengthening the global response to the threat of climate change, sustainable development, and efforts to eradicate poverty. In Press. Retrieved from https://www.ipcc.ch/sr15/chapter/chapter-5/

Sava S (2012) Needs analysis for planning educational programmes. In: Needs analysis and programme planning in adult education. Verlag Barbara Budrich, Opladen, Berlin, Toronto, pp 89–118. Retrieved from http://www.jstor.org/stable/j.ctvbkjvs2.9

Sife A, Lwoga E, Sanga C (2007) New technologies for teaching and learning: challenges for higher learning institutions in developing countries. Int J Educ Dev ICT 3(2):57–67. Retrieved from https://www.learntechlib.org/p/42360/

University of the South Pacific (2018) The University of the South Pacific 2019 handbook and calendar. Retrieved from http://www.usp.ac.fj/fileadmin/scripts/HandbookAndCalendar/HandbookAndCalendar_2019_en.pdf

UNESCO (2019) Climate change education. Retrieved from UNESCO: https://en.unesco.org/themes/education-sustainable-development/cce

Wamsler CBE, Rantala O (2011) Climate change, adaptation and formal education: the role of schooling for increasing societies' adaptive capacities. IIASA interim report. IIASA, Laxenburg, Austria: IR-11-024. Retrieved from http://portal.research.lu.se/ws/files/3125552/4392161.pdf

Wellenius B (2008) Towards universal service: issues, good practices and challenges. In: Ure J (ed) Telecommunications development in Asia. Hong Kong University Press, pp 85–112. Retrieved from http://www.jstor.org/stable/j.ctt1xwgsn.10

Weller M (2014) Open educational resources. In: The battle for open: how openness won and why it doesn't feel like victory. Ubiquity Press, London, pp. 67–88. Retrieved from http://www.jstor.org/stable/j.ctv3t5r3r.7

Yoo SJ, Huang WD (2013) Engaging online adult learners in higher education: motivational factors impacted by gender, age, and prior experiences. J Continuing High Educ 61(3). Retrieved from https://www.tandfonline.com/doi/full/10.1080/07377363.2013.836823

Chapter 12
Learning *with* Idea Station: What Can Children on One Canadian Playground Teach Us About Climate Change?

Sarah Hennessy

Abstract Idea Station, the student portion of a research project to revitalize one urban Canadian school's outdated playground helps to re-envision new approaches to education, environment and climate change, shifting the education 'of children' in favour of learning 'with children'—a shift that mirrors the language of a common world pedagogies. There is benefit from learning with non-human others. There is an organic and interdisciplinary way of thinking with the world. As adults and educators we need to value and foster more of this thinking with climate change. To ignore this approach of thinking with children we run the risk of losing their valuable insights. Idea Station reveals that children already think with natureculture and non-human others and can teach adults. Experiences from Idea Station have a number of practical applications and can model partnerships for future thinking *with* education and climate change. Idea Station presents opportunities for shifts in the ways we think about children and nature. The core learnings can all be refined down to a single shift to a more inclusive way of thinking for adults. Idea Station teaches us that the inclusive shift includes making room for (1) informal, out-of-curriculum learning; (2) the values in a child-to-adult model of learning; (3) a Common worlds pedagogy that includes a lens of natureculture, and; (4) the inclusion of technology. In thinking with Idea Station we can learn to think with bats, rats, coons, trees, roots, backstops, vines, plants, birds, bugs, technology and humans.

Keywords Early childhood · Common worlds · Pedagogies · Climate change · Natureculture

A Snapshot of Idea Station[1]

Spring recess on the playground feels warm but not yet hot. It's the kind of weather and temperature that makes you bring a jacket to school only to leave it hanging on the hook as the sun settles in. We had finished eating lunch and the giggling voices and skipping, jumping excited bodies began flowing out of every door onto the playground dumping lunch bags and

---

[1] To protect and respect the children's identities, pseudonyms are used.

S. Hennessy (✉)
Faculty of Education, Western University, 1137 Western Road, London, ON N6G 1G7, Canada
e-mail: shennes5@uwo.ca

© Springer Nature Switzerland AG 2019
W. Leal Filho and S. L. Hemstock (eds.), *Climate Change and the Role of Education*,
Climate Change Management, https://doi.org/10.1007/978-3-030-32898-6_12

jackets haphazardly by the walls. We stood to the side of one door and the children paired off with brightly-coloured, rubber-clad tablets sorting out who was going where to "do the interviews" defining the playground space by activity and age with statements like "I am going to do the climber and the grounders people", "I'm going to the nets where the grade sixes are" and "I'm going to talk to those grade ones who always play four squares". They dispersed in multiple directions, chatting and pointing, fully engrossed in their mission.

The children gathered small crowds every time they began filming. The recycling trucks were collecting on the street and the loud crushing and banging sounds were forcing children to source quieter spaces to record. They scanned the playground in efforts to source "some peace and quiet". As they moved across the playground they left conversations in their midst with off-camera children now sharing comments for their Idea Station feedback.

"I am going to say we need more hiding places and an ice cream truck."

"Duh, it's not Christmas. This isn't about asking for things we can't have. You have to think about how to make it better for real. Like fixing the pavement so I don't wipe out during track."

With new interviews captured, the roving reporters returned to the shade of a tree to share their insights. We sat on the grass, heads turning in all directions considering the space in relation to our questions, "what would you do to improve the playground? What do we need to make it better?" These questions guided each interview of Idea Station. One eight-year old, Casey, spoke saying "We could keep the backstop around the home base and grow plants and vines. It could become a habitat for birds and bugs and stuff. And if we did that we could reuse it and not send all that metal to the dump. It would take all those machines and gas to pull it [fence] up and take it away and that's just more badness for the environment."

Diz followed explaining that they had two different children talking about the trees saying "We need more trees because they fix things we wreck".

Casey commented, "Why do trees have to clean up our mess?"

"They don't have to," Diz replied "they just do."

Haniya chimed in enthusiastically sharing "we didn't talk about the trees but we talked about the animals cause it's their home not just my playground."

Diz interjected "that animals play here at night, like the bats and rats and coons."

Haniya: "that habitat [backstop] could take over"

"It takes over because it's not just our playground. It's all our stuff, all over" concluded Diz as the piercing school bell blanketed the playground.

Haniya ended our discussion as we walked back to the door filing into the school saying "we should interview Adelie about the track pavement. She didn't even realize it's the tree trunks and roots that break through the pavement. Nature is kind of sick of what we keep doing. Maybe they're tripping us as a joke?"

In deep wonder I collected the tablets, forgotten lunch bags, sweatshirts and lone skipping rope and headed to the staff room for my break.

Introduction

We continue to frame children as empty vessels in need of formal education creating more generations in our own adult vision (Dahlberg and Moss 2005; Davis and Elliott 2014; Davis 2014; James and Prout 2015). Alongside this thinking we continue to perpetuate Rousseau's romantic vision of the pure child paired with the pure nature (Elliott and Young 2016; Taylor et al. 2013; Taylor 2014, 2017; Pacini-Ketchabaw and Taylor 2015). Similarly, human-centric solutions of sustainability and stewardship (Sweeney 2015) to climate change education (CCE) (Bryan 2015; Fernandez et al. 2014; Ho and Seow 2015; Hung 2014; O'Malley 2015; Stengers 2013; Taylor 2017; Yusoff 2013) remain the default models. Idea Station presents a clear opportunity to reframe thinking on climate change education to include thinking *with* children (Kraftl 2015) and climate change. Climate change education, charged with educating "to engage in critical and thoughtful inquiry" (Hung 2014, p.18) is a necessity that exists beyond the political and economic contestation debate. How can Idea Station inform this needed shift and cement an unlearning (Taylor 2017) of how we think about children and nature to one *with* children and *with* natureculture in an era of climate change education? This paper delves into a lesser researched area of CCE involving education with young children, focusing on the process of exploring CCE with young children (O'Malley 2015) outside of formal curriculum (Fernandez et al. 2014) not the separate and necessary development of institutional curriculum content (Bryan 2015). The research needs of this paper are delineating the role of children's voices and a Common Worlds (2019) approach in CCE towards "encouraging the change in attitudes and behavior needed to put our world on a more sustainable path in the future" (UNESCO 2017). How can we emulate Idea Station thinking in climate change education?

To re-envision new approaches to climate change education, we must also reject the education 'of children' in favour of learning 'with children' mirroring the language of a Common world pedagogy (Common Worlds Research Collective 2019). The benefits of learning with non-human others by learning with children is a different kind of thinking for many adults. To ignore this approach of thinking with children we run the risk of losing their valuable insights (Istead and Shapiro 2014; Kraftl 2015). Underlying this paper is the possibility that it is adults that need to rethink when considering climate change. It is possible to think with natureculture (Haraway 2008) and non-human others. Natureculture (Haraway 2008) is a perspective where nature and human culture are positioned as inextricably enmeshed all around in opposition to a romanticized, binary separation. How we educate children may be stifling an ability to think *with*. In highlighting Idea Station and the children's thinking, it is apparent that one change within education is the need to foster more learning from and with children (Clough 2002; Goodley and Clough 2004; Istead and Shapiro 2014).

There are two distinct areas for consideration with this paper and the role of education with climate change. The first area is the beneficial results of dismantling the tendency towards a human-egocentric stewardship model (where humans can fix and solve ecological problems) and making room for a more communal common worlds pedagogy of learning (Common Worlds Research Collective 2019) with non-human others. The second area is the small but vital change from the education *of* children to educating *with* children. Idea Station encapsulates a need to consider both of these as part of a shift in thinking, approach and education as we face climate change. Even in presenting these elements there is a tension and unresolved conflict.

Whether discussing adults or children, the dialogue remains human-focused leaving thoughts in an unresolved space of trouble (Haraway 2016) to consider.

This paper explores the provocations and corresponding implications for education garnered from primary school children's understanding of the common world (Common Worlds Research Collective 2019; Latour 2004) they share with non-human others. The insights provided are sourced from my Idea Station field notes. Idea Station is the student portion of a research project to revitalize the school's outdated playground. During the video tablet conversations of Idea Station narratives of various, interwoven knowledges revealed their ability to think with materials and non-humans. Their comments and insights provoked thinking on the distinct and relational space of the playground as it relates to Haraway's (2008) natureculture. Engaging a common world's pedagogy (Common Worlds Research Collective 2019; Pacini-Ketchabaw 2013; Taylor 2017) to learning that pedagogically acknowledges moments when understanding (Savransky and Stengers 2018) of the entangled reality of experiences intersects with what is already happening in the world around and with them.

This paper builds thinking and comments from the children applying a critical reflective narrative of researching lived experiences (Van Manen 1997) to address concepts of natureculture (Haraway 2008) and Common Worlds pedagogies (Common Worlds Research Collective 2019) in thinking *with* children. First the paper positions Haraway's (2008) natureculture within a common world's pedagogy (Common Worlds Research Collective 2019). Drawing on the idea of thinking with nature the paper explores thinking with children. The narrative of Idea Station is explored within these concepts. The paper continues by addressing the distinct place of the urban playground—a much richer space than its simplified stereotype of free-playing children at recess. Finally, the paper ends with provocations and practical applications for education in relation to climate change.

The playground and backstop, 2016. Photo courtesy of the author

Context of Idea Station

Idea Station was the student component of community feedback that took place in the spring of 2016 at an urban primary school in Ontario Canada. The playground revitalization project had separate audience research for parents, teachers and the neighbourhood community in conjunction with requirements from the municipal school board. The children's (aged 6–10) video tablet conversations of Idea Station were initially designed to capture their ideas on ways to revitalize the school's urban, outdated playground. Over the course of multiple lunch hours of free play on the playground, children participated as roving reporters interviewing their peers using the questions *what would you do to improve the playground? What do we need to make it better?*

The Idea Station playground is located in the downtown urban core of Toronto, Canada. The primary school and its playground are surrounded on all four sides by residential, semi-detached single family homes in a mixed socio-economic neighbourhood. Both a large urban thoroughfare and a major urban highway are located less that a kilometre away. The playground is a blend of open grass areas, climbers, mature trees and asphalt with basketball nets and a climber. The brick school was built in 1929 and with the exception of the introduction of fences, backstop, basketball nets and a climber has not been "upgraded". Playground improvements were the result of childcare and school needs with the move to Provincial program of Full Day Kindergarten (Ontario Ministry of Education 2010) over the last decade. Idea Station received no funding.

Of the 124 children in the school, only fourteen children did not participate in Idea Station. The children collected, sorted, analyzed and presented the information to the school community highlighting their findings into recommendations for permanent fixtures, natural equipment, general equipment, visual aesthetic elements and alternative ideas.

Reflecting on Lived Experience

The children's work with Idea Station was important as a voice in change. It was their thoughtful conceptions that resonated and prompted reflection on the hidden nature of meanings (Van Manen 1997) in comments. It is the meanings assigned to the nature all around them and a shared space with non-humans that provoked my reflective thinking. The narrative anecdote is sourced from April 2016 but has continued to resonate with me over time (Todd 2001). It has fueled exploration into the Common worlds approach to learning (Common World Research Collective 2019); natureculture (Haraway 2008); thinking with, and; the space and place of the playground. This paper is evidence of how "living is full of encounters that intrigue and provoke us" (St. Pierre 2013, p. 226). What began as adults and educators looking for a simple list of needs from children for their playground revealed complex, considerate and empathetic thinking beyond themselves or human needs to consideration of a shared urban space of natureculture (Haraway 2008)?

This methodology was employed because it allows the provocations from a single anecdote as an educator to be valued in relation to the understanding of practice, children and the planet. The paper applies reflective hindsight to determine why this experience continues to resonate with me as an educator. It is in reflecting on the deep impact of experiencing children thinking with their world of non-human others that this paper was conceived. It is based on my professional journals from field practice as an Early Childhood Educator and engages a critical reflective narrative of researching lived experiences (Van Manen 1997).

As people, specifically educators, there is an orientation to pedagogy in "living with children that requires constant practical acting in concrete situations and relations" (Van Manen 1997, p. 2) and affords learning from unforeseen anecdotal experiences. The value in learning with Idea Station is derived from the experience of living with a research stance and the thought provocations that come from this way of living. Experiencing Idea Station refocused an understanding of children and their abilities to be, not just become (Clarke and McPhie 2016; Istead and Shapiro 2014). This view is a subtle but powerful shift in the understanding of education with children suggesting that the approach "does not require that children are 'educated' to become like the adults who educate them" (Van Manen 1997, p. 159). As a result of this process of writing it becomes clear that the learner in this paper is the adult and the teacher is the child – an all-too-often forgotten part of being an educator.

Natureculture as Part of a Common World's Pedagogy

The theoretical frameworks for the thinking with Idea Station children encompasses a Common worlds pedagogy (Common Worlds Research Collective 2019) grounded in Haraway's (2008) natureculture. The tangled quality of Haraway's (2008) natureculture relates directly to the pedagogy of Common worlds (Common Worlds Research Collective 2019) by implementing a posthumanist lens, that "reposition[s] childhood and learning within inextricably entangled life-worlds, and seek[s] to learn from what is already going on in these worlds" (Taylor 2017). In fusing nature with culture the tendency towards binary thinking and divisions is replaced with entanglements. This pedagogy furthers natureculture thinking to learning and children. It is this combination that lends itself to the children of Idea Station's thinking with their playground, a space of natureculture.

Haraway's (2008) natureculture is a more organic, interwoven concept that dismantles the binary of nature and humans as separate suggesting an entangled all togetherness placing humans as simply a part of nature, alongside non-human (both living and non-living) others (Pacini-Ketchabaw et al. 2016; Pacini-Ketchabaw and Taylor 2015). For the purposes of this paper natureculture is the understanding children have of being part nature all around everything, always. "Nature" is not located in a designated and romanticized place titled "nature", "park", "forest" or "wild". Haraway's (2008) more abstract concept of natureculture becomes a more concrete idea through the language of these children.

The Common Worlds pedagogy (Common Worlds Research Collective 2019) is an interdisciplinary, international collective concerned with the relationship between humans and the more-than-human world. A focus on children's existence within this common world (Latour 2004) goes beyond the ego, hero and conceit of humans to an interwoven, and inclusive model unconfined by intellectual tendencies towards simple dualism—one that leaves us with either nature- or culture-styled thinking. In pursuing children's common worlds the collective focuses on children's relations with place, materials and other species. Idea Station captures children's unfiltered experience with all three of these areas.

With increasing discussions of climate change and use of the term Anthropocene (the new unstable geologic epoch defined by the human impact on the planet) natureculture and common worlds pedagogy also become increasingly important and present an alternative model to the traditional human-egocentric stewardship pedagogy (Haraway 2008; Kopnina 2014; Latour 1993, 2011; Taylor 2017) where humans are positioned as the only ones capable of solving these problems. Thinking with the playground as an alternative pedagogical route levels humans as part of nature and removes the egotistical thinking of "we will fix this". The two concepts of natureculture (Haraway 2008) and Common Worlds pedagogy (Common Worlds Research Collective 2016) came together on the playground. As an urban space and childhood haven, the playground dismantles the separation of nature from culture and shows children's ability to think with the world, its materials and other species without defaulting to an ego-filled supreme position in conceiving of solutions.

Thinking *with*

There is a great deal of thinking in this paper and while it covers important topics it is the nature of thinking, specifically thinking *with* that is most important. There are children thinking with their playground, non-humans and climate change. There is also my thinking with children which represents a fundamental professional change from educating children to thinking *with* them.

The first hints of thinking with are revealed in my field note reference to 'hiding places'. Playing hide and seek and tag are not exclusive to humans. They are natural tendencies that connect young to their environment. The need for hiding places suggests a primal urge to blend with their world. Unfortunately this is often forbidden as it impedes the adult's ability to monitor and control. Hiding is inherently natureculture as children attempt to become one with their world to avoid detection.

The question of "what would you do to improve the playground?" was innocently conceived. Upon reflection it is apparent that it did not include "for us" as part of the question. This complex difference was understood by the children. It allowed their thinking an opening and invitation to think beyond themselves to materials and non-human others. The term "improve" is left undefined allowing a similar opening for thinking of improvement beyond the self or humans. What role did this invitation to think broadly factor in children like Casey's answers?

The backstop habitat to be, 2016. Photo courtesy of the author

Backstop and Habitats

We could keep the backstop around the home base and grow plants and vines. It could become a habitat for birds and bugs and stuff. And if we did that we could reuse it and not send all that metal to the dump. It would take all those machines and gas to pull it [fence] up and take it away and that's just more badness for the environment. (Field journal, April, 2016)

Casey's comment is complex and suggests an understanding of natureculture. In suggesting that the human-made backstop stay and that we think with it, is considerate and displays an understanding of waste and fossil fuels. It is also indicative of thinking with material, non-living things. It is more considerate to work with what is there already in the playground world than "create more badness" to clear the space for a new human imposition. Casey conveys an understanding of contradictory concepts like sustainable consumption (Kopnina 2014)—suggesting the illogic of using gas and machines to pull the backstop up and send it to a dump to make a playground more natural. In this natureculture thinking Casey understands that nature is right there with opportunity for a cooperative model with the 'plants and vines' to re-envision the backstop as a 'habitat for birds and bugs and stuff'. It is natureculture thinking with because nature is not separated and isolated but experienced all around Casey where all the "actors become who they are *in the dance of relating*" (Haraway 2008,

p. 25, author's italics). The cooperation with nature becomes collective thinking (Stengers 2012; Taylor 2017). We, as humans, could leave the backstop; the plants and vines could grow with it; the birds and bugs and stuff could make it their home. Furthermore it situates the human role as one of many and without superiority. The only role of the human is to step back, reiterated in the undertones from another child's comment about "making it better for real". Together these children ground an understanding of how the world around them works: It is not an answer on a test about the impact of fossil fuels. The field notes quote "better for real" is not simply to avoid wiping out during track (a human need) but long-term, big-picture considerate of others thinking. It is the "sometimes joined" (Haraway 2008, p. 25) thinking of natureculture with non-human others whether they are trees, backstops or animals. Casey thinks about others, in particular non-human others. The comment contains no human or individual need: It is selfless and collective in essence (Stengers 2012; Taylor 2017).

There was much consensus with the children on the lack of need for the backstop as the dimensions of the playground made baseball unrealistic—the student's bating skills leave too many balls on the school roof. It would be easy to see this thinking as the backstop no longer mattering to them (Rautio 2013). In thinking with the backstop, it reveals that it continues to "matter" in a new inter- and intra-action (Barad 2003, 2007; Taylor et al. 2013) in the space.

The inclusion of plants and vines with insects and animals suggests that concerns over plant blindness (Wandersee and Schussler 1999), the socialization of children to prioritize animals over plants, is absent. For these children the playground is a space of things, living and non-living, plant, animal insect, and structure alike, blended with collective thinking and roles absent of hierarchy.

Later in my field notes, Haniya comments that the "habitat could take over" suggesting an understanding of the interwoven ways of natureculture thinking; the presumption of human controlled architecture of the planet is absent in this thinking. The backstop would change from being a backstop to becoming a habitat in a way not prescribed and controlled by humans. Like the tree roots that trip people, it suggests an understanding that we are impossibly and inextricably interwoven with 'common worlds' (Latour 2004) as part of nature and the impacts of change have a ripple effect, whether initiated by human or more-than-human.

Bats and Rats and Coons

The understanding of shared space and thinking with non-human others continues with Haniya and Diz's discussion of the animals that also call the playground home. Building on the ability to acknowledge a more-than-human agency with 'plants and vines' and the 'birds and bugs and stuff', suggests this idea of connectivities (Taylor 2014), "foregrounding the way children understand themselves as enmeshed within complex networks of relations" (p. 124). Sometimes these relations appear as human to human in a giggle-filled game but sometimes they appear as thoughts about who uses the space when they aren't around. This network (Barad 2007; Taylor et al. 2013), in one dialogue, covers relations with an array of place, materials and other species with metal, concrete structures, plants, insects and animals as members of Latour (1993) Parliament of Things.

The separation of nature is absent in Haniya's thinking with the animals. There is also a lack of proprietary thinking with the playground. With Haniya and Diz's comments changes to the playground are not limited to humans; it is understood as a shared space as it is the home of 'bats and rats and coons'. These children think with their more-than-human others in this space even without physically interacting with these primarily nocturnal species. In this network, they think with them without interacting with them.

Thinking with natureculture is again visible as Diz and Haniya continue referencing the ability of non-living materials like the backstop and plant, insect and animal species to take over begging the question of power and the inextricable connections these children have with their world and its boundaryless space. To 'take over' can imply power and superiority to 'conquer' others. Haniya's use of the term is a shift to a more collective model with humans as one of many players in the network (Barad 2007; Taylor et al. 2013; Latour 1993). This dialogue and thinking suggests the world is not lost (Haraway 2004; Taylor 2014) with these children. They think in an enmeshed way. Their dialogue also captures a collective spirit of thinking (Stengers 2012; Taylor 2017) that runs through the dialogue, not limited to a single individual, but woven through the comments of many children beyond the limits of topics of trees, animals and the backstop.

Understanding Trees

The children's comments in the field notes frame an understanding of trees as capable, contributing, interwoven members of the world (Barad 2007; Taylor et al. 2013; Latour 1993) well beyond the simplified classroom knowledge of photosynthesis and gas exchanges. Aside from the humour of considering trees intentionality in tripping humans, trees, like habitats frame the subject of "all our stuff". Use of the term "our" is not a human-limited reference. It is a collective natureculture reference that suggests an understanding of how individuals emerge from within relationships in intra-actions (Barad 2003, 2007). This means the children understand trees and humans as constituents of action as they reconfigure the entangled space of the playground. Diz uses "our" to speak with (not for) the natureculture of the playground reiterating an earlier comment that trees 'don't have to clean up our mess they just do' suggesting a complex understanding of intra-actions (Barad 2007; Taylor et al. 2013).

This understanding is summarized by Diz saying "it's not just our playground. It's all our stuff, all around" removing a human agency in favour of an existing and evolving relationship where thinking and action are not separate but enmeshed like the children with their playground. In defining the playground as "all around", we are invited to reject the limits of border thinking and linear time for simultaneous thinking within a relationship these children have with the elements of the playground space.

The Distinct Space of the Playground

As an example of natureculture the playground is also an intermediate, peripheral zone that isn't a formal, adult-dominated space –rather it is a space regulated by children where they have a voice. In the thinking and play in this space, children reveal the distinct nature of this place. Children have a relationship with this place (Nxumalo 2015; Pacini-Ketchabaw 2013; Taylor 2017), the playground. In the complexity of thinking *with* that occurs, the playground network evolves from being a place to becoming a space.

In using the term space instead of place a more inclusive model that allows thinking *with* is included. While the playground has history, and three dimensional elements, both living and non-living it is also a less-definable collection of energies that include thinking. With Idea Station the thinking that happens with and on the playground is distinct (Clarke 2018; Gruenewald 2003a, b; Hamm 2015; Yahya and Wood 2017). With Idea Station, the playground is a space to think *with*: It is more than just a place. The playground is a place of almost exclusive informal learning (Fernandez et al. 2014; Yahya and Wood 2017). It is an intermediary space where children are in charge and the result is a different kind of thinking. The playground is the most collaborative space for these children because of its intermediary nature and their dominance in the space. Because of its designation as a space "for children" it is where they are most collaborative. The space of the playground factors in defining a different kind of child, capable of a more agenetic thinking (Clarke 2018; Istead and Shapiro 2014; Kraftl 2015). The playground is one of the least regulated spaces in these children's lives. It has the most flexible curriculum, often absent of institutional and intended curriculum. It is the space where adults play the least significant role. The minimal nature of formal curriculum enables a more organic form of thought. In the absence of formal curriculum, testing, evaluation and adults comes constant and limitless experimentation (Higgins 2009) for children. It is a space regulated by children not adults.

The simplistic understanding of children, nature, playgrounds have no place in this paper, replaced instead with the complexity of each of these as contribute to thinking with climate change. We as adults and educators continue to frame children in a limited way (Istead and Shapiro 2014; Kellett et al. 2004; Kraftl 2015). Similarly we continue to frame nature in a limited way, an idealized, romanticized, wild and distant place. The playground space for these children presents a different understanding of children and nature.

The edges of classification are diminished on the playground. Conceptually it is not quite "School" but it isn't home. It isn't really a place of adult control regardless of what adults believe. Similarly it is not experienced as nature or urban but a true natureculture, as evidenced with the children's comments. It is a soft-edged space where childcare, community, school, adults, children, vines, plants, insects, backstops, rats, bats and coons merge. This lack of defined boundaries and overlap is a factor in the generating the children's ability to think *with*.

The space of the playground invites the question of what kind of learning is happening. In addition to the child-to-adult and peer (Francis and Davis 2014) elements of the project there is an informal, complementary (Falk and Dierking 2010; Fernandez et al. 2014) learning happening. Informal learning, (Duerden and Witt 2010a, b; Fernandez et al. 2014; Kola-Olusanya 2005) is the learning that happens outside of the institutional and intended curriculum and is not taught by educators, instead modelled on open-ended learning environments. The role of informal learning builds on a stark reality that only 5% of an American's (similar results can be inferred for Canada) lifetime is spent in the classroom with the remaining 95% spent outside of classroom (Falk and Dierking 2010) paired with the Harvard Family Research Project (2007) which states that "the dominant assumption behind much current educational policy and practice is that school is the only place where and when children learn. This assumption is wrong." The informal learning with the playground is vital in their thinking and allows genuine flexibility and an openness to original thought. Informal learning is valuable, under-acknowledged and a contributing factor to insights from Idea Station.

While research recommends increases to complementary education in relation to environmentalism this negates two important areas for generating solutions. The first is that much complementary education occurs in human-controlled nature models like aquariums, zoos, and museums (Dunkley 2016; Reis and Ballinger 2018; Schweizer et al. 2013) furthering the importance of humans, separation from nature and conceit of human solutions only for environmental realities. The second is that complementary learning is devalued by suggesting it is only a 'complement' to formal learning. Both of these factor in valuing thinking with natureculture. Its value is independent of the support role afforded with a 'complement' titling.

Results and Implications

A number of results for Climate Change Education (CCE) can be garnered from Idea Station. These results have a number of practical applications and can model partnerships for future thinking *with* education and climate change. These applications present opportunities for shifts in the ways we think about "children" and "nature" in light of CCE. The core learnings can all be refined down to a single shift to a more inclusive way of thinking for adults, specifically educators. Idea Station teaches us that the inclusive shift includes making room for (a) informal, out-of-curriculum learning; (b) the values in a child-to-adult model of learning; (c) a Common worlds pedagogy (Common Worlds Research Collective 2019) that includes a lens of natureculture (Haraway 2008), and; (d) the inclusion of technology. Idea

Station shows that thinking with children, bats, rats, coons, trees, roots, backstops, vines, plants, birds, and bugs also includes technology (whether we want it to or not).

(a) **Value of the informal nature of playground learning**
Idea Station suggests that thinking *with* a playground and all its existing relationships (Barad 2007; Taylor et al. 2013; Latour 1993) happens organically distanced from adults and formal curriculum. The freedom from both adults and curriculum generates different thinking worth further exploration and deserving of consideration. The playground, as a distinct space where thinking *with* happens organically suggests we consider programming fewer field trips to human-controlled idylls and presented natures in favour of real, lived natureculture. The implications of this result indicate that CCE must include urban naturecultures and value the learning that happens here.

(b) **Child-to-adult learning**
While there is increasing advocacy and supporting research for counter narratives in education curriculum there remains a void of advocacy in favour of reconceptualising children's capabilities in generating solutions—they are able to think with their playground and we need to support this. This reconceptualization includes valuing what we, as adults, can learn with and from children. While adults are increasingly capable of acknowledging children's funds of knowledge (González et al. 2005) and abilities to learn there remains a reticence to conceive of their abilities to create solutions and original thought. Their thinking is different and valuable and essentializing their abilities (Fuller 2007) and imposing a glass ceiling on those abilities (Kellett et al. 2004) only limits solutions and thinking (Ballantyne and Packer 2009; Ballantyne et al. 2001; Goodley and Clough 2004; Istead and Shapiro 2014; O'Malley 2015; Vaughn et al. 2003). Idea Station carries a message of holding attention to learning from all directions not simply the adult-to-child model. The implications suggest CCE include approaches to learning from and with nature that implement a model of the least-adult role approach (Warming 2011) to generate different thinking.

(c) **Common Worlds pedagogy**
If children think with nature and we can think with children we can also make space for alternative pedagogies that allow an enmeshed, more-than-human world as a foundation instead of a human-centred lens. The environmental sustainability and stewardship models of addressing climate change don't seem to be working (Taylor 2017). Thinking with natureculture and thinking with children presents a viable alternative in education. The Common world's pedagogy (Common Worlds Research Collective 2019) holds potential for change in considering education's role in addressing climate change. The implications of this connection suggest more energy should be afforded in this direction of CCE?

(d) **The tablet factor**

As dangerous as limiting thinking that separates nature from culture is the romanticized tendency to limit belief that "environmental learning is best nurtured through direct and "unmediated" sensorial contact with non-human others" (Greenwood and Hougham 2015). This is problematic as it creates a model of best practice that excludes. Unjustified exclusionary thinking limits possibilities unnecessarily. Idea Station happened with children employing the technology afforded by tablets, blurring boundaries of how children can and do think with natureculture. In the case of Idea Station the thinking *with* included technology. Idea Station continues to model inclusion by involving technology in natureculture. The tablet factor is natureculture in its role of infolding and interfacing (Haraway 2008) where the tablet is embodied and enmeshed in the thinking with materials not just species and therefore should be valued. The tablet is as much a part of the encounter as the child, the backstop or the rats, bats and coons.

How we construct nature has social and ecological effects (Taylor 2011) so we need to consider the tracks we leave as educators (Pacini-Ketchabaw and Taylor 2015; Smith et al. 2019) when we construct childhood and nature in fragile, essentialized bubbles. The implications of this knowledge recommends a CCE lens that lets go of binary tendencies in favour of the messy intra-reality of natureculture that defaults us as educators to divide children from adults; nature from culture; backstops from bats, and; humans from tablets, when all are members of the Parliament of Things (Latour 1993).

Conclusion

Children are intelligent. They think differently and they can conceive of valuable solutions and approaches to the developing field of CCE (O'Malley 2015). They can naturally think in an organic and interdisciplinary way and as adults and educators we need to value and foster more of this. These children, in this particular moment show how they think in an interdisciplinary way. They show us a way to collaborate. They can tell us something about climate change. These children run and play and socialize daily in the "expected" ways of children on a playground. To demote their thinking ability because of this play is a repeated mistake of adults. Idea Station reveals the depth of thinking, thinking *with* and genuine ability to generate solutions. Some of the thinking was prompted by listening deeply to this thinking. It is not lost with me that the first three letters of Earth are "ear" and remind us to listen. Climate Change Education requires more bottom up, informal approaches that go beyond adult-led, sustainability and stewardship models can continue to purposefully place the human in the centre. Future development in the field of CCE can expand to include a less human-centric model using a common worlds approach.

Idea Station has created a way to understand children as knowledge makers and teachers *with* the world. They can help us redistribute and reorganize the way we think with climate change education. The formalities and simplified dualism of adult-child, educator-student melt away with Idea Station and there is only learning and thinking *with*. As the children think *with* natureculture, their playground and climate change, I became more conscious and capable of thinking *with* children.

References

Ballantyne R, Packer J (2009) Introducing a fifth pedagogy: experience-based strategies for facilitating learning in natural environments. Environ Educ Res 15(2):243–262

Ballantyne R, Fien J, Packer J (2001) Program effectiveness in facilitating intergenerational influence in environmental education: Lessons from the field. J Environ Educ 32(4):8–15

Barad K (2003) Posthumanist performatively: toward an understanding of how matter comes to matter. Signs 28(3):801–831

Barad KM (2007) Meeting the universe halfway. Duke University Press, Durham, NC

Bryan A (2015) Development education, climate change and the "imperial mode of living": "thinking institutionally about the ecological crisis". Policy Pract Dev Educ Rev 21:1–10

Clarke DAG (2018) Place in research. Theory, methodology, and methods. Environ Educ Res 24(1):146–149

Clarke DAG, Mcphie J (2016) From places to paths: learning for Sustainability, teacher education and a philosophy of becoming. Environ Educ Res 22(7):1002–1024

Clough P (2002) Narratives and fictions in educational research. Open University Press, Buckingham, UK

Common Worlds Research Collective (2019) Common world research collective website. www.commonworlds.net

Dahlberg G, Moss P (2005) Ethics and politics in early childhood education. Routledge Falmer, New York, NY

Davis J, Elliott S (eds) (2014) Research in early childhood education for sustainability: international perspectives and provocations. Routledge, New York, NY

Davis JM (2014) Young children and the environment. Cambridge University Press, Port Melbourne, Australia

Duerden MD, Witt PA (2010a) An ecological systems theory perspective on youth programming. J Park Recreation Adm 28(2)

Duerden MD, Witt PA (2010b) The impact of direct and indirect experiences on the development of environmental knowledge, attitudes, and behavior. J Environ Psychol 30(4):379–392

Dunkley RA (2016) Learning at eco-attractions: exploring the bifurcation of nature and culture through experiential environmental education. J Environ Educ 47(3):213–221

Elliott S, Young T (2016) Nature by default in early childhood education for sustainability. Aust J Environ Educ 32(1):57–64

Falk JH, Dierking LD (2010) The 95 percent solution. Am Sci 98(6):486–493

Fernandez G, Thi TTM, Shaw R (2014) Climate change education: recent trends and future prospects. In: Shaw R, Oikawa Y (eds) Education for sustainable development and disaster risk reduction. Springer, Tokyo, Japan, pp 53–74. https://doi.org/10.1007/978-4-431-55090-7_4

Francis J, Davis T (2014) Exploring children's socialization to three dimensions of sustainability. Young Consum 15(2):125–137

Fuller B (2007) Standardized childhood: the political and cultural struggle over early education, 1st edn. Stanford University Press, Stanford, CA

González N, Moll LC, Amanti C (2005) Funds of knowledge. Routledge, New York, NY

Goodley D, Clough P (2004) Community projects and excluded young people: reflections on a participatory narrative research approach. Int J Inclusive Educ 8(4):331–351

Greenwood DA, Hougham RJ (2015) Mitigation and adaptation: critical perspectives toward digital technologies in place-conscious environmental education. Policy Futures Educ 13(1):97–116

Gruenewald DA (2003a) Foundations of place: a multidisciplinary framework for place-conscious education. Am Educ Res J 40(3):619–654

Gruenewald DA (2003b) The best of both worlds: a critical pedagogy of place. Educ Res 32(4):3–12

Hamm C (2015) Walking with place: storying reconciliation pedagogies in early childhood education. Can Child 40(2):56

Haraway D (2004) The haraway reader. Routledge, New York, NY

Haraway D (2008) When species meet. University of Minnesota Press, Minneapolis, MN

Haraway DJ (2016) Staying with the trouble: making kin in the Chthulucene. Duke University Press

Harvard Family Research Project (2007) Findings from HFRP's study of predictors of participation in out-of-school time activities: fact sheet. Retrieved from http://www.hfrp.org/content/download/1072/48575/file/findings_predictor_OSTfactsheet.pdf

Higgins P (2009) Into the big wide world: sustainable experiential education for the 21st century. J Experiential Educ 32(1):44–60

Ho L-C, Seow T (2015) Teaching controversial issues in geography: climate change education in Singaporean schools. Theor Res Soc Educ 43(3):314–344. https://doi.org/10.1080/00933104.2015.1064842

Hung CC (2014) Climate change education: knowing, doing and being. Routledge, New York, NY. https://doi.org/10.4324/9781315774923

Istead L, Shapiro B (2014) Recognizing the child as knowledgeable other: intergenerational learning research to consider child-to-adult influence on parent and family eco-knowledge. J Res Childhood Educ 28(1):115–127

James A, Prout A (2015) Constructing and reconstructing childhood. Routledge, New York, NY

Kellett M, Forrest (aged ten) R, Dent (aged ten) N, Ward (aged ten) S (2004) Just teach us the skills please, we'll do the rest?: empowering ten-year-olds as active researchers. Child Soc 18(5):329–343

Kola-Olusanya A (2005) Free-choice environmental education: understanding where children learn outside of school. Environ Educ Res 11(3):297–307

Kopnina H (2014) Revisiting education for sustainable development (ESD): examining anthropocentric bias through the transition of environmental education to ESD. Sustain Dev 22(2):73–83

Kraftl P (2015) Alter-childhoods: biopolitics and childhoods in alternative education spaces. Ann Assoc Am Geogr 105(1):219–237. https://doi.org/10.1080/00045608.2014.962969

Latour B (1993) We have never been modern. Harvard University Press, Cambridge, MA

Latour B (2004) Politics of nature: how to bring the sciences into democracy. Harvard University Press, Cambridge, MA

Latour B (2011) Politics of nature: east and west perspectives. Ethics Glob Polit 4(1):71–80

Nxumalo F (2015) Forest stories: restorying encounters with "natural" places in early childhood education. In: Pacini-Ketchabaw V, Taylor A (eds) Unsettling the colonial places and spaces of early childhood education. Routledge, New York, NY, pp 31–52

O'Malley S (2015) The relationship between children's perceptions of the natural environment and solving environmental problems. Policy Pract Dev Educ Rev 21:87–104

Ontario Ministry of Education (2010–2011) The full-day early learning–Kindergarten program (draft version). Queen's Printer, Toronto, ON

Pacini-Ketchabaw V (2013) Frictions in forest pedagogies: common worlds in settler colonial spaces. Glob Stud Childhood 3(4):355–365

Pacini-Ketchabaw V, Taylor A (2015) Unsettling the colonial places and spaces of early childhood education. Routledge, New York, NY

Pacini-Ketchabaw V, Taylor A, Blaise M (2016) Decentring the human in multispecies ethnographies. In: Taylor C, Hughes C (eds) Posthuman research practices in education. Palgrave MacMillan, London, UK, pp 149–167

Rautio P (2013) Being nature: interspecies articulation as a species-specific practice of relating to environment. Environ Educ Res 19(4):445–457

Reis J, Ballinger RC (2018) Creating a climate for learning-experiences of educating existing and future decision-makers about climate change. Mar Policy. https://doi.org/10.1016/j.marpol.2018.07.007

Savransky M, Stengers I (2018) Relearning the art of paying attention: a conversation. SubStance 47(1):130–145

Schweizer S, Davis S, Thompson JL (2013) Changing the conversation about climate change: a theoretical framework for place-based climate change engagement. Environ Commun J Nat Cult 7(1):42–62

Smith LT, Tuck E, Yang KW (2019) Indigenous and decolonizing studies in education: mapping the long view. Routledge, New York, NY

St. Pierre E (2013) The appearance of data. Cult Stud Crit Methodol 13(4):223–227

Stengers I (2012) Cosmopolitics: learning to think with sciences, peoples and natures. Presented at the situating science knowledge. St. Marys, Halifax, Canada

Stengers I (2013) Matters of cosmopolitics: on the Provocations of Gaïa. (in conversation with Heather Davis and Etienne Turpin). In: Architecture in the anthropocene: encounters among design, deep time, science and philosophy. Open Humanities Press, London, UK, pp 171–182

Sweeney J (2015) Climate change and development education: new opportunities for partnership. Policy Pract Dev Educ Rev 20:11–30

Taylor A (2011) Reconceptualizing the 'nature' of childhood. Childhood 18(4):420–433

Taylor A (2014) Situated and entangled childhoods: imagining and materializing children's common world relations. In: Bloch M, Swadener B, Cannella G (eds) Reconceptualizing early childhood care and education: a reader: critical questions, new imaginaries and social activism. Peter Lang Publishing Inc., New York, NY

Taylor A (2017) Beyond stewardship: common world pedagogies for the Anthropocene. Environ Educ Res 23(10):1448–1461

Taylor A, Blaise M, Giugni M (2013) Haraway's 'bag lady story-telling': relocating childhood and learning within a 'post-human landscape.' Discourse Stud Cult Polit Educ 34(1):48–62

Todd S (2001) "Bringing more than i contain": ethics, curriculum and the pedagogical demand for altered egos. J Curriculum Stud 33(4):431

UNESCO (2017) UNESCO strategy for action on climate change. In; Presented at the UNESCO general conference, Paris, France: UNESCO. Retrieved from https://unesdoc.unesco.org/ark:/48223/pf0000259255

Van Manen M (1997) Researching lived experience. Althouse Press, London, ON

Vaughan C, Gack J, Solorazano H, Ray R (2003) The effect of environmental education on schoolchildren, their parents, and community members: a study of intergenerational and inter-community learning. J Environ Educ 34(3):12–21

Wandersee JH, Schussler EE (1999) Preventing plant blindness. Am Biol Teach 61(2):82–86

Warming H (2011) Getting under their skins? Accessing young children's perspectives through ethnographic fieldwork. Childhood 18(1):39–53

Yahya R, Wood EA (2017) Play as third space between home and school: bridging cultural discourses. J Early Childhood Res 15(3):305–322

Yusoff K (2013) Geologic life: prehistory, climate, futures in the anthropocene. Environ Plann D-Soc Space 31(5):779–795. https://doi.org/10.1068/d11512

Chapter 13
Using a Masters Course to Explore the Challenges and Opportunities of Incorporating Sustainability into a Range of Educational Contexts

Alison Fox, Paula Addison-Pettit, Clare Lee, Kris Stutchbury and Together with Masters Students on the Module EE830: Educating the Next Generation

Abstract Issues in the environment drive societal change (Brown in Seven global issues to watch in 2018, blog post. United Nations Foundation, New York, 2018); and hence challenge the purpose of education. Participants on a Masters course, representing informal and formal education for children aged 4–19, discussed the question "How should the purpose of education change in the light of environmental change?" The course introduces education in the context of the UN Sustainable Development Goals and the UNESCO range of competences for children forming the next generation. Participants were challenged to characterise the knowledge and skills needed by children to evaluate evidence on climate change, as well as identify and respond to local environmental issues. They also reflected on the most challenging aspects of putting Education for Sustainable Development (ESD) into practice and gave examples of integrating sustainability into educational practice. This paper reports an analysis of the responses of students who wanted to contribute, according to six dimensions of ESD (biosphere; spatial; temporal; critical; creative and active). Findings suggest that students are clear that pedagogic changes are needed and they embrace a holistic view of curriculum, pedagogy and assessment, but need support if they are to show the agency they want to support children's understanding of the impact of environmental change and how to find solutions.

Keywords Agency · Education for sustainable development · Competences for sustainability · Education practice · Curriculum change

---

The following Masters students on the module EE830: Educating the Next Generation contributed the data on which this paper is based: Anita Bjelke, Susannah Burrell, Elizabeth Edwards, Ross Griffin, Sofia Idrissi, Charlotte Jukes, Anne Kagoya, Sarah Kemsley, Karen Maddison, Jo McMullen, Llinos Nelson, James Nevin, Manka Rangar, Damian Rayner, Chris Saunders, Melanie Tickle, Emma Wilkinson, Ibrahim I. Yahaya.

A. Fox (✉) · P. Addison-Pettit · C. Lee · K. Stutchbury ·
Together with Masters Students on the Module EE830: Educating the Next Generation
School of Education, Childhood, Youth and Sport, The Open University, Walton Hall, Stuart Hall Building (Level 3), Milton Keynes MK7 6AA, England, UK
e-mail: alison.fox@open.ac.uk

© Springer Nature Switzerland AG 2019
W. Leal Filho and S. L. Hemstock (eds.), *Climate Change and the Role of Education*,
Climate Change Management, https://doi.org/10.1007/978-3-030-32898-6_13

Background and Warrant for the Paper

Climate Change, United Nations Responses and the Role of Education

We live in a world where climate change has been identified as a key, global concern (Brown 2018; Khor 2018; Sawin and Smith 2019). Living sustainably in our world, the core of sustainable development can be attributed to the Brundtland Commission's development strategy in the late 1980s, which aimed to:

> meet the needs of the present without compromising the ability of future generations to meet their own needs. (WCED 1987, p. 43)

Education has been placed centrally in United Nations (UNESCO, UNCED, UNECE) strategies ever since, particularly encapsulated in Agenda 21 (UNCED 1992) and Education For All (EFA)—the framework adopted by UNESCO as part of the Millennium Development Goals for the period 2000–15. During this period a shift in thinking meant that the nature and quality of education began to be considered as an undisputable part of planning for sustainability. A UNESCO report (Wade and Parker 2008) asked for EFA principles to be related to those of Education for Sustainable Development (ESD) and informed discussions about, and generation of, the current 2015–30 Sustainable Development Goals (SDGs)—in particular SDG 4 Education, which underpins all the other goals (UNDP 2015).

A diverse range of interpretations of ESD are offered by academics, development organisations and policymakers (including McKeown 2002; Rieckmann 2017; Vare and Scott 2007; Weik et al. 2011). All have in common three overlapping domains of sustainability: the interests of the environment, society and economy. 'The four thrusts of ESD are improving access to quality basic education, reorienting existing education programmes, developing public understanding and awareness, and providing training' (Little and Green 2009, p. 172).

Vare and Scott (2007) propose two forms of ESD are needed:

> ESD1 as the promotion of informed, skilled behaviours and ways of thinking, useful in the short-term where the need for this is clearly identified and agreed, and ESD2 as building capacity to think critically about what experts say and to test ideas, exploring the dilemmas and contradictions inherent in sustainable living. (Vare and Scott 2007, p. 191)

Whilst policymakers prefer the quick-fixes of ESD1, they argue educators should push for something more fundamental and longer-lasting, like ESD2. UNESCO (Rieckmann 2017) published eight key competences needed for sustainable living: systems thinking; anticipatory; normative; strategic; collaboration; critical thinking; self-awareness and integrated problem-solving.

Whilst much of the ESD literature is set in university rather than in primary or secondary school contexts, studies suggest that schools can contribute to this debate. A programme of ESD activities was integrated into the curriculum in 200 German secondary schools between 1999 and 2004 (de Haan 2006). By 2004:

75–80% of all pupils believe they have learned how to think with foresight and to understand complex facts in the context of sustainability, how to work with others as part of an interdisciplinary team on problems of (non-) sustainable development, and how to evaluate various solutions to problems. (de Haan 2006, p. 25)

A position piece by Pearson and Degotardi (2009) also proposes that thinking holistically about human growth and development and nurturing the development of skills for productive contributions to the sustainability of environments, through situated local practices, match well with existing effective early childhood principles and practices.

The global situation therefore poses challenging questions for those involved in planning for and providing education to the next generation: Will all global citizens have access to an education for sustainability? And How will the education they experience help citizens to live in ways which recognise and balance the interests of the environment, society and economy? This paper responds to the second of these questions, in relation to the education of 3–19 year olds, by reflecting on eighteen Masters students' views of how education, given the inevitability of environmental change, might respond to promote sustainable living. Placed as they are in a range of educational settings and roles, they have much to say about the realities of whether and how ESD might offer ways forward.

Impact of ESD for Curriculum, Pedagogy and Assessment

On the Masters module, participants studied work originating from the University of Plymouth Centre for Sustainable Futures. Warwick's (2016a) Butterfly model provides a conceptual framework for thinking about ESD, and the '7 steps to embedding sustainability into student learning' provide practical advice (Winter et al. 2015).

Conceptually, as shown in Fig. 13.1, ESD involves consideration of three relational dimensions (biosphere, spatial and temporal) and three pedagogical dimensions (critical, creative and active learning). The seven steps to implementing such a vision for ESD are:

1. Understand the principles of sustainability and ESD
2. Identify key sustainability issues
3. Develop sustainability literacy and competencies
4. Enhance teaching through sustainability pedagogies
5. Use the campus as a learning resource
6. Link the curriculum and informal learning
7. Become part of the local sustainability community

(Winter et al. 2015).

Drawing on their own experience and knowledge of their settings, students were asked to respond to these ideas, identifying changes needed to curriculum, pedagogy and assessment if ESD is to become a reality.

Fig. 13.1 The education for sustainable development Butterfly model, Figure 1 in Warwick (2016a, p. 106)

ESD focuses on how the purposes for education may be reappraised and reimagined to give young people globally a future in which they can flourish. As captured in Fig. 13.2, through developing relevant knowledge, skills, values and attitudes ESD addresses 'healthy food and ecosystems' to meet environmental agendas and incorporates 'living within our means', 'sustainable development' and living in 'equitable societies'.

Educational responses will depend on the views of sustainable development held by educators and their paradigmatic conceptions of the environment (Sauve 1996). Some alternative conceptions and possible pedagogical responses are summarised in Table 13.1.

All pedagogical responses include the relational themes of collaboration, participation, dialogue and enquiry. Developing such themes means aspiring towards pedagogies of civic compassion, through compassionate conversations, enquiry and action in which everyone works for the benefit of one another (Warwick 2016b), dialogic approaches (Vare 2018) and ESD communities of practice (Wade 2012). Vare (2018) argues that, constructive alignment of educators' curriculum and pedagogical choices (Biggs 2003), would include dialogic assessment approaches. If divergence and local appropriateness is to be valued, such approaches must be developed in consultation with partners and with one another. Underpinning these pedagogies is a view that student voice has a key role to play in ESD.

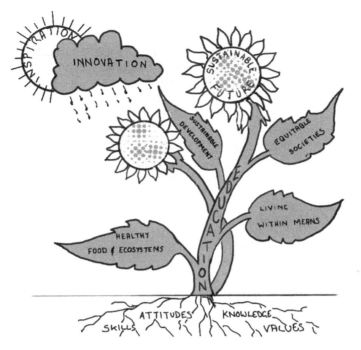

Fig. 13.2 Vision for ESD as represented by the international not-for-profit, non-governmental organisation (Adapted from Education4Sustainability http://www.education4sustainability.org/)

Table 13.1 Conceptions of the environment and possible pedagogical responses

Environment as …	Possible pedagogical responses …
Nature to be appreciated, respected and preserved	Nature exhibitions, immersion in nature
Resource to be managed	3Rs campaigns (replacement, reduction, refinement), audits of energy consumption
A problem to be solved	Problem-solving strategies, case studies
A place to live to know and learn about, to plan for and to take care of	Environmental stories of locations, eco-gardening projects
The biosphere in which we all live together, into the future	Case studies of global issues, storytelling and illustrating different world views
A community project in which to get involved	Integral action-research, environmental issues forums

Adapted from Sauve (1996, p. 13)

Impact of SDGs on Teacher Education

The SDGs provide the content and context for ESD, but repurposing education in the ways outlined above requires educators with appropriate competences (Barth and Rieckmann 2012; Sleurs 2008; Vare 2018; UNECE 2012). It is insufficient simply to expect educators to be able to reimagine and repurpose their practice. Rather, underpinning principles need to be assimilated to be enacted. Vare (2018) proposes a teacher educator framework, with three dimensions—thinking holistically, envisioning change and achieving transformation—enacted through three stages for each—integration, involvement and practice.

This study contributes to a conversation about what educators feel are ways forward, by harnessing their views when debating ESD on an applied postgraduate education programme.

The Evidence-Base for the Paper

Participants on a Masters course, studying informal and formal education for children aged 4–19, discussed the question "How should the purpose of education change in the light of environmental change?" The open course (material accessible to the wider public, not just registered students) *Looking globally: The future of education* introduces the SDGs and UNESCO competences for the next generation in order to problematize accepted views of education. Over three weeks, participants from a wide range of contexts were challenged to characterise the knowledge and skills needed by children to evaluate evidence on climate change, as well as identify and respond to local environmental issues and find solutions. On the associated Masters course, these issues were explored in more depth, with the support of tutors and access to academic publications over six weeks. This included identifying the most challenging step of Winter et al.'s (2015) framework, as well as giving examples of integrating sustainability into educational practice. Posts were extracted for those wishing to be contributors from the open course week 3 activity 1 and 4a/4b course forums, and module activity 3.10 and 3.27 module forums. The data set is formed of 101 posts, with an average of 5.6 posts per contributor and range of 1–27 posts.

This dataset is based on the posts of eighteen students who chose to contribute their forum posts for analysis against two research questions:

- What are the environmental concerns of a group of diversely located educators?
- What are the views of these educators about how education should respond to environmental concerns?

The academic team's approach to the students was approved by the University's student research project panel and the human research ethics committee in November 2018 (HREC/3067/Fox and SRPP/2018/119). This resulted in students becoming invited contributors, respecting their rights to their intellectual property and involving them in how they would be represented. Advice from these panels, which was acted upon, related to ensuring clarity about whether consenting students wanted only first or full names to appear in connection with their data. All those who consented chose for their full name to be included and offered clarification of the wording for which they wished this to appear. Advice was to ensure that the collaborative nature was prominent in the information letter provided and that students had opportunities to ask questions and offer comments both before consenting and before the paper was submitted and published. A summary of the participants' contexts and their roles forms Table 13.2.

The data were analysed by three researchers, each focusing on one of the following foci: key environmental concerns, the relational dimension (initially using the three themes from the Butterfly model, Warwick 2016a) and the pedagogical dimension (again using Warwick's three themes). Each researcher identified discrete parts of contributed posts which could be coded for each theme against emergent identified subthemes. Non-mutually inclusive coding was used so that a coded extract could relate to multiple themes or even subthemes. Inductive coding was also used under the broad foci to generate additional sub-themes. The unit of analysis is referred to as a 'coded extract' in the subsequent sections. Researchers came together, with an additional researcher, to review the coding and contribute to the synthesis.

Table 13.2 Participant contexts, on which responses are drawn (numbers of participants): total n = 18

National contexts			Educational contexts (and age range of learners)		Educational roles	
UK	England	12[a]	Primary state school (3–11 or 5–11 years)	7[b, c]	Classroom teacher	11[b]
	Wales	2	Primary private school (4–16 years)	1	Supply teacher and private tutor	1
	Scotland	0	Home education (11 years)	1[b]	Teaching support roles (classroom, special education needs and ICT)	3
	Northern Ireland	1	Secondary state school (11–19 years)	6	Home educator	1[b]
Oman		1	Private secondary school (11–18 years)	1	Senior school leader	3[d]
China		1	Special educational needs setting (11–16 years)	1	Examiner	1[d]
Peru		1	College (14–19 years)	2		
India		1[a]				

[a]One participant represented both Indian and English educational context perspectives
[b]One participant represented both home education and state primary school perspectives
[c]One participant worked in a Junior school (7–11 year) setting
[d]One participant represented both senior leader and examiner roles

Main Findings of the Paper

Introduction Participants Concerns

The local environmental concerns raised by the participants are summarised in Table 13.3.

A variety of issues were at the forefront of the participants minds, as may be expected from the global reach of the participants. Air pollution was raised globally, as was recycling plastic, thus preventing damage caused to the ecosystem by plastic waste.

It may be expected that participants reported feeling education was positioned to effect positive change within young people for local and global benefit. As the participants are studying an applied Masters course, they are, of necessity, interested and involved to some extent in the education of 4–19 year olds. What was more surprising, was the sense of powerlessness several participants expressed, in relation to teaching about and mitigating environmental challenges. The analysis reported a perceived lack of agency to affect change in their institution's or context's environmental response. This was shown through a distance from the issues expressed by use of pronouns such as "they" rather than "we" in many responses, such as; *"how can local education systems respond to wider, global environmental challenges? They can respond by teaching pupils that we are all part of one planet."* There were clear statements that any move to change depended on someone else's role, as in;

> "Awareness of constraints needs to be handled by government bodies and bigger campaigns need to happen"

> "Surely the Government should be promoting and encouraging Head Teachers to ensure that pupils know what will happen" and

> "… teachers are not generally trained in or particularly aware of the importance of ESD and often feel too constrained by pressures on other areas of learning to research and trial this."

The need for more support was commented on:

Table 13.3 The range of local environmental concerns raised by the participants

Environmental concern	Number of coded extracts
Clean water	1
Air pollution	6
Environmental damage (quarrying, fracking)	3
Environmental damage caused by eating meat/intensive farming	2
Recycling/plastics and the damage they cause in the environment	3
Renewable energy	1

"Until educators are aware of their role in ESD, they will not be able to successfully support children in developing their own skills, knowledge and attitudes. I think that educators are not being pro-active in seeking out sustainability literacy and a national resource making it clear how to develop these skills in both practitioners and pupils is the first step" whilst warning that "… people in authority that make these decisions would be very resistant to this change."

Alongside the statements above were expressions that positive environmental changes may not be made for their own sake:

"Certainly providing families and households with financial incentives to make the right environmental choice would be hugely popular and successful" and "People still believe that the system has to give them the possibility of economic growth and it is the 'system' which should do something about it."

Set against these statements of lack of agency were statements from other participants who saw the local and global as intertwined, change in one could not be made without change to the other. Participants stated clear actions that they saw as the responsibility of schools, such as designing a class project about local quarrying proposals, involving field trips and debates. These, they felt, could help young people fully understand possible consequences and hold and act on reasoned views themselves. Tackling air pollution and climate change through "turn-off zones", where car engines were not left idling outside school and "walk to school" initiatives were seen by some as a readily actionable and a necessary part of educating young people to act sustainably. The potential for technology to contribute was also mentioned as part of these initial visions for education for a sustainable future. "*Technology now allows us to share more easily with schools around the world, exploring issues in a real global context*" and "*By being part of these global campaigns, children learn that they can have an impact globally.*" The role of media, such as television programmes and videos, was considered important. The impact of the recent BBC Blue Planet II programmes on how quickly plastic bag use has been reduced in the UK was used as an example.

Who financed initiatives was discussed as part of these posts. Whilst it was considered possible for schools to model the use of renewable energy, the funding of start-up costs was acknowledged as problematic; "*Education provision is run like a business with budgetary constraints meaning freedom of choice is not always viable*". Money would also be needed for increasing understanding of sustainable education in the workforce;

… to achieve this requires training and investment to ensure that teachers are able to deliver the content required.

Despite many of the participants feeling more than willing to advocate changes towards education which would enable young people to live sustainably, one participant warned of the potential disempowering effect of some curricula; "*in an already overcrowded curriculum, finding time, resources and expertise to link aspects of a somewhat prescriptive curriculum to initiatives embracing the challenges of ESD could be a challenge.*" Another important message was given by one participant "*it* seems sometimes that the issue is too great to deal with on a personal and local level." Where what needs to be done seems to be beyond the agency that lies within the hands of enthusiastic and caring educators such as those that make up the participants in this study, where is support to be found?

Relational Dimensions

106 coded extracts were identified representing points which could be attributed to biosphere (n = 30), spatial (n = 46) and temporal (n = 30) aspects of the relational dimension to ESD.

The Biosphere

The biosphere dimension to Warwick's (2016a) model relates to connections between humans and their environment to consider the wellbeing of both. In concrete terms, seven active projects were mentioned engaging children with their environment in ways which linked with using the 'campus', as associated with their settings involving 4–19 age children, as a learning resource. Three extracts listed media such as television, films and news headlines as having the power to inspire engagement with the environment through seeing and hearing, rather than reading about ecological issues.

Students were aware of the barriers to fully being able to recognise the relational dimension to ESD in their current contexts. Firstly, local projects were considered too small to contribute significantly to the bigger issues (three extracts). Secondly, it was considered that settings had developed an inertia which acted against progress (two extracts). For example:

> I would like my institution to respond by moving away from paper-based materials and become completely electronic. However, the pen and paper is an emblem of education, so moving away from wasting excessive materials will be a difficult step.

Thirdly, and as picked up in Sect. 3.3, ESD was reported as usually not extending beyond the informal curriculum of a setting (two extracts).

Spatiality

Connection to the environment has a spatial dimension: the ways in which people are connected with one another within localities, within nations and across national borders. Twelve active projects linking local with global issues were cited, exhibiting the kind of 'glocalism' (Globus et Locus 2019) advocated as a concrete way of showing activism. Eight extracts referred to learners studying their own cultures and environment. One particularly powerful contribution referred to some work with local and global benefits:

> Frustrated by the fact that I remembered learning about [deforestation] myself as a child, I wanted to ensure that what we did would have a real impact. We therefore joined the Survival Campaign and learned how they are protecting indigenous people who in turn protect the forest. We wrote to the minister in Brazil and were then delighted to learn that, having been part of a global community, our pressure had forced action to be taken to prosecute loggers and protect a specific tribe.

Two extracts spoke of relevant guidance being offered from the government to local schools, which, although little represented, offered examples of external support for ESD. However, students also explained their view that local projects are insufficient and that bigger campaigns are needed, handled by governments (two extracts).

Considerations of spatiality and relationships in and as part of the biosphere, raise questions about who should be involved with ESD? Who are the stakeholders? Should this involve everyone? We all, young and old, here or there, have a 'stake' in the future and therefore in education for the future of our planet. Students identified businesses, non-governmental organisations and those with formal responsibilities for education as agents who could support ESD. One example was:

> A recent session chosen by the children was to clear and prepare ground for wildflower seed planting. Two children in the group had won the money through a national enterprise competition. This was a fabulous example of combining business and ESD.

Temporality

Exploration of what can be achieved now as well as for the future raises the notion of temporality. Arguably thinking in terms of time, requires considering the role of historical analysis and learning from the past, from previous experiences (of the current population and of their forebears) (Pigozzi 2003; Wade 2012).

Most of the contributions which could be considered temporal focused on the future, rather than the past and were couched as recommendations, rather than examples of citable practice. Notably however, two extracts, spoke of observing the sustained environmental activity of their learners beyond specific school-initiated projects they had been involved in, for example:

> We were encouraged this year to see that students were still bringing in recyclable waste. Even though the project was no longer officially running, they were still involved and they continued to recycle without being prompted.

These evidence changes of behaviour towards the aspired sense of sustainability. A further contributor drew our attention to the present imperative:

> These are issues which we need to be address (urgently in some cases). The answers may become apparent within our future generations, but the important of the issues need to be placed high on the [current] school agenda.

On the course, participants had studied three possible purposes for education (the human capital, the rights-based and the capabilities models (Robeyns 2006). Some concluded these helped provide a rationale for a focus on ESD, for example:

> In my opinion, these thoughts are closely related to individual choices, and freedoms, and the intrinsic value of Education to bring about changes that have long term individual, local and global benefits. These are all characteristics of the capabilities model

Others reflected on how they had come to conclude that the current dominant human capital model was inertial to ESD, and therefore educators must re-evaluate values and challenge assumptions about the current educational system in which they are situated (two extracts).

Pedagogical Dimensions

Analysis of pedagogical considerations started with those suggested in the Butterfly model (Warwick 2016a) to help address research question 2: What are the views of educators about how education should respond to environmental concerns? Responses to two activities were considered:

- In what ways do you think education can play a role in forming a collective global response to a sustainable future?
- Post a comment on the forum about an experience in your own context which demonstrates one of the suggested steps for putting ESD into practice.

Together, these yielded 52 coded extracts from the original posts and discussions that ensued. Whilst participants' contributions did not clearly separate creative, critical and active learning, their responses fell into a number of categories, which embraced pedagogies related to formal, informal and hidden curricula.

If 'pedagogy' is taken as a set of values and beliefs about learners and learning, rather than simply what teachers do in the classroom (Moon and Leach 2008), then it is entirely appropriate to analyse the data by taking a holistic view of the school curriculum expressed as the values and beliefs held by teachers in relation to both informal and formal curricula. The categories identified are in Table 13.4.

The formal and informal curricula were fairly evenly represented (23 and 26 coded extracts respectively). Whereas the examples provided to illustrate how the informal curriculum supports ESD were detailed and often included some critique, those which referred to the formal curriculum were less clear and were often represented as aspirations about what should happen, rather than specific examples. This may be a further outplaying of the perceived lack of agency reported earlier in the responses to question 1. Educators could recognise and exert their agency in the informal curriculum, rather than in the formal one.

Many of the students described whole school projects (most of which were part of the informal curriculum) designed to raise awareness and make a practical contribution. For example:

Table 13.4 The range of ideas proposed by participants for progressing pedagogy through the curriculum

Ideas as to how education can support an ESD agenda	Number of coded extracts
Through the formal curriculum through subjects, and cross-discipline activities including off-timetable days	12
Through the formal curriculum: curriculum pedagogy including active learning approaches and experience of problem-solving, critical thinking	11
Taking part in externally sponsored projects	2
Through the informal curriculum including extra-curricular activities, school councils, assemblies and whole school projects	26
Collaboration with groups external to the school	2

> Children from the newly formed eco-council walked around school and recorded the number and types of taps available. They then asked each class how often they found taps left on and decided to create a competition to create posters reminding the other children to turn off the taps.
>
> Whole school assemblies provide a chance to discuss global issues and steps we can take to achieve greater sustainability.
>
> … to teach students how to tend for and look after plants. They would then sell them every few weeks to raise money in order for the next class to grow their own.

There was also a recognition that appropriate pedagogies (as part of the formal curriculum) were likely to have a significant impact on children's understanding of the issues surrounding sustainable development.

> … there is a lot of focus on collaborative working; a recent new initiative at our school has involved using the Magenta principles which require a lot more group work.
>
> Our goal as educators isn't just to fill our students to the brim with knowledge, but to get them to actively question the world around them.

Some subjects from the formal curriculum were considered to support ESD, especially when environmental issues were viewed as being a problem to solve (Sauve 1996).

> Children will also require the opportunity to learn and practise the knowledge and skills needed to tackle these issues; a solid understanding of science, geography, maths, computing and languages.

Looking across the responses, there is a considerable amount of pessimism about being able to integrate the sort of active pedagogies compatible with ESD into everyday teaching. The relentless focus on examination results and what is perceived to be an over-crowded curriculum in the educational contexts represented by these educators were reported as issues which prevent teachers embracing ESD. This was a further manifestation of the lack of agency identified in Sect. 3.1.

> I'm afraid that the UK curriculum is more geared towards academic attainment and more exam-oriented; the schools are equally motivated and gravitated towards the results and academic performance.

... exam preparation takes precedence over everything else, to the point of squeezing projects like this out of the curriculum.

There was also some concern that teachers do not have the required skills in this area and recognition that training is required.

... incorporating more global awareness and PSHE into the curriculum is not overly burdensome providing teachers have sufficient training can take place.

... teachers are not generally trained in or particularly aware of the importance of ESD and often feel too constrained by pressures on other areas of learning to research ...

Linking back to the notion of temporality, highly correlated with notions of sustainability, students brought a critical lens to bear on the ephemerality of many pedagogical interventions, illustrated as:

All too often we tend to focus on something like a sustainability project for a short period in order to 'tick a box'. We need to learn to embed it so that it becomes second nature.

Yes, there is no use in implementing a project for a few weeks, then going back to all the bad habits we used to have.

Discussion

The evidence in this paper has addressed the research questions, What are the environmental concerns of a group of diversely located educators? and What are the views of these educators about how education should respond to environmental concerns? This revealed the range of environmental concerns held by this group of 18 educators who represented diverse locations, educational settings and roles; and a diverse range of suggested responses. All were able to cite current educational practices which were aimed at addressing these concerns and all had views about how education could and should contribute to a more sustainable way of living in the world.

The voice of the participants covers a range of views of environmental issues underpinning the pedagogical approaches which they selected to evidence (see Table 13.1). Whilst some referred to auditing water quality and use (Sects. 3.1 and 3.3), which might view the environment as something to be managed, others cited eco-gardening projects embodying the notion of the world as somewhere to live in and to care for (Sect. 3.2). Still others referred to collaborative projects which appear to have benefitted from technological developments which make it easier to connect local activity with global campaigns (Sects. 3.2 and 3.3).

However, from this analysis of course participant contributions, ESD still appears to be an add-on to the remit of the educational settings represented in Table 13.2. The analysis was able to go beyond a focus on the appropriate pedagogical approaches presented in Warwick's Butterfly model (2016a) to locate many of the challenges to education contributing more fundamentally to ESD as linked firstly, to its place in the curriculum and, secondly, in systemic limitations of the agency of teachers to make changes to these curricula. Both points raise wider questions for our educational systems as to how ESD might become embedded and enacted.

The idea of a teacher as an active agent within a learning context has been described by various writers as a willingness and capacity to act according to professional values and beliefs when dealing with the various situations that teachers face both within and beyond their classrooms (Sloan 2006; Turnbull 2005). The commitment and potential for creative, critical and active pedagogical practices was evidenced by this group of educators but, where they have only limited scope to influence the formal curriculum, their activism is limited to informal curriculum-related actions and behaviours. These involve one-off projects, individuals championing particular initiatives, school clubs and trips and taking up opportunities to engage only when they present themselves (It is worth noting that the home educator represented within the contributors was much more able to be proactive in determining the pedagogical environment, including its curriculum, for learners in their setting). Whilst there is a potential power to such behaviours if they come to form the basis of a hidden curriculum underpinning the formal, there was little evidence that sustainable values were underpinning the work of the educational settings represented.

The hidden curriculum refers to the norms, values, attitudes and beliefs that are learned within schools through the broader social environment. Messages that are part of the hidden curriculum are often conveyed through for example relationships between teachers and learners, and learners and learners, how behaviour is managed and the assessment system. Messages such as organising recycling, litter picking, encouraging walking to school or having policies for "turn-off zones", all represented in this paper, promote environmental awareness as normative expectations. However, these messages are not necessarily explicit to educators, as some participants expressed surprise to find that their analysis of practice placed them already at the forefront of environmental pedagogies:

> Examples that have inspired me so far are analysing water use on the school campus, recycling resources, clearing rubbish and growing plants. Most of these we were already doing without realising all its significance!

It is difficult to think about the pedagogical implications of ESD for education without thinking about the curriculum. A focus on curriculum, requires questions to be asked concerning teachers' agency in curriculum development as well as delivery. If teachers' agency is to effect fundamental change, the educators reflected that much more support would be needed to equip teachers with the skills and understanding to make confident contributions. This also raises questions about who else should be involved in agenda setting through curriculum design and the wide range of stakeholders who have a role to play in the future of the planet.

Conclusions and Recommendations

This paper has demonstrated how it is useful to apply frameworks for thinking about the need for and practice of ESD. This approach supported educators to reflect on their beliefs around, commitment to and realisation of their existing engagement with ESD. Frameworks can be applied from setting to setting. Warwick (2016a) and Winter et al.'s (2015) frameworks were designed for a Higher Education audience, yet had utility in transferring to a wide range of school-age settings. The consensus of these participants is that ESD needs to become 'good practice' as part of an intended 'hidden' or overt curriculum, rather than as an extra (Pearson and Degotardi 2009). They consider ESD should be integrated into all aspects of the curriculum. Whilst the Butterfly model (Warwick 2016a) offers helpful views of relational and pedagogical principles, a broader view of 'pedagogy', made it possible to reveal the importance of curriculum.

Educators are calling for more support, training and leadership across the system to help them in overcoming feelings of powerlessness, futility or limited gains from their actions. It is hoped that the higher-level courses educators engage in, such as the one represented in this paper, offer chances not only to develop the criticality central to Masters level study, but also to support activism beyond an educator's study of the programme. Applied Masters study is intended to offer sustainable outcomes (e.g. Ion and Iucu 2016). There is evidence in this paper that students had changed their thinking as a result of exposure to these ideas and begun to appreciate the value of these kinds of reflection. This reflection is absent from students' accounts as part of their practice as professionals. This is an issue which should be addressed if they are to lead 'a quality education [which] understands the past, is relevant to the present and has a view to the future' (Pigozzi 2003 cited in Wade and Parker 2008, p. 149).

Recommendations can be made from this analysis focusing on advocating the values of ESD, enabling educators to show their agency to develop pedagogical practices and allowing stakeholders to work together for a sustainable world for all.

Drawing on the participants' comments, at a national or regional scale there needs to be:

- a political agenda into which educational ESD reform is set, one which recognises collective responses as the approach to meeting global challenges
- a change in the mindset of the commercial sector to value ESD
- schools that work in partnership across sectors towards ESD agendas
- valuing actions beyond the immediate and short-term to collectively work towards more sustainable changes in behaviour
- support for educators to enable them, and the learners they support, to demonstrate sufficient autonomy to enact their values.

The curriculum enacted in the learning environments for our young people needs to support:

- activation of personal responsibilities, beyond simply awareness-raising

- the development of competence in making 'good' choices as responsible capitalist consumers
- an awareness of how thinking of others leads to demonstrating 'civic compassion'
- an appreciation that global impact can be realised from local actions, both positively and negatively
- the appetite for civic activism as future citizens prepared to use their democratic rights

The authors recognise the partiality of the data reported and indeed the inferences drawn. However, they are authentic and have allowed a voice for a group of educators who otherwise would have not been able to publicly engage with this important debate about 'the challenges and uncertainty of climate change and the role of education in developing solutions'.

References

Barth M, Rieckmann M (2012) Academic staff development as a catalyst for curriculum change towards education for sustainable development: an output perspective. J Clean Prod 26(1):28–36

Biggs J (2003) Aligning teaching and assessment to curriculum objectives. Learning and Teaching Support Network, York. Retrieved from https://www.heacademy.ac.uk/system/files/biggs-aligning-teaching-and-assessment.pdf

Brown M (2018). Seven global issues to watch in 2018, blog post, 4 Jan 2018. United Nations Foundation, New York. Retrieved from https://unfoundation.org/blog/post/7-global-issues-watch-2018/

De Haan G (2006) The BLK '21' programme in Germany: a 'Gestaltungskompetenz'-based model for education for sustainable development. Environ Educ Res 12(1):19–32

Globus et Locus (2019) What is glocalism? New mobilities, new languages and new ways of aggregation: this is the result of a dramatic change provoked by technological innovation. Retrieved from http://www.globusetlocuseng.org/About_Us/Glocalism/What_Is_Glocalism.kl

Ion G, Iucu R (2016) The impact of postgraduate studies on the teachers' practice. Eur J Teach Educ 39(5):602–615

Khor M (2018) Critical issues to watch in 2018, blog post from Penang, Malaysia, 2 Jan 2018. Retrieved from http://www.globalissues.org/news/2018/01/02/23830

Little AW, Green A (2009) Successful globalisation, education and sustainable development. Int J Educ Dev 29(2):166–174

McKeown R (2002) Education for sustainable development toolkit v2. Center for Geography and Environmental Education, University of Tennessee, Knoxville, TN

Moon B, Leach J (2008) The power of pedagogy. Sage, London

Pearson E, Degotardi S (2009) Education for sustainable development in early childhood education: a global solution to local concerns? Int J Early Child 41(2):97–111

Pigozzi MJ (2003) Reorienting education in support of sustainable development through a focus on quality education for all. Paper presented at the GEA conference, Tokyo, Japan, 25 Oct. UNESCO, Paris

Rieckmann M (2017) Education for sustainable development goals: learning objectives. UNESCO, Paris. Retrieved from https://unesdoc.unesco.org/ark:/48223/pf0000247444?posInSet=3&queryId=d0626201-9a8e-4919-88c8-f322a300fa49

Robeyns I (2006) Three models of education. Theory Res Educ 4(1):69–84

Sauve L (1996) Environmental education and sustainable development: a further appraisal. Can J Environ Educ 1(1):7–34

Sawin E, Smith N (2019) New thinking for a new year: 2018 has been full of grim climate news. 2019 could be the year cities turn this around. Commentary 31 Dec 2018 US news: a world report. Retrieved from https://www.usnews.com/news/cities/articles/2018-12-31/2019-requires-a-new-approach-to-climate-change-leadership-in-cities

Sleurs W (ed) (2008) Competencies for ESD (education for sustainable development) teachers. A framework to integrate ESD in the curriculum of teacher training institutes. United Nations Economic Commission for Europe (UNECE), Brussels. Retrieved from http://platform.ue4sd.eu/downloads/CSCT_Handbook_11_01_08.pdf

Sloan K (2006) Teacher identity and agency in school worlds: beyond the all-good/all-bad discourse on accountability-explicit curriculum policies. Curric Inq 36(2):119–152

Turnbull M (2005) Student teacher professional agency in the practicum. Asia Pac J Teach Educ 33(2):195–208

United Nations Conference on Environment and Development (UNCED) (1992) Agenda 21—program of action for sustainable development: Rio declaration on environment and development, United Nation conference on environment and development, Rio de Janeiro, Brazil. UNCED, New York. Retrieved from https://sustainabledevelopment.un.org/content/documents/Agenda21.pdf

United Nations Development Programme (UNDP) (2015) Sustainable development goals. UNDP, New York. Retrieved from http://www.undp.org/content/dam/undp/library/corporate/brochure/SDGs_Booklet_Web_En.pdf

United Nations Economic Commission for Europe (UNECE) (2012) Learning for the future: competences in education for sustainable development. UNECE, Geneva. Retrieved from: https://www.unece.org/fileadmin/DAM/env/esd/ESD_Publications/Competences_Publication.pdf

Vare P (2018) A rounder sense of purpose: developing and assessing competences for educators of sustainable development. Form@re Open Journal per la formazione in rete 18(2):164–173

Vare P, Scott W (2007) Learning for a change: exploring the relationship between education and sustainable development. J Educ Sustain Dev 1(2):191–198

Wade R (2012) Pedagogy, people and places. J Teach Educ Sustain 14(2):147–167

Wade R, Parker J (2008) EFA-ESD dialogue: educating for a sustainable world. UNESCO, Paris. Retrieved from https://unesdoc.unesco.org/ark:/48223/pf0000178044

Warwick P (2016a) An integrated leadership model for leading education for sustainability in higher education and the vital role of students as change agents. Manag Educ 30(3):105–111

Warwick P (2016b) Education for sustainable development: a movement towards pedagogies of civic compassion. Forum 58(3):407–414. Retrieved from http://dx.doi.org/10.15730/forum.2016.58.3.407

Weik A, Withycombe L, Redman CL (2011) Key competencies in sustainability: a reference framework for academic program development. Sustain Sci 6(2):203–218

Winter J, Sterling S, Cotton D (2015) 7 steps to embedding sustainability into student learning. Educational Development, Plymouth University, Plymouth

World Commission on Environment and Development (WCED) (1987) Our common future. Oxford University Press, Oxford

Chapter 14
Capacity Building Itinerary on Sustainable Energy Solutions for Islands and Territories at Risk for the Effects of Climate Change

Lara de Diego, María Luisa Marco and Mirian Bravo

Abstract This paper presents the "Online Capacity Building and Certification Program on Sustainable Energy Solutions for Islands and Territories in the Pacific, Caribbean, Africa and Indian Ocean", an itinerary to foster access to a high-quality technical scientific knowledge. This program has been developed by CIEMAT, the Spanish Center for Research in Energy, Environment and Technology as an initiative of the Global Network for Regional Sustainable Energy Centers supported by the United Nations Industrial Development Organization within the Sustainable Energy Island and Climate Resilience Initiative. The program has been conceived to meet the urgent need for affordable training and certification programs in sustainable energy on islands. The training of qualified professionals along the complete value chain of the sustainable energy sector is crucial for the deployment of sustainable energy and low-carbon climate technologies. This can be achieved by fostering the development of indigenous and renewable energy solutions to address economical and industrial productivity and competitiveness, energy security, affordable energy access and negative externalities of conventional energy (greenhouse gases emissions, local pollution). This educational and training project is aimed at meeting the general and specialized training needs required to establish a critical mass of personnel with broad skill levels. This includes persons from the general public to the public and private stakeholders in the sustainable energy sector (experts, engineers, project managers and financers, policy makers, etc.) capable of identifying, designing and implementing effective sustainable energy, climate change issues and disaster risk management measures.

Keywords Capacity building · Sustainable energy development · Small Island Developing States (SIDS) · Renewable energies · Climate change resilience

---

L. de Diego (✉) · M. L. Marco · M. Bravo
Centro de Investigaciones Energéticas, Medioambientales y Tecnológicas, CIEMAT, Avda. Complutense 40, 28040 Madrid, Spain
e-mail: lara.dediego@ciemat.es

© Springer Nature Switzerland AG 2019
W. Leal Filho and S. L. Hemstock (eds.), *Climate Change and the Role of Education*, Climate Change Management, https://doi.org/10.1007/978-3-030-32898-6_14

Introduction

Small Island Developing States (SIDS), i.e., territories in the Pacific, Caribbean, Africa and Indian Oceans, target regions of the present work, face particular critical challenges associated with the generation, distribution and use of energy. They also are especially vulnerable to climate change, natural disasters and other external impacts, e.g. hurricanes or earthquakes. Their vulnerability is exacerbated if their capacity to adapt or their ecosystem services deteriorate (IMF 2016). The energy paradigm of SIDS is mainly conditioned to their external dependence on costly imports of fossil fuels for energy production. This dependence harms the environment, contributes to climate change, affects these island's financial budgets, thereby perpetuating poverty and inequality, swelling national debt, hampering investments, and blocking sustainable socio-economic development (Dornan 2015).

At present, energy remains a major constraint to the sustainable economic growth and development of many SIDS. In contrast, indigenous renewable energy (RES) resources and energy efficiency (EE) measures can reduce their dependence on imports while simultaneously stimulating an environment for the creation of local businesses and hence providing employment opportunities. Moreover, islands are the best scenarios to demonstrate that isolated communities can meet 100% of their energy demand without greenhouse gas emissions (Couture and Leidreiter 2014). It is therefore necessary that SIDS develop alternative RES resources.

Building the momentum for the transition of SIDS to a blue-green economy requires a substantial redirection of investment, this being needed to promote the development of renewable technologies and sustainable energy solutions (UNEP 2014b). In particular, technical know-how and knowledge to implement this investment is needed for successful incorporation of renewable power options onto insular power grids. A variety of skills are needed to plan, finance, manage, operate and maintain a power grid effectively, safely, reliably and economically (IRENA 2014). From the perspective of regional climate and energy challenges, capacity building and training programs are crucial to train qualified professionals along the full value chain of the sustainable energy sector. This will enable the creation of a critical mass of skilled persons capable of driving the development of sustainable energy solutions (UNESCO 2014).

Renewable technologies, energy efficient systems, measures for climate change mitigation and resilience, mature and favorable policies, as well as appropriate energy and innovation markets all need to be strengthened in order to achieve both local and regional energy targets as we move towards the urgent transition to sustainable energy models and carbon free economies. The growth of sustainable energy markets, industries and innovation in small island countries and territories is highly dependent on the strengthening of domestic capacities (UNIDO and GN-SEC 2018). However, due to their geographic isolation, small size and limited resources, their access to capacity-building resources is limited.

Nonetheless, the promotion and boosting of the development capacities of the next generation of tradespersons to learn about climate change adaptation techniques and sustainable energy technologies, as well as their application, is a framework for action that will affect millions of people living on the frontlines of climate change around the world. A primary aspect is to also ensure the integration of gender equality issues and promote the empowerment of women to help accelerate sustainable development, economic growth and development worldwide (UN-Women 2014). An increased participation of women is needed. This is crucial to accelerate sustainable development, as well as to boost innovation and technology. This also converges with the international appeal of the United Nations Sustainable Development Goals (SDGs) to foster mechanisms that increase effective planning and management capacity in relation to climate change in the least developed countries and SIDS, focusing in particular on women, youth as well as local and marginalized communities (UN 2018).

Following the European Commission's legislative proposals in the Clean Energy for all Europeans Package, islands, more than ever, have a key role in ensuring achievement of its 2030 climate and energy targets. The final aim of this package is to accelerate a transition to carbon neutral generation and supply of energy while continuing to ensure security of supply and access, tackling climate change and helping to address economic/industrial productivity as well as competitiveness and externalities of conventional energy systems (e.g. greenhouse gases emissions, local pollution) simultaneously and in an integrated way (European Commission Website).

In 2014, at the Third International Conference on Small Island Developing States, the SAMOA Pathway (SIDS Accelerated Modalities of Actions) was adopted in order to foster collective action and coordinate United Nations organizations. SAMOA established a framework for an integrated approach to achieving the SDGs in SIDS and a model of assistance required on various issues. These include climate change, sustainable energy, disaster risk reduction plus oceans and seas (UN General Assembly 2014). Several initiatives with specific measures have been put forward to address the common challenges. These can be addressed by the regions' tremendous potential for sustainable energy solutions by progressively up-taking sustainable and clean energy systems that respect the special environment, communities and living conditions on islands.

Addressing increased vulnerability and responding to climate threats must be central to SIDS policies. However, the increased supply and use of RES and EE products and services continues to be hindered by a broad range of barriers and shortcomings. These are related to policy and regulation, fiscal and non-fiscal incentives, technical limitations, economics, finance, capacities, quality infrastructure, research, development and innovation (R&D&I) frameworks, knowledge and awareness (UNEP 2014a). For example, the United Nations "Sustainable Energy for All" initiative emerged to provide a framework for SIDS to undertake expansion of their RES sector. Thus, by 2030, they should increase the deployment, penetration, and efficiencies of renewable sources using existing cost-effective technologies (Mori et al. 2014). In the spirit of the SAMOA Pathway, and with sea-level rise in constant evolution, many SIDS have introduced targets and plans to promote RES and EE deployment as a vital component of their sustainable development and their climate mitigation and resilience efforts.

The Sustainable Energy Island and Climate Resilience Initiative (SIDS-DOCK) was established in 2009 for a similar purpose, i.e., to facilitate the development of a sustainable energy economy within the SIDS. SIDS-DOCK provides a collective institutional mechanism to assist SIDS in transforming their national energy sectors for sustainable economic development and in generating financial resources to address adaptation to climate change (SIDS DOCK Website). In 2015, SIDS DOCK, jointly with the United Nations Industrial Development Organization (UNIDO) and in close coordination with the regional organizations, launched the Global Network of Regional Sustainable Energy Centers (GN-SEC). This partnership assists regional organizations with the creation of sustainable energy centers that focus on accelerating energy and climate transformation in developing countries. This is done by complementing national efforts in the areas of policy and regulation, capacity development, knowledge and data management, awareness raising, as well as the promotion of investment, innovation and entrepreneurship (UNIDO and GN-SEC 2018). Capacity building and skills certification are important areas of the work undertaken by these centers. Without a considerable strengthening of capacities, most small-island developing countries and territories will not achieve their sustainable energy targets in the Intended Nationally Determined Contributions (INDCs) (UNIDO 2017).

Within the framework of GN-SEC, a joint initiative of the ECOWAS Centre for Renewable Energy and Energy Efficiency (ECREEE), the Caribbean Centre for Renewable Energy and Energy Efficiency (CCREEE) and the Pacific Centre for Renewable Energy and Energy Efficiency (PCREEE), CIEMAT has developed the "Online Capacity Building and Certification Program on Sustainable Energy Solutions for Islands and Territories in the Pacific, Caribbean, Africa and Indian Ocean". This has been done to provide the above centers with a training instrument that can strengthen local and regional capacity development throughout the SIDS and territories within the project regions. This Online Capacity Building Program is composed of nine online modules developed using an e-learning methodology and provides a technical approach to each technology and energy issue. The Program aims at meeting the urgent need for affordable training and qualified postgraduate certification programs on sustainable energy. It is expected, through this training program, to contribute to local capacity and to support the growth of R&D&I, and to facilitate "train the trainers" and local know-how in order to implement regional energy roadmaps and foster the development of indigenous and RES solutions. The centers are the owners and direct beneficiaries of the Program. They play a key role in its dissemination, implementation and sustainability at the regional level, together with universities, other public and private centers and institutions in their region that will cooperate to achieve a positive impact of the project.

Given the lack of training courses and educational programs at all levels that focus on renewable technologies and EE (e.g. higher education, vocational training), as well as on economic, environmental and social benefits, the Online Capacity Building Program is a potential candidate to be part of the curricula of existing sustainable energy master programs of universities, specialized energy centers, as well as technical and vocational training institutions, in order to provide accreditation based on a reference qualification framework.

Conception and Development of the "Online Capacity Building Program"

The Online Capacity Building Program has been conceived to contribute to access to high-quality scientific and technical knowledge on sustainable energy solutions for island countries facing the challenge of 100% renewable energy. This is done through the use of a solid program which describes and analyses, using a deep and technical approach, different technologies and energy issues. The program has been developed by fulfilling CIEMAT's quality criteria in terms of scientific and technical expertise, Information and Communication Technologies (ICT) tools, methodological and pedagogical resources.

Project Justification

Within the framework of the GN-SEC and SIDS-DOCK Initiative, to strengthen local and regional capacity development, UNIDO subcontracted CIEMAT in 2017 to develop, test and deliver "Online Capacity Building and Certification Program on Sustainable Energy Solutions for Islands and Territories in the Pacific, Caribbean, Africa and Indian Ocean". The project was conceived from an initial joint initiative of ECREEE, CCREEE and PCREEE. The project completion time has been one and a half years. Since its initiation in 2017, CIEMAT has provided the services requested by UNIDO within the scope of the program's activities and deliverables. For this it has executed the activities in close partnership with UNIDO, SIDS DOCK, CCREEE, PCREEE and ECREEE. The assignment also required further consultations with energy units of the Caribbean Community, Pacific Community, University of the South Pacific, University of Cape Verde and European Union Pacific Technical and Vocational Educational and Training in Sustainable Energy and Climate Change Adaptation project (PacTVET). Future collaborations are planned with the University of West Indies and other regional and international partners. The activity is implemented with financial support of the Spanish and Austrian Governments through the Austrian Development Agency and the Spanish Agency for International Development Cooperation. The project will be completed in spring 2019.

The Online Capacity Building Program has been designed on the basis of regional capacity needs and on their large renewable energy potentials and sustainable solutions deployment. However, it is still in an early stage of development. The selection of the energetic and technological areas to be addressed by the training program was made according to priorities of the Regional Centres and to CIEMAT's expertise in education and training capabilities in energy, environmental and technology matters. The training program is composed of the following nine online modules. It provides a technical review of each energy area and technology, as well as practical examples of applications focused on the target islands regions. Their titles are:

- General Introduction to Island Energy and Climate Change Mitigation and Resilience
- Solar Thermal Systems and Applications for Water Heating and Industrial Process Heating
- Grid-connected and Decentralized Photovoltaic Systems
- Efficient Energy use and Thermal Optimization in Buildings and Industry
- Geographic Information Technologies and Renewable Energy
- Bioenergy. Anaerobic Digestion of Organic Waste to Energy Solutions
- E-mobility
- Minigrids, Grid Stability in Insular Power Systems and Energy Storage
- Ocean Energy

Complying with the requirements of UNIDO and the Regional Centers, the modules have been prepared in Spanish, English and Portuguese, in accordance with the principal languages of project areas.

Objectives of the Project

The main objective of this project is to support regional sustainable energy centers in the implementation of a qualified training and certification program. It is done to respond to the urgent need for affordable training and certification programs on sustainable energy for islands. Its target is to promote sustainable energy solutions that can lead to sustainable energy and industrial development while achieving the SDGs.

The training project also seeks to facilitate access to knowledge and specialized data in order to promote the development and adoption of RES initiatives in the targeted regions. Diversifying the energy mix will increase access to a more secure and sustainable energy supply and will contribute to achieving global energy access and resilience to climate change. Other objectives of the resulting training program include promoting awareness, alerting about opportunities within EE and sustainable energy solutions fields, as well as encouraging R&D&I in the target regions. Finally, strengthening of capacities will help to achieve SIDS' sustainable energy targets in the Intended Nationally Determined Contributions.

Training Methodology

The design of the training program, the level of scientific and technical content and the selection of methodology used has been undertaken in line with the requirements of the regional Centers. These were based on the training requirements needed to meet current and future technology demands and also on the development of such technologies in the regions.

A variety of skills are needed to plan, finance, manage, operate and maintain a power grid effectively, safely, reliably and economically. Technical knowledge is required to train qualified professionals along the full value chain of the sustainable energy sector, thereby enabling the creation of a skilled and critical mass necessary to reach the development of sustainable energy solutions. The uptake of sustainable energy markets, industries and innovation in SIDS and territories is highly dependent on the strengthening of domestic capacities. However, due to their geographic isolation, small size and limited resources, their access to capacity-building resources is limited.

The "Online Capacity Building Program on Sustainable Energy Solutions for Islands" has been developed using e-learning methodology and a self-study modality. This implies complete autonomy for learning and provides the benefits and virtues of online training as a greater scope bringing knowledge to isolated and remote areas. In addition it has been developed at two speeds of learning with two different approaches in order to cover the needs for knowledge of the main stakeholders and target audience. These are a First Speed Option, aimed at anyone interested in having an understanding of each renewable technology, and a Second Speed Option, that provides a specific and specialized approach to each technology and energy issue. The latter focuses on users and stakeholders involved in the energy sector who are seeking to deepen their knowledge on sustainable energies systems and to obtain an insular view of their applications in island regions.

Both options are suitable for delivery through either a self-directed studying or a guided tutoring modality. Training is performed on an e-learning platform with an evaluation system and a final certificate of achievement. These provide an interface to successful presentation of learning contents to learners as well as tracking and evaluation systems. The latter allows assessing participant performance, which is needed for receipt of a certificate of participation. The open source e-learning platform Moodle has been selected. This selection, which has been based on accessibility, interactivity, security and availability criteria, is being widely tested with a broad user community and support worldwide.

Based on a "competency" and "skill-set" approach, the "Online capacity building program on Sustainable Energy Solutions for Islands" bases its structure and objectives on learning outcomes that can be equated to knowledge, skill and competence scopes corresponding to levels higher than 5 in the European Qualifications Framework, which is used here as reference. It starts from a base level of students with a science or technology degree onwards. This validation allows to form part of curricula of existing sustainable energy master programs at universities, specialized energy centers and technical or vocational training institutions as it provides accreditation based on a reference qualification framework (EU PacTVET 2015).

Human Resources

The content development has been carried out by experts from different research departments at CIEMAT, as well as at the Spanish Office of Climate Change, the University of Alcalá de Henares and the Technological Institute of the Canary Islands, all these being R&D&I institutions of excellence in the field of energy, EE and the environment. Material production has been undertaken by a multidisciplinary team composed of experts in e-learning methodologies, management and development capacity building projects in the field of energy, graphic designers as well as audiovisual and multimedia material producers. The online capacity building program has been led by the Knowledge Management and Training Division of CIEMAT. This division has been responsible for the coordination and technical direction of the project as well as for the methodological and didactic development of the capacity building program, the quality control and the e-learning modules final production. Finally, computer support has also been provided by CIEMAT.

Material Resources

The developed e-learning modules contain multimedia packages, one of the principal content pills, which contain the necessary elements (texts, images, animations, and voice-over, etc.) to convey the main contents of the course. Multimedia packages are self-contained web-based contents created following the Sharable Content Object Reference Model (SCORM), which defines a set of standards and specifications for sharing, reusing, importing, and exporting e-learning contents. Such e-learning pills have, as well as other benefits, full accessibility through Web technologies and the ability to be reused in multiple contexts. For audiovisual content processing, animation development, and SCORM packaging, both free and commercial software has been used.

The e-learning modules have been implemented in a Learning Management System (LMS), or virtual learning environment, that enables an effective learning process, student performance monitoring and study in a self-formative modality. This environment must be suitable, simple, easy and functional in order to undertake the learning process: communication tools, tracking tools, repositories, tools for visualizing online material, help tools, etc. For the work presented here, the LMS selected is Moodle, a open source platform. Moodle was chosen as it is a highly developed and tested LMS with a consolidated community of developers and users. Also, it is one of the platforms most widely used by higher education institutions and training centers worldwide. Moodle also presents an excellent platform for resources and communication tools, and its interface is quite intuitive, thus allowing easy navigation. It also has high usability which confers the advantages of being easy to learn and remember; efficient, visually pleasing and fun to use.

Results and Discussion

The outlined actions lead to the birth of the "Online Capacity Building Program on Sustainable Energy Solutions for Islands and Territories in the Pacific, the Caribbean, Africa and Indian Ocean".

1. Covering a total of nine sustainable energy technologies and energy issues to respond to the urgent need for capacity building and competence development of qualified professionals along the value chain of renewable energy and sustainable solutions sector. This will foster the uptake of sustainable energy markets in the target island regions, and will also contribute to achieving their sustainable energy targets in the INDCs and SDGs (Illustration 14.1).
2. The resulting capacity building program has been prepared in English, Spanish and Portuguese, the main languages of the beneficiary regions. Moreover these languages are widely spoken worldwide. This, along with the online format via internet access, maximizes the dissemination of knowledge and attracts other regions, thereby amplifying the area of scope and sustainability of the project.
3. High quality learning materials form the online modules. A good quality and well-structured program is needed for building competences and knowledge in

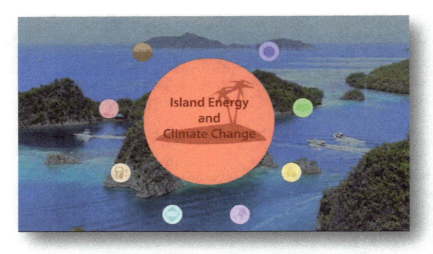

Illustration 14.1 Screenshot Online capacity building program promotional video

RES, EE and climate change mitigation. Contents have been elaborated by qualified researchers and experts with a high specialization in the different knowledge areas, together with an extensive teaching experience. E-learning treatment and virtual settings for creating an effective and successful learning program contribute to comply with the learning objectives of future target audiences. Finally, for optimal development, e-learning and ICT skilled professionals directed module production.

4. The online program is articulated in two versions with different learning speeds and with different, but complementary, approaches and learning outcomes. Both meet the criteria and standards of e-learning methodology. Also, they are offered as a self-study modality, where the student is responsible for learning while, in terms of flexibility, studying at his/her own pace thereby breaking the space and time barriers. These features permit a wide dissemination among key stakeholders, for the public and private sectors and for civil society of island regions.

General Overview Module: 1st Speed

In this version, a global overview of each technology is offered via multimedia and audio-visual material (Table 14.1).

This First Speed Option provides an overview of each technology and energy topic. It is targeted towards anyone interested in obtaining a general knowledge of such technologies and it is focused to raise awareness for a rational use of energy.

14 Capacity Building Itinerary on Sustainable Energy Solutions … 247

Table 14.1 1st speed option main features

Target audience	General public
Objective	To provide an overview of each technology
Learning material	Video presentation and multimedia content
Development site	Temporarily on the CIEMAT's servers, on the LMS Moodle 3.1.6
Final servers	ECREEE, CCREEE and PCREEE websites

Learning materials consist of multimedia and video presentations, with free access on a regional centre's website in order to achieve a wide local and regional dissemination (Illustration 14.2).

Illustration 14.2 1st speed option web appearance

Specialised Training Module: 2nd Speed

This second option is targeted towards more specialized sectors that are seeking to deepen scientific and technological aspects. Thus it provides a more technical and detailed view of each module. The teaching materials are comprehensive and complex. They contain case studies with a practical local approach and a final self-assessment test aimed at providing practical experience for each sustainable renewable energy solution. In this modality, the modules are integrated into an e-learning platform. As noted previously, the selected learning platform is Moodle, which requires student registration and the subsequent enrollment in the modules. Each module is a different course and is accessible individually (Table 14.2).

This second modality requires good study planning. This will be appropriately explained in the didactic guide provided to assist in the objectives, mode of study and expected outcomes of each module. For the correct development of the activity and to ensure the quality and effectiveness of the learning process, the following indicators must be considered and effectively achieved: access to information in the training program, student enrolment, platform access, familiarization with the platform, the identification and downloading of materials, final examination and accreditation (Illustration 14.3).

Each module includes the different training materials: an educational guide, a video presentation, synthetic content (multimedia format) and extensive content (pdf. format), practical exercises (solving a case study and its resolution) and a final self-assessment test. The estimated study time varies from twenty to forty hours, this depending on the length of each module. After completing each module, and once the evaluation criteria have been passed, a digital Certificate of Use of the module will be delivered to the participant. The following describes the established requirements for obtaining it:

Table 14.2 2nd speed option main features

Target audience	Professionals in the sector, academics and postgraduates
Objective	To provide technical and detailed view and quality specialized training for qualify and skilled personnel
Learning material	Complete learning materials including: • Educational didactic guide • Video presentation • Multimedia content • Extensive documentation • Case study • Final assessment test • Achievement certificate
Development site	Temporarily on the CIEMAT's servers, on the LMS Moodle 3.1.6
Final servers	ECREEE, CCREEE and PCREEE e-learning platform

Illustration 14.3 2nd speed option web appearance

- Displaying 100% of the content and achieving 80% in the evaluation test associated with it.
- Performing the case study and correctly answering 100% of the questions associated with it.
- Exceeding 80% in the final self-assessment test.

Learning Outcomes and Certification

Learning outcomes are defined with regard to what a learner knows, understands and is able to do after the completion of a learning process. Usually these are defined in terms of knowledge, skills and competences.

Knowledge implies the outcome of the assimilation of information through learning. It is the body of facts, principles, theories and practices that is related to a field of work or study.

Skills mean the ability to apply knowledge and know-how to complete tasks and resolve problems.

Competence means a proven ability to use knowledge, skills and personal, social and/or methodological abilities, in work or study situations and in professional and personal development.

Based on a "competency" and "skill-set" approach, the "Online Capacity Building and Certification Program on Sustainable Energy Solutions for Islands" bases its structure and objectives on learning outcomes. These can be summarized in the following points.

- Obtain fundamental knowledge and skills for understanding and implementing different technologies and sustainable energy systems
- Acquire the ability to apply knowledge acquired in the energy and environmental field
- Be able to communicate verbally, and in writing, such as technical reports, using the terminology of specialized personnel in the energy sector
- Develop learning skills for acquiring knowledge can be produced continuously and autonomously
- Be able to compare and select technical alternatives
- Understand the functional design of equipment and facilities
- Make specific calculations in the field of energy
- Have an overview of the technology and sustainable energy solutions developed in the module

The above learning outcomes can be equated to the knowledge, skill and competence scopes corresponding to levels above 5 within the European Qualifications Framework, this being taken as reference. The start point is from a base level of students with a scientific or technological degree onwards.

National and Regional qualification frameworks around the world allow for a better understanding and comparison of the qualifications levels in different countries and in different education and training systems (SPC 2015). These become the fundamental reference for alignment of the "Online Capacity Building and Certification Program on Sustainable Energy Solutions for Islands" objectives and learning outcomes in order to harmonize training and certification among SIDS recipients of this project and international standards. Regarding the Certificate of Use for the online training modules, the objective is to recognize the acquisition of knowledge and skills contained in the learning outcomes. It is in consonance with the suitability of including the modules as part of training programs of official educational and training institutions (higher education, vocational training) within the target regions, adapting the programs to offer official accreditation.

Conclusions

Capacity Development enables acquisition of the skills and knowledge required to support all aspects to implement sustainable energy technologies and to accelerate the transition of SIDS and territories towards a future with sustainable energy. A major barrier to sustainable energy and low-carbon technologies deployment in small island countries is the lack of expertise and qualified personnel along the full value chain of low-carbon technologies. Technical knowledge is required to establish a critical mass of policy makers, project financiers and engineers who will be able to manage all aspects of sustainable energy development and implementation (UNIDO 2017).

Although there is a considerable level of awareness of the benefits of deploying sustainable energy systems, educational and training programs are still lacking in small island countries, while the few that exist tend to be located in the larger islands. Moreover, due to their geographic isolation, small size and limited resources, access to capacity-building programs and materials is limited for such states.

The Online "Capacity Building and Certification Program on Sustainable Energy Solutions for Islands and Territories of the Pacific, Caribbean, Africa and Indian Ocean" responds to the urgent need for affordable training and certification programs for sustainable energy on islands. E-learning methodology, plus access through the Internet, provide added value. This is accomplished by offering broad universal access to qualified scientific and technical training at two learning speeds with the scope of areas that generally have difficulty accessing education training, and thus democratizing knowledge. In this sense, the online modules meet three basic conditions for users: quality, certification and affordability. The regions will also benefit from having the program available in three of the most widely spoken languages, i.e., Spanish, English and Portuguese.

Its structure and materials are targeted to different audience profiles, from the general public to professional and qualified persons working within the sustainable energy sector. The program is designed to train the qualified and competent personnel needed to boost this energy sector. The program will be widely promoted through Regional Centres (CCREEE, ECREEE and PCREEE), the owners and direct beneficiaries of the program. Final recipient training needs and sectors of impact determine the character and objectives of the trainings. In order to amplify the impact and sustainability of the program, the modules are expected to complement high-level education programs and vocational training and certification programs thus enabling formal accreditation, which can be organized in partnership with other regional institutions, universities, vocational training programs, R&D&I centres.

Acknowledgements This work has been carried out within the framework of the network of Regional Sustainable Energy Centers in the Pacific, Caribbean, Africa and Indian Ocean supported by UNIDO and the SIDS DOCK Initiative, and has received funding from the Spanish and Austrian Governments through the Austrian Development Agency and the Spanish Agency for International Development Cooperation.

References

Couture TD, Leidreiter A (2014) How to achieve 100% renewable energy. In: Policy handbook. The World Future Council

Dornan M (2015) Renewable energy development in small island developing states of the Pacific. Resources 4(3):490–506

EU PacTVET (2015) Synthesis report 2015. European Union Pacific technical and vocational education and training project. PacTVET

European Commission Website https://ec.europa.eu

IMF (2016) Small states' resilience to natural disasters and climate change-role for the IMF. Staff report. IMF, Washington DC

IRENA (2014) Network renewable islands: settings for success. Global Renewable Energy Islands. IRENA

Mori E, Stuart F, Ashe J, Hongbo W (2014) Our planet. Small island developing states. United Nations Environment Programme (UNEP), Nairobi, Kenya

SIDS-DOCK Website https://sidsdock.org/what-is-sids-dock

SPC (2015) Pacific qualification framework. Educational quality and assessment programme. Pacific Community (SPC), Suva

UN (2018) Sustainable development goals report 2018. https://unstats.un.org/sdgs/files/report/2018/TheSustainableDevelopmentGoalsReport2018-EN.pdf. United Nations, New York

UNEP (2014a) Emerging issues for small island developing states. Results of the UNEP foresight process. United Nations Environment Programme (UNEP), Nairobi, Kenya

UNEP (2014b) GEO small island developing states outlook. United Nation Environment Programme (UNEP), Nairobi, Kenya

UNESCO (2014) Roadmap for implementing the global action programme on education for sustainable development. United Nations Educational, Scientific and Cultural Organization (UNESCO), Paris, France

UN General Assembly (2014) Resolution 69/15. SIDS accelerated modalities of action (SAMOA) pathway. Agenda item 13(a). UN General Assembly

UNIDO (2017) Terms of reference for contracts for services and work. Project reference: SAP ID: 130200 first operational phase of the Caribbean Centre for Renewable Energy and Energy Efficiency—CCREEE. UNIDO

UNIDO, GN-SEC (2018) Regional cooperation to accelerate the uptake of common and inclusive sustainable energy and climate technology markets in developing countries. https://www.se4allnetwork.org/sites/default/files/event/files/210918_draft_conference_discussion_paper.pdf. First draft discussion paper. UNIDO, GN-SEC, Vienna

UN Women (2014) World survey on the role of women in development 2014. Gender equality and sustainable development. Sales no. E.14.IV.6. United Nations Publication, New York

Chapter 15
Taking Current Climate Change Research to the Classroom—The "Will Hermit Crabs Go Hungry in Future Oceans?" Project

Christina C. Roggatz, Neil Kenningham and Helga D. Bartels-Hardege

Abstract Climate change and its consequences at environmental, social and economic level will affect all of us, in particular the children of today who are the world's citizens of tomorrow. However, the causes, consequences and mitigating measures to counteract climate change are not currently part of the regular primary or lower secondary school curriculum in the UK. With the evident lack of practical climate change-based school activities for the UK curriculum in mind, this report describes an outreach project that takes authentic up-to-date research to the classroom with the aim to provide an example to cover this topic. The project focuses on the effects of ocean acidification and the drop of ocean pH on the foraging ability of hermit crabs. Besides a detailed description of the project set-up, this report highlights scientific as well as educational outcomes. The classroom-based experimental sessions yielded a significant scientific result, showing that the hermit crabs' ability to locate food is significantly impaired by pH conditions expected for the year 2100. Combining theoretical and practical parts, the project reached the pupils through different channels and therefore made every child take home the message in their own way, at the same time adding to their key skills in teamwork and effective communication. We could further observe a clear gain in knowledge and confidence with regards to the scientific skills obtained through this project. Professional scientists delivering the

---

Electronic supplementary material The online version of this chapter (https://doi.org/10.1007/978-3-030-32898-6_15) contains supplementary material, which is available to authorized users

---

C. C. Roggatz (✉)
Energy and Environment Institute, University of Hull, Cottingham Road, Hull HU6 7RX, UK
e-mail: C.Roggatz@hull.ac.uk

N. Kenningham
Newland St John's CE Academy, Beresford Ave, Hull HU6 7LS, UK
e-mail: deputyhead@nsj.hull.sch.uk

H. D. Bartels-Hardege (✉)
Department of Biological and Marine Sciences, University of Hull, Cottingham Road, Hull HU6 7RX, UK
e-mail: H.Hardege@hull.ac.uk

© Springer Nature Switzerland AG 2019
W. Leal Filho and S. L. Hemstock (eds.), *Climate Change and the Role of Education*, Climate Change Management, https://doi.org/10.1007/978-3-030-32898-6_15

sessions alongside school teaching staff also served as positive role models to foster the children's future aspirations for science.

Keywords Science outreach · Ocean acidification · Animal behaviour · Fostering aspirations · Authentic science

Introduction

Scientific Background

Ocean acidification, also named the 'evil twin of climate change', presents the main threat of climate change to our oceans besides global warming. The man-made excess carbon dioxide (CO_2) emitted into the atmosphere in increasing concentrations is partly absorbed by our oceans. Once absorbed, the CO_2 reacts with water and shifts the oceans' carbonate buffer system. Water and CO_2 form carbonic acid, which disintegrates (in chemistry terms: dissociates) releasing hydrogen ions. The concentration of these ions, also called protons, determines the pH of water and increased concentrations make the water more acidic. Since preindustrial times the average surface pH of our oceans already decreased from pH 8.2 to pH 8.1 and is predicted to decrease another 0.4 units to pH 7.7 by the end of this century (IPCC 2014). This equals a fourfold increase in acidity within an unprecedented short time scale compared to the stable average pH of 8.2 that has dominated the ocean over at least the past 800,000 years (Ellis et al. 2017).

At first, scientists focussed on effects of this ocean acidification process for calcifying organisms due to the shift in the oceans' carbonate system and revealed significant consequences for corals, microorganisms and the world's carbon cycle (Fabry et al. 2008). Then physiological effects and consequences for the fitness of the organisms inhabiting our oceans were unravelled (Pörtner et al. 2004). Lately, evidence is mounting that not only everyday behaviours such as foraging and homing but also important events like mating and settlement are impacted by ocean acidification (Clements and Hunt 2015). We recently uncovered that the chemical signals mediating many of these behaviours and interactions are at serious risk to be altered by a reduction in ocean pH (Roggatz et al. 2016). The consequences of these impacts for marine ecosystems and ultimately us humans are yet to be established.

Education Background

In recent years young people, as well as the general public have become more concerned about the effects of climate change and increasingly "Climate change is at the forefront of our cultural conversation about science ..." (Hawkins and Stark 2016), although the level of knowledge of the basic scientific concepts are still fairly low

(Corner et al. 2015). However, the topic is covered more and more in school curricula all over the world. UNESCO published a 'Climate Change Starter's Guidebook' (Deeb et al. 2011) and recently a special issue of the 'International Research in Geographical and Environmental Education' journal was published, bringing together articles on climate change education in different countries. In the editorial Chang and Pascua (2017) concluded that "it is the geography and environment educators' job to continue working on research that will impact the way the topic is taught and learned, with a view to helping children succeed in a climate changing world". It is important, not just to assume the interest of the young people but to actively engage them using their own interests and values, for example through social media and games (Wu and Lee 2015; Corner et al. 2015).

The English Secondary School National Curriculum covers the influence of CO_2 on the earth's climate in the subjects of Science as well as Geography (Department of Education 2013a), but the topic is missing completely from the Primary Curriculum, which focusses mainly on basic knowledge of animals and plants and their ecological needs, as well as the importance of scientific enquiry. However, children will be encouraged to learn about and reflect on the changing environment through human impact and how this can influence the natural world, positively as well as negatively (Department of Education 2013b). This section of the curriculum has been designed in a way to give teachers more freedom to highlight their preferred topics during classes (CBI 2015).

Despite this wide coverage of the climate change problem in the curriculum and in educational literature, the focus is mainly on global warming and extreme weather events like droughts or flooding, topics that directly influence the human population. Our project focusses on Ocean Acidification, which influences the whole marine ecosystem and only indirectly affects humans, for example by potentially reducing fish stock. Ocean acidification is not mentioned in any UK GCSE or A-level specification (Westgarth-Smith 2018). Even the above-mentioned 'Climate Change Starter's Guidebook' (Deeb et al. 2011) only contains one sentence and one figure about ocean acidification in a 70 page long document. However, some good examples are starting to emerge not only in Australia and the United States, but also in Europe, where the Helmholtz Centre for Ocean Research published a guide to ocean acidification for teachers with theoretical background and experiments specifically for schools (Riebesell 2012). There is also a summary for policy makers of a 'Symposium on the Ocean in a High-CO_2 World', which contains valuable information on the scientific background as well as economic and societal implications (IGBP et al. 2013).

The English Primary School National Curriculum states in its introduction that "Science has changed our lives and is vital to the world's future prosperity, and all pupils should be taught essential aspects of the knowledge, methods, processes and uses of science" (Department of Education 2013b). In spite of this statement, science seems to become less important at primary schools, with teachers reporting that children spend too little time studying the subject (CBI 2015). With the abolition of SATs testing for Science in 2009, but not for English and Maths, the focus of many schools shifted to these subjects to improve the league table results (CBI 2015).

Another reason for the lack of time spent with science topics is a shortage of primary school teachers with science related degrees, which in England only accounts for 3% (Royal Society 2010). The profession "seems to attract people into primary teaching who fear science rather than those who love it" (Appleton 2003). These teachers lack confidence in teaching science and either avoid it as much as possible (Appleton 2003), or use safe, prescribed activities that often do not lead to much enthusiasm in children (Harlen and Holroyd 1997).

An effect of this decline of science teaching in primary schools can be a lack of aspiration of children for science. Whilst most young people have high aspirations for professional careers at the age of 10–14, only few (15%) aspire to be a scientist (Archer et al. 2013). Children mostly find science interesting and have a high regard for scientists, but many see them as stereotypically brainy, nerdy, geeky people and consider themselves not 'brainy' enough to consider a science career (DeWitt et al. 2013). There are also significant differences within gender and ethnic origin, with especially highly feminine ('girly') females and black students being the least likely to aspire to be a scientist (Archer et al. 2013). These findings show how important it is to intervene at an early age to debunk these myths, by introducing interesting, authentic activities delivered by 'real' scientists for primary school and lower secondary school pupils.

Universities and industry can help to counteract this trend by collaborating with schools to set up outreach and training opportunities (CBI 2015). Siemens for example provides lesson plans for several age groups on topics like clean energy, but also about girls in STEM (CBI 2015). Many UK universities promote school projects with access and outreach grants, for example 'Meet the Scientists' sessions as part of a project on diets and lifestyle for 11–16 year old pupils at the University of Southampton (Woods-Townsend et al. 2014).

A number of those activities incorporate 'authentic science' rather than 'school science', where pupils are introduced to what practicing scientists actually do, thereby bringing real science with real scientists into schools, acting as role models and conveying a different level of enthusiasm for science to the children (Braund and Reiss 2006; Feldman and Pirog 2011). In these 'authentic science' projects at primary as well as secondary school level, pupils participate in 'actual' scientific research, for example testing how honey bees avoid predators (Robinson et al. 2012) or investigating the behaviour-changing ingredients in soft drinks with the help of flat worms (Judge et al. 2017). Hands-on working with live animals as well as experience of field work have been shown to be more motivational for children than traditional lessons or watching documentaries (Scott et al. 2012; Sammet et al. 2015; Prokop and Fančovičová 2017). However, in addition to these hands-on activities, instructions on the theoretical background of the topic are still vital to achieve an overall increase in learning (Sammet and Dreesmann 2017).

The Project

With the lack of practical climate change-based school activities for the UK curriculum in mind, the Hermit Crab outreach project combines all the above-mentioned elements. Authentic up-to-date research on the effects of ocean acidification is brought into the classroom, featuring a theory session, an experimental session with live animals and a field trip, all convened by a team of 'real' scientists. In the classroom-based theory session pupils learn about the concept of pH, why CO_2 changes the pH in the ocean, and how marine animals find their food. This leads to the formulation of the research question, whether ocean pH affects marine animals' sense of smell for finding their food. During the experimental sessions, the children perform experiments with hermit crabs and discuss their class results. They also learn that the more data are gathered, the more accurate the overall result will be, and that their own results will be part of a 'real' scientific study on ocean acidification. The field trip enables the pupils to experience the natural habitat of the hermit crabs and realise the consequences of ocean acidification and other human influences on the environment.

The aims of the project are to

- explain the reasons and possible consequences of ocean acidification.
- improve knowledge on the basic principles of pH and the chemical processes leading from human CO_2 emissions to ocean acidification.
- illustrate how marine animals find their food using chemical signalling cues.
- give children the opportunity to participate in a current scientific research project, train scientific skills and experience how "real scientists" work.

Project Structure, Organization and Realisation

The project "Will hermit crabs go hungry in future oceans?" originated from research carried out at the University of Hull on how ocean acidification affects the ability of different marine animals to smell (see for example Roggatz et al. 2016). Originally designed for a science festival, the bioassay procedure proved so popular and captivating for children that we decided to take the activity to schools. We created an outreach project, in which pupils aged 9–13 could equally participate. We have been able to run this project since 2015 on a yearly basis thanks to initial set-up funding through the University of Hull's outreach fund for the basic equipment required. The basic project structure is outlined in Fig. 15.1 together with the intended learning outcomes (ILOs) and the respective aspects of the curriculum they correspond to. The first stage included meetings to prepare participating school staff and undergraduate student assistants. The second stage consisted of a theory session, one or more experimental sessions per school and an optional field trip.

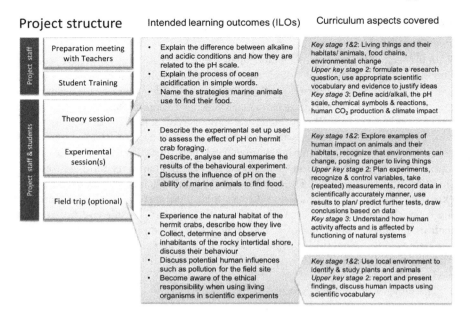

Fig. 15.1 **Structure of the hermit crab project.** Detailed learning outcomes are described in the mid-section and the corresponding aspects of the curriculum they cover are stated on the right

Preparation and Organisation

Prior to the project start within the schools, we invited the participating teachers for a preparation meeting to outline the project idea, introduce the contributing academic staff and gather feedback on work sheets we intended to use for the sessions. This facilitated valuable communication and allowed all participating staff to exchange questions.

In addition to the academic staff, a number of final year students at the University of Hull, who selected the 'Biology in Education' module within their course, were trained to help deliver the school sessions. During their training they received theoretical background knowledge on pH and ocean acidification, including the corresponding literature, and were instructed how to run the experiments. This helped them to assist in the preparation for the experimental sessions, and gave them confidence to guide the children through the project.

To structure the newly gained knowledge and mirror the work of 'real scientists', we asked all children to complete a 'Scientist report sheet' (see Supplementary Data) in the style of a scientific report. This work sheet was optimized with the help of the teachers during the initial meeting and contained four parts:

- An introduction with a gap text summarizing the theoretical aspects.
- A material and method section with set-up scheme and gaps to name the main components and experimental steps.

- A results section with a data table for times and observations.
- A discussion with questions prompting the children to think about the results in detail and in a broader context.

Theory Session

The theory session was delivered by a trained undergraduate student or a member of the project staff assisting the schoolteacher and usually lasted for one period (45–60 min). The topics covered were ocean acidification as well as animals' senses and olfaction (sense of smell). Firstly, the basic concept of pH was explained, followed by a simple demonstration using sea water with a few drops of pH indicator and a straw to bubble CO_2 into to water when exhaling. This visualized the effect of CO_2 on the pH of seawater and acted as a starting point to explain the process of ocean acidification. Potential consequences of ocean acidification for marine organisms (effects on calcification and animals' physiology/fitness) were briefly described, depending on the pupils' age. The second part of the theory session discussed the foraging strategies of marine animals and the senses involved in finding food. The study organism 'hermit crab' (*Pagurus bernhardus*) was introduced as an example and the importance of the sense of smell was highlighted. Combining both parts at the end of the theory session led to the formulation of the research question: "Will hermit crabs still be able to smell and therefore find their food or go hungry in future ocean conditions?" At the end of the theory session the pupils were given time to complete the gap text of the introduction section of the 'Scientist report sheet'. This was used to assess the pupils' learning with regards to the ILOs stated in Fig. 15.1.

Experimental Session(s)

The main focus of this part of the project was on performing feeding experiments with the hermit crabs after a short instruction given by the student or project staff and the completion of the methods section of the 'Scientist report sheet'. The children were asked to distribute and organise the individual tasks within their team (3–6 pupils) and to prepare their set-up according to the instructions. After handing out the animals, the pupils conducted the experiments either synchronized with continuous instructions (primary school level) or independently (secondary school level). Observations and times were recorded by the pupils in a data table on the report sheet. Depending on the time available, the experiments could be repeated multiple times, emphasizing the use of replication in scientific approaches. After returning the animals and cleaning up, the pupils filled in the discussion part of the 'Scientist report sheet'. The six questions within this part prompted them to describe and analyse their data quantitatively and qualitatively. The discussion part helped them to think about the meaning of their specific results and putting it into words, before asking them to think about the wider implications with respect to the research question. In a plenary session at the end, the results were discussed and findings compared across the whole class. Using differing group results as an example, the requirement of replication within scientific investigations was stressed and its implications were discussed. Similar to the theoretical part, the report sheets were later used to assess the pupils' performance with regards to the formulated ILOs detailed in Fig. 15.1.

Field Trip

The field trip was an optional component of the project due to time restrictions by tides and school schedules and took place after the experimental sessions. The location was the rocky southern tip of the beach of Scarborough South Bay (54° 16′ 09.1″ N 0° 23′ 26.8″ W), where the hermit crabs were initially collected prior to the project. This enabled us to take the hermit crabs 'home' and let the children participate in this event to emphasise the ethical use of experimental animals. The class was split into small groups of 6–8 pupils supervised by one member of the school staff. During low tide each group was then taken out for a field session separately, accompanied by a member of the academic staff. After a safety briefing the group ventured out equipped with small nets and buckets to collect animals and plants. After 15–20 min the group gathered to show the results to everyone. Some specimen of interest were transferred to a small plastic aquarium and taken back to the beach for further observations. Basic species determination guides were used by the children and staff jointly to find out which species had been collected. Based on the findings, which usually included some plastic waste or rubbish, a discussion was raised by the project staff member on how humans influence this place in the sea. The children were also prompted to think again about potential 'invisible' pollution, for example by CO_2.

The Scientific Aspect

The effects of ocean acidification on the behaviour of marine organisms has only become the focus of research over the past decade. Coral reef fish, such as clown fish, were found to lose their ability to correctly identify different odours, and, in some cases, became unable to detect them at all (Munday et al. 2009, 2010; Dixson et al. 2010). In other fish species and invertebrates, including sponges, molluscs, echinoderms and crustaceans, other behaviours are likely to be affected by olfactory impairment with decreasing pH conditions (see review by Clements and Hunt 2015).

Common hermit crabs (*Pagurus bernhardus*) are ideal study organisms. They inhabit the intertidal zone of rocky shores and are therefore used to changing environmental conditions. Spending most of their time in tide pools, hermit crabs are regularly exposed to fluctuating pH conditions ranging from 8.2 to 7.4 (de la Haye et al. 2012). They are common on UK shores, easy to collect and recognise. With their relatively small size and small claws, they pose little risk of harming the children during handling of the animal. Hermit crab behaviour has previously shown to be impacted when exposing the crabs to ocean acidification conditions. Foraging behaviour in the presence of a fish extract as a food cue was found to be significantly disrupted when tested at pH 6.8 compared to tests in current pH conditions of 8.2 (de la Haye et al. 2012). In fact, de la Haye et al.'s (2012) publication inspired us to use hermit crabs as study organisms and the bioassay design is based on their study, although noticeably simplified.

Experimental Set-Up and Procedure

The experimental set-up consisted of a small transparent plastic tank (6 L) lined on the outside with black foil (5 cm height from bottom) to reduce reflection and distraction of the animals. The bottom of the tank was marked with lines resembling a football pitch, indicating the middle line (starting point) as well as the goal areas, where the filter paper with/without food cue would be placed. A grey plastic ring (65 mm diameter) was used as starting area and constrain the hermit crab during the acclimatisation period. For every experiment the tank was filled with 1 L of pH-adjusted artificial sea water. The pH was adjusted on the day prior to the experiment by bubbling CO_2 from a pressured cylinder through the water and stored in 10 L canisters to allow for pH stabilisation. A hand-held pH meter (AP115 pH/ORP meter of the accumet Portable Laboratory by Fisher Scientific) was used to measure the pH during this process. Experiments were conducted in normal pH (pH 8.2 ± 0.1, $n = 23$) or low pH (7.6 ± 0.2, n = 14). Blue mussel (*Mytilus edulis*) extract was prepared on the day of the experimental session by crushing one mussel using pestle and mortar, dissolving the biomass in 10 mL of sea water and filtering the suspension (Whatman filter No. 1). 30 µL of extract per experiment was used.

Hermit crabs were collected several weeks prior to the experimental sessions and kept in culture at pH 8.2 (±0.1). They were fed twice a week with blue mussels, but starved for one week prior to experiments. After the experiments the normal feeding rhythm was restored before animals were released at their collection site at the end of the project during the field trip (see above).

The experimental procedure in all experiments was as follows (Fig. 15.2): The tank was filled with water and the plastic ring circumventing the starting position was put in place. The test pH was noted. Two pieces of filter paper (Whatman No. 1, approx. 4 cm^2) were prepared, one was left blank while 30 µL of mussel extract was added to the other.

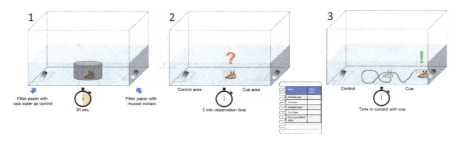

Fig. 15.2 Experimental procedure to observe the behavioural response of hermit crabs to a food cue (mussel extract) presented on filter paper. After a 30 s acclimatisation period (1), the hermit crab was allowed to move freely within the tank for 5 min (2) and observations of the time spent to find the food cue, time spent in either half of the tank or in contact with the cue were noted (3)

(1) The hermit crab was placed into the starting area on the middle line and filter papers were lowered on either side of the tank. A timer was started to stop 30 s to allow for acclimatisation of the hermit crab and the development of an odour signal trail.

(2) After 30 s the plastic ring was lifted and the hermit crab was free to move. The animal's movement and behaviour were observed for 5 min. The time spent in either half of the tank, the time spent in contact with the cue, as well as the time it took the hermit crab to locate the food cue (filter paper with mussel extract) was recorded. Additional observations, for example antennular flicking, were also recorded.

(3) After 5 min the observation was terminated, all times were summed up and noted down. The time the hermit crab spent without movement or in the centre of the tank was calculated by difference. After the experiment, the crab was returned to the holding tank and all set-up components were rinsed thoroughly.

For the analysis of the data all successfully completed runs ($n = 23$ for pH 8.2 and $n = 14$ for pH 7.6) were combined in one spreadsheet. As experiments were run with different children from different schools on different days instructed by different students, we did not manage to obtain an equal number of replicates for each treatment. We further could not monitor if one hermit crab may have been tested multiple times. Therefore, we do not apply any statistical methods for analysis here other than the basic calculation of average and standard deviation (SD)/standard error mean (SEM). If the piece of filter paper with the cue was not detected after the full duration of the assay (5 min), a time of 300 s was noted for how long it took them to reach the cue paper for the first time.

Experimental Results

Results Obtained by the Pupils During the Class-Room Experiments

The hermit crabs responded differently in different pH conditions (Fig. 15.3a). In normal pH (8.2), the hermit crabs spent most of their time in the food smell area and almost 20% (56 s on average) in contact with the filter paper containing the mussel extract. In contrast to this, in low pH (7.6) most time was spent in the centre of the tank without movement. The time spent in the control area and in contact with the control paper was similar in both pH conditions (on average 60 s in normal and 57 s in low pH conditions). This also becomes obvious when the times spent in the different areas are stacked up as percentage of the whole bioassay duration (Fig. 15.3b). The amount of time spent in the food cue area or in contact with the food cue paper (pink colours) relative to the total bioassay duration is much greater in normal pH than in low pH. The children further repeatedly observed (as noted down on their report sheets) that in normal pH the hermit crabs appeared "more active" and "waved their antennae very often". In low pH conditions the hermits appeared "sleepy" and "did not sniff with their antennae as much as before".

In 15 of the 23 runs (65%) conducted in normal pH, the hermit crabs found the food smell paper and interacted with it. On average it took hermit crabs in normal conditions 196 ± 26 s (mean ± SEM) to locate the paper with the foraging cue for the first time. In low pH the same contact with the food smell paper could only be observed in 3 of the 14 runs (21%) and location of the paper with the foraging cue for the first time was observed within an average 260 ± 51 s (mean ± SEM), slower than the hermit crabs in normal pH and with a much larger variability (Fig. 15.4 left).

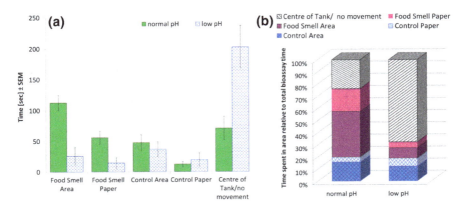

Fig. 15.3 **Response of hermit crabs to food cue under different pH conditions.** a Average time (± standard error mean (SEM) in seconds) the hermit crabs spent in the different areas or without movement in pH 8.2 (n = 23, green) compared to pH 7.6 (n = 14, blue dashed). b Relative amount of time spent in respective areas over the full duration of the bioassay in % in normal pH 8.2 (left) and low pH 7.6 (right)

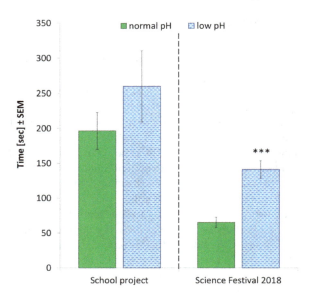

Fig. 15.4 **Average time it took the hermit crabs to locate and interact with mussel extract doped filter paper** (in seconds) in normal pH 8.2 and low pH 7.6 expressed as mean ± standard error of mean (SEM). Data on the left shows results obtained during the school project (n = 23 for pH 8.2, n = 14 for pH 7.6). Data on the right was collected during Hull Science Festival 2018, analysed statistically using a paired *t*-test and the response time was found to differ significantly with pH (***$p < 0.001$)

Results in Comparison to Experiments Performed at the Science Festival

We were able to replicate these results at the Hull Science Festival 2018 with participating children and families at the University of Hull. 20 hermit crabs were first tested in normal pH (8.2) and then in low pH (7.6) in a slightly simplified setting. In both conditions we only stopped the time it took the hermit crabs to locate and interact with the filter paper bearing the mussel extract. If the paper was not located within 3 min, we noted 180 s and stopped the experiment. Due to the clearly paired data and normally distributed differences between the pairs, we could perform a paired *t*-test in this case using the R software package v3.4.0 (Fox and Weisberg 2011; R Core Team 2016). In low pH, the hermit crabs took more than twice as long as in normal pH (Fig. 15.4 right) and the difference in response time between the pH conditions was highly significant ($p < 0.001$; $t = 8.1584$, df = 19). The hermit crabs took on average 75.8 s longer in low pH than in normal pH conditions.

Both sets of results show clearly that pH affected the ability of the hermit crabs to respond to and locate a food smell source. Reduced pH significantly impaired the location of the food cue source and rendered the hermit crabs more inactive.

Discussion of the Scientific Results

These experiments show that a filtered extract of crushed blue mussel (*Mytilus edulis*) in sea water can function as a foraging cue for hermit crabs. This is not a surprising result as hermit crabs are scavengers and often feed on dead animals or leftovers from trawlers (Ramsay et al. 1997; Groenewold 2000). Blue mussel beds are further very common in the area inhabited by the hermit crabs (Dittmann 1990).

However, in their natural habitat hermit crabs are also subject to predation (Lancaster 1988) and hence very sensitive to vibrations, shadows and light reflection hinting at potential predators. This may imply the need for a fully controlled set-up such as described by de la Haye et al. (2011, 2012) with a sheltered box, one-way mirror and light control. But despite the very simplified test set-up in our school and science festival experiments, the obtained results show a very clear trend. In normal pH the crabs were clearly more active than in low pH, which corresponds to findings of de la Haye et al. (2011) on decision making in hermit crabs. In normal pH the crabs also spent considerably more time in the area of food smell, although both treatments were subject to slight variation of cue amount, peeking children (shadows), vibrations and other distractions. This context makes our obtained result even more remarkable.

There are 4 different possible explanations for the observed reduced response of hermit crabs in low pH: (1) the chemosensory structures (antennae) of the crabs are damaged, (2) the chemical structure of the feeding cue or the receptor to receive them are changed, (3) neurotransmission of the signal to the brain and so the 'reaction' to the feeding cue is affected, or (4) the hermit crabs are physiologically stressed and less fit. Hypothesis (1) can be rejected for hermit crabs by analyses performed by de la Haye et al. (2012) and the mechanism for the disruption of neurotransmission proposed by Nilsson et al. (2012) (hypothesis 3) does not apply to crabs whose haemolymph chloride concentration increases in acidified conditions in contrast to fish and molluscs (Henry et al. 2012; Dodd et al. 2015). With the experimental design chosen in this study we are unable to differentiate between effects caused by either hypothesis (2) or (4). Physiological stress could explain the reduced movement in low pH conditions. In addition, feeding cues attracting scavengers often originate from body fluids or degrading protein from decaying animal tissue, for example amino acids and peptides. These cues contain chemical functional groups sensitive to changes in pH, which makes their molecular structure susceptible to ocean acidification conditions, potentially rendering them unrecognisable (Roggatz et al. 2016). Effects of pH on the receptors may further add to a reduction in response (Tierney and Atema 1988), and it has to be noted that although the mussel extract is a mixture of different molecules, this did not prevent the cue from becoming 'inactive'. The implications of this result could be far reaching and the scientific question "Will hermit crabs go hungry in future oceans?" has to be answered with "Potentially, yes" based on the bioassay results of this project. This result stresses the serious implications of ocean acidification and the importance to act against climate change and take measures to reduce CO_2 emissions.

The Educational Aspect

Bringing climate change and its consequences into the classroom using an authentic research experiment was at the heart of this project. The educational gains through the project were therefore as important as the scientific result. By exposing the pupils to the concept of ocean acidification and making them conduct their own experiments, we covered curriculum aspects from a different angle and with a different methodology.

Achievement of the intended learning outcomes was evaluated by reviewing the report sheets completed by the pupils as well as a set of pre- and post-experimental questionnaires. Both approaches will be described and outcomes discussed and mapped against the curriculum aspects covered.

Evaluation of the Pupils' Understanding of the Underlying Scientific Concepts

Analysis of Achievement of the Project Aims Based on the 'Scientist Report Sheet'

To assess the pupils' understanding of the concept of ocean acidification and its causes they completed a gap text in their 'Scientist Report Sheet' after the theory session. This gap text formed the introduction of the report (see Supplementary Data) and was missing key scientific terms, such as 'carbon dioxide', 'CO_2', 'ocean acidification' and 'chemicals'. The pupils were also asked to add missing words (for example 'lower') to put the statements in the correct context. The work was checked immediately and answers corrected and discussed in the plenary session. The evaluation revealed that more than 90% of the initial answers given by the pupils were correct. This indicates a significant progress in understanding the process of ocean acidification as well as the foraging strategies of marine organisms. The ILOs were fully met and the children were proud to be able to go home and "tell mum and dad that hermits smell their food" (original quote of a Year 5 pupil), which was one of the interesting new facts they had learned.

Evaluation of Additional Output and Coverage and Enhancement of UK Curriculum Aspects

Some schools extended the project over several lessons, so the children had more time to grasp the concept of ocean acidification in different ways. Year 5 pupils at a participating primary school, for example, produced posters explaining ocean acidification, with some impressive results (Fig. 15.5). The school further used the

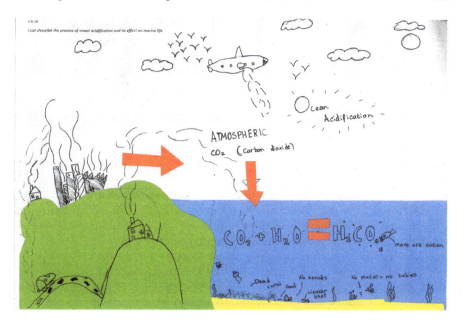

Fig. 15.5 **Example of poster explaining ocean acidification, its causes and consequences.** This poster was produced by a year 5 pupil

project as inspiration to stimulate wider learning during which the children produced poetry, narrative writing and beautiful artwork/sculptures.

As indicated in Fig. 15.1, the intention of the theory session was to educate the children on the process of OA, the difference between alkaline and acidic conditions and relate this knowledge to the pH scale. We further aimed to introduce the pupils to the main foraging strategies of marine animals including the importance of the sense of smell. These aspects relate to a range of topics mentioned in the curriculum for key stages 1 and 2 (Year 4–6 Primary School) and key stage 3 (Year 7/8 Secondary School). The theory session of the project does not only highlight the connection between 'Living things and their habitat', but also touches upon the topics 'food chain' and 'environmental change' (all key stage 1 and 2). Acid-base chemistry is only specified in the curriculum for secondary schools (key stage 3), but introduced here at a basic level focussing on the pH scale. The use of chemical symbols and writing the reaction of CO_2 with water to explain the ocean acidification process covers further aspects of the key stage 3 curriculum (for detailed mapping see Fig. 15.1).

Figure 15.5 shows an example of a poster produced by a Year 5 pupil and demonstrates that even at this young age children were able to process, understand and replicate the chemical aspects of ocean acidification based on the theory introduction. This shows that they are beginning to understand higher-level work that is not covered by the curriculum until a later stage. They also start to develop the idea that emissions by human activities cause climate change and specifically ocean acidification, fostering an understanding of how human activity affects and is affected by the functioning of natural systems (Department of Education 2013a, b).

The scientific skills improved by the hermit crab project map onto a variety of curriculum aspects. It allows the pupils to explore one example of human impact on animals and their habitats and enables them to recognize that environments can change, posing danger to the living things inhabiting them. The project further advances their skills to plan experiments and to recognize and control variables. By combining and comparing all group results, the children get further introduced to the concept of repeated measurements and learn how to record data in a scientifically accurate and comparable way.

Apart from the content-based learning there is also substantial coverage of 'working scientifically', which overarches the upper key stage 2 curriculum, where pupils are encouraged to 'formulate a research question' and 'use appropriate scientific vocabulary' as well as 'use scientific evidence to justify their ideas' (upper key stage 2). Within the discussion section of the report sheet, they use their results to plan or predict further tests and draw a conclusion based on the data. This helps them to think like a scientist and build the logical thinking and problem-solving skills associated with scientific work, skills that are easily transferable to other subjects and everyday life.

Project Outcomes as Evaluated Through Pupils' Self-assessment

Methodology of Self-assessment

In order to further assess the pupils' progress and attitude, we asked them to complete an anonymous self-assessment questionnaire before and after the experimental session. Towler and Broadfoot (1992) discussed that self-assessment is not just useful to find out about the pupils' attitudes, but is also a valuable learning tool promoting the children's own reflection of their learning. The questionnaire used a 4 point Likert scale, which has been shown as an efficient method to rate learning self-efficacy (Croasmun and Ostrom 2011). The children were asked to rate statements from "Yes", "Maybe", "Not really" to "No", which was also expressed in pictures using a thumb emoji either pointing up, 45° upwards, 45° downwards or down. The statements of the questionnaire were:

- "I can describe the experiment in my own words."
- "I feel confident to set up an experiment by myself using instructions."
- "I know how to discuss the results."
- "I can describe the effect of pH on the hermit crabs."
- "I want to go to university and become a scientist."

15 Taking Current Climate Change Research to the Classroom … 271

68 pupils completed the questionnaire in Year 5 (upper key stage 2) in two separate sessions in 2017 and 2018. The results before and after the experimental session are shown in Fig. 15.6 and expressed as percentage of pupils (%).

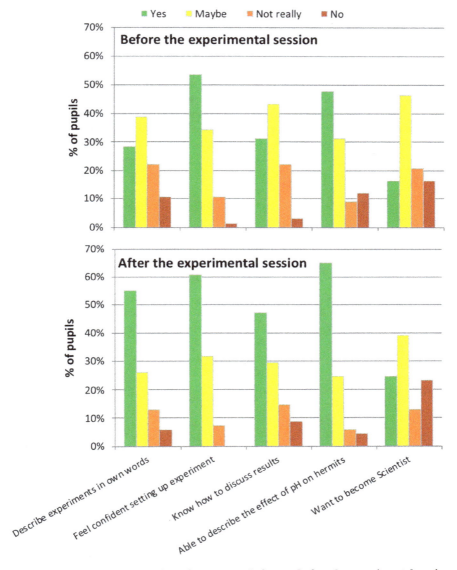

Fig. 15.6 Self-assessment questionnaire outcome before and after the experimental session for year 5 pupils (n = 68) asked to rate whether they agreed with the given statements or not in four steps "yes" (green), "maybe" (yellow), "not really" (orange), "no" (red)

Results for Pupils' Confidence with Regards to Acquired Knowledge and Skills

The majority of the children were confident in the topics they had discussed during the theory lesson and practical instructions, which already raised their interest prior to the start of the experimental session (personal observation of the project team). Consequently, they mostly stated "yes" for setting up the experiment and explaining the effect of pH on the hermit crabs. For the other statements "maybe" was the most frequent answer.

After the experimental session this changed with more than 55% stated that they feel able to describe the experiment in their own words and more than 60% felt confident setting up the experiment using the instructions given; no pupil replied "no" to the set-up question. Generally, all statements directly related to the experiment were answered with "yes" by the majority after the experiment. This indicates that most children feel they have progressed and improved their scientific skills, including experimental description and set-up, result discussion and summarising the new knowledge concluded from the experiment.

Although the percentage of pupils choosing "maybe", "not really" or "no" clearly decreased after the experimental session, the number of pupils answering "no" increased for the statement "I know how to discuss the results". This could be attributed to the fact that only after the session some pupils became aware of the complexity of the discussion and that there are a multitude of factors to think about. However, this result also points to how the project could be improved in the future, for example with an additional session dedicated to a more in-depth discussion of results. In the end, almost 90% of children stated that they can or maybe can describe the effects of pH on the hermit crabs and therefore answer the primary research question of the project.

Results Obtained with Respect to Pupils' Aspirations

Before the experimental session most children were unsure if they would like to become a scientist and answered this question with "maybe" or "not really". After the experimental session 15% of previously undecided children replied with "yes" or "no" instead. This indicates that bringing authentic research to the classroom helps them to understand the profession of a scientist and make a more distinct decision whether being a scientist fits their future aspirations. This has also been shown by Robinson et al. (2012) and Judge et al. (2017), who conducted similar projects with primary and secondary school pupils. The academic staff, delivering the sessions alongside school teaching staff and students, also served as role models. They were approached throughout the project by a number of pupils with a wide range of more general questions. The lead scientists of this project being female also helped to debunk the myths of the 'nerdy male scientist' (Archer et al. 2013) and potentially made girls feel more empowered in their aspiration to become a scientist.

The experimental session was very positively received by the children and 90% affirmed that they enjoyed the experience in the final section of the questionnaire and added comments like: "I liked this session because I learned a lot of new facts" or "I loved learning about the hermit crabs and observing them move around". One child stated "I feel more confident that hermit crabs will be extinct in the future", which sums up the main point of the project and likewise a concern many children expressed and discussed with the academic project staff during the project. It reportedly resulted in classroom lights being switched off when possible and other power-saving measures to reduce CO_2 at home (lights, heating and even fridges were reported to be targets). One of the girls completing the questionnaire stated "I am hoping to be a marine scientist. I am not very good at science so I will try to work harder to accomplish my dreams", indicating that this project has the power to inspire and motivate young people and especially girls into STEM subjects.

Field Trip Feedback

Over the duration of the project we led four field trips to Scarborough, facilitated by school staff. We did not formally assess the delivery of the ILOs for these trips, but the oral feedback received for all trips was positive throughout. To actually experience the natural habitat of the hermit crabs encouraged the pupils' natural curiosity. Collecting, observing and identifying the organisms of the intertidal and reporting and presenting them to their friends and teachers helped them to engage even more with the topic. This has also been shown by Scott et al. (2012), who describe a higher motivation in children through the experience of field work. The pupils also started to discuss human impacts, such as ocean acidification or pollution of Scarborough Bay by litter, using the scientific vocabulary they had learned during the project, all aspects covered by the National Curriculum (Department of Education 2013a, b). The highlight of each field trip was the release of the hermit crabs back to their natural habitat. This not only emphasized the ethical and respectful treatment of living organisms when using them for scientific observations, but also stressed the responsibility we have to ensure the animals' wellbeing. The 'Scarborough experience' inspired the children far beyond the trip. Several children expressed a strong interest in becoming marine biologists and continue to do so, showing significantly increased engagement in science-related topics.

Additional Skills Delivered by the Project

Based on joint observations of the project staff and the teachers, there are a range of skills beyond working scientifically that the pupils acquired through this project. Pupils showed remarkably positive attitudes towards the challenges faced in the experimental session, which fostered the pupils' level of responsibility and independence by caring for animals they would normally not encounter. When performing experiments independently, pupils held responsibility for accuracy and reliability of results. The teachers expressed that usually disruptive pupils, who often influence the learning environment, showed mature behaviour and cared for the animals throughout the session. This is backed up by Zasloff et al. (1999) who found that working with animals in a primary school setting could motivate young students with learning and behaviour difficulties. Conducting the experiments in small groups also required effective team work, and pupils engaged actively in discussion using reasoning and questioning skills in an attempt to draw conclusions about the experiment. Basic leadership skills were developed through the group coordination and distribution of different tasks (observation, time keeping, taking notes).

Conclusions

This Hermit Crab outreach project has been designed to educate children about the causes and consequences of ocean acidification, a significant consequence of climate change, which is frequently missed in media coverage and not covered by the current school curriculum in the UK. This project further covers a topical area that only recently has caught the attention of researchers world-wide to be the focus of a more thorough investigation. It therefore allows the children not only to gain insight into very state-of-the-art ongoing research, but to actively contribute to knowledge acquisition within this field. The project structure was designed to deliver theoretical knowledge as well as practical experience and allowed to foster the children's scientific skills as well as their independence, responsibility, communication and teamwork skills. It encompasses several aspects of the current National School Curriculum from key stages 1 to 3 and allowed the children to contribute to an ongoing genuine scientific investigation. The project yielded a significant scientific result by showing that hermit crabs reduce their movement and are impaired in their ability to locate a food source by smell in future oceanic pH conditions of pH 7.7. The pupils' reports and self-assessments of their skills showed a significant theoretical and practical knowledge gain. Besides delivering on the ILOs, the Hermit Crab project was found to provide valuable insights into the profession of a scientist for the children and foster their aspirations to study and engage in science. By reporting the concept, execution and observed scientific as well as educational outcomes of this novel and unique outreach research project, we hope it can serve as an example and inspire more authentic climate change research to be taken to the classroom. Proven to be so popular over the past years, we aim to continue this project in the years to come and to educate and inspire more cohorts of pupils for the future.

Acknowledgements The project was funded by the University of Hull and we would like to acknowledge the help of Dr. Jörg D. Hardege, who allowed us to use his laboratory and equipment during project development and for preparations prior to the experimental session. We further thank

Prof. Mark Loch for his valuable input to the theory session, the assisting students throughout the project and the diligent helpers during the Hull Science Festival 2018. Special thanks to Melissa Hardege for help throughout the project sessions as well as comments to this manuscript and the staff and pupils of all participating schools; you were brilliant and a pleasure to work with.

References

Appleton K (2003) How do beginning primary school teachers cope with science? Toward an understanding of science teaching practice. Res Sci Educ 33:25

Archer L, Osborne J, DeWitt J et al (2013) ASPIRES: young people's science and career aspirations, age 10–14. King's College London, London

Braund M, Reiss M (2006) Towards a more authentic science curriculum: the contribution of out-of-school learning. Int J Sci Educ 28:1373–1388

CBI (2015) Tomorrow's world—inspiring primary scientists. Brunel University, London

Chang C-H, Pascua L (2017) The state of climate change education—reflections from a selection of studies around the world. Int Res Geogr Environ Educ 26:177–179

Clements J, Hunt H (2015) Marine animal behaviour in a high CO_2 ocean. Mar Ecol Prog Ser 536:259–279

Corner A, Roberts O, Chiari S et al (2015) How do young people engage with climate change? The role of knowledge, values, message framing, and trusted communicators: engaging young people with climate change. Wiley Interdiscip Rev Clim Change 6:523–534

Croasmun JT, Ostrom L (2011) Using Likert-type scales in the social sciences. J Adult Educ 40:4

de la Haye KL, Spicer JI, Widdicombe S, Briffa M (2011) Reduced sea water pH disrupts resource assessment and decision making in the hermit crab *Pagurus bernhardus*. Anim Behav 82:495–501

de la Haye KL, Spicer JI, Widdicombe S, Briffa M (2012) Reduced pH sea water disrupts chemo-responsive behaviour in an intertidal crustacean. J Exp Mar Biol Ecol 412:134–140

Deeb A, French A, Heiss J et al (2011) Climate change starter's guidebook: an issues guide for education planners and practitioners. UN Educational, Scientific and Cultural Organization, Paris

Department of Education (2013a) The national curriculum in England framework document for geography program of study: key stage 3. Department of Education, London

Department of Education (2013b) The national curriculum in England—key stages 1 and 2 framework document. Department of Education, London

DeWitt J, Archer L, Osborne J (2013) Nerdy, brainy and normal: children's and parents' constructions of those who are highly engaged with science. Res Sci Educ 43:1455–1476

Dittmann S (1990) Mussel beds—amensalism or amelioration for intertidal fauna? Helgoländer Meeresunters 44:335–352

Dixson DL, Munday PL, Jones GP (2010) Ocean acidification disrupts the innate ability of fish to detect predator olfactory cues. Ecol Lett 13:68–75

Dodd LF, Grabowski JH, Piehler MF et al (2015) Ocean acidification impairs crab foraging behaviour. Proc R Soc B Biol Sci 282:20150333

Ellis RP, Urbina MA, Wilson RW (2017) Lessons from two high CO_2 worlds—future oceans and intensive aquaculture. Glob Change Biol 23:2141–2148

Fabry VJ, Seibel BA, Feely RA, Orr JC (2008) Impacts of ocean acidification on marine fauna and ecosystem processes. ICES J Mar Sci J Cons 65:414–432

Feldman A, Pirog K (2011) Authentic science research in elementary school after-school science clubs. J Sci Educ Technol 20:494–507

Fox J, Weisberg S (2011) An R companion to applied regression, 2nd edn. Sage, Thousand Oaks, CA

Groenewold S (2000) Effects on benthic scavengers of discards and damaged benthos produced by the beam-trawl fishery in the southern North Sea. ICES J Mar Sci 57:1395–1406

Harlen W, Holroyd C (1997) Primary teachers' understanding of concepts of science: impact on confidence and teaching. Int J Sci Educ 19:93–105

Hawkins AJ, Stark LA (2016) Bringing climate change into the life science classroom: essentials, impacts on life, and addressing misconceptions. CBE Life Sci Educ 15:fe3

Henry RP, Lucu Č, Onken H, Weihrauch D (2012) Multiple functions of the crustacean gill: osmotic/ionic regulation, acid-base balance, ammonia excretion, and bioaccumulation of toxic metals. Front Physiol 3

IGBP, IOC, SCOR (2013) Ocean acidification summary for policymakers. In: Third symposium on the ocean in a high-CO_2 world. International Geosphere-Biosphere Programme, Stockholm, Sweden

IPCC (2014) Climate change 2014: synthesis report. Contribution of working groups I, II and III to the fifth assessment report of the intergovernmental panel on climate change. Intergovernmental Panel on Climate Change (IPCC), Geneva, Switzerland

Judge S, Delgaty L, Broughton M et al (2017) Behaviour-changing ingredients in soft drinks: an experiment developed by school children in partnership with a research scientist. J Biol Educ 51:79–96

Lancaster I (1988) *Pagurus bernhardus* (L.)—an introduction to the natural history of hermit crabs. Field Stud 7:189–238

Munday PL, Dixson DL, Donelson JM et al (2009) Ocean acidification impairs olfactory discrimination and homing ability of a marine fish. Proc Natl Acad Sci 106:1848–1852

Munday PL, Dixson DL, McCormick MI et al (2010) Replenishment of fish populations is threatened by ocean acidification. Proc Natl Acad Sci 107:12930–12934

Nilsson GE, Dixson DL, Domenici P et al (2012) Near-future carbon dioxide levels alter fish behaviour by interfering with neurotransmitter function. Nat Clim Change 2:201–204

Pörtner HO, Langenbuch M, Reipschläger A (2004) Biological impact of elevated ocean CO_2 concentrations: lessons from animal physiology and earth history. J Oceanogr 60:705–718

Prokop P, Fančovičová J (2017) The effect of hands-on activities on children's knowledge and disgust for animals. J Biol Educ 51:305–314

Ramsay K, Kaiser MJ, Moore PG, Hughes RN (1997) Consumption of fisheries discards by benthic scavengers: utilization of energy subsidies in different marine habitats. J Anim Ecol 66:884

R Core Team (2016) R: a language and environment for statistical computing. R Foundation for Statistical Computing, Vienna, Austria

Riebesell U (ed) (2012) The other CO_2-problem, ocean acidification. Eight experiments for students and teachers. BIOACID/GEOMAR Helmholtz Centre for Ocean Research Kiel, Kiel

Robinson JW, Nieh JC, Goodale E (2012) Testing honey bees' avoidance of predators. Am Biol Teach 74:452–457

Roggatz CC, Lorch M, Hardege JD, Benoit DM (2016) Ocean acidification affects marine chemical communication by changing structure and function of peptide signalling molecules. Glob Change Biol 22:3914–3926

Royal Society (2010) "State of the nation" report on 5-14, science and mathematics education. Royal Society, London

Sammet R, Dreesmann D (2017) What do secondary students really learn during investigations with living animals? Parameters for effective learning with social insects. J Biol Educ 51:26–43

Sammet R, Kutta A-M, Dreesmann D (2015) Hands-on or video-based learning with ANTicipation? A comparative approach to identifying student motivation and learning enjoyment during a lesson about ants. J Biol Educ 49:420–440

Scott GW, Goulder R, Wheeler P et al (2012) The value of fieldwork in life and environmental sciences in the context of higher education: a case study in learning about biodiversity. J Sci Educ Technol 21:11–21

Tierney AJ, Atema T (1988) Amino acid chemoreception: effects of pH on receptors and stimuli. J Chem Ecol 14:135–141

Towler L, Broadfoot P (1992) Self-assessment in the primary school. Educ Rev 44:137–151

Westgarth-Smith AR (2018) Ocean acidification needs more publicity as part of a strategy to avoid a global decline in calcifier populations. J Mar Biol Assoc UK 98:1227–1229

Woods-Townsend K, Christodoulou A, Byrne J et al (2014) Meet the scientist: the value of short interactions between scientists and secondary-aged students. In: E-book proceedings of the ESERA 2013 conference: science education research for evidence-based teaching and coherence in learning. Part 10. European Science Education Research Association, Nicosia, Cyprus, pp 1821–1832

Wu JS, Lee JJ (2015) Climate change games as tools for education and engagement. Nat Clim Change 5:413–418

Zasloff RL, Hart LA, DeArmond H (1999) Animals in elementary school education in California. J Appl Anim Welf Sci 2:347–357

Chapter 16
Why Is Early Adolescence So Pivotal in the Climate Change Communication and Education Arena?

Inez Harker-Schuch

Abstract This paper explores the characteristics that make young adolescents (12–13-year olds) ideal 'change agents' in the climate science communication arena. We argue that this age group is at a pivotal age for cultivating public opinion, broadening awareness of the science and leveraging this knowledge to promote climate-friendly policy and governance. We examine the physiological and social characteristics that make young adolescents such an ideal age group. These characteristics involve intellectual development, cultivation of self-determination, and emergence of the adolescent into society—and how these characteristics can be utilised to create better communication and education tools, methods and strategies. We hope that this paper will help educators and communicators ensure climate science communication is tailored to be cost-effective, accurately designed and appropriately scaled to this key demographic. This work contributes to climate science communication and advances existing understanding of climate science communication frameworks for this specific audience.

Keywords Climate education · Early adolescence · Intellectual development · Science communication · Worldview development

Introduction

To motivate people to revise their behaviour in order to live more sustainably, communication is key—of which knowledge is the single most pivotal factor. Previous research has suggested that knowledge (specifically, the 'knowledge deficit' model as a knowledge-behaviour intervention) is largely ineffective in the climate science communication arena (Corner et al. 2015; Kahan et al. 2011). Prior knowledge, is however, the foundation of intellectual constructs regardless of how accurate or reliable that knowledge or information may be. Knowledge depends on communication,

---

I. Harker-Schuch (✉)
Fenner School of Environment and Society, The Australian National University, B141, B48, B48A Linnaeus Way, Acton ACT, Canberra 2601, Australia
e-mail: inez.harker-schuch@anu.edu.au; inez@earthspeople.org

© Springer Nature Switzerland AG 2019
W. Leal Filho and S. L. Hemstock (eds.), *Climate Change and the Role of Education*,
Climate Change Management, https://doi.org/10.1007/978-3-030-32898-6_16

as argued by Wittgenstein (1953)—for how else can we know or understand anything, particularly of a socio-cultural nature, unless we communicate? We obtain our knowledge through experience, inherited worldviews, exposure and access to information, formal and informal education environments, and via individuals who act as mentors, icons, or examples: we inherit these views in the sub-culture of our community and through close friends and family. Worldview, particularly in older adolescents and adults, plays a significant role in an individuals' attitude to or engagement with climate change (Wolf and Moser 2011), and on their knowledge development (Lewandowsky et al. 2012). It is often a stronger predictor of climate change denial than how much an individual understands the scientific basis of climate change. Socio-political identity has also been shown to 'entrench' climate change denial in individuals who identify with political parties that refute the science and evidence of climate change.

Recent research is emerging which suggests that *specific* climate science knowledge may be a more useful tool in cultivating engagement than previously thought, and may correct worldview biases (Shi et al. 2015; Stevenson et al. 2014). There is an urgent need to develop more focused methods that can adequately communicate about climate change to influential groups (Karpudewan et al. 2014; Robinson 2010). Coupled with this, we need to develop effective teaching materials that can straddle the disciplines of climate science, communicate effectively and include a willingness to build emotion into the climate science topic (Carmi et al. 2015; Moser and Dilling 2007; Rice et al. 2007; Roeser 2012).

Of even greater importance, recent research is suggesting that early adolescent may be (by virtue of their current physical and social development) the pivotal group upon which true social and behavioural change can occur. They are a group, who for the most part, share the following characteristics:

- An openness to cultivating and/or reviewing their worldviews
- The ability to affect the greatest change with the longest-term effect
- Are the easiest age group to reach
- Possess characteristics that facilitate effective communication
- Have the potential for scaling, i.e., into other groups

Although targeted and tailored climate change strategies have offered many useful and insightful perspectives (Bostrom et al. 2013) they have so far been unable to provide educators or communication strategists with a clear way forward. Our approach provides a new pathway—one that focuses on the intrinsic, biological and physiological characteristics of early adolescence that preempt, and ultimately, transcend socio-cultural and worldview biases.

The Adolescent—A Potential Candidate

In 2012, adolescents (those age was between 12 and 18) made up 0.75 billion (18%) of the world's population (United Nations (DESA) 2012). They are the largest group of climate-vulnerable people on Earth (UNICEF 2015); particularly female adolescents (Swarup et al. 2011). They are also the demographic group which the onus of responsibility for managing climate change falls on—a burden they will acquire by default (Case 1985). This obligation requires them, without exception, to confront the worst impacts of climate change as it unfolds throughout their lifetimes—and to bequeath the same legacy to their own children (Corner et al. 2015; Ferkany and Whyte 2012). A burden which they are now intellectually able to process and comprehend (Jensen and Nutt 2015). These burdens, aside from the responsibility of the physical impacts of climate change, include psychological, physical, intellectual and emotional impacts (Berry et al. 2010; Ojala 2012a; Roeser 2012; UNICEF 2015) and will involve making choices that will be far-reaching and non-retractable (Lazarus 2009). Examples of these challenges include: 'end of the world' fears (Tucci et al. 2007), increased health problems and higher mortality risks from climate-related impacts (UNICEF 2015), decreased water and food security, reduced access to education (particularly for females) (ibid.), and a general sense of worry about the future (Corner et al. 2015; Ojala 2012b, 2013). The future of their children and climate stability depends on the efforts and actions they undertake; the sheer complexity of the problem—and its 'wickedness' (Levin et al. 2012)—will require new forms of knowledge, new international cooperatives and new skills that have not existed before. These are damaging psychological burdens; particularly as they have no precedent (Ojala 2015). Within the wickedness of climate change, there is also no way of knowing if the actions they undertake will work to mitigate the detrimental effects—perhaps even within their lifetimes.

The vulnerability of adolescents to the future (and their expectations of coping with the future) are unusually acute, as they have little power to effect change. They depend on their socio-economic circumstances for the framework of their worldview. A worldview (or *weltanschauung* in the original by Dilthey (Makkreel 1975)) that can be misleading, poorly constructed and destructive (Ojala 2015; Vollebergh et al. 2001). Today's adolescent is in a precarious position, as the issue of climate change appears to hold no clear, tenable public position (Brulle et al. 2012). Currently, everyone has a steadfast, entrenched opinion, and it seems everyone is entitled to it, no matter how tenuous, idiosyncratic or unreasonable the argument may be (Ayer and Marić 1956; Postman 1985; Stokes 2012). Adolescence is also the time when each person establishes their own *weltanschauung*—a development that begins in early adolescence and slowly 'cements' before early adulthood (Case 1985; Vollebergh et al. 2001). Not only must she or he choose sides on which to stand, there will be spatial and temporal consequences of those choices that will have a real impact on our environment and future (Gowers 2005; Wray-Lake et al. 2010). It is disturbing that many adolescents are becoming increasingly less concerned about their environment—expecting governments or 'technofix' to solve the myriad of issues which currently plague our natural environment (Stevenson et al. 2014; Wray-Lake et al. 2010). It is most worrisome as their attitudes and decisions will affect every other species and habitat on our planet. No other generation has had this responsibility, nor the awareness of just what that responsibility entails.

Aside from the ethical considerations we hold for our emerging adults, there are other aspects that make adolescents essential players in climate change mitigation. Adolescents are poised on the edge of our society—a few years short of assuming the responsibilities of adulthood that will, by definition, launch their political, economic, intellectual, vocational and social identities (Checkoway et al. 2003; Gowers 2005; Quintelier 2014). These aspects include:

- Their proximity to participating in elections
- Their financial wherewithal to purchase goods and services
- The formation of personal opinions
- Their employability and gradual introduction into the workforce
- The beginnings of secure intimate and social relationships

Preparing adolescents for these tasks—and their emergence into society—is one of the main goals of society and education systems (Ghysels 2009; Shanahan et al. 2002). Adolescents also possess certain socio-cultural attributes that make them prime candidates for climate change communication. They are very adept at communication and digital technology (which are the primary avenues of information and social exchange today) and are, fortuitously, legally obliged to be enrolled in public education systems; making them a very accessible group.

Of all of these features, there are three key conditions of adolescence (and early adolescence in particular) that make this age group so ideal for climate science communication: (1) the intellectual development that takes place during adolescence; (2) as a consequence, the emergence of self-determination/self-esteem and (3) their introduction into the broader social community. These major factors are essential for developing meaningful worldviews, understanding the mechanisms of climate science, and engaging and participating in emission reduction activities towards a carbon-neutral future.

Physiological Changes—Intellectual and Abstract Reasoning Development

Similarly to very early childhood, adolescence is a critical developmental period both cognitively and socially. Cerebral changes in the human brain undergo drastic alteration during this time in order to prepare for maturation and adulthood (Steinberg 2005). Alongside augmented communication with other brain areas, the prefrontal cortex undergoes a drastic 'pruning' phase—known as the 'second critical period' of learning—signifying the traverse from childhood to maturity (Jensen and Nutt 2015). It is, according to Jensen, a 'golden age' of intellectual development (ibid.).

Throughout childhood grey matter has been increasing in various regions of the brain. This transformation peaks in the early stages of puberty, whereupon it begins the second critical 'pruning' period. This pruning results in a reduction of grey matter, and simultaneously, the production of white matter via myelination; heralding the commencement of scientific reasoning, executive function, social cognition and planned control behaviour, among others (Blakemore and Choudhury 2006; Burnett and Blakemore 2009; Dumontheil 2014; Field et al. 1997; Kuhn and Franklin 2008). Myelination acts as a sheath around axons, increasing the speed (up to a 100-fold) of neural impulses, insulating transmitting impulses, and reducing 'recovery time between neural firings'.

The resultant 'synaptic plasticity' reinforces exercised and practiced neural pathways. In its simplest 'use it or lose it' form, synaptic plasticity describes the weakening or strengthening of synapses over time. It results (according to Damon and Lerner 2008) in 'fewer, more selective, but stronger, more effective neuronal connections than they had as children' (Damon and Lerner 2008; Kuhn and Franklin 2008; Munz et al. 2014). This 'second critical period' opens up essential intellectual and social development pathways. This development allows scientific reasoning and critical thinking abilities to flourish and launches the adolescent into adulthood (Case 1985; Jensen and Nutt 2015; Piaget 1972).With regard to intellectual development, abstract reasoning arises in the 11–13 year old as a consequence of the development of the ability to coordinate 'vectorial operations' (being able to summon abstract constructs and execute simple scientific reasoning). This skill is a precursor to reasoning with multidimensional problems and highly complex concepts (Case 1985). The work by Case is strongly aligned with the Piagetian theory on intellectual development. This theory suggests that by exposing young adolescents to multidimensional and complex concepts, we effectively 'train' and 'exercise' these neurons in preparation for synaptic pruning. This in turn, promotes the preservation of important neural pathways (Case 1985; Piaget 1972). This learning period, therefore, becomes crucial; as early adolescents are now ready for constructs and concepts that employ their intellectual development. This challenges their expanding processing capabilities and prepares them to confront and engage with such issues.

For climate science communicators this intellectual development phase is, arguably, the single most important factor in communicating with early adolescents (12–14-year olds). A young adolescent is building the scaffold on which many of their intellectual constructs will later be built. T he more equipped the brain is to conceptualise these constructs, the higher the capacity to respond, process and cope with the various elements of climate science (Jensen 2015; Ojala 2013; Stevenson et al. 2014).

The physiological changes of the adolescent brain show extraordinary potential for learning of all kinds; and could be particularly useful for education with significant social dimensions (Cheshire 2017).

Social Allegiances, Perspective-Taking and Community

As synaptic plasticity influences executive function and intellectual development, so too is social cognition and perspective-taking equally affected by this synaptogenesis. These changes launch 'self-awareness', 'theory of mind', 'perspective-taking' and the 'ability to understand other minds by attributing mental states such as beliefs, desires and intentions to other people' (Blakemore and Choudhury 2006; Frith and Frith 2003). It is theorised that these profound changes in the human brain allow an individual to empathise and vicariously participate in another's experience (Blakemore and Choudhury 2006). For example, similar brain regions are stimulated both in first person and second person perspective as well as between first and third person perspective (e.g., seeing an image or hearing an explanation). These 'perspective-taking' experiences connect us to others. They help to form familial, social and romantic attachments that reinforce or revise our *weltanschauung,* and as a result, have far-reaching consequences both for society and governance.

As social cognition develops, individuals become aware of ethical and moral considerations, aside from their own survival and well-being. The nature of climate change (and the imperative towards mitigation) will very likely depend on empathetic and perspective-taking responses, as first-person perspective and experiences will be virtually impossible to attribute to the effects of climate change. For example, understanding that 'similar' or 'vulnerable' others may suffer illness, death or hardship due to climate change may motivate an individual to take action. Perspective-taking also promotes awareness of poor governance and corruption; social movements in the youth community can affect enormous social and cultural change (Checkoway et al. 2005; Ho et al. 2015). Further social cognition factors that favour communicating with this age group include example-setting and ambassadorship (youth leadership and peer-pressure for climate-friendly behaviour), community building (reinforcing group behaviours), knowledge and information sharing amongst peers, the commonality of school, and the potential of this age group to form strong social awareness constructs at the very onset of social cognition. It is, however, the maturity of the adolescent that will affect the greatest change in the world. Their emergence into society as fully matured consumers, voters, and workers ensure that their role as change agents is a very powerful one indeed.

Self-determination in Adolescence

As the brain matures, the adolescent slowly acquires the self-awareness and confidence that not only can one manage to live within one's life, but that one can also determine how it is coordinated and exercised (Blakemore and Choudhury 2006). This self-determination is fostered through an emerging self-esteem (developed from 'knowing oneself, rational/critical reasoning, and valuing and accepting the worthiness of your strengths, rights and responsibilities) that encourages us to exercise our own skills and to contribute to our own lives, and the lives of others (Gowers 2005). Recognising and achieving these self-determined contributions is a vital component in the developmental pathway of adolescence (Field et al. 1997).

There is an implicit implication in self-determination, that one can choose to act or not act. T his behaviour then results in consequences from which we learn and gain experience. We see here, the shift from cognitive processes to the realisation that one's actions [or equally, lack of them (Kuhn and Brass 2009)] have a clear and effective presence in reality, society and our environment. It is at this time that we begin to perceive the individual power that we effect or create.

Dilthey defined weltanschauung as constituting 'an overall perspective on life that sums up what we know about the world, how we evaluate it emotionally, and how we respond to it volitionally' (Dilthey, quoted in Audi 2015). Following this definition, we transpose the volition into self-determination—the exercise of one's beliefs, ideals, ambitions, and goals into physical or semi-physical realities. However, these exercises are not limited to the individual, but have enormous social and community implications (Blakemore and Choudhury 2006; Burnett and Blakemore 2009). By recognising our place in the world, and our effect upon it, we become aware of our identity and equally, our culpability. As we discuss below, recent research in social cognition in adolescents suggests that this age-range is a critical period for social development; one that has significant implications for those investigating pro-environment behavioural change and emission reduction activities.

New Approaches for Education

For those researchers and educators involved in teaching climate science, there is a general ennui with regard to education and the benefits we can obtain from structured learning environments, particularly information-based interventions (Bliuc et al. 2015; Corner et al. 2015; Stamm et al. 2000). It is generally accepted that we have 'failed' to communicate the problem effectively, or impart the knowledge and intellectual constructs necessary to process the problem (Shi et al. 2015). Rather than questioning the educational approach and 'going back to the drawing board' in light of these findings (or to methodologically reconsider to whom we are communicating), the conclusions drawn by many researchers has been that we should embark on wholly different directions, based on indications that 'knowledge deficit' is *not* one of the most significant factors in climate science understanding (Kahan et al. 2011; Potter and Oster 2008). There is, of course, no doubt that researchers are replicating findings that poor understanding and limited improvement prevail across the board; no more so than in the adolescent age group (Corner et al. 2015; Harker-Schuch and Bugge-Henriksen 2013). However, many of those findings have emerged from studies based on limited interventions (e.g. a 45 min lecture in climate science) or from correlations between 'general' science and denialism instead of *specific* climate science and denialism (Shi et al. 2015; Stevenson et al. 2014). Rather than dismissing the key messages of previous findings, in relation to climate science communication, they serve to highlight the need for communicating about climate science to early adolescents in the beginning stages of their *weltanschauung* cultivation. It also emphasises the need to give education the attention it deserves, particularly in this age group.

It is important to note, that this paper explores how climate literacy interventions may improve both the understanding of climate change, and the necessary actions to reduce emissions in the early adolescent age group. While the potential for education is significant, both at the individual level and in the broader public arena, we examine only a part of the wider and more complex issues related to climate communication, and emission reduction activities. For example, worldview bias will persist in older age groups, and attempting to overcome denialist attitudes with education may be counter-productive (Kahan et al. 2012). In addition, although we have distinguished between the physiological changes in adolescence (cognitive development, sociali-sation and self-determination) and society, in reality these cannot be isolated from each other and are functions of each other. This is manifested, for example, in the development of worldview and the context of the individual's socio-cultural environment: we are the products of the habitats that we occupy which are, in turn, shaped by our presence. However, our behaviours are typically controlled by us, particularly those related to social interactions and communication. It is therefore useful, when we explore pathways of human behaviour, physiology and society, to break the elements down into their component parts in order to assess their function and potential for alteration.

Conclusion

We have, then, a clearly defined societal group that meets vital criterion for climate science communication. Their intellectual development is at its most receptive to cultivating and adapting worldviews, they have the greatest potential to affect long-term change by virtue of their age and social position, they are highly accessible both in the educational setting and as a policy and governance requirement (through public education and the minimum education levels), and they are founding their early social allegiances that will form the basis of their adult communities: characteristics and circumstances that are found at no other time in human development.

When it comes to an individual's preconceptions, opinions, beliefs and/or ideals the effect that worldview has cannot be understated. This is particularly true in climate science, more so than any other arena. What must be examined (for the security and longevity of our civil society depends on it) is whether such worldviews can be cultivated or adapted. Particularly, whether these 'tweaks' can be made through targeted instruction and exercises aimed at fostering public engagement and better climate science understanding, at an age when worldviews are being formed and cultivated. The receptiveness of the adolescent brain is at its greatest in the early stages of puberty, and the importance of authoritative wisdom (teachers, experts, elders) diminishes as the adolescent grows older (Case 1985; Piaget 1972). As adolescents age, their willingness to adopt new concepts, broaden their worldviews and challenge their preconceptions gradually decreases and become less amenable to adapting and informing an existing worldview. It is, therefore, the younger portion of this demographic (12–14 years) that possesses the ideal physiological and social characteristics (and the strongest compatibility) for communication and education strategies. In essence, by improving climate literacy in early adolescence we are concentrating on and reaching the most biologically, physiologically, intellectually and societally ideal age group that is available. Future research endeavours may benefit from examining which aspects of climate science are cognitively easy to grasp for this age group. Studies should also evaluate how the more complex aspects of this topic can be built on top of those easier to understand constructs. Current science communication research also offers many tantalising avenues for teaching climate science that may be uniquely suitable for this age group. Evaluating these avenues (such as serious gaming) may further enhance our understanding and efficacy in communicating science to this age group.

Endeavouring to engage with climate change is a self-deterministic challenge. Accepting such a challenge involves maturity and civic courage that is a hallmark of adulthood. In addition, engaging with climate science could train cognitive function, aid in intellectual development, and promote abstract reasoning processes; further preparing an emerging adult for the tasks and responsibilities ahead. Finally, exposure to the concepts of climate change (and the science underlying the premise of climate change) may improve an individual's confidence in responding to the threat, and by default, decrease their fear and aversion to engage with it. Such exposure may provide new intellectual concepts and pathways for developing constructive and civic-conscious worldviews, or revising pre-existing ones. By teaching climate science to emerging adults, we hope to foster the development of worldviews that encompass the dimensions of civic responsibility (informed opinions, broader socio-cultural dimensions, and considerations), and embolden our youth to approach climate change confidently and effectively so that the forecast of their tomorrow looks brighter than the forecast of our today.

References

Audi R (2015) The Cambridge dictionary of philosophy. In: Audi R (ed) 3rd edn. Cambridge University Press, Cambridge. https://doi.org/10.1017/CBO9781139057509

Ayer AJ, Marić S (1956) The problem of knowledge, vol 6. Macmillan London

Berry HL, Bowen K, Kjellstrom T (2010) Climate change and mental health: a causal pathways framework. Int J Public Health 55(2):123–132. https://doi.org/10.1007/s00038-009-0112-0

Blakemore S-J, Choudhury S (2006) Development of the adolescent brain: implications for executive function and social cognition. J Child Psychol Psychiatry 47(3–4):296–312. https://doi.org/10.1111/j.1469-7610.2006.01611.x

Bliuc A-M, McGarty C, Thomas EF, Lala G, Berndsen M, Misajon R (2015) Public division about climate change rooted in conflicting socio-political identities. Nat Clim Chang 5(3):226–229. https://doi.org/10.1038/nclimate2507

Bostrom A, Boehm G, O'Connor RE (2013) Targeting and tailoring climate change communications. WIREs Clim Chang 4(5):447–455. https://doi.org/10.1002/wcc.234

Brulle RJ, Carmichael J, Jenkins JC (2012) Shifting public opinion on climate change: an empirical assessment of factors influencing concern over climate change in the US, 2002–2010. Clim Chang 114(2):169–188. https://doi.org/10.1007/s10584-012-0403-y

Burnett S, Blakemore S-J (2009) The development of adolescent social cognition. Ann N Y Acad Sci 1167(1):51–56. https://doi.org/10.1111/j.1749-6632.2009.04509.x

Carmi N, Arnon S, Orion N (2015) Transforming environmental knowledge into behavior: the mediating role of environmental emotions. J Environ Educ 46(3):183–201. https://doi.org/10.1080/00958964.2015.1028517

Case R (1985) Intellectual development: birth to adulthood. Academic Press, Orlando, CA

Checkoway B, Allison T, Montoya C (2005) Youth participation in public policy at the municipal level. Child Youth Serv Rev 27(10):1149–1162. https://doi.org/10.1016/j.childyouth.2005.01.001

Checkoway B, Richards-Schuster K, Abdullah S, Aragon M, Facio E, Figueroa L, White A (2003) Young people as competent citizens. Community Dev J 38(4):298–309. https://doi.org/10.1093/cdj/38.4.298

Cheshire B (2017) Autism and teens: Sydney father James best's radical experiment to help son, Sam. ABC News. Retrieved from http://www.abc.net.au/news/2017-07-24/sydney-dad-james-bests-radical-approach-to-autism-parenting/8723172

Corner A, Roberts O, Chiari S, Völler S, Mayrhuber ES, Mandl S, Monson K (2015) How do young people engage with climate change? The role of knowledge, values, message framing, and trusted communicators. WIREs Clim Change 2015 6(5):523–534. https://doi.org/10.1002/wcc.353

Damon W, Lerner R (2008) Child and adolescent development: an advanced course. Handbook of child psychology, vol 2. Cognition, perception, and language, 1st edn. Wiley, New Jersey. https://doi.org/10.1300/J019v04n01_01

Dumontheil I (2014) Development of abstract thinking during childhood and adolescence: the role of rostrolateral prefrontal cortex. Dev Cogn Neurosci 10:57–76. https://doi.org/10.1016/j.dcn.2014.07.009

Ferkany M, Whyte KP (2012) The importance of participatory virtues in the future of environmental education. J Agric Environ Ethics 25(3):419–434. https://doi.org/10.1007/s10806-011-9312-8

Field S, Hoffman A, Posch M (1997) Self-determination during adolescence a developmental perspective. Remedial Spec Educ 18(5):285–293

Frith U, Frith CD (2003) Development and neurophysiology of mentalizing. Philos Trans R Soc B: Biol Sci 358(1431):459–473. https://doi.org/10.1098/rstb.2002.1218

Ghysels M (2009) Will students make the grade in an education for the world ahead? The erroneous dilemma between testing and creativity. J Qual Particip 32(1):20–24

Gowers S (2005) Development in adolescence. Psychiatry 4(6):6–9. https://doi.org/10.1383/psyt.4.6.6.66353

Harker-Schuch I, Bugge-Henriksen C (2013) Opinions and knowledge about climate change science in high school students. Ambio 42(6):755–766. https://doi.org/10.1007/s13280-013-0388-4

Ho E, Clarke A, Dougherty I (2015) Youth-led social change: topics, engagement types, organizational types, strategies, and impacts. Futures 67:52–62. https://doi.org/10.1016/j.futures.2015.01.006

Jensen F (2015) The teenage brain: scaffolding the brain for lifelong learning. The Huffington Post Australia Pty Ltd., 1–2. Retrieved from http://www.huffingtonpost.com/smart-parents/the-teenage-brain-scaffol_b_7242344.html?ir=Australia

Jensen FE, Nutt AE (2015) The teenage brain: a neuroscientist's survival guide to raising adolescents and young adults, 1st edn. Harper, New York

Kahan DM, Jenkins-Smith H, Braman D (2011) Cultural cognition of scientific consensus. J Risk Res 14(2):147–174. https://doi.org/10.1080/13669877.2010.511246

Kahan DM, Peters E, Wittlin M, Slovic P, Ouellette LL, Braman D, Mandel G (2012) The polarizing impact of science literacy and numeracy on perceived climate change risks. Nat Clim Chang 2(10):732–735. https://doi.org/10.1038/nclimate1547

Karpudewan M, Roth W-MM, Abdullah MNSB, Bin Abdullah MNS (2014) Enhancing primary school students' knowledge about global warming and environmental attitude using climate change activities. Int J Sci Educ 37(1):31–54. doi.org/10.1080/09500693.2014.958600

Kuhn D, Franklin S (2008) The second decade: what develops (and how)? In: Damon W, Lerner RM (eds) Child and adolescent development—an advanced course, vol 19. Wiley, Hoboken, New Jersey, pp 517–590

Kuhn S, Brass M (2009) When doing nothing is an option: the neural correlates of deciding whether to act or not. Neuroimage 46(4):1187–1193. https://doi.org/10.1016/j.neuroimage.2009.03.020

Lazarus RJ (2009) Super wicked problems and climate change: restraining the present to liberate the future. Cornell Law Rev 94:1153–1234

Levin K, Bernstein S, Auld G, Cashore B (2012) Overcoming the tragedy of super wicked problems: constraining our future selves to ameliorate global climate change. Policy Sci 45:123–152

Lewandowsky S, Ecker UKH, Seifert CM, Schwarz N, Cook J (2012) Misinformation and its correction: continued influence and successful debiasing. Psychol Sci Public Interes 13(3):106–131. https://doi.org/10.1177/1529100612451018

Makkreel RA (1975) Dilthey-Philosopher of the human studies, 2nd edn. Princeton University Press, New Jersey

Moser SC, Dilling L (2007) More bad news: the risk of neglecting emotional responses to climate change information (Chap. 3). In: Moser SC, Dilling L (eds) Creating a climate for change communicating climate change and facilitating social change. Cambridge University Press, Cambridge. doi.org/10.1017/cbo9780511535871

Munz M, Gobert D, Schohl A, Poquerusse J, Podgorski K, Spratt P, Ruthazer ES (2014) Rapid Hebbian axonal remodeling mediated by visual stimulation. Science 344(6186):904–909. https://doi.org/10.1126/science.1251593

Ojala M (2012a) How do children cope with global climate change? Coping strategies, engagement, and well-being. J Environ Psychol 32(3):225–233. https://doi.org/10.1016/j.jenvp.2012.02.004

Ojala M (2012b) Regulating worry, promoting hope: how do children, adolescents, and young adults cope with climate change? Int J Environ Sci Educ 7(4):537–561

Ojala M (2013) Coping with climate change among adolescents: implications for subjective well-being and environmental engagement. Sustainability 5(5):2191–2209. https://doi.org/10.3390/su5052191

Ojala M (2015) Climate change skepticism among adolescents. J Youth Stud 18(9):1135–1153. https://doi.org/10.1080/13676261.2015.1020927

Piaget J (1972) Intellectual evolution from adolescence to adulthood. Hum Dev 51(1):40–47

Postman N (1985) Amusing ourselves to death: public discourse in the age of show business, 1st edn. Viking books, New York

Potter E, Oster C (2008) Communicating climate change: public responsiveness and matters of concern. Media Int Aust 127:116–126

Quintelier E (2014) The influence of the Big 5 personality traits on young people's political consumer behavior. Young Consumers 15(4):342–352. https://doi.org//10.1108/YC-09-2013-00395

Rice JA, Levine LJ, Pizarro DA (2007) "Just stop thinking about it": effects of emotional disengagement on children's memory for educational material. Emotion 7(4):812–823. https://doi.org/10.1037/1528-3542.7.4.812

Robinson K (2010) Changing education paradigms. In: RSA animate. The Royal Society of Arts, London, pp 1–4. http://www.youtube.com/watch. The Royal Society for the encouragement of Arts, Manufactures and Commerce, United Kingdom. Retrieved from www.theRSA.org

Roeser S (2012) Risk communication, public engagement, and climate change: a role for emotions. Risk Anal 32(6):1033–1040. https://doi.org/10.1111/j.1539-6924.2012.01812.x

Shanahan MJ, Mortimer JT, Krüger H (2002) Adolescence and adult work in the twenty-first century. J Res Adolesc 12(1):99–120. https://doi.org/10.1111/1532-7795.00026

Shi J, Visschers VHM, Siegrist M (2015) Public perception of climate change: the importance of knowledge and cultural worldviews. Risk Anal 35(12):2183–2201. https://doi.org/10.1111/risa.12406

Stamm KR, Eblacas PR, Clark F (2000) Mass communication and public understanding of environmental problems: the case of global warming. Public Underst Sci 9:219–237. https://doi.org/10.1088/0963-6625/9/3/302

Steinberg L (2005) Cognitive and affective development in adolescence. Trends Cogn Sci 9(2):69–74. https://doi.org/10.1016/j.tics.2004.12.005

Stevenson K, Peterson N, Bondell H, Moore S, Carrier S (2014) Overcoming skepticism with education: interacting influences of worldview and climate change knowledge on perceived climate change risk among adolescents. Clim Chang 126(3–4):293–304. https://doi.org/10.1007/s10584-014-1228-7

Stokes P (2012) No, you're not entitled to your opinion. The Conversation. Retrieved from http://theconversation.com/no-youre-not-entitled-to-your-opinion-9978

Swarup A, Dankelman I, Ahluwalia K, Hawrylyshgn K (2011) Weathering the storm: adolescent girls and climate change. London

Tucci J, Mitchell J, Goddard C (2007) Children's fears, hopes and heroes: modern childhood in Australia. Ringwood

UNICEF (2015) Unless we act now—the impact of climate change on children

United Nations (DESA) (2012) World population monitoring. New York. Retrieved from http://www.un.org/en/development/desa/population/publications/pdf/fertility/12_66976_adolescents_and_youth.pdf

Vollebergh WAM, Iedema J, Raaijmakers QAW (2001) Intergenerational transmission and the formation of cultural orientations in adolescence and young adulthood. J Marriage Fam 63(4):1185–1198. https://doi.org/10.1111/j.1741-3737.2001.01185.x

Wittgenstein L (1953) Philosophical investigations. Wiley

Wolf J, Moser SC (2011) Individual understandings, perceptions, and engagement with climate change: insights from in-depth studies across the world. WIREs Clim Chang 2(4):547–569. https://doi.org/10.1002/wcc.120

Wray-Lake L, Flanagan CA, Osgood DW (2010) Examining trends in adolescent environmental attitudes, beliefs, and behaviors across three decades. Environ Behav 42(1):61–85. doi.org/10.1177/0013916509335163

Chapter 17
Developing a Climate Literacy Framework for Upper Secondary Students

Inez Harker-Schuch and Michel Watson

Abstract While changes to our climate become more apparent and prevalent, adequate understanding in the broader public arena of the causes behind this phenomenon continue to be poor and incomplete. Education remains an essential tool for establishing an informed opinion as well as ensuring students are ready to enter society as adults. To this end, this paper explores the development of a climate literacy framework tested at a secondary school in each Australia and Norway. The pedagogy and curriculum are designed to provide a science-based overview of knowledge relating to climate science and climate change in the upper secondary age group. Measuring existing climate literacy at 53.62% ($n = 99$) we show that, not only are knowledge deficits shared across borders (culture, language, education systems) there are distinct differences in the understanding of domain-specific aspects in climate science. While no difference between the knowledge domains for the domains related to climate science, there were statistically significant differences in those domains related to drivers of natural climate variation and the impacts and consequences associated with climate change. Furthermore, our findings suggest that, while climate change is taught in secondary school in these two countries, very little improvement in understanding regarding the scientific is made during the secondary school period. We argue that too much focus is put on the consequences and impacts of climate change in the secondary classroom and too little effort if made to improve climate literacy in the physical science basis of natural climate variation and the physical causes and mechanisms underlying climate science. Our study supports previous research which shows that the consequences and impacts of climate change should be only taught after the physical science basis of climate change has been learnt. Our findings also support previous research that recommends that climate literacy interventions should

---

I. Harker-Schuch (✉)
Fenner School of Environment and Society, The Australian National University, B141, B48, B48A Linnaeus Way, Acton Act, Canberra 2601, Australia
e-mail: inez.harker-schuch@anu.edu.au; inez@earthspeople.org

M. Watson
Research School of Population Health, The Australian National University, 62 Mills Road, Acton Act, Canberra 2601, Australia
e-mail: michel.watson@anu.edu.au

© Springer Nature Switzerland AG 2019
W. Leal Filho and S. L. Hemstock (eds.), *Climate Change and the Role of Education*, Climate Change Management, https://doi.org/10.1007/978-3-030-32898-6_17

begin in early adolescence and traverse the climate science curriculum in the following order: KD1: Earth in the solar system; KD2: Greenhouse Gas molecules; KD3: Albedo; KD4: Earth's atmosphere before introducing KD5: natural climate variability; KD6: Feedbacks and climate instability; and KD7: Anthropogenic emissions. By teaching climate science in this way, we will ensure our youth are adequately informed about the issue prior to their emergence into society and we may reduce worldview bias and post-fact rhetoric.

Keywords Climate literacy · Secondary climate education · Knowledge domains · Climate change curriculum · Climate change pedagogy

Introduction

> You cannot hope to build a better world without improving the individuals
> —Marie Skłodowska-Curie

There is now robust scientific evidence and certainty that anthropogenic climate change is occurring and will have long-term consequences to the Earth's systems (e.g. atmospheric temperature increase, sea-level rise, increases in extreme weather events, and changes to the hydrological cycle, the radiative balance, and atmosphere-ocean interactions, to name only a few), as well as to human enterprises and endeavours (e.g. food production, water availability, climate-related conflicts, human health etc.) (Barnett and Adger 2007; Godfray et al. 2010; Le Treut et al. 2007; New et al. 2011; Pearson 2006). These changes demand collective international action and cooperation, particularly in relation to greenhouse gas emission reduction. Unfortunately, this collective action is impeded by many factors; most notably, a lack of public engagement on the issues of anthropogenic climate change that could drive political and social impetus for action (Moser 2004). This lack of engagement is associated with humankind's dependence on fossil fuels and an ongoing ambivalence toward climate change. This ambivalence is driven by several factors including the 'wickedness' nature of climate change (Lorenzoni et al. 2007), the effect of individual 'worldview bias' (Kahan et al. 2011), a lack of adequate scientific understanding of the problem in the broader public arena (Corner and Groves 2014), a reluctance to communicate the science of climate change (in part due to worldview barriers) (Shi et al. 2015) and an ongoing struggle to find an effective communication strategy that can straddle the barriers that impede collective action (Feinberg and Willer 2013; Moser and Dilling 2012). This communication impasse is further exacerbated by sub-factors including a tendency for those in the minority to speak out more vociferously than those in the majority (Porten-Cheé and Eilders 2015), the difficulty of 'seeing' climate change (Crayne 2015), the increased emotional burden presented by the issue (Ojala 2015) and it's inter-relatedness with other disciplines and phenomena (Crayne 2015). If we are to manage the task of mitigating climate change and its disastrous social, political, economic and environmental effects, education must form a crucial part of that solution. It will be necessary to create a public conversant with climate science in order to overcome the barriers to public action and constructive communication.

This study proposes a recently-tested approach to climate education that serves to address the main impediments to collective action and engagement, offering an effective communication and education strategy to overcome global inaction towards climate change. We propose that effective climate change education and communication depends on a fact- and cause-based approach (i.e. the mechanisms and processes that describe and drive the climate system), that should be introduced in early adolescence before worldview development; thereby embedding climate change in the natural science arena where it belongs. Therefore, we offer a pedagogical and curriculum framework for teaching climate science to secondary school-aged students based on knowledge domains (learning units) that build climate science mastery, competence and ensure that opinions regarding climate change are based in well-established climate science theory. In this study, we test the application of a climate literacy framework on older adolescents to investigate the usefulness of such a pedagogy and its potential as a climate education tool.

Barriers to Climate Change Education and Communication

Climate change has been a public issue since the mid-1980s (Jaspal and Nerlich 2014; Schäfer et al. 2016; Weingart et al. 2000) as has the awareness of the need for global action to reduce emissions. Therefore, it is no longer necessary to justify the need for such action to take place; rather, efforts to ensure a conversant public must now take priority. However, even though the issue has become a household term, the underlying understanding of climate change as a natural science phenomenon remains poor and inadequate (Corner et al. 2015; Harker-Schuch and Bugge-Henriksen 2013). While it could be argued that effective action on climate change does not necessarily depend on a conversant and informed public, the very foundation of governance in the Western world (especially that established in a democratic system driven by public opinion and elections) depends on an equally-informed public that can, as a collective, make an informed decision regarding their political and social stability both now and in the future (Wetters 2008).

When we ignore the necessity of a conversant and informed public, we divest from education to empower a post-fact and post-reason society. In doing so, we undermine the legitimacy of science as a public good; removing the evidence and increasing, according to Higgins (2016) the '*tolerance for inaccurate and undefended allegations*' that favour post-fact rhetoric. More alarmingly, we erode our capacity for resilience, adaptation and mitigation as the findings from the scientific community are invalidated, silenced and refuted. Even now, global efforts to address climate change are being handicapped as scientists are forced to excessively justify their work and/or are called upon to respond to denialist resistance both at the governance and professional level. Other scientists have felt obligated to downplay (Brysse et al. 2013), understate or underreport their findings (Lewandowsky et al.

2015; Waldmann 2018). Research also indicates that while scientists cope with the psychological burden of climate change by increasing their objectivism and suppression of their emotions (Head and Harada 2017) their psychological burden is likely to be larger due to their increased understanding of climate change as a problem. Signs of 'pre-traumatic stress', argue De Bourmont and Martindale (2015), are *'increasingly evident among those who stare at the problem of climate change head on: climate scientists, climate journalists and climate activists'*. With the pressure to find meaningful solutions in a short time-period, these additional demands on scientific expertise may have unwelcome future consequences. There is a clear need to increase understanding in the broader public arena, not least to decrease emission reductions, but also to increase public trust in the value of scientific endeavour.

Unfortunately, the task of informing the general public is thwarted by many factors including (1) the post-fact politicisation of climate change, (2) the 'wickedness' of climate change, (3) worldview bias, (4) a lack of adequate scientific knowledge related to climate science (particularly in relation to the causes and mechanisms that drive the climate system) and (5) a reluctance on the part of the science communication community to treat climate change as a natural science which, in part, can lead to a reliance and overemphasis on fear appeals. These factors are discussed below.

(1) Post-fact: The politicisation of climate change and the power of doubt

Due to the dependence of society (socio-economic/-political system) and industry on fossil fuels for continued economic growth, the vested interests in maintaining production without a significant alteration to existing systems (electricity production, personal and industrial transport, logistics, manufacture, food and material production) over more-sustainable systems (renewable energy, electric cars, recycling, upcycling, reusing) prevails. While the rationale to transition to a sustainable-resource system makes long-term economic, social and environmental sense, there is a concerted effort from industry (particularly those invested in a fossil-fuel-based economy) to destabilise the cognitive logic that drives this rationale (Oreskes and Conway 2010). Even as some countries (and political leaders) recognise the necessity for emission reduction, groups and industries within those countries often prevent greater political action taking place. This destabilisation, known as 'post-fact', is evident in the politicisation of climate change (Bryce and Frank 2014; Jaspal and Nerlich 2014) as fossil-fuel-invested and -financed politicians and industry leaders deny the scientific evidence and repudiate the findings provided by the scientific community. Such tactics undermine public trust in scientists and foster an environment whereby public opinion dominates the climate change discourse. While science must maintain scientific and methodological rigour, politicians are not required to tell the truth or maintain any intellectual honesty (Higgins 2016), making constructive discourse on this issue virtually impossible. For a lay audience, there is clear evidence of a conflict between these groups of 'trusted messengers' (Corner et al. 2015; Porter et al. 2012), thereby creating a foundation for doubt and mistrust in relation to climate change. Those that have an established trust in the scientific community are therefore more likely to support the findings of the scientists while those that distrust scientists are more likely to reject and deny the findings. The power of doubt as an impediment to, and function, of climate action is illustrated by the many studies investigating public awareness of climate change that focus, almost exclusively, on an individual's 'belief' of climate change—that is, whether they believe that climate change is happening and whether they believe it is caused by humans. Very little research into climate science knowledge i.e. an individual's understanding of the scientific processes that describe the climate system, has been undertaken. This focus on public opinion as a function of climate engagement, unfortunately, undermines the evidential basis of climate science and reinforces post-fact ideologies.

(2) A wicked problem

The wickedness of climate change also affects the dissemination of climate change information both in formal education environments (schools, universities, etc.) and the broader public arena (Lorenzoni et al. 2007; Nicholson-Cole 2009). According to Rittel and Webber (1973) wicked problems are problems outside the boundaries of 'normal' science and they possess certain, unique, characteristics that make their resolution and mitigation very difficult. Climate change is wicked insofar that: (1) the process of climate change has no defined end in which it will cease or go away; (2) it will engender effects that are both beneficial and damaging at the same time, it is neither good nor bad nor possesses a true-false dichotomy; (3) there is no way to pre-test solutions or know if the consequences of those solutions will be beneficial or damaging; (4) there will be no way to retract solutions once they are implemented; (5) there is an almost inexhaustible array of solutions—and it's impossible to know which to choose; (6) there is no precedence for tacking this issue, having never encountered it before; (7) it could be considered a symptom of another problem (e.g. population growth or the current economic model); (8) discrepancies in research findings can be interpreted in many way i.e. people base their opinions on the discrepancies that suit them best; (9) solutions will not be tolerated if they fail and those that implement them will be subject to vilification and outrage; (10) climate change does not have a definitive formulation i.e. there is no structure that defines climate change with which we can meaningfully orient ourselves to define the problem. This difficulty is exacerbated by other 'super-wicked' factors related exclusively to climate change, as described by Lazarus (2009) in that: (11) time is running out to deal with the problem of climate change; (12) those that have caused climate change are also involved in the solutions e.g. oil companies are funding research into climate change including into think-tanks that promote denialist rhetoric; (13) there is no central authority that can meaningfully manage or control the problem; and (14) 'irrational discounting' takes place that delays engagement to a distant time in the future. The wickedness of climate change makes it very difficult to teach climate change due to its complexity/vastness. This creates an ongoing challenge for individuals to perceive and decide upon actions that could be meaningful or constructive.

(3) Worldview bias

While the wickedness of climate change prevents individuals from conceptualising and establishing a strategy to engage with climate change, the issue of socio-cultural/-political worldview plays a far bigger impediment to climate change education and communication (Kahan et al. 2011). Worldview, according to Dilthey, '*is an overall perspective on life that sums up what we know about the world, how we evaluate it emotionally, and how we respond to it volitionally*' (as translated by Makkreel 1975). In essence, each individual has an ingrained attitude and perspective on life that they have inherited from their socio-cultural/-political circumstances that defines their perception of the world and dictates how they interact with it. Studies have shown that people with strong individualistic worldviews (those with strong anthropocentric attitudes who value the individual over society; whereby personal actions and interests are more important than frameworks of family, community and governance) are less likely to support environmental issues such as climate change in comparison to other worldviews such as hierarchy, fatalism, egalitarianism (Kahan et al. 2011). Lima and Castro (2005) suggest this is because individualists have lower levels of global concern and believe in man's dominion over nature. While individualistic worldviews usually represent only a small fraction of all worldviews, the nature of their worldview suggests strong self-empowerment narratives which pre-dispose them to governance and leadership roles in an industrial society. Worldview is a pernicious problem with regard to education as it is difficult to alter or revise once it is established and efforts to alter another's worldview once established—such as through education—are often met with strong resistance and further entrenchment of existing worldviews (Kahan 2013). For example, attempts by educators or communicators to alter worldviews in others often elicits strong counter-reactions—even when the efforts involve established science and fact-based information (Kahan et al. 2011, 2012). Research from Chapman and Anderson (2013) shows that individuals experience the same visceral and disgust reactions to ideas that conflict with their existing opinions as they do to things they are physically disgusted by (such as 'unpalatable foods, filthy restrooms, and bloody wounds'). The effect of worldview and its entrenchment, therefore, makes education efforts very difficult to implement, let alone establish broader outreach programmes.

(4) Poor scientific knowledge, fear appeals and the focus on impacts and consequences

General climate science knowledge also presents a significant barrier in the climate change education arena. Aside from a public that is poorly-conversant with climate change, teachers also lack the skills and knowledge necessary to effectively communicate climate change to their students—particularly those in secondary school. Misunderstandings about simple environmental processes or conflating climate change with ozone depletion are ongoing issues in climate change education. While this conflation may seem irrelevant in the broader climate change issue, not distinguishing each environmental problem could have larger communication and engagement implications. For example, as discussed previously, research has indicated that those with individualistic worldviews are likely to entrench their existing climate attitudes if challenged. Individualists may, however, consider emission reduction activities if approached with an argument regarding, for example, ocean acidification—particularly if they live, as three-quarters of the population does, near the ocean. Research has shown that when those with individualistic worldviews are confronted with local threats or concerns, they are as likely as other worldview groups to become engaged and take action (Lima and Castro 2005). More importantly, knowledge plays a role in civic engagement. A study across 28 countries in 1994 (Torney-Purta et al. 2001) investigating knowledge about civic engagement as a motivator to participate in civic life showed '*that equipping young people with knowledge of basic democratic principles and with skills in interpreting political communication is important in enhancing their expectation that they will vote*'. While this applies to civic responsibility, we can assume—given appropriate information and knowledge—students might demonstrate the same association for climate change engagement and emission reduction activities.

Other challenges to communication and engagement include the ongoing focus on the consequences and impacts of climate change as a departure point for climate change education rather than on the causes of the mechanisms that describe climate science as a scientific system. While employing future doom scenarios of consequences and impacts may capture an audience initially, research tells us that fear appeals and disaster narratives may have a polarising and paralysing effect in the long term (Potter and Oster 2008; Stern 2012). Focusing on the scientific basis of climate causes and mechanisms may reduce this effect by removing 'emotionally-charged information (specifically that related to impacts, feedbacks and consequences as well as the human role) from the climate discourse' may increase an individual's sense of competency. This is supported by Clark et al (2013), who state that 'mechanistic information' is a critical component of knowledge development and '*highly germane science information can clearly change the public's understandings and opinions*'.

The use of fear appeals needs further consideration, also. As described in the previous section on wickedness, the implementation of communication strategies that include fear appeals may have already exercised an effect in society (Nicholson-Cole 2009; Stern 2012). This effect may cultivate denialist worldviews and, as a consequence, individuals may be resistant to revision or new information. Although the ethical considerations are beyond the scope of this paper, the implications of an education approach that employs fear appeals is one that will perpetually undermine the validity of the science. By appealing to the emotions and perception of risk in an audience, we eliminate a rational and reasoned foundation for problem-solving. In effect, we load climate change into the public consciousness as a socio-cultural/-political construct instead of as a system-driven problem that has functions that can be observed, monitored, modelled, explained, explored, and solved.

(5) Climate communication worldview bias

While many in the scientific and research community consider themselves immune to worldview influences, the effect of worldview persists in all of us—this is the very nature of worldview. Communication practices that are deemed to be valuable and meaningful in the climate communication arena are supported and utilised while those that confront existing practice are challenged and, often, rejected. Challenges and resistance to the role of education to improve climate literacy is one of the most prevalent in climate change communication. Research articles continue to reject the 'knowledge deficit' model, in spite of evidence supporting the validity of knowledge-based interventions (Clark et al. 2013; Harker-schuch et al. 2019c; Shi et al. 2015, 2016). It could be argued that the accepted practice in the climate communication community regarding worldview and 'knowledge deficit' has driven the default 'fear appeal' strategy in an effort to overcome the socio-cultural/-political driver of worldview. Presented with new research supporting the validity of education interventions, however, has not shifted the worldview of climate communicators to revise their attitudes to education. Therefore, worldview bias presents a valid barrier to attitude and engagement regarding climate change within the communication and education community also. Due to the lack of research now in the role of education, we are presented with an over-abundance of research on public opinion on climate change but a dearth of research related to climate literacy and knowledge. We lack any clear insight into (1) the effect of knowledge on worldview bias, (2) how prior knowledge (of any kind, not only climate change) affects the development of worldview bias (which is known to form in adolescence), and (3) if knowledge can be leveraged at critical development stages to provide science-based information to aid in the development of an informed worldview. Of enormous significance to this discussion is the importance of treating climate change as a natural science—and how this treatment will affect worldview development, especially once the socio-cultural/-political aspects (including rhetoric and emotional content such as fear appeals) are removed.

Clearly, there are some substantial impediments to teaching climate change. These impediments, though significant, are not insurmountable and provide, if anything, a clear way forward. We need to establish climate change as a science and remove it from the public opinion arena. We need to find a pathway to circumvent the wickedness of climate change and the influence bias. We need to improve scientific understanding of climate change in the public-school system as a font of present and future public knowledge; assisting teachers and students to develop meaningful knowledge about climate change without the rhetoric of fear appeals and indoctrination. Lastly, we need to revisit the role of scientific knowledge and its potential for emission reduction and long-term global governance.

The Role of Education in Climate Literacy and Emission Reduction

In spite of the aforementioned trend to dismiss education, the importance of improving knowledge in the arena of sustainable development and climate change is undeniable (Bengtsson et al. 2018). Education improves our ability to think, it enhances our capacity to participate in public life, it stabilises democracy and public policy development, delivers many essential social and economic benefits to both the individual and society at large that have allowed enormous progress and innovation (Bengtsson et al. 2018). There is, for example, clear evidence that knowledge improves the mind. According to Bengtsson et al. (2018), *the learning processes which are associated with education lead to increasing synaptic density in our brains and thus permanently change the physiology of our brains with respect to such fundamental aspects as the ability for abstract thinking, imagining the not yet experienced, and the expanding time horizon of planning'*. Aside from the duty of preparing children and youth for the tasks of adulthood (Bobbitt 1918), public education ensures that each member in our community becomes sufficiently—and to some degree, equally—knowledgeable about issues, events, and phenomena. This results in a public that are not only capable of constructively participating in society (Bengtsson et al. 2018), they will also drive many social and individual benefits. These benefits include enhanced knowledge and skill, higher income potential, better health outcomes, enhanced social capital, entrepreneurship and innovation (Carvalho et al. 2019), a stronger adaptive capacity and overall resilience. *'The fate of empires'*, argued Aristotle, *'depends on the education of youth'*. For climate change, and its inherent connection to society and public policy, the question of whether we will be able to limit warming to within a habitable threshold may pivot on our ability to inform and prepare as many of the world's citizens as we are able. This is particularly relevant to public education where our children and youth are conveniently amassed into their respective age groups. Not only are we able to construct age-specific learning material as is done for all other school subjects, it can be designed to assist our youth in worldview development (Harker-schuch 2019) and to foster hope (Ojala 2015), develop climate-adaptive skills and cultivate resilience-thinking. Lastly, while critics of the knowledge deficit theory may challenge the importance of education and knowledge, it must not be forgotten that no climate literacy framework (curriculum and pedagogy) had ever been proposed, yet alone tested, until very recently (Harker-Schuch et al. 2019a, b, c; Milér and Sládek 2011). That worldview plays a role in denialist rhetoric and post-fact may have less to do with failures of knowledge deficit theory, but a lack of education prior to the development of worldview (Harker-schuch 2019). With or without worldview bias, education must be included in the adaptation and mitigation suite of strategies if we hope to adapt to climate change. This knowledge will be necessary if we are to have a public ready to respond to the anticipated transitions and upheavals that climate change will cause. Therefore, the urgency to construct a reliable and comprehensive climate science curriculum and pedagogy for use in the secondary public-school system is now acute.

Climate Change Pedagogy and Curriculum

Pedagogy

Target

While climate change is embedded in the science curriculum of most secondary schools in the developed world, it is frequently introduced after worldview development is well under way (Harker-Schuch et al. 2019a, b, c) or, perhaps, even established. In addition, climate change is a relatively new subject that is frequently tagged-on to existing science subjects. Therefore, efforts to introduce it to the classroom are often disjointed, incoherent or insufficiently communicated (Corner et al. 2015; Harker-Schuch et al. 2019a, b, c; Hess and Collins 2018; Shepardson et al. 2010). Recent research suggests that the topic of climate change should be introduced earlier to the classroom; at the age of 12 or the first year of secondary school (Harker-Schuch et al. 2019a, b, c). A 2017 study involving 401 12–13-year olds demonstrated that, not only was concern about climate change particularly high in this age group (Harker-Schuch et al. 2019b), they were able to cognitively-process aspects of climate science that would allow them to form a scientific understanding of climate science (Harker-Schuch et al. 2019a, b, c). This age group has recently begun the 2nd critical stage of intellectual development (Jensen and Nutt 2015) and, as such, are capable of higher-order executive reasoning (Case 1985; Piaget 1972). This is important for climate science education as many of the processes that underlie the physical science depend on a capacity for higher-order executive reasoning; particularly those related to process-dependent systems such as the albedo effect, radiative balance and feedbacks.

Introducing climate science to an earlier age group comes with important considerations. The first is that education and communication material involving consequences, impacts and future scenarios needs to be removed from the curricula. The second is that there must be a clear shift of climate change as an opinion to climate change as a science.

Shifting the Curriculum and Narrative from Consequences to Causes

Early adolescence marks the beginning of puberty and brings with it enormous social, emotional and physical upheaval. They will undergo a radical change in their relationship with society and significant others; shifting respect from parents and elders to their peers (Piaget 1972). They will be experiencing drastic emotional and psychological changes related to pubescence and changes to their physiology (including the afore-mentioned intellectual changes) (Case 1985; Jensen and Nutt 2015). They will be undergoing physical changes of an order that they have never cognitively experienced before (Case 1985; Jensen and Nutt 2015; Piaget 1972). These changes render them extremely vulnerable; particularly as they relinquish the innocence of childhood and transition to adolescence. Burdening this age group with impacts of future climate change would be foolish, not least because of the unknown effect such information might engender on their emotional and psychological well-being, but also because of the risk of polarisation and paralysation such fear appeals may foment (Kahan et al. 2012).

Shifting the Curriculum and Narrative from Opinion to Facts

Although there is a clear interest in public opinion and climate change from a political viewpoint (i.e. that related to elections and policy-development), we argue that there is little to be gained from public opinion on climate change from an education and communication standpoint. Invariably, public opinion surveys approach the opinion about climate change from an idiosyncratic belief perspective. They ask, 'Do you believe that climate change is happening?', rather than, 'What do you know about climate change?' (Leviston et al. 2014, 2015; Shao 2017). The acceptance that an idiosyncratic opinion on climate change will suffice over any useful knowledge is implicit the very moment we pose the question. By expecting a simple answer of yes or no to an opinion on climate change—or even one on a Likert scale with a range between yes and no—we relegate climate change to the very system we are trying to avoid: post-fact. We may blame lobbyists and pro-oil groups for exploiting this system, but it is a system that we have, as educators and communicators, also established and employed. Critically, from a political perspective, the focus on opinion, rather than any useful knowledge about climate change, undermines the foundation of democracy. A democratic system, argues Wetters, is '*not the vote alone, but the quality of the public forum out of which the vote emerges*' (Wetters 2008). Without a public adequately conversant and informed about climate science, we cannot hope to employ the democratic institutions that are in place to address it. Simply believing in climate change is also not a sufficient motivator to undertake actions to address climate change. In a study of 75 students in the United States, McNeill and Vaughn (2012) demonstrated that '*believing a scientific theory (e.g. climate change) is not sufficient for critical science agency; rather, conceptual understandings and understandings of personal actions impact students' choices*'; recommending that '*future climate change curriculum focus on supporting students' development of critical science agency by addressing common misconceptions*'.

In short, we argue that a pedagogy for formal education to improve climate science knowledge should begin in the early adolescence age group as, at this period in their development, they are able to cognitively process the science of climate change but have not yet formed their worldview. While consequences and impacts should form a part of the latter curriculum and pedagogy, only the physical science basis that underpins the drivers of climate science should be introduced in the early adolescent age group. By providing this age group with a solid foundation in cause-based climate science, we may reduce or circumvent the influence of worldview in later adolescence and adulthood. From an ethical standpoint, educating students about the drivers and causes of natural climate variability, we remove emotional appeals and diminish idiosyncratic biases that are known to affect attitudes and engagement. Furthermore, by focusing on cause-based climate science, we evade the influence of wickedness. This is avoided because the factors that define wickedness all occur in the future and relate to consequences, impacts and our subsequent reactions. For older adolescents, the consequences and impacts of climate change can be taught, but only once they have acquired sufficient knowledge related to climate science causes and processes.

From Lower Secondary to Upper Secondary

With a prototype climate literacy framework for early adolescence already constructed and tested (Harker-Schuch et al. 2019a, b, c), it is necessary to extend the framework to include upper secondary classes in order to have a comprehensive and age-appropriate curriculum and pedagogy for all of public secondary school. Since we have a baseline for climate literacy in early secondary school, a baseline for upper secondary will allow us to organise learning material (both curriculum and pedagogy) between these two positions and provide a better understanding of the learning distances relative to each age group—and to prepare learning material for the age groups in between. By obtaining a climate literacy measurement in this age group in comparison to pervious research done in earlier secondary years via the same research instrument, we can resolve the effect of current education interventions between the lower and upper secondary years. Although differences in cognitive ability exist between these age groups, the material for overall climate literacy stays the same. Since 12–13-year olds can cognitively process the concepts and material presented in this present study, the upper secondary students will not have cognitive difficult in processing this material, either. While we may expect upper secondary students to have well-established worldviews (Corner et al. 2015; Harker-schuch 2019), this study explores their climate literacy in order to construct learning materials between the pre-worldview age group and the established-worldview age group. Furthermore, not only are upper secondary students closer to adulthood and, hence, may effect social change sooner, their potential as change agents within their peer-community is one that may affect younger students. Providing a climate literacy result will elucidate their current climate understanding, its relation to younger years and provide context in relation to older adolescent worldview development.

Curriculum

As with any science topic, climate science can be broken down into smaller learning units, or 'knowledge domains' (Guskey 2015; Harker-Schuch et al. 2019a, b, c). For climate science, the knowledge domains (KDs) relate to the categorisation of scientific method and '*to specific, physical realms where certain phenomena take place*' (ibid.). KDs allow teachers to construct an intellectual boundary around '*what a learner is expected to understand in relation to a particular learning unit over a set time frame that can be taught in one grade or age group*'. Aside from providing a defined scope for learning, the subject matter within each KD can be scaled from easy to difficult, thereby providing navigable learning objectives as described by Biggs and Collis (1982) which then forms the grading rubric. As part of a previous doctoral research project examining climate literacy in early adolescence, we constructed a world-first curriculum for climate literacy based on these defined KDs and the Structure of Observed Learning Outcomes (SOLO) taxonomy according to Biggs and Collis (1982). In total there are 7 KDs related to climate science. These are: (KD1) Earth in the Solar System; (KD2) Greenhouse Gases as molecules; (KD3) Albedo; (KD4) Earth's Atmosphere; (KD5) Natural drivers of climate change; (KD6) Feedbacks and climate instability; (KD7) Anthropogenic emissions and their consequences. The first five KDs (1–5) relate explicitly to causes and processes that drive natural climate variability. The final two KDs (KD6 and KD7) relate to causes, consequences and the anthropogenic drivers of climate change. We distinguish between these two groups of KDs as 'climate science KDs' (KD1–KD5) and 'climate change KDs' (KD6 and KD7). The first four climate science KDs were tested for the early adolescent age group (early secondary) but KDs 5, 6 and 7 have not been tested.

This study, therefore tests KD1-KD4 and parts each of KD5 and KD7 in order to investigate the application of a climate literacy framework in the secondary school system. Due to ethical considerations, KD6 was not included in this stage of the research project as the researchers anticipate a greater public awareness of the consequences of anthropogenic emissions (e.g. sea level rise/increasing global temperatures/loss of the cryosphere) than an understanding of feedback and climate instability (such as tipping points or the runaway greenhouse effect). Aside from the concerns regarding the respondent's emotional well-being, increasing an awareness of KD5 may also trigger unwanted worldview responses.

Objectives and Aims

This study explores the climate literacy of upper secondary students in two high schools in Australia and Norway in order to inform the construction of a climate literacy education framework for upper secondary school. We investigate their prior knowledge in 5 knowledge domains (KDs: KD1-KD4 and parts of KD5 and KD7) and the relationship between these KDs. We examine whether there are differences in gender in relation to KDs. We further investigate how self-assessed knowledge of climate science corresponds to their prior knowledge performance. Lastly, we discuss the climate literacy results in relation to how it impacts the effectiveness of a pedagogy and curriculum framework previously developed by the principal investigator to teach principles of climate science and climate change at specific secondary stages to create an environment that diminishes the effects of individual worldview bias (Harker-Schuch et al. 2019a, b, c).

As such, this study has the following aims:

- To quantify the current knowledge of climate literacy in upper secondary students in KD1-KD4 and parts of KD5 and KD7
- To examine the relationship between KD1-KD4 and parts of KD5 and KD7 for the upper secondary participants
- To investigate the relationship of gender on KD1-KD4 and parts of KD5 and KD7
- To construct a climate literacy framework for upper secondary students
- Results used to inform curriculum and pedagogy framework that aims to mitigate factors that impede climate change education and communication strategies and overcomes public inaction

Methods

To quantify the climate literacy of the upper secondary age group a climate science questionnaire used in a previous research project was administered to upper secondary Norwegian and Australian students between October 2018 and February 2019. All permissions were obtained from the students and their teachers and all data was collected anonymously in accordance with Australian data privacy requirements and the General Data Protection Regulation (GDPR) of the European Union and to prevent misuse of personal information. Permission to conduct research was obtained from the participating schools, their teachers and the students. Due to ethical research protocols and privacy concerns, the regions and the schools cannot be identified. The questionnaire was provided in their native language (Norwegian for Norwegian students and English for Australian students) and during the class period (under monitoring by their teacher in exam conditions). Of the 108 students who took part, 99 were eligible for final inclusion in the analysis (permission granted and a fully-completed questionnaire). In total there were 40 Norwegian (40.4%) participants and 59 (59.6%) Australian participants. Of these, 54 (54.5%) were females, 41 (41.4%) were males and 4 (4.0%) were participants who preferred not to say. The research was conducted in accord with ANU ethics protocol 2015/583. No funding was obtained for this project.

Questionnaire

The research instrument for this study was based on a questionnaire used in a doctoral thesis project investigating climate literacy in 12–13-year olds. The same 19 questions (questions 1–19) from this questionnaire were included (corresponding to KD1–KD4 of the climate science KDs) and another 3 questions (questions 20–22) were added to include parts of KD5 and KD7 (Appendix 1). Due to the difficulty of the questionnaire, it was decided not to include the entire contents of KD5 and KD7 or any of the contents of KD6. Initially, the original questionnaire of 19 questions took more than 30 min to complete which, when including the additional KDs, may have demanded more class time than was available. Secondly, in order to establish a foundation for a climate literacy framework, the KDs that were tested in the younger age group were necessary in order to make a comparison. Lastly, the questionnaire was intellectually strenuous for the students and minimising the risk of students skipping through difficult sections (or becoming fatigued from the effort), it was logical to add fewer questions. While we will not be able to provide learning objectives for these KDs, an overall climate literacy score will be discernible.

Analysis

Climate literacy was achieved by combining the score of each question, dividing by the number of participants ($n = 99$) and summing the results for all question for each knowledge domain. The result was then divided by the number of questions in each knowledge domain. The sum of each knowledge domain score divided by the number of knowledge domains then provided the overall climate literacy.

To investigate the differences between KDs in each country, we conducted an independent T-test for paired samples (IBM SPSS statistics 23.0) between each country for each KD (KD1–KD5 and KD7). Oneway ANOVA was used to test for differences between KDs (KD1–KD5 and KD7) as well as differences in gender.

The overarching analysis approach, therefore, consisted of the following stages:

1. Descriptive values for overall climate literacy as defined by the knowledge domains (KDs).
2. Independent T-test was conducted to determine differences between countries
3. Oneway ANOVA was conducted to determine differences between each KD.
4. Oneway ANOVA was conducted to determine the differences gender and each KD.

Results

Overall Climate Literacy

Descriptive values are presented to provide an overview of climate literacy in the upper secondary age group, as defined by the KD learning units. Data is presented as mean $+95\%$ confidence interval (CI). The mean KD score in each column (Table 17.1) corresponds to the mean value of each KD as derived from the 19 climate science questions: 5 for KD1, KD2 and KD3; 4 for KD4 and for the 3 climate change questions: 2 for KD5 and 1 for KD7. The Mean KD score in each

Table 17.1 Descriptive values for each knowledge domain (sum of multiple questions within that KD) and overall climate literacy in the upper secondary student group (mean value for 22 questions); 95% CI

	Climate science KDs					Climate change KDs
	Earth in the solar system: KD1 mean (95% CI)	GHGs' as molecules: KD2 mean (95% CI)	Albedo: KD3 mean (95% CI)	Earth's atmosphere: KD4 mean (95% CI)	Natural climate change: KD5 mean (95% CI)	Anthropogenic emissions: KD7 mean (95% CI)
Individual question scores for each KD	62.37 (58.1–66.7)	42.68 (35.3–50.1)	49.24 (42.9–55.6)	25.25 (20.1–30.43)	56.57(52.5-60.6)	61.87(57.9–65.8)
	55.30 (51.3–59.3)	54.55 (47.8–61.3)	56.27 (46.6–56.5)	38.79 (31.2–46.3)	54.8(50.7–58.9)	
	51.3 (48.0–54.6)	66.41 (62.8–70.0)	45.20 (38.3–52.1)	62.88 (59.1–66.7)		
	68.94 (64.8–73.1)	47.48 (40.5–54.5)	56.82 (49.4–64.2)	20.91 (18.5–23.3)		
	66.67 (63.0–70.4)	51.0 (47.1–54.9)	64.4 (57.3–71.5)			
Mean KD score (KD CL level):	60.91 (58.9–62.9)	52.42 (49.7–54.9)	54.44 (50.9–58.0)	36.55 (33.6–39.5)	55.53 (52.7–58.3)	61.87 (57.9–65.8)
	Climate literacy (CL)		53.62			

Table 17.2 Comparisons of mean KD scores for each country

		Australia (mean = 51.7)		Norway (mean = 56.5)		Mean		
		Mean	SD	Mean	SD	Diff.	t	p
Climate science KDs	KD1	59.75	0.106	62.63	0.094	−2.87	1.39	ns
	KD2	51.95	0.138	53.13	0.130	−1.17	0.425	ns
	KD3	54.32	0.159	54.63	0.206	−0.31	0.082	ns
	KD4	37.41	0.150	35.24	0.147	2.13	−0.699	ns
	KD5	50.78	0.119	62.53	0.143	−11.74	4.44	<0.001*
Climate change KDs	KD7	56.36	0.171	70.00	0.206	−13.64	3.58	<0.001*

column represents the Climate Literacy level for each KD (KD CL level). Our analysis indicates that KD Climate literacy (KD CL) is 53.62% ($n = 99$) in the upper secondary high student group (Result 1). There were no outliers and the data was normally distributed for each group with skewness between −0.119 and 0.809 (SE = 0.243) and kurtosis between −1.171 and 315 (SE = 0.481). There was homogeneity of variance as assessed by Levene's test of Homogeneity of Variance ($p = <0.000$) when all domains were compared to one another.

Differences Between Countries

Using the independent T-test for paired samples, we analysed differences between countries. When equal variances were assumed, there was no difference between countries for KD4, KD3, KD2, and KD4 (Table 17.2). Analysis showed (Result 2) that there were significant differences in the last climate science KD (KD5) and the climate change KD (KD7). Norwegian students performed significantly better in KD5 (Mean = 62.53, SD = 0.143) than students in Australia (Mean = 50.78, SD = 0.119) and also in KD7 (Mean = 70.00, SD = 0.206) than students in Australia (Mean = 56.36, SD = 0.171). There were, however, outliers in KD1 and KD4 when analysed for country. A sensitivity analysis was conducted by modifying the outliers by replacing their KD score outcome with the next largest value. The T-test was re-run with no variation to the results as that obtained when the outliers were included.

Differences Between KDs

Using Oneway ANOVA, differences in KDs were analysed. Results from the Welch ANOVA were used to check for significant differences in the population due to heterogeneity of variance. Welch ANOVA indicated there were statistically significant differences between KD scores (Table 17.3); Welch's F(5, 272.172) = 38.604, p < 0.001. Results from the Games-Howell post hoc analysis (Result 3) were used to define where the differences lay; revealing statistically significant differences (p > 0.5) for results from KD4 (Mean = 36.96, SD = 0.148) to all other KDs; KD4 (Mean = 36.96, SD = 0.148), KD3 (Mean = 54.44, SD = 0.179), KD2 (52.42, SD = 0.134), KD1 (Mean = 60.91, SD = 0.102), KD5 (Mean = 55.80, SD = 0.141), KD7 (Mean = 61.87, SD = 0.197). Other significant differences between KDs include KD3 (Mean = 54.44, SD = 0.179) to KD1 (Mean = 60.91, SD = 0.102) and KD7 (Mean = 61.87, SD = 0.197); KD2 (52.42, SD = 0.134) to KD1 and KD7; KD5 (Mean = 55.81, SD = 0.141) to KD1.

Differences in Gender

Oneway ANOVA analysis of KDs was undertaken to determine differences in gender. The assumption of homogeneity of variances was violated for KD4 and KD3, as assessed by Levene's test for equality of variances (KD4: p = 0.024; KD3: p = 0.046). Results from the Welch ANOVA were therefore used to look for significance of differences in gender groups. There were no statistically significant differences for the effect of gender on KD2, KD1, KD5 and KD7. For KD4 and KD3 (Result 4), there were statically significant differences with males (KD4: Mean = 40.71, SD = 0.157; KD3: Mean = 57.07, SD = 0.149) scoring significantly better than females (KD4: Mean = 33.37, SD = 0.123; KD3: Mean = 53.15, SD = 0.201). Due to outliers, a sensitivity analysis was conducted by modifying the outliers by replacing their KD score outcome with the next value. The Oneway ANOVA was re-run with no variation to the results as that obtained when the outliers were included.

Discussion

This study explored the climate literacy in upper secondary students in a small sample of students in Norway and Australia. We quantified how well 15-17-year-old students understood climate science and aspects of climate impacts as a base on which to construct a climate literacy framework.

Overall, we found that climate literacy (Result 1) in this age group is 53.62%. This result describes the prior knowledge of climate literacy in the upper secondary age group and matches similar findings of climate literacy in this age group (Corner et al. 2015; Stevenson et al. 2014). Previously, Harker-Schuch and Bugge-Henriksen (2013) reported a climate literacy of 48.5% in 16–17 year olds in Denmark whereas Bodzin and Fu (2013) reported a climate literacy score of 40.8% in 13–15 year olds in the US. While this climate literacy score is low, it is generally low within the broader public arena (Clark et al. 2013).

Table 17.3 Descriptive values for each knowledge domain (sum of multiple questions within that KD) and overall climate literacy in the upper secondary school groups (mean value for 22 questions); 95% CI

KD (score)	KD4		KD3		KD2		KD1		KD5		KD7	
	Mean diff.	SE	Mean diff.	SE	Mean diff.	SE	Mean diff.	SE	Mean diff.	SE	Mean diff.	SE
KD1 (60.91)	−23.95*	0.018	−6.46*	0.021	−8.48*	0.017	–	–	−5.10*	0.017	0.96	0.022
KD2 (52.42)	−15.47*	0.020	2.02	0.022	–	–	8.48*	0.967	3.38	0.020	9.44*	0.024
KD3 (54.44)	−17.49*	0.023	–	–	−2.02	0.022	6.46*	0.021	1.36	0.021	7.42	0.027
KD4 (36.96)	–	–	17.49*	0.023	15.47*	0.023	23.95*	0.018	18.85*	0.020	24.91*	0.025
KD5 (55.80)	−18.85*	0.020	−1.36	0.023	−3.38	0.020	5.10	0.017	–	–	6.06	0.024
KD7 (61.87)	−24.91*	0.024	−7.42*	0.027	−9.44*	0.024	−0.96	0.022	−6.06	0.022	–	–

*$p < 0.5$

With regard to the country results, this study shows (Result 2) that Norwegians perform better than Australians in the KD related to natural drivers of climate change (KD5) and anthropogenic emissions (KD7). We might expect Australian students to perform slightly less well compared to the Norwegian students as they are approximately 1 year younger. In addition, a higher understanding of the natural drivers of climate change in Norwegian students might be explained, also, by a drive to improve climate literacy amongst the general public (Ryghaug and Skjølsvold 2016), a strong political advocacy towards emission reduction (Arnold et al. 2016), considerably lower levels of denialisms in the general population (Ryghaug and Skjølsvold 2016) and their geographical location (auroras, proximity to the North pole). In regards to the higher performance in knowledge related to the climate change (KD7) aspects of the questionnaire by Norwegian students, this may be explained by the physical presence of the season (visible snow cover, earlier onset of spring and later onset of winter) in comparison to the lack of a visual presence of seasons felt or perceived by students in Australian habitats. That Australia students are less familiar with the drivers of natural climate variability (KD5) and anthropogenic emissions deserves further investigation—especially in relation to both denialist rhetoric and impediments to engagement.

Comparison of the KDs to each other (Result 3) supports previous research in this area showing that knowledge domains do persist across language, culture and nationality in early secondary school (Harker-Schuch et al. 2019a, b, c). While several KDs showed no statistically significantly differences to one another, the performance at both the question- and KD-level (Figs. 17.1 and 17.2) indicates that there are knowledge commonalities that warrant further investigation. Aside from the practicalities

Fig. 17.1 Performance in country per question

Fig. 17.2 Performance in country per KD

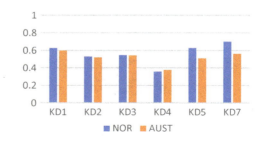

associated with ordering a topic into learning units and knowledge domains in order to teach it, the results offer a 'cumulative construction' (Biggs and Collis 1982) framework that could be employed for knowledge development. For example, they can be arranged into levels of complexity (from easy to difficult) that provides a foundation for mastery; that is, learners begin their learning journey at a level which is easy to master and gradually progress into increasingly difficult levels of learning. Lastly, although these tested upper secondary students show that they are conversant with some important aspects of climate science, large gaps in their overall understanding of the topic is lacking.

When we analyse the results for gender, we see an influence of gender (Result 4) on KD4 and KD3. Similarly to previous findings for KD1 in early adolescents, we see that the knowledge domain associated with Earth's atmosphere is less well understood by females than males (Harker-Schuch et al. 2019a, b, c). The gender influence on the knowledge domain associated with Albedo, though, is limited to this study. The higher performance by males may be explained, in some part, with childhood gender norms i.e. male children are more likely to receive toys associated with atmospheric environments (planes, satellites, kites, balloons) than the toys traditionally given to female children. A 2006 study found females, on average, to be less interested in science in high school, with the exception of biology (Miller et al. 2006). Within this study, it was discovered that females enrolled in a science major in order to fulfil requirements to enter into health profession. These finding have been replicated in other studies (Jones et al. 2000). Therefore, this lack of interest might play a role in the gender achievement gap in high school science knowledge domains as seen in this study and others (Bacharach et al. 2003).

By teaching climate science KDs as a pre-requirement for teaching climate change KDs, we embed climate change into the scientific realm where it belongs. This practice produces several key results that overcome barriers to climate change education and communication. These are; (1) Through re-establishing the phenomena of climate change as a science-based topic in public awareness, we hope to circumvent ineffective post-fact rhetoric and fear appeals in order to build both a public that is conversant with the underlying scientific mechanisms of climate change and a society that can constructively approach solutions toward mitigation, adaptation and emission reduction. (2) The wickedness of climate change is also diminished when we deal with climate change in an abstract and rational manner. In essence, we develop an intellectual relationship with the phenomena itself before we attempt to deal with it as a physical manifestation. This allows for a clearer path to action as the previous 'insurmountable' problems, as determined by climate change's wickedness, seem more achievable to solve. (3) With regard to worldview, individuals who are informed on the physical science basis are less likely to be influenced by climate change denialist hyperbole. Being adequately informed about the science of climate change also provides an intellectual resistance to denialist rhetoric as individuals will be better-equipped to cognitively process new information. This may improve their resilience against misinformation, misunderstandings and post-fact discourse. (4) By focusing on adolescents in formal and public education environments, we have the ability to teach these aspects of climate change in a safe, non-threatening environment that has both a mass communication effect and the potential to have immediate policy consequences. By shifting the narrative from consequences and impacts to the basics of climate science, we eliminate emotional and threat-laden appeals and build a narrative that further reinforces the topic in the scientific realm where it belongs. By teaching in public schools, scientific knowledge about climate change in the broader public arena will improve and have implications for attitudes, motivation and engagement. From an individual engagement perspective, not only do children and young people affect the attitudes and behaviour of familiar others (friends, parents, siblings), they have an important role in society as 'agents of change'(Checkoway et al. 2003; Checkoway et al. 2005; Ho et al. 2015)—particularly in the realm of social and digital media (Thackeray and Hunter 2010). (5) Finally, we show that the worldview barriers that arise from climate communicators and educators in relation to the knowledge deficit model are in need of revision and examination. For a start, an important critique on the underlying premise for disregarding the knowledge deficit model show that 'domain-specific science' knowledge (Shi et al. 2016) was not a measured but, rather, a general understanding of science. Remarkably, very little research has been undertaken on domain-specific climate literacy so the rejection of the knowledge deficit model was, at best, premature— and, at worst, a serious and alarmingly risky oversight on the part of educators and communicators. Furthermore, the very purpose of public education institutions is to prepare our youth for adulthood and the anticipated challenges and tasks that they will be expected to manage. Dismissing education as a valid pathway for climate engagement removes a critical function of the education system—and leaves our youth at a distinctly unique disadvantage in comparison to previous generations.

Climate Literacy Framework for Upper Secondary

Based on these findings we can now construct a climate literacy framework for upper secondary students (Table 17.4). Although KD7 has the highest level of prior knowledge, it should be the last to teach as it is related to the 'climate change KDs' which explain future events, impacts and consequences but does not describe the phenomena that cause climate change. These findings show a similar pattern of climate literacy in early adolescents (Harker-Schuch et al. 2019a; b, c). KD1 (Earth in the Solar System) should be the first KD taught as it is the KD with the highest prior knowledge from the 'climate science' KDs with a score of 60.91%. KD5 (Natural climate change) should be the second KD taught with a score of 55.53%—this KD was not included in the early adolescent age group due to the overall complexity of the topic and a risk of survey fatigue. KD3 (Albedo) should be the third KD taught with a score of 54.44%. KD2 should be the fourth KD taught with a score of 52.42%. KD4 should be the fifth KD taught with a score of 36.55%. KD7 would be, as described above, the last domain taught. Due to the similarity of the scores between KD3, KD2 and KD5, these might be swapped for any other KD in that

Table 17.4 The climate literacy framework for upper secondary: the order of KDs from easy to difficult

Climate science KDs					Climate change KD	
KD1 (earth in the solar system) KD CL: 60.91%	KD5 (natural climate change) KD CL: 55.53%	KD3 (Albedo) KD CL: 54.44%	KD2 (GHGs as molecules) KD CL: 52.42%	KD4 (Earth's atmosphere) KD CL: 36.55%	KD6 (feedbacks/climate instability) KD CL unknown	KD7 (Anthropogenic emissions) KD CL: 61.87%
Low difficulty	Increasing level of complexity			High difficulty		Taught after KD1–5
Age 11–12	Age 12–13	Age 13–14	Age 14–15	Age 15–16	Age 16–17	Age 17–18

range. Since only part of KD5 and KD7 were investigated in this study, we may find that the position of these two KDs may change in future investigations.

If this topic is introduced without any prior lessons in earlier grades, we recommend it follows this order for upper secondary. However, since worldview has already exerted an influence on this age group, we recommend that each full KD is introduced in each age group from 11–12-years onward, culminating in KD6 (climate instability and feedbacks) and KD7 (anthropogenic emissions) in the final years of secondary school. While KD6 was omitted from this study, the position of this KD will remain in the second-to-last position as it describes the causes of instability and the impacts and consequences of KD7.

Limitations

As this was initially designed as a pilot comparison study, a limited sample size was used to determine an estimate of climate change literacy in upper-secondary school students. This was, in part, to prevent disruption of classes. As explained previously in the methods, due to the difficulty of the questionnaire, it was decided not to include the entire contents of KD5 and KD7 or any of the contents of KD6 due to time pressures, potential for participant fatigue and the overall complexity of the subjects. Future projects arising from this study will include a wider and more comprehensive study cohort from across Europe and Australia to determine and compare current climate literacy levels. The sample of the participants may also not be indicative of the secondary population; both in their respective countries or in the broader global population. Further research is required to determine the significance of this study and its application in global public education secondary curricula.

Conclusion

For those concerned with climate change, we are rapidly nearing an event-horizon that will ultimately determine humankind's capacity to both cope with an environmental problem that we have engineered and are forced, through its wickedness, to overcome. If we are to manage this task, education must form an essential part of that solution. Our study shows that more efforts need to be made in secondary schools with regard to climate literacy as our findings highlight the appearance of little improvement throughout the secondary public education system and, in some cases, a gender gap between KD achievement levels in these schools tested in Norway and Australia. These results indicate that current practice (and the associated curriculum and pedagogical approach) is in urgent need of revision and improvement—particularly as a counter-measure of post-fact rhetoric and in relation to the knowledge deficit model. Knowledge gaps in crucial aspects of climate science also show that the students in these schools are still not adequately conversant with climate science to make an informed and rational decision about climate change. For policy development, resilience-building and effective public engagement in association with their proximity to public participation (voting, consumer-power, career-choice, etc.), this signal is worrisome. In response to these issues, our research offers a structured methodology for a climate literacy framework via knowledge domains; providing a solid foundation on which to establish a sound curriculum and pedagogy.

Future directions of this project will include a broader survey outreach to confirm these results in the global population and further research on the full domains of KD5 and KD7. KD6 will also need to be included, but the survey instrument will need to be split into smaller sections (e.g. one KD for each associated age group) in order to prevent survey fatigue. Through confirmation of this research, we anticipate the proposed pedagogical and curriculum framework presented in this paper to become a standard for effective climate change education. By adequately preparing our youth for tomorrow we provide a benefit to these people as individuals, and a benefit to society and the world at large. By teaching climate science KDs as a pre-requirement for teaching climate change KDs, we embed climate change into the scientific realm where it belongs; building a public conversant with the physical science basis of climate change. Not only may this prepare and empower an entire generation of game-changers, we may reinvigorate the validity of climate science and the public education system whose task it is to disseminate that knowledge.

References

Arnold A, Böhm G, Corner A, Mays C, Pidgeon N, Poortinga W, Tvinnereim E et al (2016) European perceptions of climate change (EPCC): socio-political profiles to inform a cross-national survey in France, Germany, Norway and the UK. Oxford. https://doi.org/10.1016/C2013-0-10375-3

Bacharach V, Baumeister A, Furr RM (2003) Racial and gender science achievement gaps in secondary education. J Genet Psychol 164(1):115–126

Barnett J, Adger WN (2007) Climate change, human security and violent conflict. Polit Geogr 26(6):639–655. https://doi.org/10.1016/j.polgeo.2007.03.003

Bengtsson SEL, Barakat B, Muttarak R (2018) The role of education in sustainable development. Adv Educ 32. https://doi.org/10.4314/ai.v32i1.22298

Biggs JB, Collis KF (1982) Evaluating the quality of learning—the SOLO taxonomy. Academic Press, Switzerland: New York. https://doi.org/10.1016/C2013-0-10375-3

Bobbitt F (1918) The curriculum. Houghton Mifflin Company, Chicago, US. https://doi.org/10.1017/CBO9781107415324.004

Bodzin AM, Fu Q (2013) The effectiveness of the geospatial curriculum approach on urban middle-level students' climate change understandings. J Sci Educ Technol 23(4):575–590. https://doi.org/10.1007/s10956-013-9478-0

Bryce S, Frank N (2014) What is climate change scepticism? Examination of the concept using a mixed methods study of the UK public. Glob Environ Change 24:389–401. https://doi.org/10.1016/j.gloenvcha.2013.08.012

Brysse K, Oreskes N, O'Reilly J, Oppenheimer M (2013) Climate change prediction: erring on the side of least drama? Glob Environ Change 23(1):327–337. https://doi.org/10.1016/j.gloenvcha.2012.10.008

Carvalho LC, Rego C, Lucas MR, Sánchez-Hernández MI, Viana ABN (2019) New paths of entrepreneurship development. Springer International Publishing AG, Switzerland

Case R (1985) Intellectual development: birth to adulthood. Academic Press, Orlando, CA

Chapman HA, Anderson AK (2013) Things rank and gross in nature: a review and synthesis of moral disgust. Psychol Bull 139(2):300–327. https://doi.org/10.1037/a0030964

Checkoway B, Allison T, Montoya C (2005) Youth participation in public policy at the municipal level. Child Youth Serv Rev 27(10):1149–1162. https://doi.org/10.1016/j.childyouth.2005.01.001

Checkoway B, Richards-Schuster K, Abdullah S, Aragon M, Facio E, Figueroa L, White A et al (2003) Young people as competent citizens. Commun Dev J 38(4):298–309. https://doi.org/10.1093/cdj/38.4.298

Clark D, Ranney MA, Felipe J (2013) Knowledge helps: mechanistic information and numeric evidence as cognitive levers to overcome stasis and build public consensus on climate change. In: Proceedings of the 35th annual meeting of the Cognitive Science Society, pp 2070–2075

Corner A, Groves C (2014) Breaking the climate change communication deadlock. Nat Publ Group 4(9):743–745. https://doi.org/10.1038/nclimate2348

Corner A, Roberts O, Chiari S, Völler S, Mayrhuber ES, Mandl S, Monson K (2015) How do young people engage with climate change? The role of knowledge, values, message framing, and trusted communicators. WIREs Clim Change 6(5):523–534. https://doi.org/10.1002/wcc.353

Crayne JA (2015 Teaching climate change: pressures and practice in the middle school science classroom. University of Oregon Graduate School. https://doi.org/10.1017/CBO9781107415324.004

De Bourmont M, Martindale D (2015) Coming of age in the age of extinction. In These Times 39(9):20. Retrieved from http://search.ebscohost.com/login.aspx?direct=true&db=8gh&AN=8204265&site=ehost-live

Feinberg M, Willer R (2013) The moral roots of environmental attitudes. Psychol Sci 24:56–62. https://doi.org/10.1177/0956797612449177

Godfray HCJ, Beddington JR, Crute IR, Haddad L, Lawrence D, Muir JF, Toulmin C et al (2010) Food security: the challenge of feeding 9 billion people. Science 327(5967):812–8. https://doi.org/10.1126/science.1185383

Guskey TR (2015) Mastery learning. International encyclopedia of the social & behavioral sciences, vol 14, 2nd edn. Elsevier. https://doi.org/10.1016/B978-0-08-097086-8.26039-X

Harker-schuch I (2019) Why is early adolescence so pivotal in the climate change communication and education arena? In: Climate change and the role of education. Lincoln, UK

Harker-Schuch I, Bugge-Henriksen C (2013) Opinions and knowledge about climate change science in high school students. Ambio 42(6):755–766. https://doi.org/10.1007/s13280-013-0388-4

Harker-Schuch I, Mills F, Lade SJ, Colvin R (2019a) CO2peration—structuring a 3D interactive digital game to improve climate literacy in the 12–13-year-old age group (currently under review). Comput Educ

Harker-Schuch I, Mills F, Lade SJ, Colvin R (2019b) Opinions of 12 to 13-year-olds in Austria and Australia on the worry, cause and imminence of climate change (currently under review)

Harker-schuch I, Mills F, Lade SJ, Colvin R et al (2019c) Toward a climate literacy framework: developing knowledge domains and a cause-based climate science curriculum in the 12–13-year age group (currently under review)

Head L, Harada T (2017) Keeping the heart a long way from the brain: the emotional labour of climate scientists. Emot Space Soc 24:34–41. https://doi.org/10.1016/j.emospa.2017.07.005

Hess DJ, Collins BM (2018) Climate change and higher education: assessing factors that affect curriculum requirements. J Clean Prod 170:1451–1458. https://doi.org/10.1016/j.jclepro.2017.09.215

Higgins K (2016) Post-truth: a guide for the perplexed. Nature 540(7631):9. https://doi.org/10.1038/540009a

Ho E, Clarke A, Dougherty I (2015) Youth-led social change: topics, engagement types, organizational types, strategies, and impacts. Futures 67:52–62. https://doi.org/10.1016/j.futures.2015.01.006

Jaspal R, Nerlich B (2014) When climate science became climate politics: British media representations of climate change in 1988. Publ Underst Sci 23(2):122–141. https://doi.org/10.1177/0963662512440219

Jensen FE, Nutt AE (2015) The teenage brain: a neuroscientist's survival guide to raising adolescents and young adults, 1st edn. Harper, New York

Jones MG, Howe A, Rua MJ (2000) Gender differences in students' experiences, interests, and attitudes toward science and scientists. Sci Educ 84(2):180–192

Kahan DM (2013) Ideology, motivated reasoning, and cognitive reflection. Judgem Decis Mak 424(107):407–424. https://doi.org/10.2139/ssrn.2182588

Kahan DM, Jenkins-Smith H, Braman D (2011) Cultural cognition of scientific consensus. J Risk Res 14(2):147–174. https://doi.org/10.1080/13669877.2010.511246

Kahan DM, Peters E, Wittlin M, Slovic P, Ouellette LL, Braman D, Mandel G (2012) The polarizing impact of science literacy and numeracy on perceived climate change risks. Nat Clim Change 2(10):732–735. https://doi.org/10.1038/nclimate1547

Lazarus RJ (2009) Super wicked problems and climate change: restraining the present to liberate the future. Cornell Law Rev 94:1153–1234. Retrieved from https://scholarship.law.georgetown.edu/facpub/159

Le Treut H, Somerville R, Cubasch U, Ding Y, Mauritzen C, Mokssit A, Prather M (2007) Historical overview of climate change science. In: Solomon S, Qin D, Manning M, Chen Z, Marquis M, Averyt KB, Miller HL et al (eds) The physical science basis. Contribution of Working Group I to the Fourth Assessment Report of the Intergovernmental Panel on Climate Change, pp 93–127. Cambridge University Press. https://doi.org/10.1016/j.soilbio.2010.04.001

Leviston Z, Greenhill M, Walker I (2015) Australian attitudes to climate change: 2010–2014. CSIRO, Australia

Leviston Z, Price J, Malkin S, Mccrea R (2014) Fourth annual survey of Australian attitudes to climate change: interim report, Jan 2014

Lewandowsky S, Oreskes N, Risbey JS, Newell BR, Smithson M (2015) Seepage: climate change denial and its effect on the scientific community. Glob Environ Change 33:1–13. https://doi.org/10.1016/j.gloenvcha.2015.02.013

Lima ML, Castro P (2005) Cultural theory meets the community: worldviews and local issues. J Environ Psychol 25(1):23–35. https://doi.org/10.1016/j.jenvp.2004.11.004

Lorenzoni I, Jones M, Turnpenny JR (2007) Climate change, human genetics, and post-normality in the UK. Futures 39(1):65–82. https://doi.org/10.1016/j.futures.2006.03.005

Makkreel RA (1975) Dilthey—philosopher of the human studies, 2nd edn. Princeton University Press, New Jersey

McNeill KL, Vaughn MH (2012) Urban high school students' critical science agency: conceptual understandings and environmental actions around climate change. Res Sci Educ 42(2):373–399. https://doi.org/10.1007/s11165-010-9202-5

Milér T, Sládek P (2011) The climate literacy challenge. In: Procedia Soc Behav Sci 12:150–156. https://doi.org/10.1016/j.sbspro.2011.02.021

Miller PH, Blessing JS, Schwartz S (2006) Gender differences in high-school students' views about science. Int J Sci Educ 28(4):363–381

Moser SC, Dilling L (2012) Communicating climate change: closing the science-action gap. In: Oxford handbook climate change and society. https://doi.org/10.1093/oxfordhb/9780199566600.003.0011

New M, Liverman D, Schroeder H, Schroder H, Anderson K (2011) Four degrees and beyond: the potential for a global temperature increase of four degrees and its implications. Philos Trans Ser A Math Phys Eng Sci 369(1934):6–19. https://doi.org/10.1098/rsta.2010.0303

Nicholson-Cole S (2009) "Fear won't do it" visual and iconic representations. Sci Commun 30(3):355–379. https://doi.org/10.1177/1075547008329201

Ojala M (2015) Hope in the face of climate change: associations with environmental engagement and student perceptions of teachers' emotion communication style and future orientation. J Environ Educ 46(3):133–148. https://doi.org/10.1080/00958964.2015.1021662

Oreskes N, Conway E (2010) Merchants of doubt: how a handful of scientists obscured the truth on issues from tobacco smoke to global warming. Bloomsbury Publishing, New York

Pearson RG (2006) Climate change and the migration capacity of species. Trends Ecol Evol 21(3):111–113. https://doi.org/10.1016/j.tree.2005.11.022

Piaget J (1972) Intellectual evolution from adolescence to adulthood. Hum Dev 51(1):40–47

Porten-Cheé P, Eilders C (2015) Spiral of silence online: how online communication affects opinion climate perception and opinion expression regarding the climate change debate. Stud Commun Sci 15(1):143–150. https://doi.org/10.1016/j.scoms.2015.03.002

Porter D, Weaver AJ, Raptis H (2012) Assessing students' learning about fundamental concepts of climate change under two different conditions. Environ Educ Res 18(5):665–686. https://doi.org/10.1080/13504622.2011.640750

Potter E, Oster C (2008) Communicating climate change: public responsiveness and matters of concern. Media Int Aust 127:116–126

Rittel HW, Webber MM (1973) Dilemmas in a general theory of planning. Policy Sci 4:155–169

Ryghaug M, Skjølsvold TM (2016) Climate change communication in Norway. In: Storch H (ed) Oxford research encyclopedia of climate science. Oxford University Press, New York. https://doi.org/10.1093/acrefore/9780190228620.013.453

Schäfer MS, Fox E, Rau H (2016) Climate change communication in Germany 1:1–29. https://doi.org/10.1093/acrefore/9780190228620.013.459

Shao W (2017) Weather, climate, politics, or God? Determinants of American public opinions toward global warming. Environ Polit 26(1):71–96. https://doi.org/10.1080/09644016.2016.1223190

Shepardson DP, Niyogi D, Choi S, Charusombat U (2010) Students' conceptions about the greenhouse effect, global warming, and climate change. Clim Change 104(3–4):481–507. https://doi.org/10.1007/s10584-009-9786-9

Shi J, Visschers VHM, Siegrist M (2015) Public perception of climate change: the importance of knowledge and cultural worldviews. Risk Anal 35(12):2183–2201. https://doi.org/10.1111/risa.12406

Shi J, Visschers VHM, Siegrist M, Arvai J (2016) Knowledge as a driver of public perceptions about climate change reassessed. Nat Clim Change 6:759–762. https://doi.org/10.1038/nclimate2997

Stern PC (2012) Psychology: fear and hope in climate messages. Nat Clim Change 2(8):572–573. https://doi.org/10.1038/nclimate1610

Stevenson K, Peterson N, Bondell H, Moore S, Carrier S (2014) Overcoming skepticism with education: interacting influences of worldview and climate change knowledge on perceived climate change risk among adolescents. Clim Change 126(3–4):293–304. https://doi.org/10.1007/s10584-014-1228-7

Susanne C, Moser LD (2004) Making climate hot: communicating the urgency and challenge of global climate change. Environment 46(10):32–46

Thackeray R, Hunter M (2010) Empowering youth: use of technology in advocacy to affect social change. J Comput Mediat Commun 15(4):575–591. https://doi.org/10.1111/j.1083-6101.2009.01503.x

Torney-Purta J, Lehmann R, Oswald H, Schulz W (2001) In: Wagemake P (ed) Citizenship and education in 28 countries. Eburon Publishers, Delft

Waldmann S (2018) New climate report actually understates threat, some researchers argue. Science. https://doi.org/10.1126/science.aav7128

Weingart P, Engels A, Pansegrau P (2000) Risks of communication: discourses on climate change in science, politics, and the mass media. Publ Underst Sci 9(3):261–283. https://doi.org/10.1088/0963-6625/9/3/304

Wetters K (2008) The opinion system: impasses of the public sphere from Hobbes to Habermas. Fordham University Press, New York. Retrieved from http://site.ebrary.com/lib/anuau/home.action?force=1/docDetail.action?docID=10586791

Chapter 18
Realities of Teaching Climate Change in a Pacific Island Nation

Charles Pierce

Abstract Teaching young adults about climate change in a developing nation right at the forefront of its impacts brings particular joys and challenges. This chapter discusses a teacher's experiences in creating and delivering courses on climate change and disaster risk reduction to a diverse group of learners in 2017 and 2018. Vanuatu, widely acknowledged as the world's most vulnerable country to natural hazards, was the first in the Pacific to provide this pioneer technical and vocational programme at basic and advanced levels. The chapter considers the background to delivering the courses, discusses the teaching and learning strategies used, and evaluates the contribution of the courses to adaptation to climate change in Vanuatu. It argues that the most effective way to teach climate change in a small island developing country is to focus on vulnerability assessment and ways of becoming more resilient to the negative impacts of climate change and disasters, adopting a hands-on, student-centred, experiential approach to learning.

Keywords Climate change · Disaster risk reduction · Vulnerability · Resilience · Community resilience · Adaptation and mitigation · Sustainable development · Sea level rise · Anthropogenic factors · Climate change impacts · Agent of transformation · Traditional knowledge · Fieldwork · Risk maps · Food and water security · Technical vocational education and training (TVET) · Pedagogy · Competency-based learning · Experiential learning · Constructivism · Cooperative learning · Student-centred classroom · Formative assessment · Oral communication

Introduction

Aims and Structure of the Chapter

This chapter looks at the education of young people about climate change in the context of Vanuatu, a small island developing state in the planet's largest ocean. In

---

C. Pierce (✉)
Educational Consultant and Facilitator, Vanuatu Institute of Technology, Port Vila, Vanuatu
e-mail: charliepierce19@gmail.com

© Springer Nature Switzerland AG 2019
W. Leal Filho and S. L. Hemstock (eds.), *Climate Change and the Role of Education*,
Climate Change Management, https://doi.org/10.1007/978-3-030-32898-6_18

view of Vanuatu's exposure to a plethora of hazards, education provides a positive way for people to engage with climate change and disasters and take measures to mitigate and adapt to their effects. The training courses discussed provide learners with skill sets and experiences that they can use in their own communities to foster greater consciousness of climate change and disaster issues and provide practical strategies for addressing them. We will review how the courses were developed, their characteristics and outcomes, the pedagogical approaches applied, and their usefulness to society.

This introduction provides geographical information on Vanuatu, its vulnerability to the impacts of climate change and disasters, and awareness of climate change among the population. Section 2 explains the development and delivery of courses on this subject at Technical Vocational Education and Training (TVET) level. This is followed in Sect. 3 by a discussion of relevant approaches to teaching and learning, looking at how they have enabled development of the critical skills and knowledge envisaged by the courses and highlighting innovative methods that proved successful. Finally, Sect. 4 evaluates the contribution of the Certificate level courses to the work of climate change adaptation in Vanuatu.

An underlying theme of the chapter is that the building of greater resilience depends on the practical training of young people to become agents of transformation in their own communities, passing on their skills, attitudes and knowledge about climate change and disaster risk reduction to those in the wider society.

Vanuatu

Vanuatu[1] is a volcanic archipelago in the South-West Pacific (Fig. 18.2). As with other island nations, climate change is the most significant single threat to sustainable development, particularly because of the large proportion of the population living in coastal communities (Fig. 18.1). Additionally, Vanuatu's location on the Pacific Ring of Fire means that it is highly exposed to volcanic eruptions, earthquakes, tsunamis and landslides. According to the World Risk Report (Bündnis Entwicklung Hilft 2018), Vanuatu has the highest disaster risk out of the 172 countries covered by the World Risk Index in 2018. In terms of weather-related loss events, the Global Climate Risk Index puts Vanuatu among the top five most vulnerable countries in the world (GermanWatch 2017).

Though a tiny country by global standards, Vanuatu is relatively large among South Pacific nations, comprising a Y-shaped chain of about 80 islands and islets. Most islands are volcanic, but raised coral reefs overlie volcanic materials on the larger islands, and some islets consist only of raised reefs. The total land area of nearly 13,000 km^2 includes more than 2500 km of coastline.

---

[1] Vanuatu was known as the New Hebrides until it gained independence from Britain and France in 1980.

Fig. 18.1 Marou village, Emau Island, Vanuatu

The total population of Vanuatu in January 2019 is estimated at 285,000 (World Population Review 2019), with approximately 18% living in the two urban areas of Port Vila and Luganville and the remaining 82% in rural villages and isolated settlements, mostly on the coast. The rural population relies overwhelmingly on subsistence farming and fishing for sustenance.

Impacts of Climate Change

A major impact of climate change in Vanuatu, as in other Pacific islands, is sea level rise and the concomitant coastal erosion that occurs during storms (Fig. 18.3). Projections of sea level rise between 2015 and 2030 range from 8 to 18 cm for all emissions scenarios; between 2015 and 2090, they range from 25 to 59 cm for a very low emissions scenario and 42–89 cm for a very high emissions scenario (PACCSAPP 2015). Another significant impact of warmer atmosphere and oceans is the increased intensity of tropical cyclones, such as Cyclone Pam of March 2015, at category 5 (Figs. 18.2 and 18.4).

Additional observed and expected impacts include: an increase in the number of very hot days; flooding and landslides from heavy rainfall events; more soil erosion; degradation of coral reefs through ocean acidification and intense cyclones (Fig. 18.5); longer periods of drought (Fig. 18.6); more frequent El Niño and La Niña periods; saltwater intrusion; reduction in biodiversity; heat stress in plants and

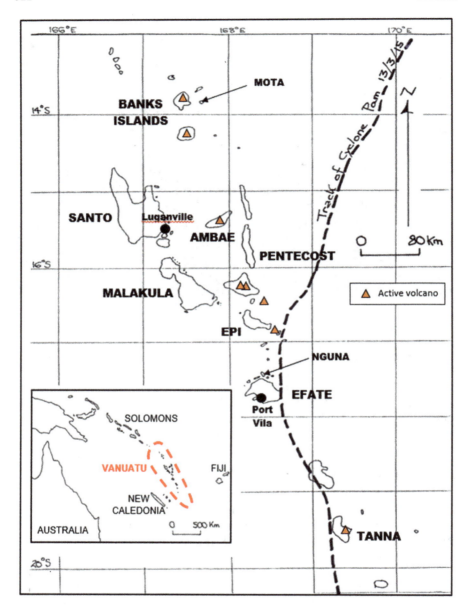

Fig. 18.2 Map of Vanuatu

Fig. 18.3 Destruction of a graveyard through sea level rise and coastal erosion at Laonamoa, Pele Island, Efate

Fig. 18.4 Destruction in South-East Epi caused by Cyclone Pam on 13th March 2015 (Hawkins 2015)

Fig. 18.5 During Cyclone Jasmine, in February 2012, broken pieces of coral reef were deposited by the storm surge on a beach near Port Vila

Fig. 18.6 During a 4-month period of drought on Mota island in 2017, people of Gamalna had to collect drinking water from a dried-up creek bed several kilometres from the village

animals; changes in the flowering and fruiting season of tree species; decline in fisheries as sea water becomes warmer; damage to food gardens and loss of food security through severe cyclones and drought; economic impacts of extreme weather events on tourism and commercial agriculture; damage to coastal infrastructures and housing through cyclones, storms and coastal erosion; loss of water security; negative effects of cyclones and droughts on education and health; increased rural-urban migration resulting from loss of subsistence livelihoods; and injuries and loss of life.

Awareness of Climate Change Among the Population of Vanuatu

Small island states such as Vanuatu are being disproportionately affected by climate change, despite being the least contributors (The Commonwealth Education Hub 2016). People in Vanuatu are experiencing the impacts of climate change and disasters at first hand, even though they may not have a clear understanding of the background science and factors responsible.

To measure perceptions of climate change, its causes and effects, the author's students interviewed a sample of 120 people on the island of Nguna in December 2018. A report on their findings (Pierce 2019) concludes that most people had heard of climate change, but only about half of them showed some understanding of its meaning, saying that it refers to changes in weather or climatic conditions; another third of the sample gave irrelevant answers, and 5% had no knowledge at all. Regarding the causes of climate change, anthropogenic factors were almost universally deemed as being responsible, with 90% of respondents identifying pollution and/or deforestation as key contributors; however, nobody could name carbon dioxide or any other greenhouse gas as being accountable for atmospheric warming, and only 2% mentioned natural factors, referring solely to the effects of volcanic activity. In terms of impacts, the three most commonly reported were changes in yields of crops and fruits (43% of respondents), sea level rise and coastal erosion (42%), and changes in temperature, seasons and other weather elements (42%). A wide range of other impacts were mentioned, reflecting the keen observations of populations reliant on natural ecosystems for their livelihoods and survival.

We should be hesitant about applying these results to the whole of Vanuatu. Nguna has benefited from efforts by aid donors and civil society organisations to promote awareness of climate change and adaptation techniques over the last decade, so there is a greater background knowledge than among populations in remoter locations. Nevertheless, these levels of awareness, and the misconceptions uncovered, have a bearing on the knowledge, skills and attitudes that should be included in any course promoting climate change education among young adults.

Delivery of Climate Change Courses

Being on the frontline of severe climate change impacts, it is not surprising that Vanuatu was the first Pacific country to establish a Ministry of Climate Change, in April 2013. Efforts to include climate change and disaster risk reduction in primary and secondary education began just prior to this. Then in 2014, the non-government organisation Vanuatu Rural Development Training Centres Association (VRDTCA) decided that a climate change course should be produced for some 35 rural training centres (RTCs) around the archipelago. Such centres provide vocational training for students unable to continue their education beyond Year 10 level. Funding was provided by Deutsche Gesellschaft für Internationale Zusammenarbait GmbH (GIZ) and the Pacific Community (SPC) through their CCCPIR (Coping with Climate Change in the Pacific Island Region) programme, and a consultant (the author of this chapter) hired to develop a curriculum and resources and deliver the course.

In keeping with the principles of Technical Vocational Education and Training (TVET), a competency-based course was designed, requiring learners to demonstrate concrete skills rather than abstract concepts. Competency-based learning refers to *"systems of instruction, assessment, grading, and academic reporting that are based on students demonstrating that they have learned the knowledge and skills they are expected to learn as they progress through their education"* (Great Schools Partnership 2014, par. 1). The emphasis is on what a person can do in the workplace as a result of completing a training program (Western Business School 2019). For a climate change course, "workplace" was assumed to be the local community. Learners were asked to demonstrate specific competencies in each unit, and generic competencies applicable to all units. An example of specific competencies is provided in Fig. 18.7. "Element" refers to the competencies to be acquired, and "Performance Criteria" the outcomes that the learner must demonstrate. For generic or key competencies, those specified by the Vanuatu Qualifications Authority (VQA) were used—initiative, communication, teamwork, information technology, problem-solving, self-management, planning, gaining new skills and knowledge, and gender equity and social inclusion.

With this competency-based approach to learning, materials were designed in ways that involve participatory learning and promote reflection and learning through experience. This conforms to constructivist theories of education and echoes Kolb's theory of experiential learning, whereby effective learning takes place in a cycle of four stages, starting with concrete experience and progressing through reflective observation, abstract conceptualisation and active experimentation to lead back again to concrete experience (Kolb 1984). The consultant wanted interactions between students, between students and teacher, and between students and communities, to be conducted with humility, with a focus on consultation between the parties concerned. Further, the curriculum stressed the fostering of unity and coherence, since the consultant is convinced that they are key factors in strengthening community resilience to hazards and climate change. Finally, learning materials were influenced by the consultant's underlying view that the aim of education is to develop capacities that are latent within a person—a belief based on these words of Bahá'u'lláh (c. 1887): *"Regard man as a mine rich in gems of inestimable value. Education can, alone,*

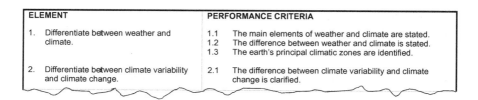

Fig. 18.7 Extract from Certificate I, Unit 2—demonstrate knowledge of climate

cause it to reveal its treasures, and enable mankind to benefit therefrom." (Tablets of Bahá'u'lláh 1988, p. 161).

It took three years for this project to be realized. Since there are three official languages in Vanuatu—English, French and Bislama (a form of pidgin English/French)—materials were produced in English and French and trialled in the outer islands. But delivery stalled after the demise of VRDTCA in 2015, and months elapsed before the Vanuatu Institute of Technology (VIT) agreed to host the programme in Port Vila. Meanwhile, VQA had its own requirements to be met, involving modification of the original content and a rigorous process for achieving official accreditation. Further funding had to be obtained from the European Union—Pacific Technical Vocational Education and Training (EU-PacTVET) project supported by the Pacific Community and the University of the South Pacific.

But in February 2017, the first-ever Certificate I course in Climate Change and Disaster Risk Reduction (CCDRR) could be launched, with 31 participants from all six provinces of Vanuatu. Twenty-seven used materials in English, four used French, and the language of delivery was Bislama. Although the course was designed for students whose education was to Year 10, those selected from over one hundred applicants included just two at this level; others had completed high school or up to two years of undergraduate courses. This variation in academic level was a challenge, as were the differences in maturity of participants, whose ages ranged from 18 to 40 and included parents of small children. There were 10 females and 21 males.

The content of the CCDRR course includes hazard risks in Vanuatu, climate and climatic variations, causes and effects of climate change, mitigation of and adaptation to climate change, vulnerability and impacts, the importance of traditional knowledge in building resilience, and the promotion of community action to prepare for climate change and disaster risk reduction. The basic aim is to empower the learners to become agents of change in their communities, able to conduct awareness programmes and demonstrate practical techniques of mitigation and adaptation.

The rationale for combining education on climate change with that on disasters is that the promotion of community resilience in Vanuatu involves coping with frequent geological as well as hydro-meteorological hazards, and that many adaptation techniques are common to both—for example, the production of village risk maps. This methodology is also recommended by Stevenson et al. (2017) in Australia. They confirm that climate adaptation should include disaster education on how to respond to bushfires, floods, droughts, prolonged heatwaves, cyclones, tsunamis and storms, adding that disaster risk reduction builds community resilience through a systematic approach to identifying, assessing and reducing risk.

An integral part of the Certificate I course was fieldwork. Learners drew risk maps for villages close to Port Vila and presented them to the communities, surveyed sources of greenhouse gas emissions in another five villages, and studied adaptation methods on the island of Pele. They stayed in six villages on the island of Emau, where they researched traditional techniques of weather prediction, food preservation and fishing (Fig. 18.8), conducted SWOT analyses, and evaluated the adaptive and coping capacities of those communities. They learnt how to present key concepts to each other, to other students at VIT and to village communities

Fig. 18.8 Investigating traditional knowledge of disaster risk reduction

(Fig. 18.9); established their own agro-forestry plot at VIT (Fig. 18.10); practised coral planting to replace degraded reefs; prepared and shared action plans for building resilience to disasters in five communities; and took a short course in First Aid. Such activities, conducted outside the confines of the normal classroom, promoted experiential, hand-on learning.

Immediately this first Certificate I course was completed, the Institute of Technology requested a further learning pathway for the graduates. The same consultant was hired to develop a more advanced course at Certificate III level, and funding was secured from two donor partners—The Asian Development Bank, and EU PacTVET in association with GIZ. While the new course was being developed, VIT took the initiative to run the Certificate I course for a second time, using a local graduate as facilitator.

Certificate III in Resilience (Climate Change and Disaster Risk Reduction) ran from August to December 2018, facilitated by the course developer. The 24 participants (11 females and 13 males) again came from all provinces of Vanuatu, ranged in age from 19 to 40 years, and had a variety of educational backgrounds. Ten came from the first Certificate I cohort, 12 from the second cohort and a further three were selected from the many applicants.

In designing this more advanced course, the developer applied feedback from the first group of CCDRR learners. Following their field experience, they wanted to upgrade their skills in writing project proposals, interpreting weather maps, drawing accurate sketch maps and graphs, responding to questions on evidence of climate change, and using percentages and statistics in surveys.

Fig. 18.9 CCDRR students presenting their assessment of community vulnerability to the people of Marou village

Fig. 18.10 Weeding the CCDRR agroforestry plot

There were huge challenges in developing and bringing this Certificate III course to fruition: pressure to produce learning materials before the start of second semester 2018; convincing aid agencies to fund the printing of books, disbursement of student fees, fieldwork costs and fees for the facilitator; mentoring two assistant facilitators appointed by VIT in the expectation that they can deliver this same course in the future; and meeting the requirements of VQA, which introduced new procedures for accreditation and delivery during this period.

Distinctive features of this Resilience course are that it builds skills in communication, mathematics and mapping, focuses on the science and measurement of climate change, and emphasizes practical adaptation techniques that strengthen food and water security. The capacities being developed are similar to those required in education for sustainable development, described by Ilisko et al. (2014)—critical thinking, problem solving, creativity and innovation skills, collaboration skills, contextual learning skills, self-direction and communication skills—but also include the ability to interact with village communities and actually demonstrate a number of adaptation strategies.

For fieldwork, students map contour lines around VIT; take photos of phenomena related to climate change and disasters and transmit them by phone to others; conduct a survey of biodiversity in undisturbed rain forest; observe and practice new agricultural techniques for adaptation to climate change (Fig. 18.11); visit a community conservation area; survey water security and sanitation in informal settlements in

Fig. 18.11 Resilience students learning a grafting technique at Tagabe Agricultural Research Station, Port Vila

Port Vila (Fig. 18.12); learn how to carry out practical adaptation techniques such as using compost made from crown-of-thorns starfish, planting vetiver grass and coral gardening (Fig. 18.13); evaluate the role of a non-government organisation; and conduct their own analysis of community resilience.

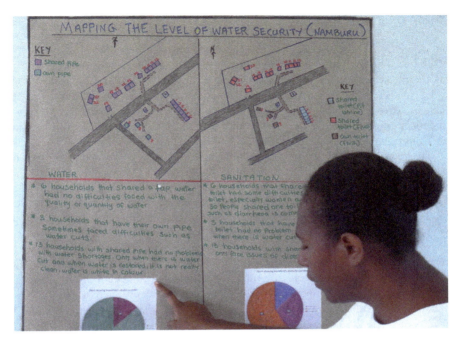

Fig. 18.12 Reporting on water security and sanitation in an informal settlement in Port Vila

Fig. 18.13 Learning to plant vetiver grass and carry out coral gardening on Pele island

Approaches to Teaching and Learning

The overarching philosophy is that learners should be able to communicate their skills, attitudes and knowledge to others, so building greater resilience among local communities and enabling populations to better cope with the challenges of climate change and disasters.

To this end, participants in the two courses must demonstrate knowledge about the science of climate change and measures to mitigate and adapt to the impacts of climate change and disasters. But much more than this, they should be able to communicate and interact closely with people in rural and urban settings in order to help build greater adaptive and coping capacity in local communities. Learners have to become articulate and sensitive to others' needs, willing to share their own understanding and at the same time learn from those who may have greater experience and traditional knowledge of resilience.

Thus certain pedagogical principles must underpin all teaching and learning. One of the foremost is constructivism, meaning that knowledge is constructed through a student's own experiences rather than received through instruction. This is linked to experiential learning, whereby learning takes place through doing and then reflecting on the results—an approach that is exemplified by field investigations and situations where learners have to master a topic and then teach it to others. Another key principle is cooperative learning—putting students into small groups in which they work together to maximise their own and each other's learning (Johnson et al. 1994). Thus the CCDRR and Resilience classrooms were organised around tables of four persons (Fig. 18.14), with the composition of the group varying according to the task. Such a student-centred classroom encourages freer discussion, promotes cooperation rather than competition, builds a team spirit, and enables students to gain new skills and knowledge from each other. It helps the learners to improve their communication skills and become more considerate of others - qualities needed when helping people in the villages to understand the nature of climate change and prepare for its impacts.

Now we will discuss the effectiveness of specific teaching and learning strategies that follow the underpinning pedagogy of constructivism, experiential learning, student-centred cooperative learning and an enquiry-based approach.

Oral Presentations

Because the overall aim of the Certificate I and III courses is to produce learners who can promote greater awareness of climate change and disaster risk reduction, they must be trained to stand up and talk to others in a clear, coherent and positive manner. In Vanuatu tradition, young people are expected to remain silent in front of their elders, and the formal education system tends to produce passive students who cannot articulate their thoughts in public. Strategies are therefore needed to help learners develop greater self-confidence when speaking in village situations.

Fig. 18.14 Resilience classroom, December 2018

One such method is through "carousel" presentations, whereby teams research different aspects of a topic and present their findings in wall charts that are pinned around the classroom. Then the members of each team re-group in new teams that circulate around the room. The member of the team who was in the group that produced the chart presents this chart to the other members of the new team. This activity engages the learners because everyone has to speak to a small group and takes ownership of the topic being presented. In the example shown in Fig. 18.15, groups researched climatic changes in the Earth's geological history. The specific competency acquired was to demonstrate that the Earth's climate has been constantly changing, and this activity clearly achieved that goal. The generic competencies gained were initiative, teamwork, problem-solving, self-management, planning, gaining new skills and knowledge, gender equity and especially an increased capacity for oral communication.

This strategy was adopted in almost every unit and proved highly popular with the learners. By the end of both courses, those who had initially been incapable of speaking in public were able to articulate their ideas in an assured manner that augurs well for the future.

Another effective strategy is the use of role plays. An example was when teams of learners in the Certificate III course used evidence from the 2013 report of the Intergovernmental Panel on Climate Change to prove that anthropogenic factors are responsible for recent changes in global surface temperatures. One team acted as visiting experts and another team represented a local community (Fig. 18.16). Learners

Fig. 18.15 Learners teaching each other about past climatic changes

Fig. 18.16 Role play of a village meeting

used their experiences of village meetings to ensure that discussions were lively and realistic. They learnt how to refute the arguments of climate-change "deniers".

Fieldwork

An axiom of the Certificate I and III courses is that participants need to be out in the field, researching the impacts of disasters and climate change and learning to demonstrate techniques of mitigation and adaptation. They need to tap into the reservoir of traditional knowledge of resilience to extreme weather events that has accumulated in Pacific island populations through millennia of experience. They must learn to assess the levels of community resilience and propose creative solutions to overcome vulnerability. They must be able to answer questions on climate science and fill the knowledge gaps revealed by the survey on climate change perceptions. All these requirements are best achieved through field experience in communities during training. This matches the conclusion reached by Kagawa and Selby (2010, last page) that "… *Looming rampant climate change calls for flexible learning and emergent curriculum approaches that embed climate change learning and action within community contexts.*"

The pivotal role of fieldwork is exemplified by Unit 9 of the Certificate III course, during which the entire class spent a week on Nguna Island to investigate community vulnerability. Each team of four lived in a different village and performed tasks linked to the desired competencies: promoting awareness on climate change; constructing detailed risk maps of the community (Figs. 18.17 and 18.18); designing and using questionnaires to interview everyone about their assets, food and water security, impacts of climate change, etc. (Fig. 18.19); and reporting to the community on its level of resilience. Back at VIT, results were refined, charts and power point presentations produced, and a final presentation given to the public in which every learner participated (Fig. 18.20). These talks were outstanding, reflecting the learners' sense of ownership of their data and their delight at having achieved such a meaningful interaction with their adopted communities.

Every generic competency in the Resilience course was demonstrated during this fieldwork. Initiative was needed to create an appropriate questionnaire for assessing community and individual vulnerability and resilience. Effective communication was essential for conducting interviews and making presentations to large groups of people. Teamwork was developed through working together on tasks and solving problems. IT skills were used when accessing Google Maps. Problem-solving was in evidence when deciding how to collect personal and household information. Self-management was shown when individuals took responsibility for carrying out tasks allotted by their team, and during daily reflections on progress. Planning was necessary to ensure that all tasks were completed on time, and in organising presentations. Learning new skills and knowledge came through the collection and processing of a mass of information from different sources, as well as in cooperating with others to complete a challenging project with minimal guidance. Gender equity and

Fig. 18.17 Constructing a risk map of Nekapa village, Nguna

Fig. 18.18 Sketch map of Utanlangi village, Nguna

social inclusion were shown through sensitivity to village culture and in ensuring that discussions included both male and female perspectives.

One student made this pertinent observation: "*Unit 9 is very very useful. I can say that ..... what we have done in this unit is the actual outcome of this course, not the certificate at the end. This unit directs us on what we must or will be doing in the future.*" (Mele 2018).

A difficulty encountered during investigations is that group members may not cooperate with each other—an issue that can escalate during residential fieldwork. If the facilitator knows his/her students well, he/she can pre-empt this by careful

Fig. 18.19 Conducting an interview in Unakap village, Nguna

Fig. 18.20 Public presentation on the findings of resilience surveys on Nguna

selection of groups. Alternatively he/she will have to spend time in the field with group members to encourage mutual tolerance and understanding.

The main challenge, however, is the cost, especially for travel to villages on Efate and its offshore islands. The first Certificate I and III courses were fortunate in that such expenditure was funded by donor partners. But when VIT decided to run the Certificate I course for the second time, without this funding, there was no finance for travelling and no fieldwork was done outside Port Vila. As a consequence those learners did not acquire the same communication and practical skills as had the first cohort, and this posed difficulties when they entered the Certificate III programme.

Practical Demonstrations

Among the many demonstrations carried out by facilitator and students were an experiment to indicate how heating causes expansion of water, and experiments to show the melting of ice on water and on land (Fig. 18.21). Both proved highly effective in helping learners to understand sea level rise.

The ice activity illustrates the four stages of experiential learning. By taking measurements (experience) and thinking about the results (reflective observation), learners discovered that the melting of ice that floats on water did not raise water levels, whereas the melting of ice that rests on a large rock in a container of water raised the water level in that vessel by about 1 cm. From this learning, they could adjust their previous understanding that whenever ice melts it leads to a rise in water levels (conceptualisation) and then apply the new understanding to the world (active experimentation) to see that the melting of land-based ice sheets in Greenland and Antarctica contributes to global sea level rise, whereas the melting of ice floating on the Arctic Ocean does not.

Fig. 18.21 Experiments to show the melting of ice in water (left) and on land (right)

Assessment as a Tool for Learning

Five methods of assessment were used—an end-of-unit summative test and four formative strategies: assessment of oral participation, completion of workbook activities, observing learners during practical and field work, and learner reflections. It was through the four formative methods that learning was more effectively enhanced.

The most innovative was the assessment of oral participation for each of the learning outcomes. During activity time, the facilitator(s) visited each table of learners for face-to-face conversations with individuals about concepts being studied. Although these interactions demanded a lot of facilitator time and energy, they clarified misunderstandings and helped the facilitators to ensure that each learner grasped a concept before proceeding further, so fulfilling the ethos of competency-based learning. This one-on-one approach proved enjoyable for everyone, with several learners saying how much they gained from such exchanges. Oral participation also included group discussions and formal presentations by learners to the class, to students in other classes at VIT (Fig. 18.22) and to people in local communities. For each learning outcome in each unit, the facilitator ticked a check-list of oral participation for every learner.

Reflections were done orally, often at the end of the day, by asking learners to talk about something they had learned or enjoyed during classes. From time to time,

Fig. 18.22 Resilience students presenting their learnings about new agricultural techniques to students of tourism at VIT

written reflections were required. The value of reflection is that when learners reflect upon the teaching and learning of an issue, they strengthen their own capacity to learn. They assimilate new learning, relate it to what they already know, adapt it for their own purposes, and translate thought into action (Te Kete Ipurangi 2019).

Innovative Activities That Were Particularly Successful

Certain activities captured learners' attention, or proved invaluable in helping them to internalize difficult concepts and explain them to others. Two examples follow, both illustrating the principles of constructivism.

Cross-cultural sharing of traditional techniques: Although Vanuatu is part of the Melanesian cultural complex, its cultural traits vary from island to island and within islands. Several opportunities were given to learners for sharing traditional techniques that build resilience to disasters and a changing climate. For example, they worked in island groups to research and share traditional methods of fishing (Fig. 18.23), cultivation (Fig. 18.24), food preservation, agro-forestry and house construction that enable a community to survive after storms, floods, droughts, volcanic eruptions and tsunamis. This aroused great excitement among the learners, who shared their cultural features with great pride. Equally effective was the one-on-one sharing that took place between students from different islands (Fig. 18.25), often on an informal basis.

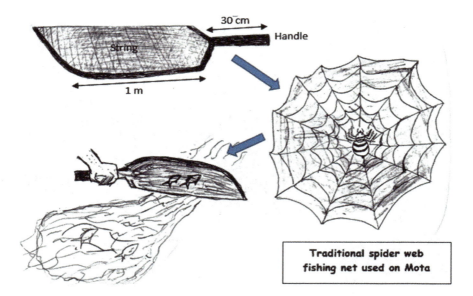

Fig. 18.23 A traditional fishing technique from the Banks Islands (Wogale, 2015)

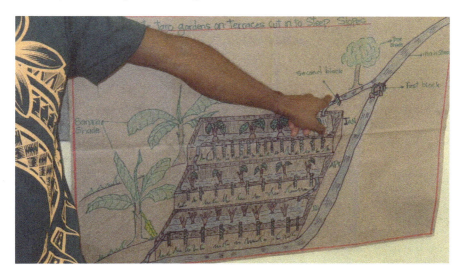

Fig. 18.24 A traditional technique of taro cultivation on the island of Santo

Fig. 18.25 Exchanging traditional techniques of adaptation

Using a diagram of the global carbon budget to explain anthropogenic causes of climate change: In Unit 4 of Certificate III, students learn how to explain the build-up of carbon dioxide in the atmosphere. An effective resource is the diagram of the global carbon cycle (Australian Academy of Science 2018) in Fig. 18.26. Through a series of questions on this resource, critical thinking and new learning is developed.

When learners drew their own version of this diagram and created a second one to show the situation before the Industrial Revolution, they found it much easier to explain anthropogenic causes of climate change to people in the villages (Fig. 18.27).

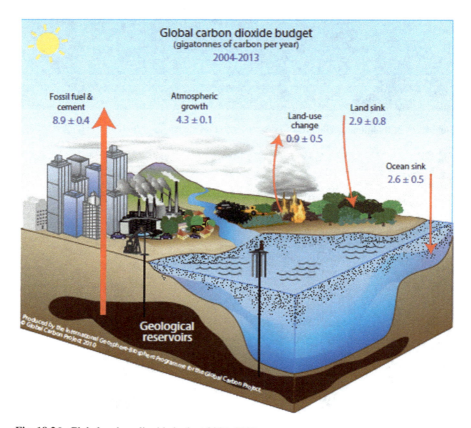

Fig. 18.26 Global carbon dioxide budget 2004–2013

Fig. 18.27 Explaining the carbon cycle to people in Woralapa village, Nguna

How the Certificate Courses Are Contributing to Adaptation to Climate Change

Because Vanuatu is a high-risk environment, with coastal populations exposed to the certainty of climatic change but the uncertainty of its impacts, it was considered essential that the Certificates I and III courses prepare learners to assess vulnerability and help communities to become more resilient. Stevenson et al. (2017, p. 2) argue that "…. *climate change involves creatively preparing children and young people for a rapidly changing, uncertain, risk and possibly dangerous future; just how dangerous depends on the actions we take today*." The author agrees, but suggests that for Vanuatu and other Pacific island states, these actions should focus on adaptation. Mitigation measures are important, but because GHG emissions from anthropogenic sources in these islands are an insignificant proportion of world totals, mitigation should focus on tree-planting and forest conservation, which are also adaptation tools.

Hence adaptation to climate change and disasters is a key learning outcome of the Certificates I and III courses. Participants learn about traditional and modern techniques, and in nearly all cases, they see actual examples of these, either visually in the classroom or through field visits. More importantly, they learn how to demonstrate these strategies to others. For example, they gain hands-on experience in spacing crops and trees during agro-forestry, in planting vetiver grass, in coral gardening and in the grafting of citrus fruit. They learn how to use charts to give simple explanations of many other strategies, both to each other and to people in the villages.

Thus the CCDRR and Resilience courses are directly contributing to efforts in Vanuatu to promote climate change adaptation. Graduates should have developed practical skills and capacities to be agents of transformation in the community - helping people to understand climate change and take measures to mitigate its causes, as well as assisting them to promote adaptation measures that reduce the negative impacts of climate change and disasters. But how do we measure success?

One way is to look at evaluations made by participants in the course:

> This course has helped me to know about climate and weather, how to identify the different causes of climate change and the impacts of climate change on the environment, and how we people can take action to slow down these effects, especially with the aid of traditional knowledge from the past. When I returned to my community I gave an awareness talk to the whole village. They were very grateful because now they know what climate change is, its causes, and actions they can take.
>
> Enory Wogale, 22 years.

> I have been a CCDRR student for the last six months. During this time, the course helps to … build up my techniques and skills of communication. In addition, I have become better equipped to do my job as an Agricultural Extension Officer. I can link agriculture with climate change.
>
> Rexly Narai, 26 years.

> My aim in taking up the CCDRR course was to build up my capacity in my work environment. I have been employed in the National Disaster Office for the past five years, and through the fieldwork projects we have undertaken, this course has boosted my understanding of climate change issues and how to reduce the impacts of climate change … Also, by taking up this course I have a better understanding of DRR issues.
>
> Brenda Williams, 29 years.

> Certificate III is much more advanced than the Cert I course. I have learnt more about different ways of adaptation to impacts, so as to become more resilient. I like the advanced skills and knowledge I have gained in mapping, biodiversity, food security and agriculture.
>
> Manuel Lishi, 25 years.

> This Certificate III course was so fantastic because it shows me many ways of how to live with the impacts of climate change and disasters …. After the completion of this course, I will make sure that all the rural communities on the west coast of Santo must be resilient to the impacts of climate change and disasters.
>
> Jerry Rojo, 22 years.

These testimonies show that the Certificate courses have indeed expanded learners' awareness of climate change and disaster risk reduction, improved their communication skills, enabled them to apply their skills in the fields of agriculture and disaster management, reinforced the value of traditional knowledge, and empowered them to take action to build greater resilience in their local communities. Participants have acquired the capacity to become "agents of change" in the field of adaptation.

In the longer term, it is too early to comment on initiatives taken by the 24 graduates of the Certificate III course, but some of the outcomes of the Certificate I course can be mentioned. One graduate returned to his village on the remote west coast of Santo and proceeded to establish a community conservation area and a Community Disaster and Climate Change Committee. Another pursued his interest in aquaculture, helping others to initiate their own projects in their communities. Another set up his own backyard food garden in an urban setting, using environmentally-friendly techniques that he had learned from the course. Yet another worked with his community to establish an agro-forestry project and is now pursuing a degree in Environmental Science at the University of the South Pacific, with the long-term aim of promoting marine biodiversity.

Conclusion

In view of Vanuatu's high level of vulnerability to the impacts of climate change and disasters, the teaching of climate change demands that the acquisition of knowledge be translated into practical action. The strategies illustrated in this chapter have proved highly effective in training learners at technical and vocational level in a tertiary institution, and could well be adopted in university education. The emphasis is on providing young people with the tools to help their communities cope with an uncertain environmental future. Such tools include the capacity to explain climate change issues in a coherent manner, the competence to assess risks, and the ability to demonstrate strategies that can build resilience. They also imply a flexibility of approach and a motivation to transform society. This chapter has tried to demonstrate that a constructivist, student-centred approach to education should be used and that learning about climate change must incorporate field experience in building community resilience, with a focus on the assessment of vulnerability and the acquisition of knowledge and skills relevant to adaptation.

The usefulness of the Certificate courses is illustrated by these comments from prominent citizens:

> As climate researcher I can confirm that Certificate III in climate change and disaster risk reduction offered by Vanuatu Institute of Technology is the best in the Pacific islands.
>
> Tigona, Vanuatu Meteorology and Geohazards Department (2018).

> This Certificate III course is the first of its kind to be run in the Pacific, and we must take pride in this. You students have learnt about such topics as food and water security, management of marine resources and the re-stocking of coral. You have learnt how to write project proposals - an important tool that has taken some of us years to learn. You are so privileged! And the way you have used tools such as risk maps and SWOT analysis with communities in the field is very powerful.
>
> Wells, Asian Development Bank (2018).

Building on this success, however, will present challenges. Firstly, delivery of the courses at VIT must be facilitated by knowledgeable, dedicated teachers with a high level of commitment to both the promotion of resilience and to the welfare and development of their students. Those learners who completed the first Certificate

I and III courses were already highly motivated, and they had the benefit of being taught by the course developer. In future, such high levels of devotion and interest may not be present, especially if there is insufficient funding to guarantee effective fieldwork and the colour printing of learning materials. Secondly, there must be ongoing moral and financial support for these programmes from community leaders, government departments, civil society organisations and overseas donor partners, ensuring that the practical and communication skills promoted are not downplayed in deference to academic content. The courses could also be delivered at the Vanuatu Agricultural College and the Vanuatu Institute of Teacher Education, and the associated knowledge, skills and attitudes are appropriate for inclusion in primary and secondary school curricula.

The CCDRR and Resilience courses are fully in line with Vanuatu's National Sustainable Development Plan, 2016–2030, and Vanuatu's Climate Change and Disaster Risk Reduction Policy, 2016–2030. Let us hope that the efforts already initiated can be developed still further, thereby enabling a greater degree of resilience to be achieved in Vanuatu's burgeoning communities.

References

Australian Academy of Science (2018) Global carbon dioxide budget 2004–2013, based on Global Carbon Project, 2010. Accessed on 24 Apr 2018 at https://www.science.org.au/ learning/general-audience/science-booklets-0/science-climate-change/3-are-human-activities-causing

Bahá'u'lláh (c.1887) Tablets of Bahá'u'lláh. Bahá'i Publishing Trust, USA, 1988, p 161. ISBN-10 0877432163

Bündnis Entwicklung Hilft and Ruhr University Bochum—Institute for International Law of Peace and Armed Conflict (IFHV) (2018) World risk report. ISBN 978-3-046785-06-4. Accessed on 5 Jan 2019 at https://reliefweb.int/report/world/world-risk-report-2018-focus-child-protection-and-childrens-rights

GermanWatch (2017) Global climate risk index 2017. Accessed on 6 Jan 2019 at https://germanwatch.org/sites/germanwatch.org/files/publication/16411.pdf

Great Schools Partnership (2014) The glossary of education reform: competency-based learning, Portland, USA. Accessed on 9 Mar 2019 at https://www.edglossary.org/competency-based-learning/

Hawkins K, Radio New Zealand International (2015) Destruction in South East Epi caused by Cyclone Pam on 13th March 2015. Accessed on 5 Apr 2015 at http://www.radionz.co.nz/international/programmes/worldandpacificnews/galleries/vanuatu-after-cyclone-pam

Johnson DW, Johnson RT, Holubec EJ (1994) Cooperative learning in the classroom. Association for Supervision and Curriculum Development, Alexandria, Virginia, USA

Kagawa F, Selby D (eds) (2010) Education and climate change: living and learning in interesting times: a critical agenda for interesting times. Taylor and Francis, New York. ISBN 0-203-86639-8 Master e-book ISBN

Kolb D (1984) Experiential learning: experience as the source of learning and development, vol 1. In: Mcleod S (eds) (2013) Kolb—Learning styles. Prentice-Hall, Englewood Cliffs. Accessed on 7 Mar 2019 at https://www.simplypsychology.org/learning-kolb.html

Mele F (2018) Comment in the evaluation section of the last page of the learner workbook for Unit 9

Pacific Climate Change Science and Adaptation Planning Programme (PACCSAP) partners (2015) Country brochures: current and future climate of Vanuatu. Accessed on 8 Mar 2019 at https://www.pacificclimatechangescience.org/wp-content/uploads/2013/06/15_PACCSAP-Vanuatu-11pp_WEB.pdf

Pierce C (2019) Perceptions of climate change in a rural community of Vanuatu. Port Vila, Vanuatu

Stevenson R, Nicholls J, Whitehouse H (2017) What is climate change education, published online in curriculum perspectives. Springer, Berlin. https://doi.org/10.1007/s41297-017-0015-9. Accessed on 7 March 2019 at https://www.researchgate.net/publication/316242167_What_Is_Climate_Change_Education

Te Kete Ipurangi (2019) Effective pedagogy, the New Zealand curriculum, p 34. Ministry of Education, New Zealand, accessed on 10 March at http://assessment.tki.org.nz/Assessment-for-learning/Assessment-for-learning-in-practice/Reflection-on-the-learning

The Commonwealth Education Hub (2016) Climate change and education: a policy brief. Accessed on 7 Mar 2019 at https://www.thecommonwealth-educationhub.net/wp-content/uploads/2016/02/Climate-Change-Policy-Brief_Draft_140416_v4.pdf

Tigona R, Vanuatu Meteorological and Geohazards Department (2018) Comment on page of Facebook group "Vanuatu Rainfall and Agro-Meteorology Outlook" on 7 Dec 2018

Wells N, Project Officer, Asian Development Bank, Port Vila (2018) Extracts from her speech given at the ceremony for the completion of certificate III in resilience, 14 Dec 2018

Western Business School (2019) Competency-based training and assessment process, Melbourne, Australia. Accessed on 9 Mar 2019 at https://www.wbs.org.au/competency-based-training-assessment-process/

Wogale E (2015) Traditional method of fishing used on Mota Island in the Banks Group, Vanuatu

World Population Review (2019) Accessed on 6 Jan 2019 at www.worldpopulationreview.com/countries/vanuatu-population/

Chapter 19
From Academia to Response-Ability

Raichael Lock

Abstract The need to tackle climate change through both mitigation and adaptation is increasingly urgent with all nations and all sectors of society needing to respond. Given this state of affairs social action is important not only to raise awareness but to stimulate appropriate responses to the climate crisis. As a consequence, social action programmes in schools have been a crucial part of learning about climate change as they provide political and practical engagement. This paper explores the experience of the Manchester Environmental Education Network (MEEN), a small UK charity that runs a social action project with schools focusing on the facilitation of inter-generational activities around climate change. Presenting three vignettes, written by myself as the MEEN coordinator, the aim is to examine the Carbon Classroom programme and discuss the ethical value of these intergenerational interactions. Rather than working with the idea of responsibility, which focuses on duty and accountability, this paper will explore climate change education through the notion of response-ability (Haraway 2016) with the aim of cultivating an ethical and open approach where responses are mutually engaging. Thus, by drawing on Barad's theory of agential realism, the vignettes will highlight instances where enactments of response-ability occur. The conclusion will discuss the recent rise in young people engaging in activism on climate change, such as the Fridays for Future, Youth4Climate and the Extinction Rebellion movements, and question, in the light of these events, how the role of climate change educators need to be reconfigured to become more response-able.

Keywords Social action · Education · Response-ability · Agential realism

---

R. Lock (✉)
Manchester Institute of Education, University of Manchester, C3.23, Ellen Wilkinson Building, Manchester, UK
e-mail: raichael.lock@manchester.ac.uk

© Springer Nature Switzerland AG 2019
W. Leal Filho and S. L. Hemstock (eds.), *Climate Change and the Role of Education*,
Climate Change Management, https://doi.org/10.1007/978-3-030-32898-6_19

Introduction to Agential Realism: Who Is Response-Able for Climate Change Education?

There are calls from climate change education researchers for there to be more emphasis on the collective response to climate change rather than pressing for individual behaviour change (Waldron et al. 2016). Such thinking describes the tendency to view the individual, or the self, as removed from the 'other' who exists out there and this duality often troubles our thinking around climate change. A common example could be the position that it is not worth acting on climate change because others are failing to do so or, at the other extreme, that one small action does not have an impact. What these examples share is that they are premised on the duality of the collective and the personal.

Barad's theory of agential realism (Barad 2003) attempts to move our thinking away from such duality so is helpful as a methodological framework for thinking through change. She proposes that rather than individuals having agency the universe *is* agency in that matter is in a constant state of being reconfigured: agency is the dynamism of the universe (Barad 2003). From this perspective there are no subjects or objects but rather enactments of matter that might or might not involve humans (Barad 2003). She suggests that the way we interpret the world is to examine the specific details being enacted and to make temporary separations, in other words by viewing the world as subjects and objects, as a means to make sense of the world.

This move beyond individualism is not unique to agential realism. There are other methodologies that focus on collectivity which would be in agreement with Barad's statement that, 'There are no singular causes. And there are no individual agents of change' (Barad 2007, p. 394). For example, there are commonalities with Latour's concept of distributed agency which, rather than ascribing agency to individuals, views it as spread across a network including both humans and nonhuman 'actants'. It also has similarities with Haraway's notion of 'sympoiesis', a term that means 'making-with' where, she states that 'critters' such as cells and organisms 'interpenetrate one another' and form 'sympoetic arrangements she describes as 'ecological assemblages' (2016, p. 58).

Such relational methodologies also point to a relational ethics. According to Barad, and in accord with Haraway, there is a need to attend to relations and this brings with it 'relations of obligation' (Barad 2010). More specifically, Barad asserts:

> First of all, agency is about response-ability, about the possibilities of mutual response, which is not to deny, but to attend to power imbalances. (Barad as interviewed by Dolphijn and van der Tuin 2012)

But what might it mean to enact response-ability, to form mutual responses and attend to power imbalances? A response is an action but a 'mutual' response affirms inter-connectivity. Barad uses the term 'intra-activity' to describe the connectivity inherent in all action stating that:

> Particular possibilities for acting exist at every moment, and these changing possibilities entail a responsibility to intervene in the world's becoming, to contest and rework what matters and what is excluded from mattering. (Barad 2003)

Therefore, the enactment of mutual responses from the perspective of climate change can be seen working in multiple ways. For example, there are mutual responses between greenhouse gases and soils as they form exchanges through biogeochemical processes; between plants and gases through the process of decomposition and through the relations between humans and fossil fuels. Although the stories of mutual response in this paper will focus largely on humans it is always in relation to the material world and the role of our collective actions in climate change. But these complex webs of inter (or intra) action are all based in very specific and detailed actions that occur moment by moment and, according to Barad, it is by attending to every mutual response that response-ability is enacted.

But, as Barad suggests, response-ability, 'is not a calculation to be performed': there is no recipe for becoming response-able: it is rather about enacting a state. 'It is a an iterative (re)opening up to, an enabling of responsiveness' (2010, p. 265). The concept of opening up or, as Haraway puts it, 'cultivating response-ability' (2016, p. 32) is about being able to respond to whatever arises or, to put it another way, it is to practise the skill, or the ability, to respond ethically.

In theory, if no one moment is quite the same as the last, in the 'ongoing flow of agency' then the process of educating, or researching for that matter, means that when attempting to define, or enact, 'response-ability' what is appropriate in one instance may not be appropriate in the next. The notion of becoming response-able also destabilizes, at least theoretically, the apparent dichotomy of free will and determinism. Barad argues that because intra-actions always include specific material connections it means that other phenomena, by necessity, are excluded. In her view this means that, 'intra-actions are constraining not determining. That is, intra-activity is neither a matter of strict determinism nor unconstrained freedom' and she further contests that, 'The future is radically open at every turn' (Barad 2003, p. 826).

Such a position may be scientifically contentious but, read metaphorically, the message is useful to the climate change educator. By most accounts climate science is pointing us towards the likelihood of catastrophic climate change whilst calling for 'collective human action' to stabilize earth 'in a habitable interglacial-like state' (Steffen et al. 2018). Yet the UK, for example, is struggling to meet its own carbon reduction targets (Le Page 2018). Given such precarity the idea of an 'open' future means that we cannot exclude the possibility of sudden breakthroughs, disruptions or turnarounds in relation to climate science, behaviour change or other current unknowns. It also means that every action matters.

However, the complexities surrounding climate change should not prevent us from attending to the material intra-actions. As educators, and the educated, it is necessary to attend to the specific enactments we are involved in and relate them to our collective knowledge. Walking the talk, so to speak, is imperative otherwise our actions and words are incongruous. Furthermore, those in the position of being educated may rightly perceive this as a power imbalance if they are told how to behave by others who are not acting on their own advice. We need to respond to our collective footprint because, as Haraway puts it, we must attend to the details that matter, because 'details link actual beings to actual response-abilities (2016, p. 115) and to act with response-ability is to act on our knowledge.

Consequently, this paper will examine instances of response-ability, presented as brief vignettes, as they have been documented in the course of my work with MEEN. The first vignette explores the details of an intergenerational intra-action in a public setting; the second how intergenerational issues play out in a school setting and the third focuses on intergenerational responses in a specific political setting. All three vignettes will be examined through the notions of response-ability and openness. But first it is necessary to attend to my role as both researcher and as the deliverer of MEEN's school activities.

The Researcher's Response-Ability

As an environmental educator devising and delivering MEEN's project-based activities this paper includes both a contextual backdrop to MEEN's work with schools and three vignettes drawn from my position as MEEN coordinator.

Whilst both the names of the schools and people involved are anonymised the vignettes have been referred to those involved for written or verbal permission for inclusion. The key players have read the vignettes in which they are cited and have agreed they can be shared. In one case a school staff member agreed to have the comments they made following reading the vignette included whilst in another a photograph is used to show the feedback from an adult who visited the Carbon Classroom. In keeping with Barad's agential realism and the iterative reconfigurings of the world all the vignettes are open-ended, in other words, the stories are not seen as conclusive but rather as ongoing.

The aim of the Carbon Classroom is to invite young people to run a stall in a public place and engage members of the public in discussion and game playing activities for learning about climate change (Brown and Lock 2017). Although the Carbon Classroom is devised as a setting for enacting social response-ability in relation to climate change it is not exclusive in this role and the final vignette draws on a research project called Save Our Soils.

All three schools have previously engaged in the Eco Schools programme and, although they may not be actively signed up to the programme now, the stories are drawn from working with groups of young people on an Eco team whose role is to attend to the environmental issues in their school communities. The stories were identified as specific examples of response-ability in action.

Although the vignettes focus on the specifics of MEEN projects, it must be acknowledged this boundary presents 'a danger of both privileging that network and rendering invisible its multiple supports' (Fenwick and Edwards 2010, p. 6). MEEN's projects participate in a much wider network of activities and the boundaries are controversial.

The School Context in Relation to Climate Change Education as Social Action

This section of the paper contextualises MEEN's work and generalises how the organisation tries to response-ably negotiate and deliver the Carbon Classroom in intra-action with the school community. It includes how policy makes a difference, particularly the curriculum, in relation to taking social action in schools, discusses intergenerational power issues and the need to address them with response-ability.

Many young people, including those working with MEEN, express deep concerns about climate change, stating they feel fear and dread on learning about the issues and they cite social action as the most positive response for addressing the problem (Gayford 2009). Meanwhile, England's National Curriculum (DfE 2011) having put the emphasis on learning about climate change education in secondary Science and Geography, may have improved young people's knowledge but it has not put the emphasis on responding with social action. Recent research in England's schools affirms that the approach to teaching climate change is, 'largely dominated by a conservative vision' where there is little 'provocation to encourage alternative thinking about environmental challenges or to genuinely encourage environmental activism' (Glackin and King 2018, p. 9). Since 2011 MEEN has been facilitating The Carbon Classroom with Manchester schools with the aim of addressing this perceived gap.

School staff generally welcome the opportunity to participate however, the intra-actions are complex. Secondary schools find creative ways of fitting the project into the school day. It might be worked in as a transition project, an after-school club, or as engagement for a leadership programme. Primary schools are more likely to identify the work with eco teams, school councils or by working with classes as a cross-curriculum project. There are as many points of access to delivering the Carbon Classroom programme as there are participating schools with each having specific material conditions, as Barad would state, for the project to materialise and unfold.

But for the process to enact mutual intergenerational responses, 'Adult partnership is needed to ensure that children are in fact allowed to make decisions and take on leadership roles' (Torres-Harding et al. 2018, p. 5). This introduces the idea of intergenerational power imbalances. For example, in setting up a carbon classroom, decisions about the types of activities to be offered are made, initially, without input from the children. Carbon Classrooms can be delivered with another school, in a public place such as a museum, shopping centre, place of worship, or, as is the most recent case, in a Student's Union. In every case such activity needs particular agreement from both the schools and organisations involved. In the spirit of openness MEEN discusses with the team why certain options have to be excluded. It may be the schools do not have staff cover or the potential venue considered too risky. In this regard the multiple layers of responsibility, from both the school and MEEN, can make intergenerational response-ability difficult.

When the project is extra-curricular it requires the involvement of staff, pupils and parents and a variety of other factors that need to come into alignment. Sessions may be crammed into quick bursts of activity as they are squeezed into lunchtime or there may be clashes with parent evenings or staff training whilst pupils need to rearrange being picked up from school or free themselves from other commitments. Primary schools are more likely to run the project during curriculum time, but pupils on the eco team can be called away for academic activities deemed more important. Parents also need to give permission for their children to participate in Carbon Classrooms and children can worry about being excluded whilst school and MEEN staff need risk assessments and photo permissions. Whatever the details of the specific scenario it is evident that for young people to engage in social action on climate change in schools it is adults who are responsible for enabling the activity. Indeed, it is the ability of MEEN and the school community to respond flexibly and collectively to complex and changing situations that enables Carbon Classrooms to happen at all.

If the school community exercises response-ability to make the project happen then MEEN's task is to be response-able to the participants. To begin any intra-action the first activity is to ask what is already known about climate change: for high school pupils this may be a repetition of the science or geography curriculum, blended with information gleaned from television, on-line or other sources, whilst primary school pupils may not have heard of climate change at all.

In the latter case MEEN runs mini-roleplay scenarios starting from what pupils already understand, such as the fact the sun gives light and heat to the earth. For example, pupils enact the role of sun and earth, whilst others become a stream of light that travels to earth only to be transformed into heat. Tasked with escaping back into space they are captured by greenhouse gases and the role-play enacts how this becomes increasingly difficult the more greenhouse gases are present. Such activity is invariably experienced as fun, but once pupils have a reasonable understanding of the climate science, its possible impacts and solutions, they are then tasked with becoming community educators. This disruption of the educator/learner boundary gives value to their knowledge and pupils generally respond well to the challenge. More importantly though it provides them with the opportunity to exercise response-ability as they are asked to initiate conversations on climate change.

In terms of agential realism and the enactment of mutual responses, this does not put the onus on the 'child as educator' (Burman 2013) but rather on each intra-action that unfolds in the entanglements of each Carbon Classroom. By inviting young people to become community educators there is a sense that intra-actions can consolidate or dissipate agency as passers-by might stop and engage but equally, some might remain passers-by and ignore or dismiss those addressing them. The fact that school children are asking for a conversation around climate change often surprises members of the public as it does actively disrupt boundaries around who is the teacher and who holds the knowledge and power.

To help with this MEEN always runs a session on how to approach people and role-plays possible responses suggesting that to cultivate 'possibilities for mutual response' there needs to be a balance, an interplay, between the propagation of openness and trust alongside a need to safe-guard young people. Obviously the context of each Carbon Classroom is different but the detail of each setting matters— organising in a shopping centre or community centre is different to setting up in a museum or a library. Whatever the context though, the pupils need to be open and response-able to the detail of each conversation, to each specific intra-action, as it unfolds. An example of this intra-action is given in the first vignette.

Vignette 1—A Response-Able Conversation

A mixed group of primary and high school pupils are running a Carbon Classroom as part of a climate change exhibition. The pupils are approaching visitors and asking if they can spare a moment to talk about climate change.

In this context a primary pupil approaches a tall man in a suit. I'm within earshot as a safeguarding measure but far enough away not to be a participant in the conversation. The pupil asks whether the man is concerned about climate change. He replies that he is not because scientists are working on carbon capture and storage. The pupil says she hasn't heard of this, so he explains how it works. She contemplates this scenario then says she's not sure it will work as a solution, firstly, because the technology is untested and we need to act now; then we would still need to extract fossil fuels which takes energy and then transport the gas to a storage site, with no guarantees the gas will not escape. She then states there are other safe renewable energy sources that could sort out the problem. He acknowledges her challenge by commenting that he hadn't thought about all the energy involved in mining, processing and transportation and he is clearly thinking through her ideas when she asks him to step into the Carbon Classroom to look at the games they are running. He accepts. His written feedback (Fig. 19.1) *states she, 'Challenged my views on*

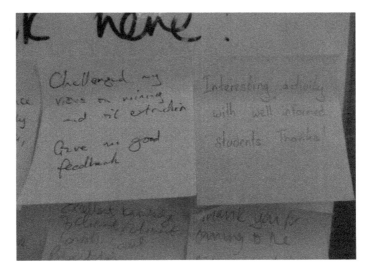

Fig. 19.1 Written feedback

mining and oil extraction. Gave me good feedback,' and he verbally stated to the MEEN coordinator that his views on climate change had been well challenged.

Being able to respond to something openly is difficult especially around issues where people, often adults, have already adopted a position, but the discourse in this instance had both participants being responsive to new ideas.

The pupil's approach was open: she had already experienced a multitude of responses from the public, having been ignored, even dismissed, by some passers-by, and having talked with those who know very little about climate change to engaging with experts. Her previous activities have given her a sense of what might happen but, she has little idea how the scenario will pan out and on her approach she was looking for signs on which to base her response.

The adult stopped to engage. It was evident he was an employee of the University of Manchester identifiable as such because he was wearing a lanyard. The pupil may not have registered this but she may well have been reading other signs such as the fact he was wearing a suit although whether this was perceived as indicating an imbalance of power is impossible to determine. In another instance pupils attending a Green Summit alongside businesses, as cited in the third vignette below, were tasked with talking to other attendees to share stories. Interestingly they chose to target people wearing suits with one pupil saying, 'There's a man in a suit. Let's get him!'. The sense was that men in suits had power, therefore, more responsibilities, and the pupils wanted to challenge them in order to prompt response-able intra-actions.

Returning to this vignette the adult engaged was very tall and bowed his head down in a gesture that showed he was giving attention to what the pupil had to say, whilst in response she craned up to engage. By moving in and creating an attentive closeness they both indicated involvement in the intra-action. When the man then presented his case for not worrying about the impacts of climate change the pupil's response to carbon sequestration could have been awkward as she had never heard of the proposed technology. However, retaining attentiveness she matched his confidence and asked him to explain what it is and how it works.

One of the consistent responses young people give when asked about the Carbon Classroom, or similar scenarios involving the public on matters of social concern, is that they experience greater levels of confidence. Pupils state that because they might know more about climate change than those they approach they can query their adults' positions. Of course, the adults' responses to the shift in power are key: they need to engage on mutual terms. When faced with a climate change denier the discourse that ensues might be short, fraught or, as in one instance, slightly antagonistic. But even in polarity there is the potential for a response-able exchange, a discussion, a debate where different opinions are aired.

In this instance, the adult was clear there is nothing to worry about and provided an accurate description of carbon capture and storage enabling the pupil to consider the proposed scenario. The ruminating silence as the pupil thought about their response to the information was surprisingly long but respectfully so as it allowed the pupil time for clarification. Finally, the pupil's response enabled a mutually agreeable outcome whereby the adult's assumptions were challenged and the pupil learned about the possibility of technological fixes.

The engagement highlights a shift in adult/child relations but also in educator/learner boundaries. Of course, even given further details it would be impossible to determine how this particular intra-action became an enactment of mutual response. As Barad asserted, response-ability, 'is not a calculation to be performed' (2010, p. 265). It could be the case that the adult is a trained educator, whose job is to listen and respond to the ideas of young people, or maybe that the pupil has been trained in debating techniques, but whatever the unknown details, this specific instance highlights how social intra-action can help to reconfigure a climate change discourse. Returning to Barad's notion that 'agency is about response-ability and the possibilities of mutual response' (as quoted in Dolphijn and van der Tuin 2012, p. 55) this instance demonstrates how mutually enacting due attention enables response-able social intra-action.

In terms of agential realism, it is impossible to comment on whether this social intra-action has left any mark on those involved. Whether the adult reviewed his position on climate change, or whether the pupil even remembers the encounter, remains unknown in this instance. However, the next vignette explores how a response-able intra-action might unfold as it plays out through a sequence of events.

Vignette 2—Community Response-Ability

A group of high school pupils have been meeting with school staff and MEEN in their lunchtime to consolidate their learning on climate change. The aim is for the group to run an afternoon session on climate change for visiting pupils from a feeder primary school. On the day the session begins with pupils and staff sharing a low-carbon lunch from the Real Junk Food project, then the high school pupils deliver an inter-active presentation before running a climate change game playing session and action planning session to help both schools respond to the climate emergency. A variety of adults from the school community attend the session including the Chair of Governors who has considerable knowledge and concern around climate change.

At the end of a busy afternoon the high school pupils are talking about what they want to do next and it's clear they want to target the school staff: they state this is important because it seems that many of the adults in the school don't understand the urgency of the problem, except for the geography teachers and the science teachers. The pupils want to organise a school day dedicated to learning about and taking action on climate change with a range of activities from game playing, tree planting, and an action planning session with the school staff.

On sending this story to a member of staff involved in the project they reply that a 'full day of climate change work' was unlikely to materialise soon because of intense academic pressures. Rather they saw the on-going nature of the project as being able to reconfigure the school's response to climate change as, 'a slow fuse project' which would 'get real lasting change' into the school culture, including the school eventually being willing to take the whole school, or even a year group, off timetable for a day.

This vignette highlights what Barad might call a 'congealing of agency' (Barad 2007) where a group of young people, engaging with staff, MEEN and others, explore response-able social intra-action and are clearly keen to continue the process. They are enthused and excited about their potential role in building a whole school response to climate change through intra-acting not only their peers but also their teachers. Once more the power imbalance between adults and children is disrupted with the pupils becoming confident enough to want to plan a day of activities and further social intra-actions.

Enfolded within this instance there are myriad back stories: the key staff member had been trying to instigate such a project for years and felt frustrated by a lack of opportunity and had waited a year for MEEN to be able to lend support. Then, following the event, the staff member was sure the project would wind down into what they called a 'slow fuse' project as they expected the school to priorities other issues. Yet having framed the response as a 'slow fuse' project when the team met again they produced an extensive action plan. They wanted to deliver an energy saving programme called Switch-off Fortnight which is a part of the Eco School programme and committed to organising assemblies and run energy saving competitions. They also planned to talk not only to school staff at a Friday staff briefing session but also talk to governors. In fact, under 'Engaging with leadership' the pupils outlined demands including for any financial savings made from energy saving to be re-invested in further environmental action. Such an action plan suggested that the pupil's sense of response-ability in relation to climate change had increased, but for the plan to be enacted the senior leaders need to also exercise response-ability, not only in relation to the members of the team but also in relation to the climate crisis. Has this happened? In all these instances there is no success or failure as the activity, as Barad would suggest, is on-going.

MEEN tries to respond by continuing to provide support in an effort to congeal the agency in any given instance. Although such a position is antithetical to conventional project management, for MEEN, in order to practise response-ability, it is necessary to find ways to continue relations with a school and to keep the material-discourse open and active. One move has been to link projects into a continuum and, in this instance, MEEN has planted trees with the team and engaged them with an intergenerational conference on climate change where they can share their knowledge, skills and actions.

These vignettes, therefore, need to be read as open-ended. They cannot be conclusive in the context of the project or in relation to climate change. In this specific scenario however, the Senior Leadership team need to respond with equal enthusiasm and attention to the issues with MEEN waiting to be updated. The third vignette extends the context of climate change education beyond personal intra-actions and beyond the school by engaging young people in the political context of a Green Summit.

Vignette 3—Becoming Response-Able

MEEN is running a project with primary school eco clubs on impoverished and contaminated soils in inner city Manchester. One school has a member of staff committed to the agenda and a team of enthusiastic pupils whose ages range from 6 to 11.

When MEEN secures last minute invitations for three members of the team to be delegates at the Greater Manchester Mayor's Green Summit the school is quick to respond. At the Summit the pupils are encouraged to mix and mingle with adult delegates inviting them to share stories and commitments around environmental activities. One pupil, who demonstrates a specific interest in climate change, is prompted by me to ask a renowned climate scientist about the connections between soil and climate change. The scientist says there are connections but they are very complex relationships still under research.

Later in the day the pupil is walking down a corridor with me and states, 'This is the most important day out.' I ask why and the response is, 'it's much better than going to Blackpool on a trip because here you really feel as though you're doing something.'

Three months later at MEEN's Intergenerational Save Our Soils conference, designed to bring schools, universities, NGOs and others together to share soil knowledge and experience, the same pupil confidently addresses an audience on the issues they see as connecting soils, trees and climate by linking the impact of floods and droughts to soils and trees.

This vignette focuses on a range of intra-actions that highlight the benefits of being response-able to emerging situations. It also attends to an on-going situation where adults in positions of power respond to the intra-actions they have with young people. In an example where, as Barad (2003, p. 822) would put it, 'agency congeals' the conference organisers agree a school can be invited and the school manages to give staff time and provide parental permissions for pupils to attend in a couple of days. Being response-able requires immediate attention and the value of this attention is evident from pupil responses to the experience of the summit. It is perceived as 'doing something' about an issue of concern, they have contributed to negotiations, contributions that reconfigure their community and their relations to that community.

Furthermore, if agency is about cultivating mutual responses it is necessary to collectively press for response-able change by attending to those in power, whether with politicians, universities or businesses and the inclusion of young people into an adult summit changed the event. The pupil stated he was engaged in something 'important' which, in my view, could indicate the perceived value of relational engagements as businesses, scientists and the Mayor took the time to have response-able intra-actions. The pupils may have struggled to pay attention to all the speakers in the vast conference centre but when it came to approaching others in conversations it disrupted conference norms. Being stopped by a ten-year-old wanting to hold a company or an organisation to account in relation to climate change is not a common phenomenon.

However, in the on-going 'flow of agency' (Barad 2003, p. 817) these intra-actions drew a response that has reconfigured the format of the summits. Some thirty schools from across Greater Manchester attended the next Summit March 2019, with three schools working with MEEN to deliver a Carbon Classroom in the conference marketplace. However, the act of involving young people in the conference is not an end in itself: how schools are integrated into the day is something that needs attention. Reading this vignette through agential realism serves as a reminder that what is included, or excluded, in any instance, is an enactment shot through with ethical implications. Those working on the conference have taken on extra work to ensure that the schools in attendance have suitable facilities and activities to engage young people with others in a response-able manner.

MEEN's method for running the intergenerational soil conference, as mentioned in the vignette, is an attempt to facilitate mutual responses. The soils conference offered opportunities for young people to present and run workshops for adults as well as attend adult run sessions and it is organised like this to attend to power imbalances and generate mutuality. It is this that marks such an event as an enactment of intergenerational social intra-action.

In the delivery of response-able intergenerational activity it also helps to be flexible with subject boundaries. Throughout MEEN's soil project pupils were encouraged to explore soil through their chosen 'matter of collective concern', whether it was concern for animals or the pollution caused by plastics. In the vignette, the pupil cited is particularly concerned with climate change and, despite being told by an expert that many of the links between soil and climate are still being researched, the pupil's presentation confidently explored relations between soil and atmosphere. This enactment was another example of agency congealing through many different material-discourses: it was a creative and useful contribution to the conference and highlighted how climate change education can be woven into other subjects.

This vignette shows that climate change education does not exist in isolation as a single subject and that all educators, of whatever age, can exercise response-ability through its inclusion. It also emphasises how specific intra-actions contribute to change where there is a political openness to act with response-ability.

Conclusions and Possible Reconfigurations

In the introduction it was stated the paper would explore climate change education as social intra-action through agential realism and response-ability. The vignettes have presented acts of mutual response at three different scales all showing how flexibility and openness are necessary for the cultivation of response-ability. The disruption of boundaries, whether age related, educator/learner related or subject boundaries, also seems to help facilitate response-ability by challenging habitual responses. However, trying to reach conclusions about how to teach response-ability does not in itself cultivate response-ability. Response-ability is an open-ended process which occurs spontaneously and through specific relations. As Barad stated, it is not a 'thing' which can be bought in or a method that can be implemented rather it is a state to be practised through enactment.

In light of this, I would like to give an example of how response-ability might be enacted through MEEN's relationship to current developments. Social action in schools is currently being reconfigured. Youth activism in relation to climate change has been transformed by the global school strikes inspired by Greta Thunberg and the international social action group, Extinction Rebellion. Manchester has seen two school strikes and more are planned. Prior to both occasions, MEEN was approached by parents, grandparents and other adults, wanting information about how they could support the action indicating the strikes have strong intergenerational support. However, the question for MEEN, and for other climate change educators, must be how to facilitate mutual responses not only with young people and our network members but also with schools.

Many Manchester schools have taken the stance that attendance is their primary responsibility and this has made it difficult for MEEN to generate a response-able link between the school strikes and young people in school settings. However, not addressing the school strikes would be antithetical to creating a mutual response. Young people want to be heard on climate change so it is imperative to listen, acknowledge and respond.

Attending to our position in relation to these different pressures is critical and leads to questions such as, if social action on climate change is now being led by young people does MEEN have a role to play in this area of activity? Should MEEN respond with even more intense and practical efforts at mitigation and adaptation? Or maybe MEEN needs to build stronger and more effective links between young people and politicians? The value of the Carbon Classroom is that it offers opportunities for social intra-action in a school context but it is worth questioning whether its continual re-enactment is a response-able way to address the climate emergency.

Whatever MEEN decides to prioritise, it is clear that climate change educators need to be response-able for creating open-ended, responsive, material-discourses, not only in relation to the science of climate change but also in the activities created in the on-going flow of mitigation and adaptation. Most importantly though, if we are to have any integrity in our responses with young people, it is imperative that we attend to, and act with, response-ability in relation to each and every specific intra-action in which we are enmeshed.

Bibliography

Barad K (2003) Posthumanist performativity: towards an understanding of how matter comes to matter. Signs 8(3):801–831

Barad K (2007) Meeting the universe halfway: quantum physics and the entanglement of matter and meaning. Duke University Press, Durham

Barad K (2010) Quantum entanglements and hauntological relations of inheritance: dis/continuities, spacetime enfoldings, and justice-to-come. Derrida Today 3(2):240–268

Brown S, Lock R (2017) Enhancing intergenerational communication around climate change. In: Leal Filho W et al (ed) Handbook of climate change communication, vol 3. Springer International Publishing, pp 385–398

Burman E (2013) Conceptual resources for questioning 'child as educator'. Stud Philos Educ 32(3):229–243

Department for Education (2011) The framework for the national curriculum. A report by the Expert Panel for the National Curriculum review. Department for Education, London

Dolphijn R, van der Tuin I (2012) New materialism: interviews & cartographies. Open Humanities Press

Fenwick T, Edwards R (2010) Actor-network theory in education. Routledge, London

Gayford C (2009) Learning for sustainability: from the pupils' perspective. A report of a three-year longitudinal study of 15 schools from June 2005 to June 2008. The Institute of Education, the University of Reading. Available at: https://www.meen.org.uk/Assets/Documents/WWF%20Pupils%20Perspective.pdf. Accessed 9 Mar 2019

Glackin M, King H (2018) Understanding environmental education in secondary schools in England: perspectives from policy (Report 1). King's College London, London

Haraway D (2016) Staying with the trouble. Duke University Press, USA

Harre N (2007) Community service or activism as an identity project for youth. J Commun Psychol 35(6):711–724

Le Page M (2018) UK is not on track to meet its own climate targets says report. New Scientist. Available at: https://www.newscientist.com/article/2172829-uk-is-not-on-track-to-meet-its-own-climate-targets-says-report/. Accessed 9 Mar 2019

Steffen W et al (2018) Trajectories of the earth system in the anthropocene. Proc Natl Acad Sci 115(3):8252–8259

Torres-Harding S, Baber A, Hilvers J, Hobbs N, Maly M (2018) Children as agents of social and community change: enhancing youth empowerment through participation in a school-based social activism project. Educ Citizensh Soc Justice 13(1):3–18

Waldron F, Ruane B, Oberman R, Morris S (2016) Geographical process or global injustice? Contrasting educational perspectives on climate change. Environ Educ Res. Available at: https://www.tandfonline.com/doi/abs/10.1080/13504622.2016.1255876. Accessed 9 Mar 2019

Chapter 20
Recognition of Prior Learning (RPL) in Resilience (Climate Change Adaptation and Disaster Risk Reduction) in the Pacific: Opportunities and Challenges in Climate Change Education

Helene Jacot Des Combes, Amelia Siga, Leigh-Anne Buliruarua, Titilia Rabuatoka, Nixon Kua and Peni Hausia Havea

Abstract Climate change adaptation and disaster risk reduction in the Pacific led to a high number of informal capacity building workshops where community members were trained on different aspects of climate change adaptation and disaster risk reduction. However, as there were no assessments at the end of the workshops, participants could not build on these certificates to continue to develop their capacities. The European Union Pacific Technical and Vocational Education and Training (EU-PacTVET) project partnered with Fiji Higher Education Commission to develop a process to recognize prior learning acquired during informal workshops. The project focused on how to assess the acquired skills and credit the learners to then complete a resilience qualification. Recognition of Prior Learning (RPL) creates opportunities for people who acquired skills during informal workshops or through their personal

---

H. Jacot Des Combes (✉) · L.-A. Buliruarua · P. H. Havea
PaCE-SD, USP, Lower Campus, Laucala Suva, Fiji
e-mail: helenejdc@hotmail.com

L.-A. Buliruarua
e-mail: leighanne.buliruarua@usp.ac.fj

P. H. Havea
e-mail: ilaisiaimoana@yahoo.com

Present Address:
H. Jacot Des Combes
National Disaster Management Office, Majuro, Republic of the Marshall Islands

A. Siga · T. Rabuatoka · N. Kua
SPC, 2nd Floor Lotus Bldg, Nabua, Private Mail Bag, Suva, Fiji
e-mail: amelias@spc.int

T. Rabuatoka
e-mail: trabuatoka2508@gmail.com

N. Kua
e-mail: nixonk@spc.int

© Springer Nature Switzerland AG 2019
W. Leal Filho and S. L. Hemstock (eds.), *Climate Change and the Role of Education*,
Climate Change Management, https://doi.org/10.1007/978-3-030-32898-6_20

experience and provides them with credits for a qualification. As a result, it increases their employability, assists in the diversification of their source of income and supports the professionalization of the resilience sector in the Pacific. One major challenge for RPL is the need to build the trust of practitioners and potential employers for the process. Another challenge is to develop fair and robust assessment practices for resilience which is a broad, cross-cutting sector, where skills cannot be directly demonstrated during an assessment session.

Keywords Technical and vocational training · Recognition of prior learning · Pacific Islands · Climate change · Resilience

Recognition of Prior Learning (RPL)

Recognition of Prior Learning (RPL) is the process to assess the skills obtained out of formal education, mostly from life experience, work experience, and informal and non-formal education so that they can be credited in formal education (English 2005). RPL, also called Prior Learning Assessment and Recognition (PLAR), accreditation of prior experiential learning (APEL), or validation of prior learning (VPL), is now commonly used in different parts of the world, including Canada, Europe, South Africa and Australia and New Zealand (English 2005). RPL's original goal was, through the recognition of different types of experience and skills obtained outside of traditional education, to improve social justice. However, there is still a gap between this expectation and the reality of the outcome of RPL, at least in Australia (Cameron and Miller 2004). More recently, RPL has been used in the context of international migration to Europe to assess the skills of immigrants (Andersson et al. 2004; Moss 2014).

Although there is a consensus to agree on the potential benefits of RPL, there is also a consensus to point the difficulties, challenges and limitations of this process. First of all, "There is no clear agreement among writers, researchers and major policy-influencing agencies regarding what RPL is, does or encompasses. Views vary from quite tightly defined notions of RPL as access to a training program or qualification, through to conceptions of it being a reflective process that can directly influence the nature of learning and the process of training" (Smith 2004, p.5).

Another main challenge in the use of RPL is trust between the different stakeholders. On one hand, traditional education, in particular tertiary education, has developed as an individual and rational way of learning as opposed to the contextualized way, based on actual application of knowledge and skills to a specific situation, of learning that is assessed through RPL (Michelson 2006). As a result, there is still a significant reluctance to use RPL to access, or for credit towards, university programs. On the other hand, the RPL process is also considered confusing, very demanding in terms of collecting the necessary evidence and finalizing the paperwork for a risky result, partly due to questions on the quality of the evidences produced and the inconsistency of the assessment process (Hargreaves 2006).

However, despite these challenges, RPL has been considered extremely important for the Resilience (climate change adaptation and disaster risk reduction) sector in the Pacific region.

RPL in the Resilience Sector: Rationale and Challenges in the Pacific Small Island Developing States

The regional Resilience qualifications developed in the Pacific Small Island developing States (PSIDS), in particular at levels III and IV, integrate many different skills, thus reflecting the cross-cutting approach of activities in this sector (EU-PacTVET and FHEC 2016). Some of these skills include the recognition of the use of indigenous knowledge and local knowledge (IKLK) to identify and implement resilience building solutions in a cost-effective and sustainable manner in communities. IKLK is an important aspect of Resilience as demonstrated by several studies that have recently discussed the benefits of using these types of knowledge for climate change adaptation and disaster risk reduction in the context of coastal and small islands communities (e.g. Hiwasaki et al. 2014; Syafwina 2014; Bich Hop et al. 2017). As a result, skills that are included in the regional Resilience qualifications can be acquired by life experience in communities.

However, in the Pacific, life experience is not the only way to acquire skills that are included in the regional Resilience qualifications. In the last decade, many projects were conducted in the PSIDS to adapt to climate change or to improve disaster risk management. Many of the se projects included a capacity building component, often via capacity building workshops (USP EU GCCA Team 2017). These workshops increased the awareness of participants of the different issues linked to climate change and disaster risk management and supported the acquisition of skills to identify issues and potential solutions for them. However, due to a lack of assessment realized at the end of these workshops, it was not possible for the participants to have these skills recognized and accredited. So, when it comes to skills related to Resilience, the PSIDS are not a blank page and a process to assess these skills and credit them under the new regional Resilience qualifications is essential.

Unfortunately, assessing these skills is not without challenges. Contrary to most, if not all, traditional TVET disciplines, Resilience skillsets are not specific but integrate different types of knowledge and skills to achieve the expected outcome. Moreover, although some of the skillsets can be broken down into more specific tasks and skills, the articulation of these skills to solve a problem is as important as the individual skills. As a result, a demonstration in a controlled environment, a process used for other TVET skills, is not an appropriate process to assess Resilience skills. To address this specific challenge, it is more relevant to use a common RPL assessment method based on the preparation of a portfolio, where evidence of the skills acquired by the applicant can be collected and collated, completed with an interview to clarify some questions identified during the assessment of the portfolio. This method is considered to have at least two potential limitations. First, it may be difficult to collect evidence, for example when skills were acquired from a project that ended a few years prior and when the staff associated with this project left. Similarly, collecting evidence related to IKLK may be difficult because some of this knowledge is considered sacred and secret and is not supposed to be shared outside the community, or even family where this knowledge is used and kept. Similarly, some of this knowledge may be constrained by specific local condition and may not be applicable to other communities, or, if it is applicable, it may have different outcomes. In addition, the collected evidence needs to be of quality, and it is generally necessary to have the evidence supported by a witness. When the skills are obtained in employed work, the witness is generally the direct supervisor. However, when the applicants provide evidence of skills acquired during work in their own community, the witness may be the community leader or a family leader who may be related to the applicant. Although this is not a problem per se, it may raise questions on potential bias in the witness statement and on the reliability of the evidence provided.

Because the assessment process is complex and requires specific documents to support the application, it is necessary to make it as clear and explicit as possible for all involved: assessor, assessed, advisor, and administrators. A clear step-by-step guide on the preparation of the portfolio and the associated interviews should be provided to the applicant. A clear role of all involved should also be available and a list of criteria for decision-making should also be made available, as well as information on the possibility to appeal the decision and how to proceed. It is important to have a clear and transparent process to (1) reinforce the trust of all stakeholders in the process and (2) decrease the nervousness of applicants who have not been in contact with formal education in a long time. For some people, the prospect of an interview during the assessment process can be very intimidating and discouraging. This is one of the reasons why advice to the applicant prior to the assessment is considered at least partly responsible for the quality of the process (Van Kleef 2014).

Another potential bias in the process is due to the low inter-assessor reliability on the decision on the competence of an applicant (Stenlund 2013 in Travers and Harris 2014). The differences in assessing an applicant between two assessors can be based on different reasons including (but not limited to): background and experience of the assessors, cultural differences, or different philosophical perspectives that can explicitly or implicitly inform assessment (Travers and Harris 2014). In a sector as broad as resilience, integrating different issues and with the potentially limited pool of accredited assessors in the PSIDS, there is a possibility that the background and specific area of expertise of the assessors influence their decision. Different solutions exist to address this challenge in the PSIDS including, but not limited to, decision by a panel of assessors instead of one; very specific guidelines on criteria for decision-making; regular refreshing workshop with analysis and discussion on case studies simulating challenges identified by both assessors and assessed. However, these solutions will increase the demand on financial and human resources to support the RPL process.

Limited human and financial resources are among the main challenges for RPL in the context of the regional Resilience qualification on the PSIDS. This is particularly necessary because RPL focuses on an individual assessment of the documents and evidence provided by each individual applicant, thus making increased demands on the education system, including the development and professionalization, if needed, of the RPL workforce, assessors and advisors if these roles are taken by different people (Travers and Harris 2014). If the demand for RPL increases, in particular from applicants from islands away from the capital city, then additional human and financial resources will be needed. To ensure a comprehensive training of the accredited assessors is especially important since some studies question the reliability of RPL assessment due to low inter-assessor reliability on the decision whether or not an applicant is competent (Stenlund 2013 in Travers and Harris 2014). In the context of the RPL assessment for Resilience qualifications in the PSIDS, this issue is especially challenging.

This is made even more critical by the way the qualifications have been developed to allow variations between countries to accommodate specific conditions. For example, one of the skillsets of the qualifications aims at participating in, or organizing, vulnerability assessment, depending on the qualification level. However, there is no standard vulnerability assessment tool for the whole region. Some countries have developed national assessment tools (e.g. Reimaanlok in the Marshall Islands— Reimaan National Planning Team 2008). In other countries, different stakeholders involved in climate change adaptation or disaster risk management projects use their own tools to assess vulnerability (e.g. McNamara et al. 2012). This makes assessment of these specific skills under RPL very context-specific and requires from the assessors the capacity to validate the applicants' ability to apply tools in their specific contexts which may be slightly different from the ones they are used to. As a result, the probability for an applicant to be assessed by someone familiar with the specific tool they used is lower, with potential influence on the decision on competence of the applicant.

Contribution from the EU-PacTVET Project

The EU-PacTVET project partnered with Fiji Higher Education Commission (FHEC) to develop an enabling and supporting environment for the TVET providers in the PSIDS interested in introducing a process to recognize prior learning acquired during informal workshops. The project focused on how to assess the acquired skills and credit the learners to then complete a Resilience qualification. This partnership, as well as the significant consultation with different stakeholders hopes to find processes to address the challenges regarding RPL in the region, first by building the trust of practitioners and potential employers for the RPL process and to develop fair and robust assessment practices for resilience.

During discussions with several trainers and assessors for the regional TVET qualifications in Resilience regarding RPL, several challenges were identified, in addition to the ones mentioned in the previous section, some contextual and some very practical. Contextually, the challenges focused on the lack of a policy framework for RPL and of commitment to propose RPL, partly due to a lack of awareness, and partly to the limited acceptance of the RPL process as a support for quality education. Practically, the absence of specific guidelines and worries about the difficulties for applicants to compile enough evidence were added to 'classical Pacific issues' such as high turnover of assessors, limited outreach to rural or isolated communities and limited access to funding (Bateman 2018). In order to address these challenges, the EU-PacTVET project and FHEC decided to work on both the context and practical aspects.

More specifically, under the EU-PacTVET project, a regional RPL policy was developed. Although this policy specifically targeted the Resilience and Sustainable Energy qualifications, there is nothing preventing this policy from being adjusted and used for other qualifications in the region. Information was shared with the EU-PacTVET partners in the PSIDS covered by the project to raise awareness about the process. On the practical side, different tools have been developed to assist higher education institutions (HEIs) in successfully implementing RPL. These tools include the RPL Work Instruction, the RPL Trainer/Assessor Guide, the RPL Applicant Guide, the RPL Application Form and the RPL Applicants Work History Form. These tools have been presented to trainers, assessors and potential employers from different countries in the region and have been made available for any HEI/country interested. It is expected that these tools will provide a standardized process to assess skills acquired during prior learning and guide both the applicants and the assessors, thus ensuring the clarity, quality, and fairness of the whole process. Such a precise and transparent process will help build the trust of the different stakeholders for the RPL process.

However, the need for more advocacy and awareness of the RPL policy and tools is still extremely high before the Pacific can get enough understanding of the RPL regional policy for the qualifications in Resilience and Sustainable Energy and of the purpose and proper usage of the RPL tools. Only then will they be confident enough to use this process and tools to develop their capacity, using their prior learning as foundation.

Conclusions

As a global first, regional TVET qualifications on Resilience (Climate Change Adaptation and Disaster Risk Reduction) have been developed in the Pacific region. Due to the high number of climate change and disaster risk management projects with a capacity building component in the region, Pacific Islanders have acquired a variety of skills that are included in these regional qualifications. In addition, traditional practices for natural resource management and disaster risk reduction, including based on indigenous knowledge and local knowledge, are currently used in the region to build resilience and are recognized by the qualifications, thus encouraging their preservation and continuous use. As a result, skills are present in the region and a robust and fair process to assess and recognize these skills, either for access or credit in the new regional Resilience qualifications is needed.

However, Recognition of Prior Learning (RPL) is not a simple or straightforward process in terms of Resilience in the Pacific. First, because of the cross-cutting aspect of Resilience, the most relevant method for RPL is a combination of a portfolio compiling evidence of the acquired skills and interviews to better assess the skills of the applicant. However, this method is complex, and the quality of the assessment relies heavily on the quality of the evidence provided. As a result, the process needs to be clear and transparent and advice and guidance should be provided to the applicant, in particular for applicants who left the formal education system a long time ago and who can be intimidated by the process. On the other hand, training of assessors and advisors needs to be comprehensive to compensate the fact that one assessor may not be familiar with all the tools used in the Resilience sector in the Pacific region. In such a broad domain, assessment by a panel of assessors rather than by only one is expected to decrease the inter-assessor difference in the final decision, although this increases the demand on human and financial resources.

The EU-PacTVET project, in partnership with other stakeholders, in particular the Fiji Higher Education Commission, supports the development of the RPL process in the region, including through the development of guidelines and the training of assessors. However, additional resources may be needed to make people aware of the RPL process and enable them to access it. Deployment of RPL assessment teams in communities in the outer islands in the PSIDS is expected to reach community members with Resilience skills and increase their employability, providing diversified source of livelihoods for the community and participating in the professionalization of the resilience sector in the PSIDS.

References

Andersson P, Fejes A, Ahn S-E (2004) Recognition of prior vocational learning in Sweden. Stud Educ Adults 36(1):57–71

Bateman A (2018) Report on the workshop for assessors: European Pacific Technical Vocational Education and Training (EU-PacTVET) project. Bateman and Giles PTY LTD, Oct 2018, Suva, Fiji

Bich Hop HT, Huu Ninh N, Thu Hie LT (2017) The role of traditional ecological knowledge in the disaster risk management strategies of Island communities in Cat Hai, Vietnam. Clim Disaster Dev J. https://doi.org/10.18783/cddj.v002.i02.a03 https://doi.org/10.18783/cddj.v002.i02.a03

Cameron R, Miller P (2004) RPL: why has it failed to act as a mechanism for social change. Paper presented to the social change in the 21st century conference. Centre for Social Change Research, Queensland University of Technology, Brisbane, Australia

English LM (2005) International encyclopedia of adult education. Palgrave Macmillan, Houndsmill

EU-PacTVET & FHEC (2016) Regional certificates in resilience (Climate Change Adaptation and Disaster Risk Reduction/CCA & DRR). Pacific Community, Suva, Fiji

Hargreaves J (2006) Recognition of prior learning: at a glance. National Centre for Vocational Education Research (NCVER), Australian Government, Canberra, Australia

Hiwasaki L, Luna E, Syamsidik, Shaw R (2014) Local & indigenous knowledge for community resilience: hydro-meteorological disaster risk reduction and climate change adaptation in coastal and small Island communities. UNESCO, Jakarta

McNamara KE, Hemstock SL, Holland EA (2012) PaCE-SD guidebook: participatory vulnerability and adaptation assessment. Pacific Centre for Environment and Sustainable Development (PaCE-SD). The University of the South Pacific, Suva

Michelson E (2006) Beyond Galileo's telescope: situated knowledge and the recognition of prior learning. In: Andersson P, Harris J (eds) Re-theorising the recognition of prior learning. National Institute of Adult Continuing Education (NIACE), Leicester, UK, pp 141–162

Moss L (2014) Prior learning assessment for immigrants in regulated professions. In: Harris J, Wihak C, Van Kleef J (eds) Handbook of the recognition of prior learning—research into practice. National Institute of Adult Continuing Education (NIACE), Leicester, UK, pp 384–406

Reimaan National Planning Team (2008) Reimaanlok: national conservation area plan for the Marshall Islands 2007–2012. N. Baker, Melbourne

Smith L (2004) Valuing recognition of prior learning—selected case studies of Australian private providers of training. Australian National Training Authority, Australian Government, Canberra

Stenlund T (2013) Agreement in assessment of prior learning related to higher education: an examination of interrater and intrarater reliability. Int J Lifelong Educ 32(4):535–547

Syafwina (2014) Recognizing indigenous knowledge for disaster management: Smong, early warning system from Simeulue Island, Aceh. Procedia Environ Sci 20:573–582

Travers NL, Harris J (2014) Trends and issues in the professional development of RPL practitioners. In: Harris J, Wihak C, Van Kleef J (eds) Handbook of the recognition of prior learning—research into practice. National Institute of Adult Continuing Education (NIACE), Leicester, UK, pp 233–258

USP, EU GCCA Team (2017) USP EU-GCCA project phase I: support to the global climate change alliance through capacity building, community engagement and applied research final report december 2010–january 2017 (2017) Pacific centre for environment and sustainable development (PaCE-SD). The University of the South Pacific, Suva, Fiji

Van Kleef J (2014) Quality in PLAR. In: Harris J, Wihak C, Van Kleef J (eds) Handbook of the recognition of prior learning—research into practice. National Institute of Adult Continuing Education (NIACE), Leicester, UK, pp 206–232

Chapter 21
The Role of Informal Education in Climate Change Resilience: The Sandwatch Model

G. Cambers, P. Diamond and M. Verkooy

Abstract This paper explores the role of informal education in enhancing climate change resilience for beach and coastal systems using a multi-stakeholder approach called the Sandwatch programme. Sandwatch is an action-oriented, volunteer network working to enhance beach environments using a tried and tested pedagogical approach called MAST—Monitoring, Analysing, Sharing, Taking action. School students and community members learn and work together to scientifically monitor their beach environments, critically evaluate the problems, share their findings, and design and implement sustainable activities to address the issues, enhance the beach environment and build resilience to climate change. Sandwatch is an informal education tool, often conducted as an extra-curricular activity in school settings. In 2014, at the end of the United Nations Decade of Education for Sustainable Development, Sandwatch was recognized as one of 25 success stories from around the world of education for sustainable development practices in action. In 2017, an impact assessment of the Sandwatch programme was conducted using a flexible tool, the "most significant change" technique. Analysis of long-running Sandwatch programmes from nine countries showed: (i) the importance of using a participatory approach that includes hands-on learning outside of the classroom; (ii) evidence of different ways in which learners had been empowered such as by adopting a natural resources management career path; and (iii) the outreach potential of the Sandwatch programme to take action for sustainable development and enhance climate resilience.

Keywords Beaches · Conservation · Education for sustainable development · Most significant change · Extra-curricular

---

G. Cambers (✉)
16 Statham Street, Suva, Fiji
e-mail: g_cambers@hotmail.com

P. Diamond
1103-271 Ridley Boulevard, Toronto, ON M5M4N1, Canada
e-mail: pdiamondskn@gmail.com

M. Verkooy
PO Box 7710, Tacoma 98417, USA
e-mail: mverkooy@gmail.com

© Springer Nature Switzerland AG 2019
W. Leal Filho and S. L. Hemstock (eds.), *Climate Change and the Role of Education*,
Climate Change Management, https://doi.org/10.1007/978-3-030-32898-6_21

Introduction

This paper presents the Sandwatch programme, which started in 1998, and in 2014, at the end of the United Nations Decade of Education for Sustainable Development (DESD), was recognized as one of 25 most successful projects from around the world showing education for sustainable development (ESD) practices in action. The paper discusses the impact of the Sandwatch programme and its role in building climate change resilience.

The Sandwatch programme provides a framework for children, youth and adults, with the help of teachers and local communities, to work together to measure the changes in their beach environments using a standard methodology, and then critically evaluate the problems and conflicts. They then design and implement activities such that beach environments become more resilient to climate change. The programme consists of a network of action-oriented, volunteers, teachers, students and communities, coordinated by the non-profit Sandwatch Foundation, and supported by the United Nations Educational, Scientific and Cultural Organisation (UNESCO) and other development partners.

Education for Sustainable Development

In December 2002 the United Nations General Assembly proclaimed the United Nations Decade of Education for Sustainable Development (2005–2014, DESD). Following that UNESCO was announced as the designated coordinator of the Decade. The vision for DESD was described as "a world where everyone has the opportunity to benefit from education and learn the values, behaviours and lifestyles required for a sustainable future and for positive societal transformation" (UNESCO 2006). The way forward to a sustainable future for the planet was intertwined with education and the DESD was planned to tackle it head on.

The DESD was characterized by two different phases. From 2005 to 2008 efforts were focused on defining and promoting ESD, identifying and working with activities already taking place and developing monitoring and evaluation mechanisms. Phase two, 2009–2014, encompassed advancing ESD in the context of quality education. UNESCO shifted the focus of ESD to three sustainable development issues which would be addressed by education. The issues were climate change, biodiversity and disaster risk reduction (UNESCO 2014).

Extensive materials exist as a result of the DESD and the wealth of knowledge gained, shared and transformed was extremely significant. In 2007, UNESCO created a Monitoring and Evaluation Expert Group and they commissioned a review, to date, of processes and learning for ESD. Tilbury's (2011) review upheld that learning within the framework of ESD was about:

- Learning to ask critical questions
- Learning to clarify one's own values

- Learning to envision more positive and sustainable futures
- Learning to think systemically
- Learning to respond through applied learning
- Learning to explore the dialectic between tradition and innovation.

In order to address sustainable development issues such as climate change, sustainable consumption, disaster risk reduction, among others, "ESD requires participatory teaching and learning methods… to empower learners to take action for sustainable development" (UNESCO 2014, p. 20).

Climate Change

Climate change is one of the most serious threats to sustainable development around the world. Evidenced by a slow, gradual upward rise in air and sea temperatures since the industrial revolution in the 1700s, now in the 21st century climate change is impacting all aspects of human society and natural systems.

Understanding and concern about climate change has been evolving since the 19th century (Bhandari 2018) as the scientific evidence pointed toward the fact that human-caused greenhouse gas emissions were influencing the global climate. But it was not until 1992, that the United Nations Framework Convention on Climate Change (UNFCCC) was adopted at the Earth Summit in Rio de Janeiro, and subsequently ratified in March 1994. A few years prior to the adoption of the UNFCCC, in 1988, the Intergovernmental Panel on Climate Change (IPCC) was created by the World Meteorological Organization and the United Nations Environment Programme. The objective of the IPCC is to provide governments at all levels with scientific information that they can use to develop climate policies. Among other, roles the IPCC produces assessment reports on changes in climate every 4–5 years.

The 5th IPCC Assessment Report presents unequivocal evidence for warming of the climate system and that many of the observed changes are unprecedented over decades to millennia (IPCC 2014). The report further states "Continued emission of greenhouse gases will cause further warming and long-lasting changes in all components of the climate system, increasing the likelihood of severe, pervasive and irreversible impacts for people and ecosystems" (p. 8).

In many small islands and coastal areas, climate change is impacting the environment through rising temperatures, varying rainfall patterns, changing wave energy, increased magnitude and frequency of extreme weather events, ocean acidification and rising sea levels.

History of the Sandwatch Programme

Figure 21.1 shows some major milestones in the evolution of the Sandwatch programme between 1998 and 2017.

Sandwatch was developed as an open access programme, and all the tools are freely available on the website so that any group can implement Sandwatch. The Sandwatch methodology (MAST) uses simple, easy to use measurement techniques that require little to no equipment and can be done by people of all ages and backgrounds.

1998–2005

Sandwatch can trace its early beginnings to an environmental education workshop held in Trinidad and Tobago in 1998, organized by UNESCO. The workshop participants, teachers and students, saw first-hand the problems facing their beaches—erosion, pollution and poorly planned development—and proposed a Sandwatch project to address the problems. At this time, one of the preliminary objectives of Sandwatch was to make science interesting for students and one way to do that was by taking principles learned in the classroom and applying them outside the classroom on the beach (Cambers and Ghina 2007). Sandwatch can be envisaged as a multi-disciplinary programme that can be integrated into any discipline, a science class, a language class, a social studies class, an art class or developed as an after-school activity.

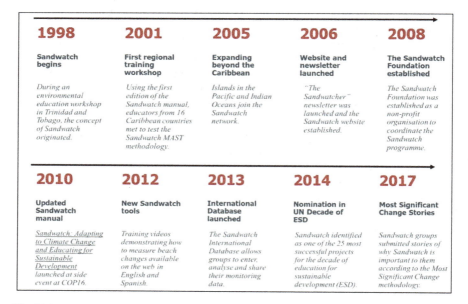

Fig. 21.1 Major Sandwatch events 1998–2017

Beginning in 2001 with a regional workshop held in Saint Lucia, teachers and students from 18 Caribbean countries were trained to use simple, standardized methods for the measurement of beach changes including erosion and accretion, wave and current action, water quality, beach ecology, pollution and human activities that impact the beach. During this first training workshop, a specific methodology involving four steps was adopted: (i) Monitoring beach changes; (ii) Analysing the data; (iii) Sharing information with communities; and (iv) Taking action to address issues affecting the beach (MAST). MAST has become the foundation of the Sandwatch programme.

An example of the use of the MAST methodology is exemplified by a Sandwatch primary school team located in the Bahamas. Students, with the help of their teachers, measured the beach over several months, and after analysing their data they concluded one of the main issues impacting the beach was tourists standing on top of a small patch reef and damaging the corals. The students then conducted a questionnaire survey to determine visitors' views about the beach and the reef and worked with a local environment group to design a brochure on proper reef etiquette, which was distributed to nearby hotels and rental properties (Cambers et al. 2008b).

As Sandwatch progressed, teachers and students addressed many different beach issues including erosion, pollution, beach littering and the illegal placement of coastal structures. Teachers also began applying some of the Sandwatch methods to their formal teaching of the curriculum, e.g. in Cook Islands, aspects of the Sandwatch MAST methodology were integrated into the science and social science curricula (Cambers et al. 2008a).

2006–2015

The Sandwatch programme, which had started as an environmental education programme, further evolved alongside two major historical developments, (i) the 4th Assessment Report of the IPCC (2007), which presented unequivocal evidence about the cause of observed climate change and resulted in the award of the Nobel Peace Prize to Al Gore and the IPCC for "… their efforts to lay the foundations for the measures that are needed to counteract such change" ("The Nobel Peace Prize" 2007); and (ii) the DESD, "a global movement which seeks to transform education policy, investment and practice" (Tilbury 2011, p 7).

The 4th Assessment Report of the IPCC and the emerging global importance of climate change inspired a revision of the Sandwatch manual to integrate climate change adaptation into all chapters and activities, and the manual was retitled: *Sandwatch: adapting to climate change and educating for sustainable development* (Cambers and Diamond 2010).

Since the revision of the manual in 2010, Sandwatch groups have explored the relationship between healthy coastal environments and the adverse impacts of climate change. An Australian Sandwatch group began exploring ocean acidification. Students collected a range of materials from the beach including shells, sand, limestone and granite rock and treated them with acetic acid (vinegar). All samples containing calcium carbonate reacted with the acid and produced bubbles of carbon dioxide gas. This prompted the discussion about how increasing ocean acidification, a result of carbon dioxide dissolving in the ocean, may affect marine organisms, local fisheries and the global food chain (Wilson 2013). An exchange with a visiting student from a small Pacific island atoll country, Kiribati, heightened the students' understanding of the impacts of climate change (including sea level rise and ocean acidification) on people living in low-lying island countries.

At the end of the United Nations Decade of ESD (2005–2014), a survey was conducted by UNESCO to identify good practices and success stories in ESD. Sandwatch was identified as one of 25 success stories of ESD practices in action and showcased at the World Conference on Education for Sustainable Development (ESD), in Aichi-Nagoya, Japan from 10 to 12 November 2014 (UNESCO 2014).

Sandwatch tools include a methods manual which was first created in 2001, revised and published by UNESCO in 2005, and further revised and published in 2010. The manual is available online in English, Spanish, French and Portuguese. Other tools include training videos in English and Spanish to accompany the methods manual, the Sandwatch website (www.sandwatchfoundation.org) and newsletter, and the Sandwatch International Database and its training videos. These tools have been created with funding from UNESCO, along with other partners such as the University of Puerto Rico—Sea Grant College Program and the Government of Denmark.

The MAST methodology has been applied to other environments e.g. deserts (Yembuu et al. 2016). In 2014 the Sandwatch methodology was incorporated into a UNESCO education course "*Climate change education inside and outside the classroom*" and delivered to participants in four regional workshops in Africa, Caribbean and Pacific regions, (UNESCO and Sandwatch Foundation 2014).

The explosive growth in communications technology and social media globally over the last 15 years helped expand the outreach of the Sandwatch programme. In the early 2000s Sandwatch launched a global communications strategy leveraging the emergence of affordable access to the internet in small communities around the world. Initially utilizing email, simple website design, and desktop publishing, Sandwatch was able to recruit and empower numerous small community organizations and demonstrate to them the ease in which they could quickly develop a truly global audience and initiate mutually beneficial partnerships.

Sandwatch has continued to expand and develop its digital communications strategies and most importantly share such skills with partners around the world via regional workshops and seminars. This includes the use of tele-conferencing, video editing programmes, the use of social media forums (Twitter, Facebook, YouTube) as well as a host of other applications and online utilities. In addition, assistance is provided with traditional media such as authoring news articles, press releases, grant applications and finding local sponsorship opportunities.

Methodology to Assess the Impact of the Sandwatch Programme

With the Sandwatch programme approaching the two-decade mark of implementation, it was timely to try and assess the long-term impact of Sandwatch so as to plan for the future development of the programme.

However, while the MAST methodology for measuring and analysing changes in the beach environment is standardised, assessing the contribution of Sandwatch to ESD in the context of climate resilience is extremely complex. The uptake of certain learning skills can be assessed, e.g. using a compass to measure and understand changes in wave direction and how this impacts the movement of the sand, and then applying that learning to the understanding of mathematical functions. But ESD is about education for living, empowerment, leadership skills, citizenship, listening to others, conflict resolution in complex situations and so much more. Standard quantitative assessment methods, such as logical framework analysis, and indicators such as the number of active ongoing programmes, or the number of students trained, just did not address the diversity of Sandwatch or the complexity of ESD.

It was important to try and assess the impact of Sandwatch on those involved in the programme, whether students, teachers or other involved persons. One major constraint was that there was no funding available for the assessment, so any method selected had to be through self-evaluation. After reviewing various methodologies, the most significant change (MSC) technique was selected to evaluate Sandwatch. This technique is a qualitative and participatory form of monitoring and evaluation, based on the collection and systematic selection of stories of reported changes from development activities. The technique was developed to meet the challenges associated with monitoring and evaluating a complex participatory rural development programme in Bangladesh, which had diversity in both implementation and outcomes (Davies 1996). Since then the technique has been extensively used especially in developing countries (Davies and Dart 2005; Serrat 2009).

The methodology involves the collection of significant change stories from the field operators, and the systematic selection of the most significant of these stories by panels of designated stakeholders or staff. The designated staff and stakeholders are initially involved in searching for project impact and the selection process progresses through several different levels so as to identify the most significant stories.

Starting in April 2017, the MSC technique was applied to the assessment of the Sandwatch programme. The Sandwatch Foundation contacted Sandwatch champions in 18 countries with individual emails to ask if they would like to contribute stories about how Sandwatch has brought about positive change in their country. They were asked to have a discussion with their Sandwatch teams to identify the most significant change brought about by Sandwatch over the past three to five years. A simple template was provided with five basic questions:

WHAT was the change?
WHO was involved?
WHEN did it happen?

WHY is it significant?
WHAT was the impact of the change and why is it significant?

Over the course of five months ten stories were received from nine countries: Australia, Bahamas, Cape Verde Islands, Cuba, Kiribati, Madeira, Puerto Rico, St. Vincent and the Grenadines, Trinidad and Tobago. A matrix was developed identifying for each story, the identified change(s), their impact, and why the group considered them to be significant. The matrix was analysed by a team from the Sandwatch Foundation. The individual stories are available on the Sandwatch Foundation website at http://www.sandwatchfoundation.org/most-significant-change-stories.html.

Results and Discussion

The most significant change stories were inspirational and provided valuable insight to the wide reach of the Sandwatch programme. Major themes emerging from the stories reflected the power of reaching out to other groups, changing attitudes to the environment, influencing the way education is perceived and delivered, and inspiring students to adopt environmental career paths. These themes all reflect realisation of the Sandwatch vision "*Sandwatch seeks to change the lifestyle and habits of children, youth and adults on a community-wide basis, and to develop awareness of the fragile nature of the marine and coastal environment and the need to use it wisely*" (Cambers and Diamond 2010).

For the purposes of this discussion, the stories are presented and discussed using the major outcomes of the DESD: participatory teaching, empowering learners, and taking action for sustainable development, as stated in the final DESD report "ESD requires participatory teaching and learning methods … to empower learners to take action for sustainable development" (UNESCO 2014, p. 20).

Participatory Teaching and Learning Methods

The Sandwatch approach requires going out of the classroom, into the field and experiencing the natural environment at the beach. In some countries, from a formal education perspective, this approach is met with resistance due to the significant extra work required for obtaining the necessary permissions, arranging transport, and consideration of safety issues.

The story from a primary school teacher in Trinidad and Tobago illustrated the importance of participatory, hands-on learning outside the classroom and that this also requires confidence building for the teachers.

> I have been exposed to this programme for just about three years now and I can say that I have changed immensely due to this programme. From my experience of doing beach clean-ups, I have learnt the importance of keeping the environment clean and the impact of

pollution to life on earth. I have also seen hands on learning taking place, as students are actively involved in activities. I now see things differently and think differently as I am more confident and empowered to participate in activities and speak out at things that are wrong. I have learnt and adapted new and innovative teaching techniques. I have also learnt that education goes beyond the traditional paper and pencil technique.

Mrs. Ali, Teacher at Mayaro Government Primary School, Trinidad and Tobago.

Many of the other stories also touched on the necessity of demonstrating to concerned parents, teachers and school principals that education goes beyond the four walls in the classroom. In the beginning many parents considered the beach as an environment to have fun and not a learning environment, but gradually this resistance was overcome as Sandwatch demonstrated the importance of new and innovative teaching techniques, particularly engaging students through hands-on learning. In some countries where field visits were too difficult to arrange, interested teachers and Sandwatch champions implemented Sandwatch activities as an extra-curricular option, e.g. through environmental clubs.

The significant change story from Australia showed how the Sandwatch approach helped students set global issues such as climate change in their local and regional contexts. At the start of the implementation of the Sandwatch approach, students became more aware of local issues relating to the coast including littering, vegetation damage from pedestrians, and erosion events relating to storms. The local issues were then used as a lead-into global issues such as climate change and sea level rise and how these could affect local beaches. The school then linked up with a school in Kiribati and with the help of a locally based Kiribati student discussed everyday life and culture in Kiribati including issues that this low-lying atoll nation faces under projected climate change scenarios where the highest point is only 3 m above sea level. This exchange strengthened cross cultural experiences and fostered a greater understanding about people whose homes and lifestyles are now under immediate threat due to climate change. It also encouraged students to think about climate change issues at the global level and how they could respond to those issues. Energy conservation ideas for the home were among the responses that could be implemented at an individual level and so contribute to the mitigation of climate change.

Overall the stories emphasised the value of direct interaction with the natural environment and the importance of interacting with other Sandwatch groups and communities from other countries. Regional and international conferences have been very popular with the Sandwatch community, and exchange visits have also been beneficial. The Sandwatch group in Trinidad and Tobago have organized exchange visits with schools in Brazil and the UK, which have enriched the students' global and cultural experience of the world as well as the reach of the Sandwatch programme.

Empowering Learners

Several of the stories from countries where the Sandwatch programme has been operational for ten or more years, including the Bahamas, Puerto Rico, St. Vincent and the

Grenadines, and Trinidad and Tobago, highlighted how the Sandwatch programme had inspired students to choose environmental conservation in their further (tertiary) studies and career paths.

One of these stories from a student in Puerto Rico showed how participation in the Sandwatch programme had influenced her career path. This story was particularly insightful, showing a depth of understanding about the need to understand psychology to motivate other people to conserve the environment.

> I have loved everything ocean-related since I was little. Once I started working on the Sandwatch programme, I saw the ocean from a different perspective. I don't simply go to the beach anymore and see how fun and beautiful it is; now I pay more attention to details that changed from one visit to the next. My love for marine ecosystems also grew, and lit a passion inside me for protecting the environment. The Sandwatch programme experience led me to become interested in oceanography as an academic career path. I entered the University of Puerto Rico at Mayagüez in the Geology major, intending to continue my studies along the marine geology path, following that passion for environmental protection. However, during my time in the university, I realized that the only way to preserve nature is by educating others. This made me want to dedicate myself to the education field, but from a different perspective. I changed majors and am currently studying psychology with the end goal of specializing in educational psychology. The objective is still the same: save the ecosystems, but this time, I will reach it by helping create a generation more conscientious about the environment. In the future, I hope to have my own school, with my own educational system, in which I'll foster education through projects like Sandwatch, in which children learn by having the most direct contact possible with nature.
>
> Nayrobie Lee Rivera Estévez, Puerto Rico.

Empowerment has many different components. The story from Trinidad and Tobago described how the Sandwatch programme has empowered many young people to become leaders and confident individuals able to take their knowledge and experience beyond their classroom environment, and beyond the context of Sandwatch, for example, having the confidence to judge the secondary schools' environmental debate competition.

The Sandwatch programme has also been taken up at the postgraduate level. The story from Cuba described how one educator from Cuba had applied his 16 years' teaching experience with Sandwatch to his Ph.D. research and had made several presentations at conferences and seminars on the role of environmental education and secondary school students

> The Sandwatch project has allowed me in these 16 years, to learn new knowledge on the pollution and erosion of the beaches, to use different instruments, to meet personalities in this field, and to initiate myself in the world of research. In addition, I have worked with groups of different students and shared the experiences so that students and parents have undertaken actions to conserve and protect the beaches, an ecosystem that the new generations must protect.
>
> Dr. C. Raudel Cuba Jiménez—Escuela Provincial Pedagógica de Matanzas.

Taking Action for Sustainable Development

The Sandwatch methodology (MAST) with its focus on taking action to conserve the beach environment based on monitoring and analysing the beach changes, and then sharing the results with the local community, has inspired generations of students, parents, teachers and community members to join young Sandwatchers in taking positive action for sustainable development.

Many of the Sandwatch groups addressed the problem of littering and pollution by conducting beach clean-ups. For example, in Madeira, in the North Atlantic Ocean, off the coast of Portugal, the Sandwatch groups' clean-up efforts inspired other schools, hotels and resorts to carry out similar activities on their own, and they were further supported by the local council and private businesses.

In the Caribbean, Sandwatch students in Bequia, in St. Vincent and the Grenadines, expanded the beach clean-up concept to community clean-ups. While in Kiribati, in the central Pacific, after seven years of focusing Sandwatch activities on primary and secondary school students, attention turned to a specific community where the beach was extremely polluted and had become a dump for scrap vehicles.

In recent years the issue of plastics polluting the oceans has received widespread publicity. More than 30 years ago a coastal clean-up programme was started in the USA and spread to become the International Coastal Clean-up programme (Ocean Conservancy 2011). Not only did this programme focus on coastal clean-ups, but also on categorizing and recording the different types of debris found, so that its origin could be determined and efforts could be focused on the cause of the problem—the polluters. This programme and its methodology was adopted to become one of the standard methods conducted by Sandwatch groups, and is included in all editions of the Sandwatch methods manual. Many Sandwatch groups also take part in the annual international coastal clean-up, as well as their own individual and national events. During the first Sandwatch competition in 2005, students from a secondary school in St. Lucia identified pollution from the nearby Choc River as a major issue resulting in high levels of debris and pollution on the beach. The group focused, not only on cleaning up the debris, but more importantly trying to influence the attitudes of the residents living near the Choc River and others who were dumping their garbage in the river.

The stories demonstrated other ways in which the Sandwatch programme had reached out to others to take action. Sandwatch in western Australia is led by a conservation organisation. This organisation recorded increased participation by families from the school involved in the Sandwatch programme in weekend "*Coastcare*" events, highlighting the level of understanding and the importance those families now place on coastal conservations actions.

Sandwatch students in the Bahamas and their coastal community saw for themselves the benefits of replanting natural vegetation such as sea oats to stabilise eroded sand dunes as several hurricanes impacted the island over a period of years.

Conclusions

Results from the Sandwatch impact assessment show that in today's advanced technological environment, Sandwatch is still relevant, perhaps more relevant then it was 20 years ago, as it requires individuals, students and teachers going outside of the controlled classroom or office setting, and experiencing and understanding first-hand the infinitely complex, natural environment in which we live and interact. The most advanced physical and mathematical climate and environmental models in the world cannot do full justice to that complexity.

The role of education for all in addressing the challenges of climate change cannot be over-emphasised. Climate change affects every person and every aspect of our lives, our health, what we eat and drink, our very survival. On our crowded planet, the framework of ESD presents a comprehensive approach to addressing the issues and making the required behavioural changes. Sandwatch, along with other programmes, has a continuing and leading role to play to creating positive change in both the informal and formal educational system such that we all become part of the action that needs to take place.

The results from the Sandwatch programme's most significant change stories have clearly demonstrated the value of participatory teaching and learning methods, and how the use of such methods is empowering learners of all ages to take action for sustainable development.

The main objective of the impact assessment was to plan for the future development of the Sandwatch programme and one of the main learnings is to harness the pool of experience and knowledge from the generations of students, who have passed through the Sandwatch programme and are now young adults, in how to move the Sandwatch programme forward.

One of the first proposed actions is to expand the collection of most significant change stories from past students who have been involved in Sandwatch over the past five years. Then, depending on obtaining the necessary funding, regional consultations can be held, to brainstorm and develop proposals on new ways to expand and sustain the Sandwatch programme over the next five years. From this pool of young adults, a Young Sandwatchers Network and Advisory Board can be formed to devote some of their time to advance the Sandwatch programme and guide the Sandwatch Foundation in its future activities.

Empowering students and young adults to conserve their beaches and the wider environment, and to use them wisely, while involving their friends, families, teachers, and communities in sustainable activities that also build climate resilience remains central to the Sandwatch vision.

References

Bhandari M (2018) Climate change science: a historical outline. Adv Agric Environ Sci 1(1):5–12. Retrieved from http://ologyjournals.com/aaeoa/aaeoa_00002.pdf

Cambers G, Ghina F (2007) The sandwatchers. World Sci UNESCO Nat Sci Newslett 5(1):16–19

Cambers G, Diamond P (2010) Sandwatch: adapting to climate change and educating for sustainable development. UNESCO, Paris, France

Cambers G, Belmar H, Brito-Feliz M, Diamond P, Key C, Paul A, Townsend G (2008a) Sandwatch: a practical, issue-based, action-orientated approach to education for sustainable development. Caribb J Educ 30(1):1–28. Retrieved from https://www.mona.uwi.edu/soe/publications/cje/article/154

Cambers G, Chapman G, Diamond P, Down L, Griffith AD, Wiltshire W (2008b) Teachers' guide for education for sustainable development in the Caribbean. UNESCO, Santiago, Chile

Davies RJ (1996) An evolutionary approach to facilitating organisational learning: an experiment by the Christian commission for development in Bangladesh. Impact Assess Proj Appraisal 16(3):243–250

Davies R, Dart J (2005) The most significant change (MSC) technique: a guide to its use. Retrieved from https://www.clearhorizon.com.au/f.ashx/dd-2005-msc_user_guide.pdf

IPCC (2007) Climate change 2007: synthesis report. In: Pachauri RK, Reisinger A (eds) Contribution of working groups I, II and III to the fourth assessment report of the intergovernmental panel on climate change. IPCC, Geneva, Switzerland, 104 pp

IPCC (2014) Climate change 2014: synthesis report. In; Pachauri RK, Meyer LA (eds) Contribution of working groups I, II and III to the fifth assessment report of the intergovernmental panel on climate change. IPCC, Geneva, Switzerland, 151 pp

Ocean Conservancy (2011) 25 years of action for the ocean. Retrieved from https://oceanconservancy.org/wp-content/uploads/2017/04/2011-Ocean-Conservancy-ICC-Report.pdf

Serrat O (2009) The most significant change technique. *Knowledge Solutions*, 25. Retrieved from https://www.adb.org/sites/default/files/publication/27613/most-significant-change.pdf

The Nobel Peace Prize for 2007 (2007) Retrieved from https://www.nobelprize.org/prizes/peace/2007/summary/

Tilbury D (2011) Education for sustainable development: an expert review of process and learning. UNESCO, Paris. Retrieved from https://unesdoc.unesco.org/ark:/48223/pf0000191442.locale=en

UNESCO (2006) UNESCO's role, vision and challenges for the UN decade of education for sustainable development (2005–2014). Connect UNESCO Int Sci Technol Environ Educ Newslett XXXI(1–2):1–5

UNESCO (2014) Shaping the future we want: UN decade of education for sustainable development (2005–2014) final report. UNESCO, Paris. Retrieved from https://unesdoc.unesco.org/ark:/48223/pf0000230171.locale=en

UNESCO and The Sandwatch Foundation (2014) Climate change education inside and outside the classroom. Retrieved from http://www.sandwatchfoundation.org/ccesd-course-materials.html

Wilson C (2013) Exploring ocean acidification in Perth, Australia. The Sandwatcher 16:11

Yembuu B, Khadbaatar S, Munkhbadar T, Uranchimeg G, Tugjamba N (2016) Desertification watch: learning to combat desertification. UNESCO

Chapter 22
A Plexus Curriculum in School Geography—A Holistic Approach to School Geography for an Endangered Planet

Phil Wood and Steven Puttick

Abstract Since 2010 English education has seen a large and rapid shift in emphasis from a skills-based curriculum to one based on the idea of 'core knowledge', aligned with and given traction by the concepts of 'cultural literacy' (Hirsch) and 'powerful knowledge' (Young), and reliant on a belief that only 'academically sanctioned' knowledge is fit to offer students. This shift has led to a far-reaching reappraisal of the curriculum, prioritising traditional views of knowledge and content which in school geography have re-established a content heavy, traditional offer. This offer may lay some foundations for further study, but fails to engage students in crucial, more complex issues facing the planet, and facing them as citizens in the present and in the future. In this paper we outline what we see as being deficient in the current 'core knowledge' agenda, and offer instead an approach we refer to as a plexus curriculum. This is based on a more holistic approach to the subject which seeks to consider how various features of the geography curriculum can be interconnected for greater effect. This includes the intertwining of academic knowledge with the everyday, and the intertwining of different elements of the subject into more holistic and interdependent lenses. By using climate change, the Anthropocene and earth systems as a core conceptual framework around which the subject knowledge base is structured and interconnected, we argue that a plexus curriculum can develop a more critical and holistic understanding of geography, as well as playing a central role in developing geographical imaginations.

Keywords Plexus curriculum · Climate change · Anthropocene · Cultural literacy · Powerful knowledge

---

P. Wood (✉)
Centre for Research and Knowledge Exchange, Bishop Grosseteste University, Lincoln, UK
e-mail: philip.wood@bishopg.ac.uk

S. Puttick
School of Teacher Development, Bishop Grosseteste University, Lincoln, UK

© Springer Nature Switzerland AG 2019
W. Leal Filho and S. L. Hemstock (eds.), *Climate Change and the Role of Education*,
Climate Change Management, https://doi.org/10.1007/978-3-030-32898-6_22

Introduction

In 2010 there was the election of a new Government in the UK. The following period has seen large-scale and rapid shifts in many areas of policy and life in the country, not least within the area of education. Prior to 2010, a long-standing 'New Labour' government, first elected in 1997, had pursued an ongoing agenda of curriculum change. The then Prime Minister, Tony Blair, put education at the centre of his vision for a progressive British social and economic revolution, the 'Third Way' (Giddens 1998). Education was deemed a national priority as it was seen as the driving force for creating a competitive workforce; education was considered within this programme to drive economic development. The role of the curriculum was to some extent redefined to suit the perceived needs of employers which led to a shift from a more 'traditional' subject-focused curriculum to one emphasising a skills agenda. This shift might also be seen as part of a wider international trend, characterised by Sahlberg (2012) as the Global Education Reform Movement (GERM); a coming together of government policy with the needs and preferences of corporatism. Agencies such as the World Bank and the Organisation for Economic Cooperation and Development (OECD) began to measure educational outcomes (e.g. Programme for International Student Assessment), prioritising big data to report on educational change. In the case of the World Bank, focusing attention on low income countries (e.g. World Bank 2010). The move towards a 'skills-based curriculum' led to a much greater emphasis on '21st Century skills' such as creativity, problem-solving and ICT-based approaches (Saavedra and Opfer 2012). As a result of this shift some have argued that a coherent body of knowledge as core to the curriculum became increasingly marginalised and replaced by politicised subjects, for example citizenship (Whelan 2007).

With the coming to power of the Coalition government in 2010 Michael Gove became Secretary of State for Education. His tenure in this position lasted only four years but brought radical and rapid changes in policy and practice, a central element of which was a shift in the curriculum. He was critical of the curriculum developments under New Labour and called for a 'renewal' of the English curriculum (the other countries of the UK each have responsibility for their own education systems and policy). Where New Labour had championed a skills-based approach, Gove wanted a return to a knowledge-led curriculum, framed as liberalism:

> The eminent Victorian, and muscular liberal, Matthew Arnold encapsulated what liberal learning should be. He wanted to introduce young minds to the best that had been thought and written... I want to argue that introducing the young minds of the future to the great minds of the past is our duty. (Gove 2011)

Gove discontinued many vocationally orientated qualifications and called for a greater degree of 'rigour' in more traditional, academic subjects. This rapidly led to new subject syllabi with expanded content, and greater input from university academics in deciding what that content should include. At the same time, a new quality assurance measure was created for schools, the English Baccalaureate (EBacc) which is now used as one measure of the quality of education provided by a school. EBacc measures the percentage of students who gain good grades in Maths, English, a Science, a Modern Foreign Language and a Humanities (either Geography or History). In much of the debate around these changes, the return to a traditional, academic and subject-based curriculum was often argued for on the basis of the work of two academics, E. D. Hirsch and Michael Young.

E. D. Hirsch and Cultural Literacy

Hirsch has become a central figure for some in the English curriculum debate. He is a professor emeritus of education and humanities at the University of Virginia who developed the concept of 'Cultural Literacy', out of which emerged the Core Knowledge movement. In the 1980s Hirsch noticed that some of his students appeared to have little background knowledge on which they could rely when attempting to understand the context of literary works. This led him to reflect that,

> During the period of 1970-1985, the amount of shared knowledge that we have been able to take for granted in communicating with our fellow citizens has also been declining. More and more of our young people don't know things we used to assume they did know. (1987: 5)

Hirsch believed that this lack of background knowledge was highly problematic because, without it, students would not be able to take part in the conversations which bind a culture together as there would be limited common understandings. Hence, as summarised by Cook (2009: 489) *'According to Hirsch, what students need is a firm foothold in the shared background knowledge of literate culture which he terms "cultural literacy"'*. Therefore, cultural literacy is argued to be essential as it determines an individual's ability to understand and fluently take part within a given culture. This led to the acid test sometimes quoted around Hirsch's ideas:

> To be truly literate, citizens must be able to grasp the meaning of any piece of writing addressed to the general reader. All citizens should be able, for instance, to read newspapers of substance. (Hirsch 1987: 12)

At the centre of Hirsch's concerns is that children from poor backgrounds who may not, on average, be exposed to the same level of knowledge as their richer peers, will have less cultural literacy. If this is the case, these children will find it more difficult to succeed in society as they will not have the same cultural resources on which to call as they grow up. He, therefore, saw the development of a core knowledge curriculum as a way of narrowing this cultural gap.

There are a number of critiques of Hirsch's work, some of which centre on a belief that his model would lead to a narrow pedagogy based on the memorisation of facts and little else. For example, Christenbury (1989:14) writes, *'cultural literacy encourages superficial notions of knowledge'*. Whilst Scholes (quoted in Cook 2009: 491) reflected that cultural literacy

> trivializes the concepts of culture and of literacy by suggesting that culture can be reduced to just 5000 bits of information and literacy to the passive possession of those bits.

Whilst these reviews tend to focus on cultural literacy as reductive and instrumental in nature, Cook goes on to develop a more nuanced reading of Hirsch. He stresses that the facts listed by Hirsch which are set out at the end of his book as 'The List' (approximately 5000 core facts all students should learn) should not be characterised as the whole curriculum and is offered only as an experiment. Scott (1988) highlights that Hirsch himself says that the list will almost certainly be misused and is only a heuristic and that the core knowledge agenda which results from Hirsch's cultural literacy should only be seen as one element of a wider curriculum. Unfortunately, this aspect of his work appears to have been lost in translation as curricula based on a conceptualisation of core knowledge with strong resonances to 'The List' have emerged.

The use of cultural literacy as a framework for geography education can be traced back to the early 1990s and the creation of the National Curriculum in England. When the Geography Working Group (GWG) was formed to create the first National Curriculum, the geography curriculum of the 1970s and 1980s was argued by some (mainly beyond the subject) as being a deficit model of the subject with little focus on knowledge coverage. Dowgill and Lambert (1992) surmise that the GWG were steered towards the cultural literacy framework when developing the original National Curriculum attainment targets in geography. They begin to interrogate the likely impact of cultural literacy on geography and conclude that it appears to predominantly focus on locational knowledge and a regional view of the subject. They also find it difficult to place cultural literacy within the wider curricular context. Is it meant as an aim in the curriculum? Does it underpin the curriculum conceptually? Or is it merely a list of interesting content? Dowgill and Lambert (1992: 151) are left wondering what the place of cultural literacy is within the subject or indeed, as they inquire at the end of their consideration, 'is geography (merely) an element in cultural literacy?'

The heavy influence of regional and locational geography can be seen in more recent developments in the Core Knowledge movement. Core Knowledge UK (http://www.coreknowledge.org.uk/index.php) have produced an outline curriculum for primary school geography which is wholly directed through the use of regions as the basis for the curriculum. Different processes are introduced within different regions, reducing place to location (Agnew 1987) with little sense of locale (a setting and scale for everyday processes and human action) and no 'sense of place' (including subjective feelings of place). This appears to relate to Dowgill and Lambert's (1992) concern that cultural literacy, if not engaged with critically, may collapse into a list of facts, and may therefore become very descriptive in nature, a medium more suited to regional geography which Walford characterises, (quoted in Wood 2009:9),

> The attraction of the regional curriculum frame was that pupils could be said to "cover the world" if they stayed with geography over five years of study; the knowledge base was complete, if superficial. (Walford 2001, p. 143)

Cultural literacy does play a useful role in curriculum thinking in that it stresses the need for consideration of the knowledge which is deemed important within a subject. But in a geographical context, this appears to translate into a fixation with locational knowledge which, whilst important, does not deal with knowledge as a network of ideas and concepts which together help develop an emerging holistic and critical understanding of major contexts and processes. Neither does it engage in any depth with questions about why some areas of knowledge are more important than others, other than a fairly vague idea of cultural discourse. This final point is, in a way, the starting point of the work of Michael Young.

Michael Young and Powerful Knowledge

Discussion around 'powerful knowledge', particularly in the context of school geography, might be understood as the meeting of two coinciding narratives. One is the journey of Michael Young's (2008) thought from 'social constructivism to social realism', involving radical changes in his epistemological beliefs, including shifts in the significance and role that disciplines and subjects ought to play in education. The other is the journey of geography education research in developing increasingly critical accounts of knowledge in school geography. The latter provided a fertile environment for Young's conception of 'powerful knowledge' to be widely engaged with in geography education. The nature of the relationship between geography as a school subject and geography as an academic discipline runs beneath both of these narratives. Different configurations of this relationship, and different conceptions of the discipline also play important roles in supporting—or potentially undermining—competing visions of what school subjects ought to be and do.

Critiques of knowledge in school geography have often used visual language of 'seeing' and 'gazes' in their claims that knowledge has been given insufficient attention. Whether this is the case or not is hard to assess: the claims are not made on the basis of empirical study of school geography or the longitudinal or comparative work that might be necessary for concluding that teachers' and schools' 'attention' had shifted focus. Nevertheless, these critiques seem to be widely accepted—including by those not writing from an explicitly 'powerful knowledge' perspective. These narratives also seem to be reflected more widely in the ways in which current Ofsted and DfE policy texts seem to construct narratives around previous 'inattention' to knowledge, and an urgent 'return' to subjects and core knowledge. One example in geography education is Morgan and Lambert's (2011) argument that 'thinking skills, learning to learn and the emotional dimensions of learning have assumed more immediate or urgent attention than a critical gaze on the material content of lessons' (p. 281). Consequently, a narrowly defined focus on pedagogy 'has marginalised knowledge in the practical day-to-day work of making the curriculum' (p. 281). In Firth's (2011) terms, 'geographical knowledge … has been marginalised by the exigencies of everyday practice and the imperatives of policy' (p. 312).

One example illustrating the readiness of geography education research to adopt the concept and associated discourse of powerful knowledge is Morgan's (2011) paper *Knowledge and the school geography curriculum: a rough guide for teachers*.

The education policy of the (then) previous 13 years of Labour government is summarised as 'the creation of numerous education 'strategies' where the emphasis was on generic 'learning', free from any sense of subject or disciplines' (p. 90). Using Young's distinction between pedagogy as the 'how' and knowledge as the 'what' of the curriculum, Morgan summarises the critique from Young and others in these strong terms: 'schools and teachers had become so focused on the 'how' of learning that the question of what was to be learned had been neglected' (p. 90). The *Importance of Teaching* white paper (DfE 2010) is then presented by Morgan as making a 'call for a return to focus on subject-based teaching and within that a concern with the core knowledge that makes up the subjects' (Morgan 2011, p. 90). The choice of the term 'core knowledge' is notable; there are close, and mutually-reinforcing relationships between Young's powerful knowledge and Hirsch's core knowledge in these discourses around knowledge and school geography.

There have been a number of critiques of powerful knowledge, including Catling and Martin's (2011) arguments for understanding children's 'everyday' knowledge or 'ethno-geographies' as powerful. This argument aims to rearrange the dichotomous conceptualization of disciplinary knowledge as being 'above' children's ethno-geographies, and instead sees the purpose of school geography as fostering a generative dialogue between a number of different types of knowledge which might—at least potentially—be equally 'powerful'. Roberts (2014) is similarly critical of Young's distinction between powerful and everyday knowledge. By engaging with academic geographers' conceptions of 'everyday' knowledge, she develops her argument by suggesting a tension between the dichotomy underpinning Young's powerful knowledge (that is, between everyday and powerful knowledges) and the ways in which disciplinary knowledge in geography has in several instances rejected this dichotomy.

Roberts' (2014) paper followed a debate at the Institute of Education (UCL) between Young and Roberts, and there have been a number of other collaborations between Young and geography education researchers (e.g. Young et al. 2014). Taking the example academic urban geographies, Roberts' critical engagement compares these approaches against Young's criterion of powerful knowledge, summarised as being:

- Conceptual as well as based on evidence and experience;
- Reliable and in a broad sense 'testable' explanations or ways of thinking;
- Always open to challenge;
- Organised into domains with boundaries that are not arbitrary and these domains are associated with specialist communities such as subject and professional associations;
- Often but not always discipline-based

(Roberts 2014, p.3)

One example of academic geographers' engagement with everyday knowledges is Cloke et al. (2005) encouragement for undergraduates to be 'aware of the human geographies wrapped up in and represented by the food you eat, the news you read, the films you watch, the music you listen to, the television you gaze at' and to 'think about how and what you read in books or articles connects or doesn't to your everyday life and why that might be' (p. 602). A later edition (Cloke et al. 2014) begins by arguing that geography is

> ...all around us, a part of our everyday lives...not confined to academic study but includes a host of more popular forms of knowledge through which we come to understand and describe our world. (p. xvi)

There are lines of enquiry to pursue around who 'we' might refer to, particularly in relation to Young's sociological conception of communities, associations and disciplines (including the intriguing condition 'often'), but not withstanding this, the boundaries between 'academic study' and 'more popular forms of knowledge' are clearly blurred. Stating that 'Geography' is 'not confined to academic study' creates a tension against dichotomous categorisations of powerful and everyday.

Another important way in which academic geography problematises knowledge, and which also creates tensions with 'powerful knowledge' is through discussion of representation, construction, and 'the real' (McCormack 2012). Young and Muller (2013) also contend that powerful knowledge is 'real', which may be tested by 'whether the world answers to knowledge claims' (p. 241), but as Roberts argues (following Daniels et al. 2008), geography is not a 'direct reflection of a straightforward reality that is out there but a social construction ... interpretations of the world differ from different vantage points in time and space' (p. 2). Arguments around representation and construction are an important part of the 'critical physical geography' introduced below.

Both Hirsch and Young offer useful insights to the curriculum debate. However, there are also serious limitations to their work when related to geography education. To see the subject as merely a 'list' of knowledge to be covered is reductive and does not explicitly make clear the need to create a more networked conceptualisation of subject knowledge where choices about both content and its internal structuring are not considered. Unlike Hirsch, Young develops a more explicit consideration as to how subject content should be chosen. However, in doing so he has often discounted experiential knowledge which, particularly within a geographical context, must be seen as a core resource for contextualising and engaging with 'disciplinary' knowledge; indeed, there are important senses in which everyday knowledge is part of disciplinary knowledge. As with Hirsch, Young's conception of powerful knowledge does not seem conducive to developing holistic understandings of the interconnections different elements of subject content. Where this holistic understanding is absent, aspects of subjects are reified and artificially treated as isolated 'bits', rather than as intrinsically and complexly linked aspects of a wider whole. It is this deficit in curriculum thinking that a plexus approach attempts to address and expand.

Defining the Plexus Curriculum

Once a curriculum is based upon the knowledge it contains there is the real threat that it will begin to be structured in a way that makes that knowledge atomistic. By this, we mean that there may be a tendency for the different elements of the curriculum to be taught in isolation from one another, leading to aspects of a subject being known, but the interconnections left implicit. This may well lead to students developing a series of uneven and poorly connected schema, giving them pools of expertise within a subject with little understanding of how the various elements fit together. To overcome this, there needs to be a more developed and explicit consideration of the various interconnections which exist across the different parts of a subject. This is the core of a plexus approach, which can be defined as,

> An approach to education which focuses on the interconnections between issues, ideas, theories and methodologies to build holistic and complex networks of educational understanding. An approach which explores connecting processes and builds multi-dimensional perspectives.

In the context of a geography curriculum this means that we need to consider how each of the elements within a syllabus is networked to create a whole. How do different elements of a programme explicitly link together? What is the nature of the connections and how do we understand them? How can these connections be built into the curriculum to encourage the emergence of holistic networks of understanding?

Within a geographical context this requires us to begin to think about the different links there are between processes, environments, issues and concepts and how we might begin to bring these various perspectives together into a coherent whole. In other words, this is to think about a 'knowledge led curriculum' whose focus is on building holistic conceptual networks which allow students to build emergent geographical imaginations.

Gieseking (2017) reviews the changing meaning of geographical imagination from its origin in the work of Prince (1962), who saw imagination within geography as a way of bridging the subjective and objective, to more recent considerations of the interconnections between humans and nature and the ways in which these are understood and developed,

> Increased discussions around climate change and the anthropocene indicate yet another shift in the geographical imagination around nature. (Gieseking 2017: 4)

This results in geographical imagination being the ways in which people think of and use space, and how they use their knowledge of these spaces to make decisions and value judgements. As Daniels (2011: 182) reflects,

> As many writers on the theme have found, the geographical imagination in its various forms and meanings is a powerful ingredient of many kinds of knowledge and communication, within and beyond geography as an academic subject, as a way of envisioning the world, experiencing and reshaping it too.

Hence, a central aim of a plexus curriculum must be to develop the interconnections within and beyond the subject which over time allow for ever greater opportunities to envisage the world in—at least some—of its complexity. At the same time, this expanding imagination should also allow students to begin to consider how the knowledge they are engaged with might point towards new and interesting ways of reshaping the world around them. A plexus curriculum therefore aims to develop interconnections within the subject content which it embraces. It does this, in part, to aid the emergence of a geographical imagination within students to help them make sense of their world whilst also demonstrating the contribution geographical knowledge can make in thinking about future issues and bringing about change.

A Conceptual Framework for a Plexus Curriculum in Geography

To develop a coherent conceptual framework for a plexus curriculum in geography, we need to develop some core concepts around which a knowledge base can be structured. We suggest that in addition to the normal geographical concepts such as place, space and scale, large substantive geographical issues can productively contribute to this conceptual core, based on their complex, synthetic nature. The concepts discussed here are climate change, the Anthropocene and earth systems.

Climate Change climate change can act as a strong unifying concept as it requires a number of interconnections across a geography curriculum to be explored to develop a holistic understanding of climate change itself. For example, to understand climate change at a planetary scale there needs to be an understanding of long-term natural processes such as orbital forcing/Milankovitch Cycles. To understand the process of anthropogenic climate change it is then important to engage with atmospheric processes, some atmospheric chemistry and sources of anthropogenic greenhouse gases. Having started to engage with these physical processes, understandings of the links to human economic activity, its relation to energy use and economic development need to be explored. There also needs to be an understanding of how societies might mitigate some of the impacts and work to minimise the sources of climate change. In this way there is clear potential to begin to synthesise elements of the pre-existing curriculum by using a climate change lens. In a number of examination specifications at GCSE (examinations sat by 16-year olds in England) and A-level (examinations sat by 18-year olds in England) climate change is studied, but always as a discrete issue. Therefore, its complex and overarching nature is lost. A plexus approach aims to overcome this.

Anthropocene including this concept as a core aspect of a plexus curriculum begins to make more explicit the links between human and physical systems. Too often, school geography treats human geography and physical geography as being essentially separate entities. However, it is now the case that humans have had such an acute and far ranging impact on the physical sphere (Schwagerl 2014), that it is no longer possible to treat these two arms of the subject as being divorced from one another (if it ever has been; see, for example, Goudie 1986). Numerous examples can now begin to be explored which bring these two areas of the subject together and interconnect them in various ways. Using the Anthropocene as a context also extends out the critical investigation and analysis of humans and their interactions with their environment beyond that of climate change. This includes considerations of biodiversity (McGill et al. 2015), relationships between humans and their environments (Angus 2016), the impacts of economic systems on the natural environment and also the processes increasingly affecting humans due to climate change such as loss of habitable land and mass migration. In all of these cases using the Anthropocene as an organising concept allows for the explicit discussion of a number of interconnections between the traditionally isolated schemes of work which tend to occur in many geography courses. Here, knowledge is developed 'in-between' traditional foci to create more connected schema amongst students.

Earth Systems much of the physical geography taught within geography curricula tends to focus on specific physical systems, most often rivers, coasts, and glacial landscapes. Whilst these are all legitimate areas of study, this approach tends to under-represent any notion of how these systems relate to planetary scale physical systems. As students begin to develop their understanding of each of these systems, it is important for them to begin to gain a clear and explicit understanding of how they are interrelated. For example, it is rare for the interconnections between rivers and coasts to be developed in any consistent manner other than to explain that delta morphology relates to the relative energy and sediment supply from the river and coast systems. If we begin to develop clearer and more critical understandings of how the various physical systems operate together in complex ways, then students can begin to build more holistic understandings of the physical systems which are responsible for planetary characteristics.

Connections are no less true of human systems and processes such as globalization, the rise of technology, mass migration and the impact of these processes on urban areas, transport systems and human land use. Given that many of these issues are also impacted by physical processes and in turn impact those processes, involving students in more explicit exploration of the interconnections involved is necessary for the curriculum to help them understand the complexity of the issues which they and their societies increasingly face and construct.

We suggest these 3 concepts as a core framework for considering how to develop interactions within the geography curriculum. We argue that these concepts and the interconnected knowledge which surround them will be increasingly important to any understanding of news and current affairs for the general reader moving into the immediate and medium-term future. This is Hirsch's acid test for cultural literacy, and hence, we argue that this approach also meets the test of a core knowledge agenda in addition to the aim of holistic understanding and complexity that we have argued for. By placing these concepts at the core of a network of knowledge, individuals have the potential to develop a holistic understanding of major issues that are occurring in the world and therefore should be at the centre of any developing geography education. By using these concepts as a core framework for encouraging an explicit consideration of interconnections, knowledge and understanding become the starting points for critical and imaginative engagement with the subject.

It is important to think about how the core concepts and knowledge might be operationalized together to give a practical basis for a curriculum. The nature of the interconnections chosen should be decided by teachers as curriculum makers deciding what would work best for their own context. Here, we give three examples of recent approaches which could serve to exemplify interconnections across geography and relate back to our chosen core concepts.

The first example of how a number of geography issues can begin to be brought together in a holistic way is through the heterodox economics model of Raworth's (2017) 'doughnut economics'. Raworth takes as a starting point the idea that we need to evolve an economic perspective which emphasises the insight that the earth has limits in terms of resource use and sustainability. This is an economic model which takes account of humans' reliance on particular environments. She highlights the need for a stable climate, fertile soils and biodiversity, and understands that there is a spectrum of activities humans required for communities to operate in a sustainable and positive manner, whilst also realising that there is an ecological ceiling to economic activity. Go beyond this ceiling and human activity begins to have adverse impacts on earth systems. Doughnut economics would therefore offer one lens through which interconnections between economic, social, environmental and physical geography could be encountered by students. Over a period of time they would begin to understand how economic activity can have an impact not only on humans but also on physical systems through a more coherent and complex engagement with the spectrum of processes involved than would be the case in more 'traditional' geography curricula.

A second lens offering exciting potential for understanding the complexities of physical systems is to introduce the model of the 'critical zone'. Sullivan et al. (2017: 1) describes critical zone (CZ) science as having,

> created a transdisciplinary nexus that seeks to understand the response of Earth's near surface processes to climatic and human perturbations. CZ science brings together researchers from geology, soil science, geomorphology, hydrology, meteorology, and ecology to study Earth's living skin from bedrock to the top of vegetation.

The critical zone is a relatively new concept identifying the boundary layer where rock, soil, water, air and living organisms interact to regulate the natural habitat. It therefore determines the health of ecosystems, biodiversity and ultimately our survival via food production through agriculture, water quality for human use, and so on. The US National Science Foundation has created a number of critical zone observatories which are interdisciplinary in nature and as such not only show the interconnections between various physical processes operating at a number of scales and their interactions with humans, but also help students begin to understand that disciplinary boundaries are porous and that many interconnections occur between disciplines in an attempt to understand complex and difficult problems. Sullivan et al.

(2017) have begun to establish some of the links between human dependence and the natural processes within the critical zone, such as the impact of human activity on processes and the nature of the critical zone, and in turn, how changes in the critical zone impact on human activities. Field et al. (2015) focus on the potential to use a knowledge of the critical zone to understand the potential of 'ecosystem services', the potential for ecosystems to supply products which can be used by society. By considering the processes in the critical zone, including how they create and regulate the zone, and how they become the basis for human activity, we are able to begin to think in terms of how weather, rivers, soils and ecosystems interact, and how these interconnected processes are linked to resource supply and exploitation. This then opens up consideration of how changes in climate, levels of human exploitation, and their interactions with physical systems create sustainable or unsustainable environments. Obviously, the detail which is covered would depend on the age of the students, but simple links could be made at lower levels within the school system, with increasing detail and criticality as students progress.

The final lens we exemplify here is 'critical physical geography' (CPG). In their discussion of CPG, Lave et al. (2014: 3) state,

> Its central precept is that we cannot rely on explanations grounded in physical or critical human geography alone because socio-biophysical landscapes are as much the product of unequal power relations, histories of colonialism, and racial and gender disparities as they are of hydrology, ecology, and climate change.

Therefore, an important aspect of physical geography focuses on attempting to understand how physical systems and human social economic processes interact to have an impact on both the physical systems themselves and the humans who operate in them. Simple examples might be to consider the 'industrial' landscapes of agriculture, and of water extraction in the case of reservoirs and dams. As well as the more obvious discussions over why reservoirs and agriculture are necessary, and how they impact on the physical environment, CPG opens additional foci for consideration, for example how agricultural policies impact on agricultural activity and how this then impacts on physical and biotic processes. An interesting spin on this focus is also to consider how physical systems begin to work with human environments to produce hybrid landforms. Dixon et al. (2018: 118) capture the development of urban weathering and the production of what they call Anthropocene regolith,

> If the stone fabric of the urban landscape is considered as an Anthropocene geological formation, then over decadal timescales weathering processes will develop Anthropocene urban regoliths as well as more intensively reshaped surfaces.

They go on to identify urban stalactites and stalagmites, the result of leaching of calcite from concrete, and at a larger scale urban sink holes, which they characterise as an Anthropocene geohazard. Including the study of these processes and landforms within the physical geography of school curricula helps to illustrate the complexity of the interconnections between physical and human environments and processes.

Towards a Conceptual Model of a Plexus Curriculum in Geography

Our conceptualisation of a plexus curriculum has at its core the idea that we need to build a more holistic engagement with the knowledge base of geography. Figure 22.1 sets out an initial conceptual model which we believe could inform curriculum thinking to develop a plexus approach.

At the core of the model are is the conceptual framework which acts as the structure for interconnecting the chosen disciplinary knowledge. In practice, the knowledge base might not be dissimilar from pre-existing curricula, but in a plexus approach, it is reworked to allow explicit links and interconnections to be explored. These interconnections can use various frameworks for structuring and exploring content. An example might be the linking of physical and urban geography through a lens of critical physical geography, exploring pseudo-karst processes which are a result of processes linked to the Anthropocene. Likewise, climate change can be explored through the lens of doughnut economics as it relates to globalisation, industrialisation and environmental degradation.

As the knowledge base of the curriculum is developed, and interconnections built, it is important to iteratively relate what is learned back to student experience and reflection. This allows students to locate themselves and their communities within their emerging understanding of geographical issues and processes. With this being an iterative process, student experience is also co-opted into the curriculum by acting as the foundation for asking questions and exploring the subject further, and through this

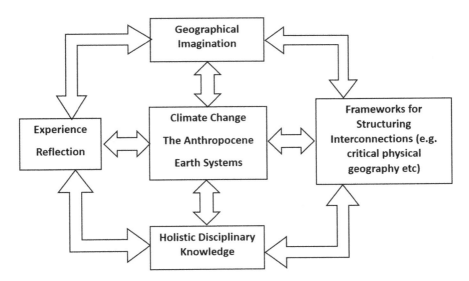

Fig. 22.1 Outline conceptual model for plexus curriculum thinking

approach personal experience and disciplinary knowledge become complementary aspects of the curriculum.

Finally, the interplay of these aspects of a plexus curriculum are the basis for developing students' geographical imaginations. As outlined by Daniels (2011) the development of the geographical imagination acts as the foundation for both understanding the world and envisioning how it might be reshaped. The powerful knowledge of geography is given its power not as abstract disciplinary content, but as an emerging core for critical and holistic engagement with geographical issues.

The concepts we have chosen as the core to the process of developing interconnected knowledge we see as being crucial to understanding the complex and shifting nature of the planet. This is why we have put them at the core of our model. However, this is in some ways a thought experiment; a way of beginning to engage with the need for developing holistic rather than atomistic knowledge and understanding through a geography curriculum. Therefore, we would encourage teachers to consider if these are the concepts which they believe should sit at the centre of the process, or whether in their own context, with their own expertise, they believe alternative concepts would work as a better foundation for interconnected thinking. As curriculum makers, teachers should have opportunity for dialogue and discussion as to what should sit at the conceptual core of their work. But whatever concepts are deemed central, they should allow for a structuring of the knowledge within the subject in such a way that the interconnected, holistic nature of the subject is explored and made explicit for students. At a time when our planet is endangered, we need individuals who can think holistically and critically—and developing this in individual students and across and between societies and cultures is a task for education that no-one should underestimate.

References

Agnew J (1987) Place and politics. Allen & Unwin, Boston, MA

Angus I (2016) Facing the Anthropocene: fossil capitalism and the crisis of the earth system. Monthly Review Press, New York

Catling S, Martin F (2011) Contesting powerful knowledge: the primary geography curriculum as an articulation between academic and children's (ethno-) geographies. Curriculum J 22(3):317–335. https://doi.org/10.1080/09585176.2011.601624

Christenbury L (1989) Cultural literacy: a terrible idea whose time has come. Engl J 78:14–17

Cloke P, Crang P, Goodwin M (2005) Introducing Human Geographies 2nd edn. Routledge, Abingdon

Cloke P, Crang P, Goodwin M (2014) Introducing human geographies, 3rd edn. Routledge, Abingdon

Cook PG (2009) The rhetoricity of *cultural literacy*. Pedagogy 9(3):487–500

Daniels S (2011) Geographical imagination. Trans Inst Br Geogr 36(2):182–187

Daniels P, Bradshaw M, Shaw D, Sidaway J (eds) (2008) An introduction to Human geography: issues for the 21st century, 3rd edn. Pearson, London

DfE (2010) The importance of teaching. The Schools White Paper 2010. DfE, London

Dixon SJ, Viles HJ, Garrett BL (2018) Ozymandias in the Anthropocene: the city as an emerging landform. Area 50(1):117–125

Dowgill P, Lambert D (1992) Cultural literacy and school geography. Geography 77(2):143–151

Field JP, Breshears DD, Law DJ (2015) Critical zone services: expanding context, constraints, and currency beyond ecosystem services. Vadose Zone J 14(1) Accessed at: https://dl.sciencesocieties.org/publications/vzj/pdfs/14/1/vzj2014.10.0142

Firth R (2011) Making geography visible as an object of study in the secondary school curriculum. Curriculum J 22(3):289–316

Giddens A (1998) The third way: the renewal of social democracy. Polity Press, Cambridge

Gieseking JJ (2017) Geographical imagination. In: International encyclopedia of geography: people, the earth, environment, and technology. Pre-print version accessed at: https://www.researchgate.net/publication/314239640_Geographical_Imagination

Goudie AS (1986) The integration of human and physical geography. Trans Inst Br Geogr 11(4):454–458

Gove M (2011) The secretary of state's speech to Cambridge University on a liberal education. Accessed at: https://www.gov.uk/government/speeches/michael-gove-to-cambridge-university

Hirsch ED (1987) Cultural literacy: what every American needs to know. Houghton Miffin Company

Lave R, Wilson MW, Barron ES (2014) Intervention: critical physical geography. Can Geogr 58(1):1–10

McCormack D (2012) Geography and abstraction: towards an affirmative critique. Prog Hum Geogr 36(6):715–734

McGill BJ, Dornelas M, Gotelli NJ, Magurran AE (2015) Fifteen forms of biodiversity trend in the Anthropocene. Trends Ecol Evol 30(2):104–113

Morgan J (2011) Knowledge and the school geography curriculum: a rough guide for teachers. Teach Geogr 36(3):90–92

Morgan J, Lambert D (2011) Editors' introduction. Curriculum J 22(3):279–287

Prince H C (1962) The Geographical Imagination. Landscape 11:22–25

Raworth K (2017) Doughnut economics: seven ways to think like a 21st-century economist. Chelsea Green Publishing

Roberts M (2014) Powerful knowledge and geographical education. Curriculum J 1–23. https://doi.org/10.1080/09585176.2014.894481

Saavedra AR, Opfer VD (2012) Learning 21st-century skills requires 21st-century teaching. Phi Delta Kappan 94(2):8–13

Sahlberg P (2012) Finnish lessons: what can the world learn from educational change in Finland?. Teachers' College Press, New York

Schwagerl C (2014) The Anthropocene: the human era and how it shapes our planet. Synergetic Press, London

Scott P (1988) Review: a few more words about E. D. Hirsch and cultural literacy. Coll Engl 50:333–338

Sullivan P, Wymore A, McDowell W et al (2017) New opportunities for critical zone science. Arlington Meeting for CZ Science White Booklet. Accessed at: https://criticalzone.org/images/national/associated-files/1National/CZO_2017_White_Booklet_Final.pdf

Walford R (2001) Geography in British schools 1850–2000. Woburn Press, London

Whelan R (ed) (2007) The corruption of the curriculum. Civitas, London

Wood P (2009) Locating place in school geography—experiences from the pilot GCSE. Int Res Geogr Environ Educ 18(1):5–18

World Bank (2010) The education system in Malawi World Bank working paper no. 182. The World Bank, Washington. Last Accessed 21st Apr 2019: http://siteresources.worldbank.org/EDUCATION/Resources/278200-1099079877269/Education_System_Malawi.pdf

Young M (2008) Bringing knowledge back in: from social constructivism to social realism in the sociology of education. Routledge, Abingdon

Young M, Muller J (2013) On the powers of powerful knowledge. Rev Educ 1(3):229–250. https://doi.org/10.1002/rev3.3017

Young M, Lambert D, Roberts C, Roberts M (2014) Knowledge and the future school: curriculum and social justice. Bloomsbury, London

Chapter 23
(Latent) Potentials to Incorporate and Improve Environmental Knowledge Using African Languages in Agriculture Lessons in Malawi

Michael M. Kretzer and **Russell H. Kaschula**

Abstract In their official language policy, nearly all Sub-Saharan African states use their indigenous language(s) as Language of Learning and Teaching (LoLT) only at the beginning of primary schools. This is also the case in Malawi. The curricula in the various school subjects are also highly dominated by 'Western' ideas and include very little Indigenous Knowledge (IK). Nevertheless, indigenous languages are frequently used during lessons. This research focused on answering the following questions: How is a meaningful Science Education for pupils in Malawi possible? Does the inclusion of IK and teaching through African Languages assist pupils in any way? Research was done in the Northern Region of Malawi. To obtain a better understanding, semi-structured interviews and ethnographic observations were conducted. The main focus of these interviews was on the subject of 'Agriculture'. Aspects such as IK, environmental education, climate change, school gardens and others were covered. Teachers saw the importance and necessity of the subject 'Agriculture' especially in rural areas, as Agriculture remains the backbone of economic activity in Malawi. To include more practical tasks like the running of school gardens, IK next to 'Western' Science and (partly) use more of indigenous languages were some of the outcomes. Teachers emphasized aspects such as deforestation and the conservation of soil and water as the main environmental aspects of the teaching of Agriculture. Examples of IK are also provided from the Global South. This is done in order to support the argument made in this chapter regarding the potential use of IK within the teaching and knowledge production process.

Keywords Malawi · African languages · Indigenous Knowledge Systems (IKS) · Agriculture · Environmental education

---

M. M. Kretzer (✉) · R. H. Kaschula
School of Languages and Literatures (African Language Studies Section),
Rhodes University, P. O. Box 94, Grahamstown/Makhanda 6140, South Africa
e-mail: m.kretzer@ru.ac.za

R. H. Kaschula
e-mail: r.kaschula@ru.ac.za

© Springer Nature Switzerland AG 2019
W. Leal Filho and S. L. Hemstock (eds.), *Climate Change and the Role of Education*,
Climate Change Management, https://doi.org/10.1007/978-3-030-32898-6_23

Introduction

> Drought, floods forecast for Malawi rainy season: MET warns of El Niño. (Nyasa Times 2018)
>
> New Malawi weather alert warns of more flooding this week. (Nyasa Times 2019)
>
> Water scarcity in Balaka and Ntcheu: Government drills High Yielding Boreholes to mitigate water shortage. (The Maravi Post 2018)

These are three very recent headlines from newspapers in Malawi. They provide an initial impression of how serious the current weather conditions are in Malawi. Additionally, various publications from the United Nations Development Programme (UNDP) highlight the vulnerability of Malawi to a variety of extreme weather events. The most common climatic hazards include floods and landslides, as well as seasonal or even multi-year droughts (UNDP 2019a).

Considering that agriculture is the economic backbone of Malawi (involving millions of smallholder farmers), any increase in climatic hazards severely affects the population. Malawi is a small, landlocked but densely populated country in South-East Africa, with a very fast growing population of currently around 18.6 million (World Bank 2019). The UNDP Human Development Report ranked Malawi in 2018 as 171 out of 189 countries and territories (UNDP 2019b). Despite an increase of 40.2% of its HDI Malawi remains in the low human development category. In recent years Malawi has faced several severe famines, and it relies continuously on donor-aid to improve the livelihood of its population.

The United Nations (UN) devotes the Sustainable Development Goal (SDG) 13 to climate change and processes of mitigation and adaptation to its severe changes and challenges. Within this context, landlocked countries are mentioned as more vulnerable regions (UN 2019). Other international documents, such as the Agenda 2063 from the African Union (AU), clearly state the unfair current situation in the following quote: "Whilst Africa at present contributes less than 5% of global carbon emissions, it bears the brunt of the impact of climate change" (AU 2015). Clear links between the goals of Agenda 2063 and the UN SDG 13 regarding climate change are visible.

Many African countries formulated and (partly) implemented various national climate-related policies. Ethiopia presented in 2011 its Climate Resilient Green Economy strategy. Contrary to this, Malawi's development policies hardly focuses on climate change and mitigation strategies. As mentioned above, Malawi faces severe challenges, including a very fast growing population and its concomitant effects on food security. This is why the second Malawi Growth and Development Strategy (2011–2016) had its emphasis on such aspects, rather than the establishment of a green economy (Maupin 2017: 139).

The education system likewise faces severe challenges common to many Sub-Saharan African countries. Nevertheless, the extent of its challenges remains more prevalent. Malawi's current political, economic and social developments are still highly influenced by the long lasting legacy of the decades-long dictatorship of Dr. Hasting Banda. Only in 1994 the first, free, fair and democratic elections took place.

During the presidency of Bakili Muluzi significant political and educational reforms took place in an effort to improve the quantity and quality of teaching and learning in Malawi. Furthermore, education systems, and specifically Science Education, still rely and focus predominantly on the Global North. Curricula are therefore very biased and largely exclude IKS. An exception is the Science Education in Ghana, which includes IKS (Anderson 2006; Shizha 2013). Further exceptions are provided later in this chapter.

Objectives of This Chapter

The research for this chapter focused on answering the following questions: How is a meaningful Science Education for pupils in Malawi possible? Does the inclusion of IK[1] help the pupils? Does the teaching through African Languages assist cognition? The remainder of this chapter is as follows: Firstly, a methodology section describes the main aspects of this empirical study, including the research area. This section will be followed by the conceptual framework and a section with further examples from the Global South. Following this, characteristics of the Malawian education system are elaborated on and the results of the field work are outlined. This chapter ends with a conclusion and some recommendations.

Methodology

The research was conducted in three districts of the Northern Region in Malawi (see Fig. 23.1). The authors selected a qualitative approach, with semi-structured interviews, in order to obtain some insight into teaching of Agriculture at schools in Malawi. Furthermore, an ethnographic research approach, as well as a document analysis of school books used in the curriculum, and associated examination documents was done. The ethnographic research focused mainly on the usage and conditions of the various school gardens. The document analysis focused mainly on the textbooks and examination papers used by the Malawi National Examination Board (MANEB). To avoid biases during the interviews as a stand-alone method, a multimodal approach was selected to assist in contextualizing statements by the teachers during the interviews (Taylor and Bogdan 1998: 56). A total of 14 teachers from six different schools were interviewed in late 2015. The field work was self-funded by the authors.

All the teachers provided their consent and agreed to be part of the research. All interviews were recorded and transcribed. All information from the interviews is

---

[1]The authors solely use the term IK for the purpose of reader's access. The authors are aware of its multidimensional body of meanings and its depiction as inferior and primitive (Kincheloe and Steinberg 2008: 136).

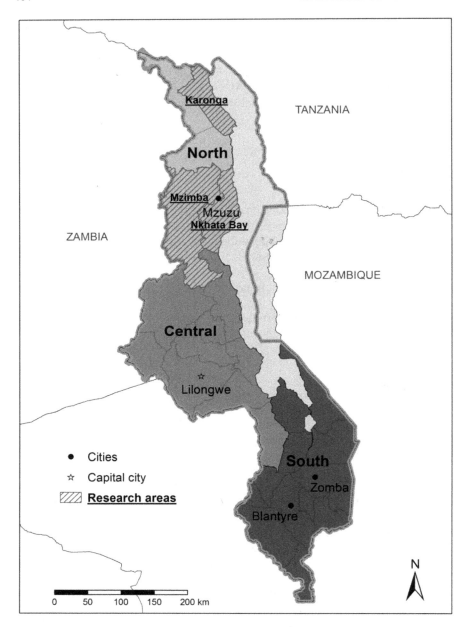

Fig. 23.1 Research area in Malawi. *Source* Kretzer et. al. 2016: 238

stored, though any quotation provided by the teachers is treated anonymously. The authors' preferred to conduct face-to-face interviews, using an interview-guide.

All interviews were conducted by one of the authors at the respective work place of the interviewee. This ensured a familiarity and comfort for the interviewees them (Longhurst 2016: 150; Helfferich 2011: 177). Interviewer related biases existed and may have influenced the interview; and perhaps the answers to some extent (Smith 2016: 98; Groves 1989: 398–404). The author was free to change the order of the questions with only one limitation. The first question was developed as a very open question in order to help start a good conversation flow and to establish a comforting interview situation (Kromrey et al. 2016: 338–340; Dörnyei 2007: 140).

Conceptual Framework

Although the term IK is frequently used in academia, and also in the popular media, no overall definition exists. There are numerous definitions that are used in regard to IK. Most common definitions focus on its orality, locality and its intergenerational transmission. Govender (2012, 112) defines IK as: "as a body of knowledge produced and owned by local people in their specific communities and passed on from generation to generation, through practice and mainly oral channels".

What is most striking are the many incorrect assumptions about IK. Often IK is seen as a stable and monolithic knowledge, and mainly used as a term for the various traditional communities in the Global South, which is problematic. Such separation between IK and Western Science is a biased, artificial and neo-colonial assumption. No knowledge is stable over space and time (a static notion of IK) but rather all knowledge is flexible, versatile and adapts to changing circumstances in space and time (Kincheloe and Steinberg 2008: 143). Orality is one of its main characteristics. Therefore, knowledge is versatile and bound to the owner of such knowledge (Zegeye and Vambe 2006). Often, IK is associated with medicine, healing, farming practices, arts, nutrition, but not limited to those disciplines. IK is still largely marginalized and stigmatized and not seen as an equal or equivalent to knowledge related to Western Science[2] based concepts (Govender 2012: 113).

> It was often referred to in a negative or derivative manner, with phrases such as primitive, backward, archaic, outdated, pagan and barbaric. This demeaning reference did not create space for IK's integration with other forms of knowledge, commonly referred to as scientific, western or modern knowledge (Ocholla 2007: 3).

IK and African worldviews are closely related to each other (Kayira 2015; Le Grange 2012; Cloete 2011). The concept of Sankofa originates from West Africa and has its origins in the Akan language. Akan is spoken in Ghana and the Sankofa consists of three words: 'San' (return), 'Ko' (go) and 'Fa' (retrieve). The Sankofa has

---

[2]By using the term Western Sciences the authors refer to Europe/North America and the usage of the term is solely for the purpose of the reader's access, without taking any political sides regarding meaning.

been symbolized as a mythical bird with its head facing backwards. For the Akan, the past is seen as a resource and it can assist by providing an understanding of the present and future (Berea College 2018; Kayira 2015: 114). Sankofa can be used as a traditional form of African (informal) education. Recently, scholarly interest in the Sankofa concept rose as an increasing number of African Americans brought attention to it. The implementation of the Sankofa concept needs the individuals to fully understand and be rooted in it and its holistic African worldview (Karenga 1995: 39).

Examples of Science Education in the Global South

In further research conducted by Dube and Lubben (2011), they focus on the existing and varying perceptions of Science teachers regarding the integration of IK into teaching in Swaziland.[3] Some do not see any necessity for the integration nor do they feel competent enough in IK. Nevertheless, the results revealed many cultural practices which are related to various aspects of Science Education. These are connected to Biology or reproduction, such as the famous Reed dance (*uMhlanga*), or environmental aspects such as the *Kukhonta* system. This indigenous land management system gave the Chief a very responsible role. His duty was to control the grazing and settlement so that an over-utilization was prevented (Dube and Lubben 2011: 78–79). Similarly to other African communities, the Swazi also have different taboo animals depending on the specific clan. This, to an extent, helps to preserve the habitat and biodiversity and is still utilized and practiced by present and hopefully future generations.

The urgent question about IK and Science Education is not only limited to countries in Sub-Saharan Africa, but rather all countries of the Global South, if not the whole world. The second example is taken from a rural area in Anchetty in India (Shukla et al. 2017). Any loss of IK can affect food security of a community significantly. This can be seen in Anchetty, as the younger generation hardly grow the small millet plant and associated crops of the region (SMAC). Shukla et al. (2017) underline in their research the unused potentials of integrating IK or Indigenous Agricultural Knowledge (IAK) into the teaching of Science. They further propose that this would help to preserve these teachings. In addition, it can help to connect the rural school with their surrounding community, and vice versa, as a place-based education. The authors highlight the obstacles for creating connections between IK and Science Education. Such obstacles exist mainly in the attitudes of the parents and pupils themselves. IK, IAK or anything related to agriculture is highly stigmatized by the young generation. Pupils see any agricultural activity as a sign of poverty, illiteracy, and backwardness. Therefore, it has a very low priority and reputation among

---

[3] King Mswati III declared on the 19th of April 2018 during the celebration of its 50th anniversary of independence that the name had changed to Eswatini (Times of Swaziland 2018). Due to the fact that the name Swaziland is still used, the authors use the names interchangeably.

them (Shukla et al. 2017). The same perception among pupils is quite common in other places and is even a topic dealt with in the mass media. For example, the popular mass media journal *Move,* in South Africa, raised this topic early in 2018. The article portrays a young school girl who is interested in agriculture and who started her own pepper plantation. Throughout the article the same perceptions as presented by the above mentioned pupils from India are prevalent (Nndeleni 2018). The same perceptions are often also applied to the learning of, or in, African or indigenous languages (Kaschula 2013).

Challenges of the Education System in Malawi in General and Teaching of Agriculture in Specific

The education system of Malawi faces many challenges, and its outcomes remains poor. Drop-out rates are very high and any meaningful teaching and learning faces severe obstacles. Alongside his so called "zasintha" (things have changed; Kayambazinthu 1999) policy, the former President Bakili Muluzi introduced Free Primary Education (FPE). FPE was one of the main promises made during his election campaign. It was a very visible and recognizable promise and with the support of donor aid, FPE was implemented. This led to a substantial increase in enrolment rates and classrooms were often extremely crowded. The quality of learning and teaching suffered significantly as the government was not able to train sufficient new teachers in such a short time (Dzama 2006: 248). To avoid even more overcrowded classrooms, many untrained teachers were employed (Edwards 2005: 26). While the net enrolment rate in primary schools in 2016 was close to 100% (and no significant gender gap was prevalent), the net enrolment rate at secondary schools was as low as 31.45%. Furthermore, the pupil-teacher ratio remains extremely high at 62.31; although significant improvement has been made (UNESCO Institute of Statistics 2019). Various international evaluation studies, such as the Southern and Eastern Africa Consortium for Monitoring Educational Quality (SACMEQ), indicate the ongoing challenges in schools in Malawi. The consistent and accurate supply of textbooks or other learning materials is still problematic. In 2007, only 24% of grade 6 pupils in schools in Malawi had access to an individual mathematics textbook (SACMEQ III 2011). Other studies demonstrate even more severe and disturbing results. Castel et al. (2010) and DeStefano (2013) expose in their research the poor teaching conditions in Science or Agriculture; in 2008 up to 200 pupils were forced to share a single textbook. This provides an overall impression of the teaching conditions in public schools in Malawi, and specifically in teaching of Agriculture. As highlighted earlier, the economic backbone of Malawi is agriculture. Therefore, the overarching goal of the subject Agriculture is to teach pupils a broad knowledge about various aspects of agriculture. This aims to equip them with the necessary skills to enable them to be employed within the agricultural sector or to improve their own smallholding production (Vandenbosch 2006: 33).

The subject Agriculture prepares pupils scientifically and practically for future meaningful participation in Agriculture. The subject defines 22 learning objectives, which consist of eight main topics and its related sub-topics. Some of the learning objectives are general and very broad, such as "understand the importance of agriculture in Malawi's economy" or "observe farm safety rules". Contrary to this, others are more specific and related to the environment and its protection. Examples are "know Malawi's environmental resources", "understand the various methods of conserving natural resources", "apply the various conservation measures to Malawi's natural resources" or "appreciate the importance of conserving natural resources for sustainable use" just to name a few. One of the eight main topics focuses on "agriculture and the environment" and another on "challenges in agriculture development" (Vandenbosch 2006: 45–47). Although climate change itself is not mentioned as one of the topic areas, it is related to many of the aims and topics. The Department of Curriculum Development (DCD) realized some weaknesses in the Agriculture syllabus and tried to address them in a curriculum review of the secondary school course content. This included cross-cutting issues such as entrepreneurship, climate change or disaster risk management, to name a few (Malawi Institute of Education 2012).

Nampota (2011: 137) highlights the importance of these cross-cutting issues and she aims for "mainstreaming" the topic of environmental impact in formal education in Malawi. Other topics such as gender equality or inclusion are examples of such a mainstreaming approach. Nampota (ibid.) believes there is untapped potential if the topic of environmental education is restricted to one subject, namely Agriculture. Such transmission education is not sustainable as knowledge is bound to a specific subject and simply transferred from the various subjects teacher to the pupils. A more inclusive and sustainable way of teaching results in a transactional education. Transactional education requires, for example, school gardens as the pupils interact with the content. This is closely related to the pedagogical ideas of constructivism. Applying constructivism inside a classroom defines or constructs learning as a cultural activity (Chinn 2007). Teaching and learning is therefore not isolated from the life and experiences of the pupils, but rather adds to the already existing knowledge. Such an inclusive and connected approach helps not only to include Indigenous Knowledge Systems (IKS), but also requires the usage of African Languages. If African Languages are utilized, it helps to learn what is locally relevant and to connect both spheres of life and learning, thereby becoming more meaningful (Trudell 2009: 76).

Recent developments in Malawi are unfortunately contradictory, and hinder such a pedagogical approach. The increased use of various African Languages in schools in Malawi after the dictatorship of Banda stopped suddenly in 2014. The previous Minister of Education, Luscious Kanyumba introduced an English-only language policy. This results in many of the children being faced with a very unfamiliar language as LoLT. Many pupils do not speak English, or never speak English at home prior to their first school day. Furthermore, even some teachers are overwhelmed and unable to teach through the medium of English. Kamwendo (2016: 224) is highly critical of this approach in the following quote:

> This is a nightmare to the majority of the young learners given that the classroom is the first place where they come into contact with English. The vast majority of learners come from homes in which English is never used at all.

Every language is culturally embedded in its environment, and language does not only function as a neutral tool for communicating. English is a language which is not related to the various different landscapes of Africa and is limited in its vocabulary to describe them in comparison to African Languages. On the contrary, African Languages have a vast variety of words and categorizations to describe African landscapes, due to their embeddedness in the region.

Like all languages, English is dynamic, flexible and inventive—but it is, from a global perspective, an intrinsically urban language—swirling in and around centres of power. In Africa, an imported language is myopic and conceptualisation of the natural environment through this language is less flexible (Cloete 2011: 42). It is therefore debatable as to whether it would be a suitable LoLT when it comes to the teaching of Agriculture in Malawi, indeed of any content subject, with the exception of teaching English as a subject in its own right.

Teaching of Agriculture at Schools in Malawi

During the interviews, the teachers highlighted the general role of Agriculture and the conditions under which the subject of Agriculture is taught. All teachers focused on the tremendous role agriculture plays within the economy in Malawi; as Malawi remains an Agro-based Economy. A sample of interview responses follows:

> Agriculture is a very important subject, ja because not everybody when he has finished education can start working, as maybe as a teacher or any work, but if a student knows what Agriculture is when he finishes his school and he cannot be at work he can be a farmer and we teach them that Agriculture is a business, so they know that in Agriculture they can do find money as well as food, ja income.

> Malawi relies on Agriculture, about 80% of Malawi's economy depends on Agriculture, so if we don't teach Agriculture in terms of economy our country will be dying, so it's like we are importing the knowledge of Agriculture to our students.[4]

Nevertheless, many teachers also saw specific obstacles for the teaching of Agriculture with regard to the educational background of the teachers. This is supported by the fact that the majority of the interviewed teachers did not study Agriculture. They teach it as non-specialists and this in turn influences the quality of the lessons. Teachers see a further obstacle in the current way of teaching the subject of Agriculture. They complain about the fact that Agriculture is taught in a way that is too theoretical. Practical tutorials or assessments are negligible and do not form a major part of the subject. This includes the often non-existent school gardens, which could help significantly in the practical understanding of the subject. Such a practical

---

[4] All interviews were slightly edited regarding formulations or repeating of words to ensure a better reading of these interviews.

approach could help the teaching of Agriculture as it would include a constructivist teaching method. The incorporation of the daily experiences of pupils in their gardens at home helps them to relate the new knowledge and skills to their previous experiences. The inclusion of school gardens could also help to engage with the local communities. This place-based education (Smith 2007) has many advantages. Through the connection with the local community and the school a reciprocal relationship can be established. Such an approach can also help to mainstream aspects of environmental education and to create synergy effects outside of the Agriculture lesson. Awareness of aspects of conservation could also be raised. Unfortunately none of the teachers reported such place-based education, but they rather referred to no involvement. They also commented on the fact that some teachers use the school garden for their own private purposes, therefore defeating any possible objective to use the gardens to enhance education.

Furthermore, earlier research in the Northern Region indicated the synergy effects on other subjects such as literacy and numeracy (Engler and Kretzer 2014: 229). The usage of African Languages as LoLT helps to overcome the current artificial and obstructive teaching situation during Agriculture lessons. As in many other African countries, teachers emphasized the use of Code-Switching (CS) between English and an African Language as a support strategy for teaching and learning during their lessons. As the official language policy in Malawi opts for English-only as LoLT from the beginning, CS depends entirely on the individual teachers' preferences and decisions as well as the overarching school setting. Such covert language practices (Schiffman 1992) are indeed common in Sub-Saharan Africa. Research from Botswana (Kretzer 2017) showed this, but also in South Africa (Kretzer 2016, 2018). The following quote from one teacher clearly shows the existing challenges regarding some agricultural tools, which might be unknown to a large number of pupils, especially in the rural areas of Malawi.

> In fact I can say that the equipment is not enough, mostly the thing that we are doing in Malawi we just teach theory, we just say this is a plough, most of the student they don't know what is a plough they even, especially in this maybe … remote areas they don't know what is a plough, you just tell them a plough has got this what? Parts, but they have not seen those things, so … I can say that the situation is not very conducive, most of the things you just teach them as a theory, but in order to see the most practical it is very difficult and those practical's that are just very few, you can talk about soil, which is still there, talk about maybe plants like tomatoes, goats, locally goats, those are some of the things that are there, but the equipment and some of the things to do practical they are not, … not really found in our schools.

Some of the teachers indicated the enormous role of food security in Malawi as very important for the teaching of Agriculture. Thus, climate change and the environment are also relevant aspects of the lessons. In some of the researched schools some awareness information leaflets about climate change where available. These posters were made available by the UNDP, but unfortunately they are presented only in English. However, statements from the Malawian government described the availability of such information materials in African Languages such as Chichewa or Chitumbuka (Malawi Government 2013). Furthermore, this strategy paper clearly speaks about 'opportunities for mainstreaming CC [climate change] in curricular' (Malawi Government 2013: X) to improve knowledge about climate change regarding its scientific explanations, its various severe impacts and possible responses to climate change (Fig. 23.2).

Fig. 23.2 Climate change information leaflet. *Source* Michael M. Kretzer, 07/2015

Interviewed teachers highlighted the various causes of climate change and its devastating effects on the economy and Malawi's population. Food security is under severe threat due to climate change and the fast growing population. Therefore, teachers include in their lessons the consequences of climate change on the different landscapes in Malawi and its plants and animals. Some also highlight the steps that can be taken in order for humanity to adapt to these changes during their lessons. These may include the incorporation of IK, for example the planting of local plant varieties to avoid a total dependency on commercial hybrid seeds. The quote below from a teacher illustrates and supports this point of view.

> The curriculum in fact, for the, there are so many in fact crops, like sweet potatoes, so the Malawians they plant local varieties, but if we talk about maize, we encourage they must, according to the book, they talk about crop improvement, whereby we show them that, you know that in order to increase the, they must plant the hybrids, … so in another way we still encourage like for maize that they should really buy those seeds from the shops, but for sweet potatoes, cassava they just use their home, the remaining ones, they take these stems and plant for another season, so this are one of the things they can do.

Agriculture school books often use the term 'conservation', when it comes to adaptation and mitigation strategies for climate change (see Fig. 23.3), or the teaching of environmental knowledge in general.

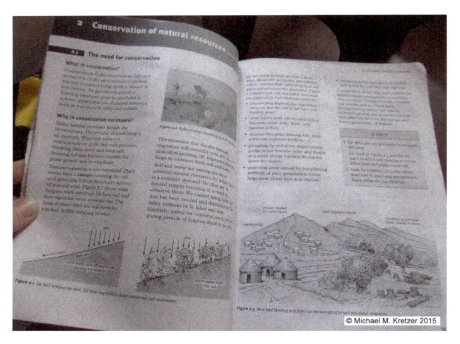

Fig. 23.3 Example of agriculture textbook page about conservation. *Source* Kretzer et. al. 2016: 247

Many teachers emphasized the importance of teaching sustainability and environmental aspects during Agriculture lessons. Figure 23.3 gives an example of a school book with its focus on erosion. Due to lack of sufficient electricity throughout Malawi, especially in remote areas, together with its increasing population, there is severe pressure on the remaining forests. Many Malawians use firewood for cooking or to produce charcoal, with all its devastating effects on the soil and the environment. Therefore, many teachers focused on soil conservation. Water usage and deforestation are also a very severe threat for the soil quality of Malawi as the two quotes below indicate.

> It is part of the curriculum, for example earlier on I said, I talked of environmental conservation which is a sustainable aspect, so agriculture promotes that, promoting environmental conservation, so I think that is part of agriculture, ja for example conserving soil, conserving forests and others, that is part of agriculture, [...] Especially so, when you conserve for example the soil, that will promote conservation of water it means that will, I think that is part of water conservation, that is of course conserving soil but at the same way it means we also promoting high infiltration rate. ... When it comes to usage you don't necessarily focus on usage, maybe we talk generally about agriculture and usage of water.

> We talk about degradation, we talk about what are the causes, cutting down trees in our country, where at the end of the day there is rain coming and that rain encourage land loss [...] So we talk about how we can conserve water and soil, so somehow there is the relationship between forestry, how we can keep our trees, care about our trees in order, so that the soil, we can still have some.

Within this context it became clear that the majority of teachers had their focus on deforestation, and not on any other aspects of environmental education. Some teachers studied forestry, which explains why their focus was on deforestation, but also the severe threat resulting from deforestation was mentioned several times during the interviews. Other teachers highlighted that the majority of teachers who teach Agriculture did not actually study the subject, but rather teach it as non-specialists. This negatively affects their confidence and the quality of teaching significantly.

> Currently in Malawi we had a problem of floods, especially in Chikwawa, it's because of cutting down of trees, the people they don't know the importance of why they shouldn't be cutting down trees, so we teach our students the importance of keeping forests, because they have to conserve the soil.

One teacher reflected on her own behaviour very critically and explained the underlying patterns of using charcoal from the sellers and buyers perspective. She is clearly aware of the negative consequences of buying charcoal, but emphasizes economic constraints which force people to sell or buy charcoal. Even though she knows how devastating deforestation is, she also does not act accordingly, and also purchases charcoal.

> People don't have jobs so they burn charcoal, they do charcoal business, so that is why trees are cut down. [...] In Malawi we don't have, everybody does not have electricity and electricity is more expensive so you compare, even me I use charcoal, because when I use electricity I found it is more expensive, so I know that cutting down of trees is bad, but I prefer to use charcoal because it is cheap so these people who are cutting, they are just

cutting because they know people are buying a lot, they cannot stop so it is a big problem and this problem can be stopped if electricity, the alternative to charcoal can be found and everybody shall have electricity.

Only very few teachers emphasized other aspects of environmental education or conservation, over and above the examples about deforestation and the soil. The example below speaks about air pollution and also gives more details about adapting water conservation to the changing seasons and the severe droughts and floods.

It is called conservation of national resources, we talk about sustainability, we talk of conservation of soil, of forestry, we talk of air pollution. We teach them how to construct different structures to conserve water, like storm draining […] and contour draining and so on and so on.

The challenge in the education system is to create suitably qualified teachers of Agriculture who fully understand and comprehend the factors that influence the teaching of this subject globally. These factors include the effects of global climate change. The lack of a more practical approach to the teaching of Agriculture also results in a highly theorised approach, which denies the role of any IKS in the education process.

Conclusion

This article showed some of the severe environmental challenges for agriculture and food security in Malawi. It also offered insight into the difficult general educational challenges in schools in Malawi, affecting the teaching of Agriculture. Not only are classroom sizes bigger, but there is a lack of physical agricultural equipment, together with a shortage of textbooks. This compounds the problems facing the teaching of Agriculture.

Furthermore, many teachers are not happy with the general teaching conditions which hinder meaningful teaching and learning. They also see the teaching of Agriculture as too theoretical. The use of English also hinders teaching and the inclusion of IK. Some policy documents suggest a need to mainstream climate change as part of the course. Many of the teachers themselves seem unaware of climate change. It was also found that the frequent absence of school gardens hinders the practical aspects of the teaching process. Even though conservation of natural resources is part of the curriculum, school books and awareness posters about climate change, they are only monolingual and available only in English. This is contrary to the policy documents which depict examples using African Languages. Regarding the teaching of conservation as environmental education, teachers focused predominantly on aspects of deforestation and the conservation of soil and water. Only one teacher highlighted air pollution. Interestingly, teachers are aware of the severe negative effects of using charcoal instead of electricity, but nevertheless due to economic constraints a teacher confessed her preference for charcoal. This example shows clearly that a pure awareness campaign or teaching of environmental aspects has little influence on the future, because behaviour remains the same if the economic situation and pressure remains the same.

This chapter has shown the need for a more practical, and at the same time, a globally informed approach to the teaching of Agriculture in Malawi. This would require the use of IK and African Languages in imparting knowledge, while at the same time taking cognisance of global climatic events. Teachers need to be properly equipped to teach this subject, both from a professional and practical point of view. This is important if one considers the link between agriculture and the economy of Malawi as being intrinsically intertwined. To enable the learning of environmental teaching as a discipline, it is necessary to improve the general teaching conditions. For example, to include school gardens so that the lessons are not focused on the theoretical and to also highlight the importance of behavioural changes from teachers. The example of charcoal emphasized that awareness and knowledge is not enough, but rather it must include a change of behaviour in order to become environmentally friendly. This must be accompanied by financial incentives, otherwise such behaviour changes will not be sustained.

Acknowledgements The authors would firstly like to thank the participating teachers and Mzuzu University for their support. Thanks also to Joshua Kumwenda (University of Witwatersrand) for his support during the data collection in the Northern Region. A special thanks also to the two anonymous reviewers who helped to improve the original manuscript. Nevertheless, all remaining mistakes are solely the responsibility of the authors. The financial assistance of the National Research Foundation (NRF) towards the publication of this research is hereby acknowledged. Opinions expressed and conclusions arrived at, are those of the authors and are not necessarily to be attributed to the NRF.

Disclosure Statement
No potential conflict of interest was reported by the authors.

References

African Union (2015) Agenda 2063. The Africa we want (popular version). Retrieved from https://au.int/en/Agenda2063/popular_version. 12 Feb 2019
Anderson IK (2006) The relevance of science education, as seen by pupils in Ghanaian Junior Secondary Schools. University of Western Cape, Kapstadt
Berea College (2018) The power of Sankofa: know history. Retrieved from www.berea.edu/cgwc/the-power-of-sankofa/. 10 May 2018
Castel V, Phiri M, Stampini M (2010) Education and employment in Malawi (working papers series no 110). African Development Bank, Tunis
Chinn PWN (2007) Decolonizing methodologies and indigenous knowledge: the role of culture, place and personal experience in professional development. J Res Sci Teach 44:1247–1268
Cloete EL (2011) Going to the bush: language and the conserved environment in southern Africa. Environ Edu Res 17(1):35–51. https://doi.org/10.1080/13504621003625248
DeStefano J (2013) Teacher training and deployment in Malawi. In: Motoko A (ed) Teacher reforms around the world: implementation and outcomes, vol 19, pp 77–97

Dörnyei Z (2007) Research methods in applied linguistics. Oxford University Press, Oxford

Dube T, Lubben F (2011) Swazi teachers' views on the use of cultural knowledge for integrating education for sustainable development into science teaching. Afr J Res Math Sci Technol Edu 15(3):68–83. https://doi.org/10.1080/10288457.2011.10740719

Dzama E (2006) Malawian secondary school students learning of science: historical background, performance and beliefs. PhD thesis, University of Western Cape

Edwards F (2005) The neglected heart of educational development: primary teacher education strategy in Malawi. J Edu Teach 31(1):25–36. https://doi.org/10.1080/02607470500043557

Engler S, Kretzer MM (2014) Agriculture and education: agricultural education as an adaptation to food insecurity in Malawi. Univ J Agricult Res 2:224–231

Groves RM (1989) Survey errors and survey costs. Wiley, Hoboken, New Jersey

Govender N (2012) Educational implications of applying the complexity approach to Indigenous Knowledge Systems (IKS). Alternation 19(2):112–137

Helfferich C (2011) Die Qualität qualitativer Daten. Manual für die Durchführung qualitativer Interviews. VS Verlag für Sozialwissenschaften, Wiesbaden

Kamwendo GH (2016) The new language of instruction policy in Malawi: a house standing on a shaky foundation. Int Rev Edu 62:221–228. https://doi.org/10.1007/s11159-016-9557-6

Karenga M (1995) Making the past meaningful: Kwanzaa and the Concept of Sankofa, Narratives, 36–46

Kaschula RH (2013) Challenging the forked tongue of multilingualism: scholarship in African languages at South African universities. In: Altmayer C, Wolff E (eds) pp. 203–222

Kayambazinthu E (1999) The language planning situation in Malawi. In: Kapman R, Baldauf R (eds) Language planning in Malawi. Multilingual Matters Ltd, Mozambique and the Philippine Sydney, pp 15–86

Kayira J (2015) (Re)creating spaces for uMunthu: postcolonial theory and environmental education in southern Africa. Environ Edu Res 21(1):106–128. https://doi.org/10.1080/13504622.2013.860428

Kincheloe JL, Steinberg SR (2008) Indigenous knowledges in education: complexities, dangers, and profound benefits. In: Denzin NK, Lincoln YS, Smith LT (eds) Handbook of critical and indigenous methodologies. SAGE Publications, Thousand Oaks, London, pp 135–156

Kretzer MM (2016) Variations of overt and covert language practices of educators in the north west province: case study of the use of Setswana and Sesotho at primary and secondary schools. South Afr J Afr Lang 36(1):15–24. https://doi.org/10.1080/02572117.2016.1186891

Kretzer MM (2017) Linguistic repertoire and language practices of pupils at secondary schools in Botswana: a case study. In: Ralarala M, Barris K, Ivala E, Siyepu S (eds) (CASAS BOOK SERIES): African language and language practice research in the 21st century: interdisciplinary themes and perspectives. Centre for Advanced Studies of African Society (CASAS), Cape Town, pp. 163–186

Kretzer MM (2018) Implementierungspotential der Sprachenpolitik im Bildungssystem Südafrikas. Eine Untersuchung in den Provinzen Gauteng, Limpopo und North West. (Unpublished PhD thesis) Justus-Liebig-University Giessen

Kretzer MM, Engler S, Gondwe J, Trost E (2016) Fighting resource scarcity – sustainability in the education system of Malawi – case study of Karonga, Mzimba and Nkhata Bay district. S Afr Geogr J 99(3):235–251. https://doi.org/10.1080/03736245.2016.1231624

Kromrey H, Roose J, Strübing J (2016) Empirische Sozialforschung. Modelle und Methoden der standardisierten Datenerhebung und Datenauswertung mit Annotationen aus qualitative-interpretativer Perspektive. Konstanz und München: UVK Verlagsgesellschaft

Le Grange L (2012) Ubuntu, Ukama and the healing of nature, self and society. Edu Philos Theory 44(2):56–67. https://doi.org/10.1111/j.1469-5812.2011.00795.x

Longhurst R (2016) Semi-structured Interviews and Focus Groups. In: Clifford N, Cope M, Gillespie T, French S (eds) Key methods in geography. SAGE Publications, Los Angeles, pp 143–156

Malawi Government (2013) Malawi's strategy on climate change learning, Deckblatt, Retrieved from https://www.uncclearn.org/sites/default/files/sg4_2-4_national_climate_change_learning_strategy_of_the_republic_of_malawi.pdf. 12 Feb 2019

Malawi Institute of Education (2012) Promoting quality education for all. DCD. Introduction. Retrieved from http://www.mie.edu.mw/index.php?option=com_content&view=article&id=19: dcd&catid=2&Itemid=257&showall=1&limitstart=. 1 Mar 2019

Maupin A (2017) The SDG13 to combat climate change: an opportunity for Africa to become a trailblazer? Afr Geogr Rev 36(2):131–145. https://doi.org/10.1080/19376812.2016.1171156

Nampota D (2011) Exploring the potential and challenges of integrating environmental issues in formal education in Malawi. Afr J Res Math Sci Technol Edu 15(3):137–152. https://doi.org/10.1080/10288457.2011.10740723

Nndeleni O (2018) Planting seeds of her future. Mahlatse Matlakane grew an interest in farming when she was just 15 years old. Move 622, May 2018, 20–21

Nyasa Times (2019) Drought, floods forecast for Malawi rainy season: MET warns of El Niño, Retrieved from https://www.nyasatimes.com/new-malawi-weather-alert-warns-of-more-flooding-this-week/. 12 Feb

Nyasa Times (2018) New Malawi weather alert warns of more flooding this week. Retrieved from https://www.nyasatimes.com/drought-floods-forecast-for-malawi-rainy-season-met-warns-of-el-nino/. 12 Feb

Ocholla D (2007) Marginalized knowledge: an agenda for indigenous knowledge development and integration with other forms of knowledge. Int Rev Inf Ethics 7:1–10

SACMEQ—Southern and Eastern Africa Consortium for Monitoring Educational Quality (2011) SACMEQ III main report, 2011, Retrieved from www.sacmeq.org/sites/default/files/sacmeq/reports/sacmeq-iii/national-reports/mal_sacmeq_iii_report-_final.pdf. 22 Feb 2014

Schiffman H (1992) Resisting arrest' in status planning: structural and covert impediments to status change. Lang Commun 12(1):1–15

Shizha E (2013) Reclaiming our indigenous voices: the problem with postcolonial Sub-Saharan African school curriculum. J Indigenous Soc Dev 2(1):1–18

Shukla S, Barkman J, Patel K (2017) Weaving indigenous agricultural knowledge with formal education to enhance community food security: school competition as a pedagogical space in rural Anchetty, India. Pedagogy Cult Soc 25(1):87–103. https://doi.org/10.1080/14681366.2016.1225114

Smith FM (2016) Working in different cultures and different languages. In: Clifford N, Cope M, Gillespie T, French S (eds) Key methods in geography. SAGE Publications, Los Angeles, pp 88–107

Smith GA (2007) Place-based education: breaking through the constraining regularities of public school. Environ Edu Res 13(2):189–207. https://doi.org/10.1080/13504620701285180

Taylor SJ, Bogdan R (1998) Introduction to qualitative research methods. John Wiley, New York

Times of Swaziland (2018) Kingdom of Eswatini change now official, Retrieved from http://www.times.co.sz/news/118373-kingdom-of-eswatini-change-now-official.html. 22 Nov 2018

The Maravi Post (2018) Water scarcity in Balaka and Ntcheu: government drills high yielding boreholes to mitigate water shortage, Retrieved from http://www.maravipost.com/water-scarcity-in-balaka-and-ntcheu-government-drills-high-yielding-boreholes-to-mitigate-water-shortage/. 22 Nov 2018

Trudell B (2009) Local-language literacy and sustainable development in Africa. Int J Edu Dev 29:73–79. https://doi.org/10.1016/j.ijedudev.2008.07.002

United Nations (2019) Goal 13: climate action. Retrieved from http://www.undp.org/content/undp/en/home/sustainable-development-goals/goal-13-climate-action.html. 12 Feb 2019

UNDP (2019a) Malawi. Retrieved from https://www.adaptation-undp.org/explore/eastern-africa/malawi. 12 Feb 2019

UNDP (2019b) Briefing note for countries on the 2018 statistical update. Malawi. Retrieved from http://hdr.undp.org/sites/all/themes/hdr_theme/country-notes/MWI.pdf. 12 Feb 2019

UNESCO Institute of Statistics (2019) Malawi, Retrieved from http://uis.unesco.org/country/MW. 22 Feb

Vandenbosch T (2006) Post-primary agricultural education and training in Sub-Saharan Africa: adapting supply to changing demand. World Bank, Nairobi

World Bank (2019) Population, total, Malawi, Retrieved from https://data.worldbank.org/indicator/SP.POP.TOTL?locations=MW. 12 Feb 2019

Zegeye A, Vambe M (2006) African indigenous knowledge systems. Review 29(4):329–358

Chapter 24
Sixty Seconds Above Sixty Degrees: Connecting Arctic and Non-Arctic Classrooms in the Age of Climate Change

Mary E. Short and Laura C. Engel

Abstract Science education literature highlights a need to better understand teachers' experiences of navigating barriers to implementing climate change education curriculum. This chapter presents findings of a qualitative study of teachers' perspectives during the facilitation of a digital environmental storytelling project called #60above60, which connects students in Arctic and non-Arctic urban contexts. The focus is on teachers' cognitive interpretations of the project and broader individuals in climate change education. The results and subsequent discussion in this paper will be useful to people and organizations interested in understanding how teachers and institutions implement curriculum incorporating climate change, sustainability, and student action-oriented instruction.

Keywords Teachers · Environmental education · Climate change education · Digital storytelling · Arctic

The Case for Connecting Arctic and Non-Arctic Classrooms

Anthropogenic activity disproportionately affects the Arctic's overwhelmingly fragile ecosystem, with implications for the health and well-being of the entire planet. As a result, there is a growing need to develop students' skills in ways necessary for studying, documenting, and addressing a growing litany of climate-related issues (Anderson 2012). Literature underscores the particular importance of teachers in facilitating and advancing student perspectives on global warming and climate change, particularly as students often lack a deep understanding of these complex issues and their local and planetary implications (Shepardson et al. 2009, 2011).

This chapter focuses on teachers' experiences in climate change education by investigating teachers' perspectives during the implementation of a globally-oriented, digital environmental storytelling project, called #60above60. #60above60 is part of

M. E. Short (✉) · L. C. Engel
The George Washington University, Washington, DC, USA
e-mail: bshort@gwu.edu

L. C. Engel
e-mail: Lce@gwu.edu

© Springer Nature Switzerland AG 2019
W. Leal Filho and S. L. Hemstock (eds.), *Climate Change and the Role of Education*,
Climate Change Management, https://doi.org/10.1007/978-3-030-32898-6_24

a National Science Foundation (NSF) funded Partnerships for International Research and Education (PIRE) project, Promoting Urban Sustainability in the Arctic. Participants in #60above60 include students and teachers from public and private elementary and middle schools located in the United States (U.S.), Norway, and Finland. In #60above60, students design and create a series of 60-second digital stories focused on unique features and environmental problems facing local city contexts, as well as proposed solutions to those problems. These digital stories are then exchanged between students above and below the 60th degree parallel (what the PIRE project defines as the Arctic), connecting students in Arctic and non-Arctic urban environments.

In our chapter, we first elucidate some of the barriers facing teachers in implementing climate change education. In our chapter, we first elucidate some of the barriers facing teachers in implementing climate change education, illustrating the importance of better understanding teachers' cognitive domains, which both guide teacher interests in climate change education and shape its implementation. We then elaborate on a set of findings related to the perspectives of two middle school teacher participants in #60above60. These teachers were located in Anchorage, Alaska and Washington, DC. While both are contextually part of the U.S., Anchorage, Alaska and Washington, DC are culturally and geographically distinctive, which allows students and teachers to identify and investigate environmental issues that are simultaneously grounded in their local communities, while transcending them to broader global environmental dynamics. Our findings on how teachers draw across cognitive constructs in their sensemaking when implementing #60above60 highlight the role that motivations of individual teachers play in climate change education.

Climate Change Education and the Role of Teachers

95% of active climate scientists attribute recent global warming to anthropogenic activity, suggesting a common understanding amongst scientists about the causes of climate change (Doran and Zimmerman 2009; Plutzer et al. 2016). This widespread recognition of climate change and its causes has intensified the need for climate change education, defined broadly as "processes aimed at improving the degree to which an education system is prepared for, and is responsive to, the challenges of climate change" (Mochizuki and Bryan 2015, p. 5). Frameworks of climate change education acknowledging the significance of anthropogenic activity frequently emphasize a number of key dimensions, such as highlighting the causes and consequences of climate change, teaching local-global interconnections, and inspiring students to take appropriate action to reduce and adapt to climate change related conditions (Mochizuki and Bryan 2015). Yet, implementation is fraught with uncertainties and challenges. In our chapter, we highlight four challenges specific to teachers: (1) Anxiety about incorporating a perceived controversial subject into curriculum; (2) A perceived lack of personal knowledge about the human causes of climate change; (3) Concerns as to whether or not climate science issues are developmentally appropriate for students; and (4) Challenges in guiding students to make local-global connections when considering and taking action on climate change (Hestness et al. 2011; Oversby 2015).

First, literature frequently points to teachers' anxieties about teaching a potentially controversial issue (Dunk et al. 2019; Oversby 2015; Sadler et al. 2006). For example, teachers report concerns about possible parent and/or student resistance to teaching climate change in their classrooms (Plutzer et al. 2016; Sullivan et al. 2014). Coupled with a distinct lack of teacher-vetted, standards-aligned climate change curriculum and professional development resources, the perception that human-related causes of climate change are scientifically controversial may result in teachers feeling obligated to teach students *both sides* of climate change, framing it as a *scientific* and *social* controversy (Sullivan et al. 2014). This is exasperating for science teachers who lack training and professional development in teaching and in addressing controversial topics (Lombardi and Sinatra 2013; Oversby 2015; Plutzer et al. 2016).

Second, teachers often do not feel that they have the adequate knowledge about human causes of climate change. For example, a study of 1500 middle and high school teachers in the U.S. found that fewer than half of surveyed teachers were aware of the overwhelming scientific agreement that "global warming is mostly caused by human activity" (Plutzer et al. 2016, p. 645). Many scholars advocate for providing explicit climate science education instruction during teacher training, including developing skills in interpreting scientific data and climate models in order to mitigate reluctance towards highlighting human-related causes in climate science curricula (Hestness et al. 2011; Lombardi and Sinatra 2013; Plutzer et al. 2016; Sullivan et al. 2014). Despite the attention to additional training, research on preservice teachers participating in a teacher education program that explicitly addresses climate change education showed discrepancies between teacher held beliefs about climate change as a scientific topic and as a curricular topic (Hestness et al. 2011). For example, one preservice teacher asserted that climate change occurs naturally, while simultaneously demonstrating conceptual understanding of human-induced climate change (Hestness et al. 2011). Therefore, it is important to continue to examine teacher perceptions and beliefs about climate change, and in particular, the human causes of climate change.

Third, research suggests a perpetual uncertainty among teachers about age appropriateness of climate change education. In Hestness et al. (2011), preservice teachers repeatedly expressed concerns about climate change being developmentally inappropriate for primary school students. Teachers frequently reported that the subject area was too complex and frightening for students with limited science knowledge (Hestness et al. 2011; Sullivan et al. 2014). This finding is consistent with Borgerding and Dagistan's (2018) report of preservice teachers citing concerns that their students were too immature to engage in robust argumentation around socially divisive topics like climate change.

Lastly, climate change education emphasizes the need for students to recognize local and global interconnections, also referred to as teaching global competence (Longview Foundation 2008; Parkhouse et al. 2016). Hestness et al. (2011) argued:

> When teacher candidates and their students can build their understandings of the local relevance of a socioscientific issue, they can make personal connections and understand the ways they may be personally impacted. But when they can understand its global relevance, they take an even more profound step, in understanding ways their lives and circumstances are interconnected with those of others. (p. 367)

Despite the emphasis on "making local/global connections," literature points out that teachers are often unaware, uncomfortable, and/or underprepared to focus on larger global issues in connection with local communities (Engel and Siczek 2018). In addition, "taking action" is similarly challenging for teachers, particularly in climate change education. Oversby (2015) noted that "many teachers are unprepared for the integration of action and content knowledge that characterizes climate change education, especially those in science where subject knowledge tends to be more factual" (p. 24).

As teachers face these notable barriers in their implementation of climate change education, it becomes important to better understand how, at an individual level, teachers make sense of their experiences while incorporating climate change education into their curriculum (Siczek and Engel 2017). While their knowledge of climate science is vital, so too are the motivations and beliefs that they bring to their own understanding and their teaching of climate change.

Teacher Sensemaking in Climate Change Education

In this chapter, we draw on Siczek and Engel's (2017) framework of teacher sensemaking and global education, built from Spillane et al.'s (2002) seminal work that elucidates how teachers "make sense" of policy changes impacting teacher practices. Spillane et al. (2002) organize teacher sensemaking into three main cognitive constructs: (1) individual cognition, (2) situated cognition, and (3) the role of representation. Siczek and Engel (2017) adapted these constructs for their study of U.S. teachers' sensemaking of global education policy reform initiatives. Figure 24.1 provides an overview of the sensemaking framework.

Individual cognition is connected to personal experiences, beliefs, and knowledge. New information and phenomena are always processed through existing schemata (Spillane et al. 2002; Siczek and Engel 2017). While previous *experiences* are important resources, most of the time, these enduring understandings may also interfere with teachers' interpretations while implementing curriculum changes (Spillane et al. 2002). In this study, individual cognition relates to the ways previous experiences and perceptions shape teachers' current interpretations of digital environmental storytelling as an approach in globally-oriented climate change education. We focus particularly on teachers' self-view and values that are applied to classrooms, as well as how teachers own individual experiences and personal views shape their ideas about students' abilities to cognitively engage with global socio-environmental problems.

Situated cognition refers to teachers' sensemaking in a "particular social context" (Siczek and Engel 2017, p. 6). Analysis of situated cognition considers teachers'

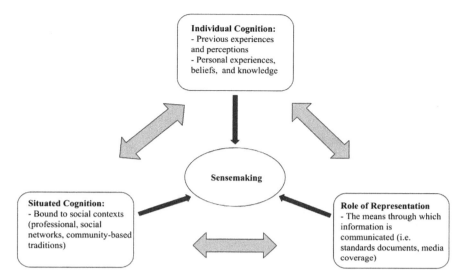

Fig. 24.1 Teacher sensemaking. Adapted from Siczek and Engel (2017)

interpretations within interconnected organizational structures, professional relationships, social networks and community-based traditions (Spillane et al. 2002). We focus specifically on teachers' interpretations as influenced by what Spillane et al. (2002) call "thought communities" in describing various social influencers (religious, political, social group membership etc.) that shape individual perspectives and worldviews (as quoted in Siczek and Engel 2017, p. 6). Therefore, our analysis of teachers' situated cognition attends to teacher values and beliefs in relation to social contexts both within and beyond their school communities. It also helps frame the kinds of expectations that teachers hold about their students' abilities to engage in analysis of global environmental problems.

Within the sensemaking model, the role of representation includes the means by which information is communicated to curriculum implementors (in this case teachers), such as through media coverage, standards documents, professional development materials, and assessment tools (Spillane et al. 2002). Therefore, the role of representation concerns the mechanisms through which specific messages are communicated to teachers charged with local-level implementation. This last frame is useful for better understanding how teachers cognitively make sense of climate change education initiatives, as well as how teachers' use these messages to develop ideas about students' roles in, and capacities for, solving local and global environmental problems.

Together, these three domains (individual cognition, situated cognition, role of representation) provide important insights into how teachers make sense of a climate change education project. In doing so, we provide a better understanding of how teacher motivations, beliefs, and experiences shape the implementation of climate change education.

Digital Environmental Storytelling Exchanges Across the 60th Degree Parallel

Launched in 2017, #60above60 is part of an NSF funded research project focused on producing an index of Arctic urban sustainability. The orientation of #60above60 is toward educational outreach, with a central objective to enhance students' skills necessary for asking and discussing critical questions about both local and global environmental human impacts. Every six months, we organize a cycle of #60above60, in which teachers in Arctic and non-Arctic regions participate in the project. During the project, students in multiple Arctic and non-Arctic locations engage in producing 60-second digital stories of local or place-based environmental problems and solutions, exchanging them with classrooms in cities located above and below the 60th degree parallel. By engaging in the exchange of student produced the aim is for students to become more fully aware of the ecological connections they have to their local environments, as well as to peers in different contexts.

In #60above60, the primary medium is digital storytelling, which is an integrated application of multiple media and software that utilizes the art and techniques of storytelling (Haigh and Hardy 2011). According to Robin (2008), digital storytelling allows students to creatively tell stories through the traditional processes of selecting a topic, conducting research, writing a script, and sharing knowledge, to then combine narrative with multimedia. The student's use of digital technologies is shown to increase student self-confidence. With its place-based environmental focus, it instills in students the capacity and motivation to act on their learning, rather than passively participate (Meadows 2003). Students follow three lines of inquiry in #60above60, creating and exchanging one 60-second digital story for each prompt: (1) What is special about your city?; (2) What is an environmental problem facing your city?; and (3) What is a solution to the environmental problem facing your city? The philosophy underlying this approach is that students are fully entitled to participate in discussions of global proportions and should be encouraged to advocate for sustainable corporate, governmental, and individual practices worldwide. Digital storytelling creates an accessible vehicle for action, as well as cultivates interconnections between multiple contexts.

24 Sixty Seconds Above Sixty Degrees: Connecting Arctic ...

Table 24.1 Teacher participant profiles

Name	Teacher profile
Amanda	Caucasian female at an independent school in Anchorage, AK. Her class is comprised of 14 sixth graders, three of whom are students of color (Alaska Natives). The school's focus is on holistic practices with an emphasis on developmental theory. There is no technology in the classroom other than a single laptop assigned to the teacher
Jennifer	Caucasian female at a public charter school in Washington, DC. Her class consists of 25 seventh graders, the majority of whom are African American or Latinx. The school has a global and STEM focus. Technology is readily available in the classroom; each student is provided a laptop

Research Methods

Our focus in this chapter is on two teachers' experiences and perspectives related to the implementation of #60above60. One teacher is based in Anchorage, Alaska and one in Washington, DC. Teacher data was generated through a series of semi-structured interviews at the beginning and end of January 2018–May 2018, during which both teachers implemented the #60above60 project. Table 24.1 provides cursory profiles of both teachers, their respective school contexts, and classroom demographics. Interviews were conducted via Skype, telephone, and face-to-face. They were digitally audio recorded and transcribed verbatim. Interview transcriptions and coding were conducted by the first and second authors, as well as a graduate research assistant. To analyze interview data, we incorporated both inductive and deductive techniques. We used the sensemaking framework as an a priori coding schema to identify common perspectives within participant interviews. This was followed by a process of open coding to identify phrases or ideas that captured apparent trends at each level of sensemaking, and where different levels of sensemaking appeared to overlap. Having identified themes through initial tentative coding, we reviewed the data again to refine and solidify main themes, developing a set of representative quotes for each. From this process, five main themes were identified that illustrate the ways participants made sense of, and adapted, a digital environmental storytelling project into their existing curriculum. Funding for travel and researcher support was provided by the NSF (Award #1545913) and our research with human subjects was approved by the George Washington University Office of Human Research. Participant's names have been changed to pseudonyms to ensure confidentiality.

Findings

Findings are organized by two main themes: (1) Individual and situated cognition in teachers' sensemaking and (2) the role of representation in teachers' sensemaking. In addition, Table 24.2 provides an overview of salient quotes organized thematically.

Table 24.2 Salient quotations of teacher cognitive constructs in sensemaking

Category	Finding	Salient quotation
Individual cognition	Self-view: ecophobia	"I really don't want the students to have climate change as a big piece of this. I find that when we tell students too young about global issues that are…they feel overwhelmed." (Amanda) [quoting the students] "Why don't you do something about it. You guys messed it all up. Why don't you – they really challenged me." (Jennifer)
	Grounding issues locally	"I think it's hopeful for us to look locally, and then add a case study on another country and then look, like more general worldwide. You know, I found that this helps them to make sense of things." (Jennifer)
Situated cognition	Perception of developmental constraints	"In the primary setting of [the school] up until eighth grade, they really only look at themselves – their education is really centered around themselves. Who they are. And once they reach the age of fourteen, they can take that cognitive leap." (Amanda)
		"That's a huge challenge, just in general socially and emotionally, for them to think about others, to think about their teacher, to think about their friend; it's all about them […]" (Jennifer)
	Perceptions the project within current curricular frame	"The next stage will be me introducing this, "OK what environmental issues do we have?" And it's really interesting for me because they have just completed projects at the end of last year […] the project is that the student looks at something that they try to contribute to their community in some way." (Amanda)
Role of representation	Grounded in authority	"I had one girl who went to carbon dioxide production, and my discussion with her was, this feels too big for me, that it will feel big for you—that you, yes you can do something—how can you in your local environment change? I wouldn't only bring in the carbon dioxide issue because we cannot see carbon dioxide, we can't. That's not something that—it's not like they've been told this issue, but it's very abstract." (Amanda)

Individual Cognition Intertwined with Situated Cognition in the Teachers' Sensemaking

For both teachers, personal experiences played a strong role in shaping interpretation and adaptation of #60above60 before and throughout implementation. Examples of prior experience attended to by teachers during interviews included previous curriculum implementation with current and past students, as well as teachers' own experiences as students.

Amanda

In the case of Amanda, statements containing aspects of individual cognition were closely tied to those tagged as situated cognition. At various times throughout interviews, Amanda used situational statements to support individualistic ones, and vice versa. For example:

> I really don't want the students to have climate change as a big piece of this. I find that when we tell students too young about global issues...they feel overwhelmed. (Amanda)

On the surface, this statement appears born from Amanda's personal observations, assumingly from previous teaching experience. However, then she expands from the "I" in the initial sentence to a broader "we," signifying that it is more than just *her* teaching practice. Amanda goes on to speak about *the practice of teaching* from an institutional or philosophical perspective:

> When we introduce the 'Amazon rainforest is being cut down at the rate of a hundred football fields,' and we are telling grade two students this, I think it's very easy for the students to feel that I can't do anything. That's it. It's done. Can't do anything. (Amanda)

In this quotation, she offers a hypothetical perspective of students' responses to environmental degradation in second grade. Here she reverts back to the first tense. While she is attending to social contexts in adopting a hypothetical student's voice, she may be simultaneously speaking to her individual cognition. As shown in both statements, she expresses particular concerns about hopelessness related to premature abstraction in education. The concept of premature abstraction is derived from the work of Piaget and "based on the assumption that elementary school children are 'concrete thinkers', whose reasoning is tied to objects and their manipulation" (Metz 1995, p. 103). Global climate change, and global issues in general, require abstract thinking (Cutter-Mackenzie et al. 2010; Sobel 1996, 2005, 2008). Therefore, Amanda argues, they are developmentally inappropriate for her sixth-graders. However, she also shares an individualized sense of overwhelm and even apathy in the face of global climate change.

Following #60above60, Amanda similarly conflated her individual perspective with that of her students. While discussing the role of global issues in her students' videos, she reiterated that she believed global issues were too large and abstract for her sixth-graders. She said:

> I had one girl who went to carbon dioxide production, and my discussion with her was, this feels too big for me, that it will feel big for you—that you, yes you can do something, but how can you in your local environment change? I wouldn't bring in the carbon dioxide issue because we cannot see carbon dioxide. It's not like they haven't been told this issue, but it's very abstract. (Amanda)

Here again, it is evident that she gauges the students' cognitive capacity based on her individualized perspective (i.e. it felt "too big"). She simultaneously reinforces her position by arguing that carbon dioxide is too abstract because it is invisible. This is remarkably aligned with a popular misinterpretation of Piaget asserting that prior to middle school students' science investigations should focus solely on that which is observable and concrete (Metz 1995); which, as she explains at various points in both interviews, connects to the school's educational philosophy. Both processes of individual cognition and situated cognition appear to play important roles in shaping Amanda's decisions about #60above60 and are arguably strong mediating factors in the learning environment.

Jennifer

Like Amanda, Jennifer's statements representing individual cognition were often closely associated with those tagged as situated cognition. However, she appears to take a different position on incorporating global learning into her classroom from the individual and situated perspectives Amanda expressed. This difference was most obvious at the start of #60above60. When asked about the role local-global interconnectivity plays in her classroom, she said the following:

> As someone who studies international ed and global citizenship, I am always incorporating global learning into my school. [I am] talking about these local examples and then expanding their views to global examples and then thinking back to 'why does this matter to me, to my community, to the world?' (Jennifer)

Jennifer's situated cognition is at the forefront of her approach to #60above60, where she closely identifies herself with the content of her academic studies, a filter she applies to her expectations of her students' learning and engagement. By the end of #60above60, her ideas about students' capacities have changed. She frames her students as somewhat disengaged in the project, which she attributes to developmental theories, implying that adolescents may not be mature enough to think beyond themselves and their local community:

> That's a huge challenge, just in general socially and emotionally, for them to think about others, to think about their teacher, to think about their friend; it's all about them. (Jennifer)
>
> I think it's helpful for us to look locally, and then add a case study on another country and then look, like more general worldwide. I found that helps them to make sense of things. (Jennifer)

At multiple points during implementation of the project, she acknowledges her personal passion for environmental justice, illustrating the interconnected nature of her situated and individual cognition. Individually, her passion for environmental justice is situated within her role as a science teacher. This is particularly evident when Jennifer explains that she has been inspired to take action in her own community because of her students' "anger" about the planet they are inheriting from generations before them. Speaking as her students she says:

> Why don't you do something about it? You guys messed it all up. Why don't you – they really challenged me and I ended up joining [deleted for confidentiality]. I actually had my colleague here today to look at what I've done since joining that. I was like I need to go, take action and even if I'm not sharing about that every day, I know that I'm being a role model in that way. I'm taking action to do clean ups, and to build a community, to educate others about waste overflow and pollution, urban runoff (Jennifer)

As the interview continues, Jennifer explains how she used the project's conceptual framework and the underlying pedagogy, but adapted deliverables for perceived situational constraints:

> I couldn't just bring the kids outside with the technology and have them take pictures of some stuff. They would be crazy. So, I decided to give it to them for homework like to do all of the preplanning stuff, and actually I had them like write out, 'this is what I want to take a video of, this is something unique.' I had to do all the brainstorming in the classroom. (Jennifer)

The assertion that her students "would be crazy" is grounded in her perception of the school's behavioral expectations, while also based on her previous experience with students outdoors. She has hypothesized that the students would be difficult to control if she allowed them the freedom to take pictures independently during class. Different from the outset of the project, Jennifer creates a new mental and curricular model for achieving the project's intention. Additionally, even though her students may not have completed all of the tasks outlined by the #60above60 project materials, she believes she has achieved the overarching objectives. For example, she describes other aspects of her curriculum that she perceives have achieved the same goals as the #60above60 project for which she is being interviewed:

> It [environmental justice] has come up in all our engineering units too, right? [...] So, it just naturally connected in that way, even though I don't know if I explicitly was doing 60above60. (Jennifer)

Jennifer implies here that she is able to use her situated knowledge to curate materials from different sources to produce the same result. But she still sees herself as a participant, even if her students did not produce digital stories in ways identical to Amanda's school.[1]

---

[1] Jennifer's students created digital responses to a local issue. Rather than films, they co-developed a Google document that included pictures and text about a local watershed.

Role of Representation

The role of representation was included in both teachers' sensemaking, and was often discussed in conjunction with other cognitive constructs in the sensemaking framework. At times, the role of representation was at odds with the situated and/or individual cognition of both teachers. As discussed above, the role of representation emerges alongside Jennifer's situated cognition when she mentions how environmental justice emerged in other engineering units. In our interpretation, this finding indicates that Jennifer's identity as a middle school science teacher is shifting as she draws on new materials and resources beyond her previous experience in international education, reminding us that situated cognition is dynamic. It shifts as people find their footing in new social contexts. In developing this new perspective, Jennifer consults information with authority in science education. Her inclusion of the popular *Engineering is Elementary* curriculum and the ways that she cites the *Next Generation Science Standards (NG)* are examples of this (NGSS Lead States 2013):

> There's eight different engineering and science practices from NGSS, and that's what I'm grading my students on, so that's something that I will need to look for. So, for example ask questions and define problems so for them to be able to do that, for them to be able to plan and carry out an investigation is another one. (Jennifer)

Clearly, Jennifer wants the students to apply the NGSS practices, reflecting the powerful role of representation of the NGSS in her consideration and implementation of #60above60. However, it also appears at odds with her reluctance to allow students to use technology to lead their own investigations, which is urged by the NGSS (NGSS Lead States 2013). Similar to Amanda, there is a predominant idea that the students are not ready to independently investigate a global problem. While talking about her students' capacity for engaging independently with peers from Alaska, Jennifer says:

> In eighth grade, I think they could handle that one. I wouldn't do that with the sixth [graders]. I would maybe do that with the seventh [graders] for next year, so I'll know the kids, but if there was more [structure]. (Jennifer)

The language used by Jennifer is strikingly similar to Amanda's:

> In the primary setting of [the school][2] up until eighth grade, they really only look at themselves – their education is really centered around themselves. Who they are. And once they reach the age of fourteen, they can take that cognitive leap. (Amanda)

As evidenced by these perspectives, both teachers seem to appeal to ideas relating to cognitive and developmental theories, at times even drawing on their teacher training and professional contexts.

---

[2]Deleted for confidentiality.

Discussion

Both of the teacher participants featured in this chapter drew on developmental theories in their experiences when implementing a global digital environmental storytelling project; yet, they drew on these developmental theories for seemingly different reasons. In the case of Amanda, she often focused on students' cognitive capacity, closely aligned to premature abstraction ideas in science education. She also highlighted her concern over what she perceived as potential hopelessness or apathy amongst students confronted too early by complex global environmental issues. Amanda spoke about specific ages when students would take a "cognitive leap," enabling them to think about issues from a global perspective. This view seems consistent with Amanda's professional community where she has worked for over a decade. Because much of Amanda's teacher training is also grounded in the cognitive developmental theories, the authority of her training holds a powerful role of representation and seems to reinforce her situated cognition.

In contrast, Jennifer did not express concerns about premature abstraction during #60above60, and often articulated the importance of making broader global connections for her students. She frequently took what Borgerding and Dagistan (2018) referred to as a "values-based advocate" approach to climate change education. Jennifer's values-based approach appears connected to her individual cognition, especially evident as she described her personal passion for participating in community environmental actions, in part driven by her own travel abroad experiences (Ortloff and Shonia 2015) and her professional community. For example, she describes her own reaction to her students' anger about inheriting a planet on the brink of collapse, with an aim to model the behavior of an environmental activist in response her students. This sentiment is deeply attached to Jennifer's personal ideas about social justice and action, as she explained, "I need to go take action [...] I know that I'm being a role model in that way". Also notable in Jennifer's perspectives was a shift over the course of the project in her thinking about what she could or could not do with her students. By the end of the project, Jennifer frequently cited limitations in students' social emotional development as a barrier to teaching climate change topics, which is consistent with findings from other studies (Hestness et al. 2011; Plutzer et al. 2016; Sullivan et al. 2014).

Interestingly, both Amanda and Jennifer's respective sets of beliefs (which we argue represent their situated cognition) tend to contradict science education research that suggests that students as young as elementary school are capable of sophisticated scientific investigations about complex phenomena that require abstract thinking (e.g., Kuhn 1989; Metz 1995; Russ et al. 2008). Nonetheless, what this suggests is the important role that cognitive and/or developmental stages play in teacher decision-making about whether and how to implement student investigations into global issues, like climate change.

These perceptions, and both teachers' sensemaking during the implementation of #60above60 appeared highly dependent on their school contexts, as well as the values associated with their perception of themselves as teachers within that context. Over the course of her first year in a new role as a middle school science teacher, Jennifer's approach to teaching and learning appeared to shift in tandem with her acclimation to teaching in a standards-dominated educational context. During the course of the project, she held herself increasingly accountable to the NGSS. Amanda, on the other hand, referred frequently to her schools' learning objectives for students at various levels in their development. There was very little change in how she discussed these issues across the life of #60above60, suggesting that her situated cognition is significantly more stable than Jennifer's. This makes sense when considering that Amanda has been working in her current school for over a decade, whereas Jennifer was a first-year science teacher.

In addition, our findings support research that discuss important barriers in climate change education, such as concerns about teaching controversial topics and managing complex global topics (Borgerding and Dagistan 2018; Dunk et al. 2019; Hestness et al. 2011; Oversby 2015; Plutzer et al. 2016; Sadler et al. 2006). Both Amanda and Jennifer voiced concerns about teaching complex global science concepts. For example, Amanda was concerned about student attempts to deal with the topic of CO_2 in her student's digital stories, which she felt was "too big" of a topic. She frequently viewed global environmental problems as cognitively inaccessible to her students. Likewise, Jennifer voiced concerns about whether her students were able to handle the independence needed to develop digital stories during class time. This echoes findings from Borgerding and Dagistan (2018) and Ekborg et al. (2013) who found it is common for teachers to want to provide structure for students during investigations into complex global environmental issues, like climate change.

Conclusion

The increased attention to global education reform initiatives, focused on sustainable development (Smithsonian Science Education Center 2018), suggests a growing need for including human induced causes of global climate change into curriculum reform (NRC 2012; NGSS 2013). Despite the common understanding amongst climate scientists on the causes of climate change, there appears to be a lack of common orientation among teachers for implementing climate change education. Our study illuminates some of the ways teachers make sense of climate change education, through their participation in the #60above60 program in their classrooms. Although we are unable to draw generalizations from our study, given the small number of participating teachers, the deeper dive into individual sensemaking of two teachers implementing climate change education points to several important lessons and implications for further research and practice.

Overall, findings from this study indicate that teachers often draw across cognitive constructs in their sensemaking when implementing global learning. Teachers who are individually drawn to teaching global competence, or who are personally interested in issues of global importance, may be more likely to take up efforts to incorporate climate change education into their classrooms. These findings suggest that individual cognition may be a strong motivating factor for teachers' responsiveness to initiatives related to climate change education. Furthermore, our findings suggest that teachers' estimations of students developmental and cognitive capabilities play a strong role in whether and how they implement curriculum/programs focused on global environmental issues. For teachers to buy into teaching and learning about places and issues far removed, it is essential that they feel students are capable of the task at hand. Although this sounds like a simple statement, it is indicative of a serious challenge facing efforts in climate change education, and one worthy of further research. The need to re-assess the dominant assertion that teaching students about global environmental challenges is beyond their cognitive and developmental capacity of students prior to the age of 15 is an important aspect which needs further investigation. This step will be an important move in addressing the growing need to develop students' skills in ways necessary for studying, documenting, and addressing a growing litany of climate-related issues worldwide.

Acknowledgements The authors wish to thank the GW Arctic PIRE project, specifically Dr. Bob Orttung, Binyu Yang, and members of the research team. This project is supported by the National Science Foundation grant (Award #1545913). Authors would like to thank the teachers participating in #60above60.

References

Anderson A (2012) Climate change education for mitigation and adaptation. J Educ Sustain Dev 6(2):191–206

Borgerding LA, Dagistan M (2018) Preservice science teachers' concerns and approaches for teaching socioscientific and controversial issues. J Sci Teacher Educ 29(4):283–306

Cutter-Mackenzie AN, Payne PG, Reid A (2010) Experiencing environment and place through children's literature. Environ Educ Res 16(3–4):253–264

Doran PT, Zimmerman MK (2009) Examining the scientific consensus on climate change. Eos Trans Am Geophys Union 90(3):22–23

Dunk RD, Barnes ME, Reiss MJ, Alters B, Asghar A, Carter BE, Conter S, Glaze AL, Hawley PH, Jensen JL, Mead LS, Nadelson LS, Nelson CE, Pobiner B, Scott EC, Shtulman A, Sinatra GM, Southerland SA, Walter EM, Brownell SE, Wiles JR (2019) Evolution education is a complex landscape. Nat Ecol Evol. https://doi.org/10.1038/s41559-019-0802-9

Ekborg M, Ottander C, Silfver E, Simon S (2013) Teachers' experience of working with socioscientific issues: a large scale and in-depth study. Res Sci Educ 43(2):599–617

Engel LC, Siczek MM (2018) A cross-national comparison of international strategies: global citizenship and the advancement of national competitiveness. Compare J Comp Int Educ 48(5):749–767

Haigh C, Hardy P (2011) Tell me a story—a conceptual exploration of storytelling in healthcare education. Nurse Educ Today 31(4):408–411

Hestness E, Randy McGinnis J, Riedinger K, Marbach-Ad G (2011) A study of teacher candidates' experiences investigating global climate change within an elementary science methods course. J Sci Teacher Educ 22(4):351–369

Kuhn D (1989) Children and adults as intuitive scientists. Psychol Rev 96(4):674–689. https://doi.org/10.1037/0033-295X.96.4.674

Lombardi D, Sinatra GM (2013) Emotions about teaching about human-induced climate change. Int J Sci Educ 35(1):167–191

Longview Foundation (2008) Teacher preparation for the global age: the imperative for change. Longview Foundation, Silver Spring, MD. Retrieved from http://www.longviewfdn.org/files/44. pdf

Meadows D (2003) Digital storytelling: research-based practice in new media. Vis Commun 2(2):189–193

Metz KE (1995) Reassessment of developmental constraints on children's science instruction. Rev Educ Res 65(2):93–127

Mochizuki Y, Bryan A (2015) Climate change education in the context of education for sustainable development: rationale and principles. J Educ Sustain Dev 9(1):4–26. https://doi.org/10.1177/0973408215569109

National Research Council (2012) A framework for K-12 science education: practices, crosscutting concepts, and core ideas. The National Academies Press, Washington, DC. https://doi.org/10.17226/13165

NGSS Lead States (2013) Next generation science standards: for states, by states. The National Academies Press, Washington, DC

Ortloff DH, Shonia ON (2015) Teacher conceptualizations of global citizenship: global immersion experiences and implications for the empathy/threat dialectic. In: Maguth BM, Hilburn J (eds) The state of global education: learning with the world and its people. Routledge, Taylor and Francis Friends Group, New York

Oversby J (2015) Teachers' learning about climate change education. Procedia Soc Behav Sci 167:23–27. https://doi.org/10.1016/j.sbspro.2014.12.637

Parkhouse H, Tichnor-Wagner A, Cain JM, Glazier J (2016) "You don't have to travel the world": accumulating experiences on the path toward globally competent teaching. Teach Educ 27(3):267–285

Plutzer E, McCaffrey M, Hannah AL, Rosenau J, Berbeco M, Reid AH (2016) Climate confusion among US teachers. Science 351(6274):664–665

Robin BR (2008) Digital storytelling: a powerful technology tool for the 21st century classroom. Theory Pract 47(3):220–228

Russ RS, Scherr RE, Hammer D, Mikeska J (2008) Recognizing mechanistic reasoning in student scientific inquiry: a framework for discourse analysis developed from philosophy of science. Sci Educ 92(3):499–525

Sadler TD, Amirshokoohi A, Kazempour M, Allspaw KM (2006) Socioscience and ethics in science classrooms: teacher perspectives and strategies. J Res Sci Teach 43(4):353–376

Shepardson DP, Niyogi D, Choi S, Charusombat U (2009) Seventh grade students' conceptions of global warming and climate change. Environ Educ Res 15(5):549–570

Shepardson DP, Niyogi D, Choi S, Charusombat U (2011) Students' conceptions about the greenhouse effect, global warming, and climate change. Clim Change 104(3–4):481–507

Siczek M, Engel LC (2017) Teachers' cognitive interpretations of U.S. global education policy. Educ Policy

Smithsonian Science Education Center. Dec 2018. https://ssec.si.edu/global-goals

Sobel D (1996) Beyond ecophobia. Orion Society, Great Barrington, MA

Sobel D (2005) Place-based education: connecting classrooms & communities. The Orion Society, Great Barrington, MA

Sobel D (2008) Childhood and nature: design principles for educators. Stenhouse Publishers, Portland, ME

Spillane JP, Reiser BJ, Reimer T (2002) Policy implementation and cognition: reframing and refocusing implementation research. Rev Educ Res 72(3):387–431

Sullivan SMB, Ledley TS, Lynds SE, Gold AU (2014) Navigating climate science in the classroom: teacher preparation, perceptions and practices. J Geosci Educ 62(4):550–559

Chapter 25
Diving Ecotourism as Climate Change Communicating Means: Greek Diving Instructors' Perceptions

Georgios Maripas-Polymeris, Aristea Kounani, Maria K. Seleventi and Constantina Skanavis

Abstract Climate change and loss of biodiversity are some of the most significant challenges in coastal areas. It is therefore imperative to design effective adaptation and mitigation strategies to achieve ecological sustainability patterns. The Diving Community could become powerful change agents to tackle and mitigate climate change; and thus marine ecosystems' extinction. This research provides an extensive literature review to link the benefits of diving ecotourism. The issue as to whether diving tourism can be a tool to promote sustainability in coastal regions is also investigated. Based on face-to-face interviews with diving instructors who work in Greek Diving Centers, this survey also revealed their perceptions of diving tourism and climate change. The most important findings of the research revealed that Greek diving instructors appeared to have the perceptions that "Diving tourism can be a means of promoting education and communication on climate change" (mean = 4.2),"Diving training promotes responsible environmental behavior" (mean = 4.02), "Diving tourism promotes the conservation of ecosystems in the area" (mean = 4), "Diving tourism promotes sustainability and can therefore promote campaigns to reduce the impact of climate change" (mean = 3.82), and "Diving training can promote strategies to reduce the impact of climate change" (mean = 3.77). Through the findings of the survey it was concluded that diving ecotourism could play a vital role in promoting mitigation and adaptation strategies in coastal areas. Consequently, Greek diving instructors are key players in raising awareness of climate change in coastal areas.

G. Maripas-Polymeris · A. Kounani (✉) · M. K. Seleventi
Department of Environment, University of the Aegean, Mytilene, Greece
e-mail: akounani@yahoo.gr

G. Maripas-Polymeris
e-mail: env12045@env.aegean.gr

M. K. Seleventi
e-mail: maria.seleventi@gmail.com

C. Skanavis
Department of Public and Community Health, University of West Attica, Athens, Greece
e-mail: kskanavi@uniwa.gr

© Springer Nature Switzerland AG 2019
W. Leal Filho and S. L. Hemstock (eds.), *Climate Change and the Role of Education*,
Climate Change Management, https://doi.org/10.1007/978-3-030-32898-6_25

Keywords Climate change communication · Diving ecotourism · Diving instructors · Greece

Introduction

Climate change is a pressing global issue, having a wide range of impacts at different levels within society and industries. Human-induced changes to the environment have started causing major socio-economic issues, capturing attention around the globe. Climate change is currently recognized as an important social and environmental issue; surpassing the issue of world population and resource depletion. Climate is a significant factor in tourism development, and is regarded as an invaluable asset in tourism globally (Siddiqui and Imran 2018). Countries such as Greece, which has a wealth of natural attractions, can observe a decline in tourism as a result of climate change (Liu 2016). Extreme weather events, sea level rise (SLR), floods, droughts, wildfires, infectious disease, etc. can influence tourist activity as well as their safety (Siddiqui and Imran 2018).

Several efforts to reduce climate change impacts (such as conservation societies, ecotourism, and the Paris Summit in 2015) require support to be implemented on a global scale. A promising sector of the tourism industry nowadays is diving tourism. Diving tourism, as one form of ecotourism, is an activity that requires effective management to protect ecological and cultural values and ensure the sustainable use of natural resources (Rangelm et al. 2014).

This research article investigates the role of diving ecotourism and diving instructors on climate change mitigation and adaptation strategies. The main aim of the survey is to explore the linkages between climate change and diving ecotourism, based on a literature search. Another objective is to assess the perceptions of Greek diving instructors on climate change, diving ecotourism, as well as the impact of climate change on coastal regions. Finally, this article looks to evaluate whether diving instructors could promote mitigation and adaptation strategies through their training pathways.

The Impacts of Climate Change on Coastal Regions

Coastal regions are some of the zones most affected by climate change on the planet. Anthropogenic climatic forces are mediated primarily by greenhouse gas emissions (predominantly by CO_2). The combination of elevated CO_2 levels and the resultant increase in global mean temperature will produce a cascade of physical and chemical changes in marine systems (Harley et al. 2006).

Climate change can affect coastal areas in a variety of ways. Coasts are sensitive to sea levels rising, changes in the frequency and intensity of storms, increases in precipitation, and warmer ocean temperatures. In addition, rising atmospheric concentrations of carbon dioxide (CO_2) cause the oceans to absorb more of the gas and become more acidic. This rising acidity can have significant impacts on coastal and marine ecosystems (USEPA 2019). Furthermore, due to the warming of seawater, the global ocean is expanding. Coupled with freshwater input from ice-melt, thermal expansion of the ocean causes the sea level to rise by 2 mm per year (IPCC 2001). Sea level rise could erode and inundate coastal ecosystems and eliminate wetlands (USEPA 2019).

Undeniably, different regions of the world are affected differently by the impacts of climate change (Siddiqui and Imran 2018). These impacts are likely to worsen problems that coastal areas already face. Some existing challenges that are already cause for concern in many areas that also affect man-made infrastructure and coastal ecosystems are shoreline erosion, coastal flooding (USEPA 2019).

Major large-scale events like El Nino can also play an important role in understanding ongoing challenges (Siddiqui and Imran 2018). Furthermore, global warming is responsible for the destruction of the coral reefs and beaches in the tropical coastal areas. As a result, this could have an impact on economies that strongly depend on tourism (Agnew and Viner 2001).

Climate Change and Tourism

The association of tourism and climate change hints at complex interactions, and can be described as a two-way relationship. On the one hand, the tourism industry is a great contributor to climate change (Nicholls 2006) since the emissions from global tourism (including transport, accommodation, and tourism activity subsectors) make up 5% of the total CO_2 emissions. According to Gossling (2002), in 2001 tourism was estimated to be responsible for the consumption of 14,080 PJ of energy, resulting in global emissions of 1400 Mt of CO_2-e. The major contributor is the private automobile, and air transport (Chapman 2007), followed by other forms of transport, and the accommodation subsector (UNWTO 2008). Per unit of energy consumed, air travel has the greatest impact on global warming (Gossling 2002). On the other hand, the tourism industry is affected considerably by the impacts of climate change. Most notably due to its effect on the attractiveness of tourist destinations flows (Amelung et al. 2007; Lise and Tol 2002).

Tourism depends on natural resources, such as water, coastlines, landscapes and biodiversity. These elements influence the potential attractiveness of a destination (Yazdanpanah et al. 2016). Studies show that the selection of the destination, season and length of stay of a tourist is determined by various climate-related factors (Siddiqui and Imran 2018). Climate change threatens the existence of some of the relevant natural resources (Yazdanpanah et al. 2016). Major pressures in terms of the tourism industry aggravate the risk of species extinction, increasing heat waves, decreasing freshwater, increasing accidents due to wildfires, growing health and life insecurities, and rising risks of diseases. Alterations in coastal zones and loss of islands can contribute to the deterioration of the situation in developing countries, due to the limited resources available (Siddiqui and Imran 2018).

Climate Change and Diving Tourism

Climate change, as a significant long-term challenge, will create new risks and opportunities for different segments of the tourism sector (Yazdanpanah et al. 2016). Marine tourism is a major contributor to local economies of both developed and developing countries. Despite its economic significance, marine tourism is based on natural resources that are extremely vulnerable to climate variation, and consequently is influenced by the impacts of climate change. Diving ecotourism (as a form of marine tourism that is based on marine ecosystems) is especially affected by the alterations in climate. Increased monsoonal weather will reduce the number of 'safe diving' days per year. According to Southeast Asia START Regional Center (as cited in Tapsuwan and Rongrongmuang 2015) monsoonal sea level rise will add another 2–3 cm to the already rising level. Sea level rise will lead to vast coastal erosion of beaches. A typical example to this is the case of Koh Tao Island. Climate predictions suggest that the average rainfall for Koh Tao Island is likely to increase by 26% due to the increased frequency of heavier rainfall in the next 30 years (Tapsuwan and Rongrongmuang 2015). The sea level is predicted to rise by 20 cm from 2008, and consequently it is going to cause vast coastal erosion of the island's flat beaches. The impact of climate change could potentially be destructive to the dive tourism industry on Koh Tao Island. Coastal communities are vulnerable to climate change because they are isolated, have small land mass, their population and infrastructure are concentrated on the coastline, and their economy is dependent on natural resources (IPCC 2012; Scott et al. 2012; Tapsuwan and Rongrongmuang 2015).

Historically, communities have adapted to change and improved their resiliency in the process (Scott et al. 2012; Tapsuwan and Rongrongmuang 2015). Climate change is a rising concern for all forms of tourism which requires attention, further study, and demands holistic management steps to reduce its future footprints. There are various ways to make tourism a sustainable industry, and this is in the best interest for those in ecotourism. This industry is also the best suited for this work as ecotourism helps to familiarise people with nature (Siddiqui and Imran 2018). Diving ecotourism, as a form of ecotourism, can be a means to promote sustainability in coastal areas (Kounani et al. 2017).

Climate Change Communication (CCC)

Communication plays a crucial role in constructing notions of climate change and its relationship to the public within a society (Fox and Rau 2017). Climate change communication (developed out of science communication) aims to bring the knowledge obtained through experimentation to the public, so that the findings can be visible and relevant to everyday life (Dulic et al. 2016). Much of the research in climate change communication focuses on the public understanding of climate change, factors that affect public understanding, media coverage and framing, media effects, and risk perceptions (Chadwick 2017). Change is more likely to occur at a local level when community perspectives are embedded in the proposed local solutions and actions. This involves a change process that is clearly linked to local experience and is compelling and interactive (Fox and Rau 2017). Most of the current practices in the communication of climate change have not been empirically evaluated for their effectiveness in raising awareness of climate change risks, and stimulating adaptation action (Wirth et al. 2014; Grothmann et al. 2017).

The Role of Diving Ecotourism in Communicating Climate Change

Ecotourism (an integral part of sustainable tourism) is a form of tourism based on the contact with nature; providing the local communities with opportunities for both financial growth and social progress. Visiting natural areas with a view to being educated, and researching or engaging in eco-friendly pursuits are the key components of this field (Kiper 2013). What is more, it aims to mitigate the adverse effects of tourism, thus supporting environmental conservation in the long term (Skanavis and Kounani 2017). Another element of ecotourism is the close contact with ecology and ecosystems. As well as the creation of up-close environment-related experiences, with the aim of making individuals more environmentally aware (Obenaus 2005).

Over the last decades, scuba diving has played a fundamental role in both marine and coastal tourism (Dimmock and Musa 2015). This sport has also emerged as a profitable business, as over a million divers are certified every year, according to the Professional Association of Diving Instructors statistics (PADI 2019). As a result of this expansion, scuba diving has attracted many investors, resulting in the multiplication of scuba diving centers, schools, resorts, equipment shops, and charter businesses (Townsend 2008; Lucrezi et al. 2017). However, despite the economic benefits derived from this industry, there are some environmental hazards that cannot be overlooked since diving, if not properly performed, can pose a great risk on the marine ecosystems (Musa et al. 2011; Lucrezi et al. 2013). These dangers increase proportionately with the number of certified divers. Climate change profoundly affects dive-tourism as it is heavily dependent on the condition and beauty of coral reefs (SDL 2017).

Changes in divers' behavior and the way they perform their diving activities can significantly contribute to the prevention of such damage to the health of the marine ecosystem (Fatt Ong and Musa 2011). Undoubtedly, diving can be a pillar of environmental conservation, due to its educational character, its close relation to the improvement of marine life and its strong interest in nature (Dearden et al. 2007; Skanavis and Kounani 2017).

The Role of Diving Instructors in Promoting the Education and Communication of Climate Change

Over the past twenty years, Climate Change and Environmental Education and Education for Sustainable Development, has taken the lead in encouraging environmental protection and sustainable development (UNICEF 2013). The contribution of environmental educators to raising awareness about environmental issues is significant.

Diving instructors, through the alternative form of environmental education they offer, should focus on discovering methods which affect tourists' attitude towards the environment. Importantly, they should not only focus on training new divers, but they should train environmentally conscious divers (Kounani et al. 2017; Cook 2019).

Once they realize the extent to which they affect the environment, divers will understand that it is their responsibility and part of their role to protect marine ecosystems. Through their passion, they can make the difference and alleviate the problems plaguing marine ecosystems (Musa et al. 2011; Mowery 2017).

The Case of Greece

The Impacts of Climate Change in Greece

According to the estimation of the Research and Policy Institute "diaNEOsis", the temperature in Greece is going to increase by 2.5 °C by 2050 (compared to 1961–1990); an increase attributed to climate change. Heat waves are going to become more frequent, a rise of 15–20 annually has been predicted and we are expected to experience more than 50 "tropic days" annually. At the same time, a 12% decline in the average rainfall has been forecast.

The combination of the expected rise in temperature and the drop in rainfall will dry forest areas out, leaving them less protected against fires. A 15–70% increase in the number of high risk days each year, and an extension of 2–6 weeks a year of the high risk season, has been forecast, phenomena that will seriously impact soil fertility (Dianeosis 2017).

A considerable rise in the sea level will take its toll on Greece, since a great number of the country's residents live in close proximity to the coast and the majority of its resorts have been constructed by the sea. According to the IPCC, tourism is one of the main sources of income for Greece (Bank of Greece 2014), Given that, a rise in the sea level will not only cause a significant loss of the country's land but also as a consequence a 2% reduction in the country's GDP (Sauter el al. 2013).

The tourism industry is a vital economic sector and occupies a dominant position in the Greek economy. As the research has revealed, the continental tourist areas of the Greek mainland will face heatwave episodes more frequently. In most of the studied regions, 5–15 more days that exceed 35 °C will occur every year. The largest increases are found in 'summer' days (>25 °C) and 'tropical' nights (>20 °C). The former may be considered a positive impact as it may prolong the tourist period. The number of 'tropical' nights per year seems to substantially increase especially in the islands. For example, Rhodes and Chania expect a sum of 40 additional 'tropical' nights. In coastal regions, such warm conditions in combination with high levels of relative humidity can result in uncomfortable conditions for foreigners and the local people (Giannakopoulos et al. 2009).

Scuba Diving in Greece

With a length of more than 15,000 km, the Greek coastline is the longest among the Mediterranean countries. The country also boasts a considerable number of islands and rocky islets, 3,000 in total, which spread all over its archipelago (Scalkos et al. 2009).

Due to its magnificent undersea sites, clear seas, rich marine life, mysteriously sunken ancient towns, well-known shipwrecks as well as its unique seabed, Greece is a mecca for aspiring divers. The varied attractions transform scuba diving from a simple leisure activity into a profitable business in Greece (Alternative Greece 2019). A further advantage of Greek seas is the high temperature and good visibility it offers; tanging from 6 to 50 m, depending on the season (Scubandros 2019).

Greece aims to attract divers with expendable income, particularly during the economic crisis which the country is currently going through, as scuba diving is a significant source of revenue (Skanavis and Kounani 2017).

With the reformation of the related law, diving is now allowed everywhere throughout the Greek sea. This is a great improvement for divers who were previously only able to dive in 620 miles of the 10,000 miles of Greek coastline (Skanavis and Kounani 2017). According to Galanopoulos (2012), in 2009 it was estimated that 186 diving centers certified by the Hellenic Ministry of Shipping were in operation. Additionally, 5 training organizations for diving instructors were certified all around Greece, boosting tourist development in their wider area.

Methodology

In May of 2018 a questionnaire-based survey supplemented with semi-structured, face-to-face interviews was administered to 36 diving instructors from diving training

Fig. 25.1 Location of diving training centers

centers throughout Greece. Figure 25.1 indicates the location of the diving centers. Each interview lasted about 1 h. During the interviews, all interviewees were asked the same list of questions in the same order. The questions, 29 in total, were based on a bibliographic review of similar research projects adjusted to the peculiarities of the specific target group. The first 8 questions supplied data about the demographics (e.g. gender, age, educational status, marital status, region origin etc.). The next 21 were developed to gauge their knowledge and perceptions on general environmental issues, scuba diving and climate change. The collected data was analyzed using Microsoft Excel 2016.

Results

In order to receive information about the sample, the authors included 8 separate questions in the questionnaire; the results of these questions are presented in a similar way. The demographics' results, characterized the sample. The second set of questions, provided information on the perceptions of Greek diving instructors towards the climate change issue, diving ecotourism, the impact of climate change on coastal regions, and their views towards the implementation of mitigation and adaptation strategies through their training pathways.

Demographics

Figures 25.2 and 25.3 illustrate the gender and age groups of respondents. As shown, the vast majority of respondents (83%) were males and 64% were over 36 years old.

The marital status of the interviewees is shown in Fig. 25.4. Overall, 36% of all respondents reported they were married at the time of the survey, 47% were single and 17% were divorced. 61% had no children, 22% had one child and 11% had two children. A small percentage (6%) reported having three children (Fig. 25.5).

Figure 25.6 indicates the respondents' place of origin. Half of the respondents (50%) are from another Greek seaside area, while 33% are from the wider area surrounding the diving centre.

With regards to the educational level of respondents, Fig. 25.7 illustrates that 64% were university graduates and a mere 3% were Ph.D. holders.

Fig. 25.2 Gender of the respondents

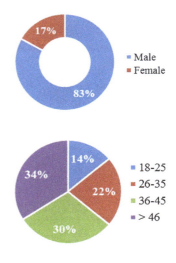

Fig. 25.3 Age groups of the respondents

Fig. 25.4 Marital status of the respondents

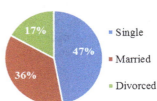

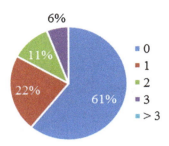

Fig. 25.5 Number of children of the respondents

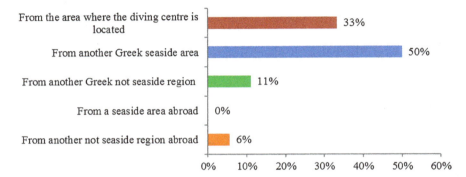

Fig. 25.6 Respondents' place of origin

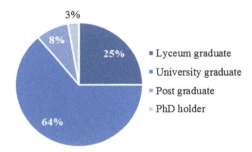

Fig. 25.7 Respondents' educational level

In addition to their studies, 75% of the respondents had received environmental education in their training as diving instructors. Figure 25.8 illustrates the reasons behind respondents' choice to practice diving as a profession. As shown below, the majority of respondents (75%) made the choice because of their love for the ocean and marine life and for the environment in general.

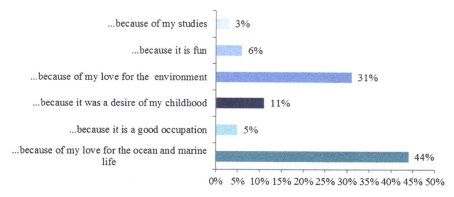

Fig. 25.8 The reason behind the choice of participants to practice diving as a profession

Knowledge and Perceptions Towards General Environmental Issues, Scuba Diving and Climate Change

Concerning the perceptions of the respondents towards the state of the environment, 72% of the participants believed that it is in poor condition, but with a lot of effort, it will possibly change for the better, while a small percentage believed that it is in good condition (Fig. 25.9).

Meanwhile, the vast majority (97%) considered themselves environmentally conscious. Additionally, participants were asked to express their opinion on the most serious environmental challenges our planet is confronted with, on a scale of 1–5, with 1 being 'not serious' and 5 being 'extremely serious'. Table 25.1 shows the results and indicates that "water pollution' (mean = 4.48) and "deforestation" (mean = 4.22) are the most serious challenges.

On a scale of 1 to 5 (1 = not serious, 5 = extremely serious), the most serious problems associated with coastal areas was expressed to be "water pollution (from

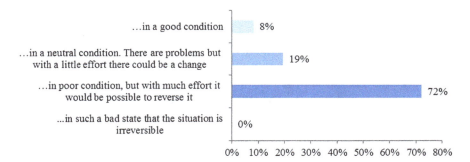

Fig. 25.9 Respondents' perceptions toward the state of the environment

Table 25.1 Participants' beliefs about the most serious environmental challenges our planet is facing [on a scale of 1 to 5 (1 = not serious, 5 = extremely serious)]

	Mean	Min	Max	St.dev
Water pollution	4.48	2	5	0.81
Deforestation	4.22	1	5	0.93
Threatened species	4.08	2	5	0.97
Air pollution	4.05	2	5	0.9
The increase of solid waste production	3.97	1	5	1.06
The depletion of ozone layer	3.77	2	5	0.92

Table 25.2 Participants' beliefs about the most serious environmental problems coastal areas are facing [on a scale of 1–5 (1 = not serious, 5 = extremely serious)]

	Mean	Min	Max	St.dev
Water pollution (from ships, urban and industrial waste)	4.48	2	5	0.77
Overexploitation of the coastal seabed	4.37	2	5	0.87
Uncontrolled residential development of coastal areas	3.85	2	5	0.94
Extreme weather events	3.65	1	5	0.95
The coastal zone's watering	3.57	2	5	0.99
Sea level rise	3.54	1	5	1.09
Coastal erosion	3.31	2	5	0.89

ships, urban and industrial waste)" (mean = 4.48) and "overexploitation of the coastal seabed" (mean = 4.37) (Table 25.2).

Subsequently, the participants were evaluated on their perceptions towards diving tourism. Firstly, they were asked whether they considered "diving tourism as a form of ecotourism", with the majority of the sample (85%) responding positively. Participants expressed strong agreement towards diving tourism supporting conservation, economic activities and environmental education. Moderate views were expressed for diving tourism supporting local communities, being more a recreational activity than a study of the natural environment, being a source of revenue for environmental protection and being a reason for declining biodiversity. Environmental degradation associated with diving activities was the least agreed on item (Table 25.3).

An evaluation of the participants' knowledge and perceptions of climate change indicate that the vast majority (97%) of the diving instructors answered that they were aware of what climate change means. Moreover, the strongest perception expressed by participants was that anthropogenic activity is the crucial factor associated with climate change (Figs. 25.10 and 25.11).

Subsequently, participants were asked to express their opinion on the state of climate change by choosing among a set of offered alternatives. Half of the respondents believe that "Earth's climate is in a delicate balance. A small rise in global warming

Table 25.3 Participants' perceptions towards diving tourism [on a scale of 1 (strongly disagree) to 5 (strongly agree), they agree with the following statements]

	Mean	Min	Max	St.dev
Diving tourism promotes the conservation of ecosystems in the area	4	1	5	0.94
Diving tourism is an economic activity that supports the local economy, since it brings great economic benefits to the region	3.97	2	5	0.9
Diving is associated with the cleanliness of the seabed in the area	3.97	1	5	1.04
Diving tourism promotes environmental education	3.91	1	5	0.98
Diving tourism is generally supported by the local community and provides jobs to local residents	3.4	2	5	0.99
Diving tourism is an activity most chosen for recreation and less for contact with the natural environment	3.05	1	5	1.06
Revenues from diving tourism benefit from environmental protection	2.94	1	5	1.11
Diving tourism causes decline to the biodiversity of an area if it is uncontrolled	2.71	1	5	1.18
Diving is an activity that causes environmental degradation in the area	1.82	1	5	1.13

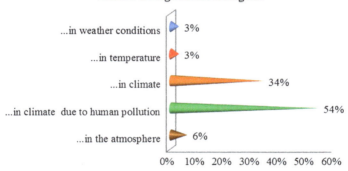

Fig. 25.10 Perceptions towards the meaning of climate change

will have devastating effects", while 25% of them believe that "The climate of the Earth is changing slowly. Global warming will gradually lead to dangerous impacts".

Overall, 86% of the diving instructors believe that Greece is particularly vulnerable to the effects of climate change. More specifically, they were asked to express their views according to some comments related to the effects of climate change on Greece, using a scale from 1 (totally disagree) to 5 (totally agree). As can be seen in Table 25.4, the answer with the highest mean was "Rising temperatures of seas, lakes and rivers resulting in the destruction of aquatic ecosystems" (Mean = 3.65).

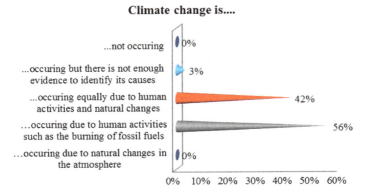

Fig. 25.11 Respondents' opinions towards CC

Table 25.4 Participants' perceptions towards the effects of climate change in Greece [on a scale of 1 (totally disagree) to 5 (totally agree)]

	Mean	Min	Max	St.dev
Rising temperatures of seas, lakes and rivers resulting in the destruction of aquatic ecosystems	3.65	1	5	1.19
Immigration of species	3.62	1	5	1.14
Rising temperature	3.57	2	5	0.87
Acidification of the oceans	3.57	1	5	1.02
Greater floods	3.51	1	5	1.02
Frequent droughts	3.42	1	5	1.1
Increase sea level	3.42	1	5	0.99
Coastal erosion	3.4	1	5	1.08
Extreme weather events	3.2	1	5	1.09

After that, they were asked to evaluate a range of options that could help mitigate and/or reduce the effects of climate change, using a scale from 1 (totally disagree) to 5 (totally agree). "Reduction of electricity consumption or use of renewable sources" and "Education and awareness programs for climate change" are the options with the highest mean (Mean = 4.37). (Table 25.5)

The vast majority of the sample (94%) believed that it is possible for residents to become aware of climate change through diving tourism. More specifically, the diving instructors were asked to express their opinion about the ways environmental awareness can be achieved, by choosing among a set of offered alternatives (Fig. 25.12). Almost half of the respondents (47%) believed that environmental awareness can be achieved through providing environmental education during the diving training process.

Participants were also asked to express their views according to some comments related to the role of diving tourism on climate change education and communication,

Table 25.5 Participants' opinions towards the ways in which the effects of climate change can be mitigated/reduced [on a scale of 1 (totally disagree) to 5 (totally agree)]

	Mean	Min	Max	St.dev
Reduction of electricity consumption or use of renewable sources	4.37	2	5	0.72
Education and awareness programs for climate change	4.37	1	5	0.87
Creating campaigns by the media	4.22	2	5	0.87
More frequent use of mass media	3.74	1	5	1.22
Reduction of products consumption	3.14	1	5	1.21
Change in diet (less meat consumption, etc.)	3.02	1	5	1.14

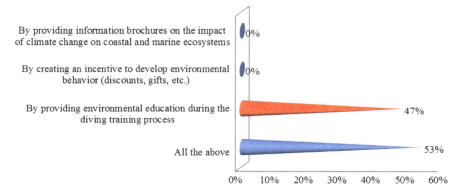

Fig. 25.12 Participants' opinions on "how environmental awareness can be achieved"

using a scale from 1 (totally disagree) to 5 (totally agree). As can be seen in Table 25.6, the answers that were presented which had the highest mean were "Diving tourism can be a means of promoting education and communication on climate change" (mean = 4.2) and "Diving training promotes responsible environmental behavior" (mean = 4.02).

Moreover, 81% of the respondents stated that they had already used their training as a means of promoting climate change communication and education. Finally, the vast majority of the diving instructors (89%) had tried to raise trainees' consciousness towards climate change issue during their training.

Discussion

In terms of the demographics, it is noteworthy that the vast majority of the participants were men (Fig. 25.2). According to PADI's statistics, this is a general representation worlds divers. Diving has always been a male-dominated activity. For many years

Table 25.6 Participants' perceptions towards the role of diving tourism on climate change education and communication [on a scale of 1 (totally disagree) to 5 (totally agree)]

	Mean	M in	Max	St.dev
Diving tourism can be a means of promoting education and communication on climate change	4.2	2	5	0.89
Diving training promotes responsible environmental behavior	4.02	1	5	1.06
The trainees' contact with the environment through diving training increases their environmental awareness	4.02	1	5	0.97
Diving tourism promotes sustainability and can therefore promote campaigns to reduce the impact of climate change	3.82	2	5	0.79
Diving training can promote strategies to reduce the impact of climate change	3.77	1	5	0.92
Diving training promotes communication of climate change	3.65	1	5	0.79

the gender ratio of 65% male versus 35% female remained constant. Nowadays, the ratio of female diving instructors worldwide is almost 40% (PADI 2017).

According to Skanavis and Kounani (2017), Greek diving instructors are individuals with a rich environmental background; they are environmentally conscious and love the ocean and diving ecotourism. Similarly, Greek diving instructors that participated in this survey are individuals that have received environmental education in their training as diving instructors; and appeared to have high levels of environmental consciousness. Furthermore, as Fig. 25.8 shows, for the most part they have chosen their profession due to their love for the ocean, marine life, and the natural environment in general. This is most likely due to the fact that the majority of the participants come from seaside areas (Fig. 25.6).

Regarding the participants' beliefs of what the most serious environmental challenges to our planet are they stated, "water pollution" and "deforestation" to be the most pressing (Table 25.1). Concerning their perceptions towards the most serious problems coastal areas are being confronted with, according to Table 25.2, "water pollution" and "overexploitation of the seabed" are the most important ones. The "sea level rise" and "coastal erosion" were considered to be less serious problems.

Diving ecotourism is considered to be a vital source of employment, as it contributes significantly to local incomes (Cinner 2014). However, it poses numerous challenges as well. High rates of visitation to coastal areas lead to an increase in the pressure to marine ecosystems. These pressures are compounded with the large numbers of often inexperienced and poorly trained divers (Wabnitz et al. 2018). According to Table 25.3, the participants in this study had a similar belief, as they said that "diving ecotourism is an economic activity that supports the local economy, since it brings great economic benefits to the region". In contrast to Wabnitz et al. (2018), the results of this study revealed that they considered diving ecotourism (as a form of ecotourism) to "promote the conservation of ecosystems in the area" and to "associated with the cleanliness of the seabed in the area". It is worth noting that the comment "diving is an activity that causes environmental degradation in the area" had the lowest mean (Table 25.3).

Regarding diving instructors' viewpoints towards climate change, research conducted in Thailand revealed that the diving instructors believed that human activities (such as the burning of fossil fuels) provokes climate change (Tapsuwan and Rongrongmuang 2015). Similarly, Fig. 25.11 shows that over half of Greek diving instructors (56%) believed that climate change is "mainly due to human activities such as the burning of fossil fuels" while 42% believed that "both human activities and natural changes contribute to climate change". A great portion of the 86% of the diving instructors believed that "Greece, as a country that has many coastal regions, is particularly vulnerable to the effects of climate change".

As Wabnitz et al. (2018) stated, increases in sea surface temperatures of between 1 and 3 °C by 2100 are projected to significantly impact ecosystems in the tropical. In a similar way, this research revealed that the respondents considered that "rising temperatures of seas, lakes and rivers result in the destruction of aquatic ecosystems (Table 25.4).

With regard to improving the effectiveness of climate change education and communication programs, increasing public participation in climate change policy making and enhancing the adaptive capacity of the dive tourism industry are broad enough to apply to many dive tourism industries around the world (Tapsuwan and Rongrongmuang 2015). Likewise, the vast majority of the participants of this survey (94%) believed that it is possible for locals to become more aware of climate change through diving tourism. Specifically, almost half of the respondents (47%) believed that "environmental awareness can be achieved through providing environmental education during the diving training process" (Fig. 25.12). Furthermore, as Table 25.6 shows, participants agreed that "the trainees' contact with the environment through diving training could increase above all their environmental awareness". Also, they supported the notion that "diving tourism could be a means to implementing sustainability and can therefore promote campaigns to reduce the impact of climate change". Consequently, they claimed that diving training can promote strategies to reduce the impact of climate change, and thus promote communication of climate change (Table 25.6).

According to Ballantyne and Parker (2011), diving instructors, having the essential role of ocean ambassadors, can increase the public's knowledge of environmental issues. While at the same time, they concentrate on ways in which the tourism sector and leisure experiences can deliberately and positively influence the environmental behavior of visitors; as well as increase environmental consciousness. In a similar way, Greek instructors can act as conservators of the ocean and promote climate change mitigation strategies. This is supported by the results of this study as 81% of the respondents stated that they had already used their training as a means of promoting climate change communication and education and 89% responded that they "have tried to raise trainees' consciousness towards climate change issues during their training".

Conclusions

To combat climate change, various essential steps must be taken worldwide. The Paris Agreement, in 2015, recognized the need to limit the impacts of climate change through different approaches that aim to decrease the global average temperature. Diving ecotourism could be a key tool in communication strategies for raising awareness of climate change in coastal areas, as it is directly influenced negatively by the impacts of climate change. Consequently, diving instructors play a crucial role to play in the implementation of such strategies. Diving tourism cannot achieve sustainability goals without the support of local communities, academic institutions, the Managing Authorities, non-governmental organizations, or without the diving instructors' contribution (Dimmock and Musa 2015). The outcomes of this research revealed that diving instructors in Greek diving centers appeared to have received adequate environmental education in order to meet the requirements needed to promote climate change mitigation and adaptation strategies. As diving ecotourism could play a vital role in promoting sustainability in coastal areas, instructors are integral in the implementation of mitigation and adaptation strategies in Greece. Effective interpretational and educational programs can have a 'transformative' influence which gives participants not only a deeper understanding of the attraction itself, but also a consequent adherence to a more environmental ethos.

References

Agnew MD, Viner D (2001) Potential impacts of climate change on international tourism. Tour Hosp Res 3:37–60

Alternative Greece (2019) Alternative tourism greece: scuba diving greece—diving centers—schools. Retrieved from https://www.alternativegreece.gr/WebForms/CategoryDisplay.aspx?ID=91

Amelung B, Nicholls S, Viner D (2007) Implications of global climate change for tourism flows and seasonality. J Travel Res 45(3):285–296. https://doi.org/10.1177/0047287506295937

Ballantyne R, Packer J (2011) Using tourism free-choice learning experiences to promote environmentally sustainable behaviour: the role of post-visit 'action resources'. Environ Edu Res 17(2):201–215. https://doi.org/10.1080/13504622.2010.530645

Bank of Greece (2014) Greek tourism and climate change: policies adaptation and new development strategy. Retrieved from https://www.bankofgreece.gr/BogEkdoseis/CCISC_Tourism%20and%20climate%20change_Fw%20Ch1.pdf

Chadwick A (2017) Climate change communication. Oxford research Encyclopedia of communication, Oxford University Press, USA, pp 1–29. https://doi.org/10.1093/acrefore/9780190228613.013.22

Chapman L (2007) Transport and climate change: a review. J Transp Geogr 15(5):354–367. https://doi.org/10.1016/j.jtrangeo.2006.11.008

Cinner J (2014) Coral reef livelihoods. Curr Opin Environ Sustain 7:65–71. https://doi.org/10.1016/j.cosust.2013.11.025

Cook N (n.d.) (2019) The importance of marine conservation as a dive professional. Retrieved from https://goo.gl/8KuaNu

Dianeosis (2017) The impact of climate change on the greek economy. Written by Georgakopoulos T. Retrieved from https://goo.gl/E6ZZki

Dimmock K, Musa G (2015) Scuba diving tourism system: a framework for collaborative management and sustainability. Mar Policy 54:52–58

Dearden P, Bennett M, Rollins R (2007) Perceptions of diving impacts and implications for reef conservation. Coast Manag 35(2–3):305–317. https://doi.org/10.1080/08920750601169584

Dulic A, Angel J, Sheppard S (2016) Designing futures: inquiry in climate change communication. Futures 81:54–67

Fatt Ong T, Musa G (2011) An examination of recreational divers' underwater behavior by attitude behaviour theories. Curr Issues Tour 14(8):779–795. https://doi.org/10.1080/13683500.2010.545370

Fox E, Rau H (2017) Disengaging citizens? climate change communication and public receptivity. J Irish Polit Stud 32(2):224–246. https://doi.org/10.1080/07907184.2017.1301434

Galanopoulos G (2012) Diving tourism in Greece. ATEI of Crete, School of Management and Economics, Heraklion

Giannakopoulos C, Kostopoulou E, Varotsos K, Plitharas A (2009) Climate change impacts in Greece in the near future. A report of national observatory of Athens, Greece

Grothmann T, Leitner M, Glas N, Prutsch A (2017) A five-steps methodology to design communication formats that can contribute to behavior change: the example of communication for health-protective behavior among elderly during heat waves. SAGE Open, pp 1–15. https://doi.org/10.1177/2158244017692014

Gossling S (2002) Global environmental consequences of tourism. Glob Environ Change 12:283–302

Harley CDG, Hughes AR, Hultgren KM, Miner BG, Sorte CJB, Thornber CS, Williams S (2006) The impacts of climate change in coastal marine systems. Ecol Lett 9(2):228–241. https://doi.org/10.1111/j.1461-0248.2005.00871.x

IPCC (Intergovernmental Panel on Climate Change) (2001) Climate change 2001, synthesis report. A contribution of working groups I, II, and III to the third assessment report of the intergovernmental panel on climate change. Cambridge University Press, Cambridge, UK

IPCC (Intergovernmental Panel on Climate Change) (2012) Managing the risks of extreme events and disasters to advance, climate change adaptation special report of the intergovernmental panel on climate change. In: Field B, Barros V, Stocker TF, Qin D, Dokken DJ, Ebi KL, Mastrandrea MD, Mach KJ, Plattner G-K, Allen SK, Tignor M, Midgley PM (eds). Cambridge University Press, Cambridge, UK and NewYork, NY

Kiper T (2013) Role of ecotourism in sustainable development. In: Ozyavuz M (ed) Advances in landscape architecture. IntechOpen, London. https://doi.org/10.5772/55749

Kounani A, Skanavis C, Koukoulis A, Polymeris-Maripas G (2017) Diving tourism as a tool for the promotion of the sustainable management of coastal areas: the role of the scuba diving instructors, the 7th Pan-Hellenic conference on management and improvement of coastal zones, Athens, pp 89–100, 20–22 Nov (in Greek)

Lise W, Tol RS (2002) Impact of climate on tourist demand. Clim Change 55(4):429–449

Liu TM (2016) The influence of climate change on tourism demand in Taiwan national parks. Tour Manag Perspect 20:269–275. https://doi.org/10.1016/j.tmp.2016.10.006

Lucrezi S, Saayman M, Merwe P (2013) Managing diving impacts on reef ecosystems: analysis of putative influences of motivations, marine life preferences and experience on divers' environmental perceptions. Ocean Coast Manag 76:52–63. https://doi.org/10.1016/j.ocecoaman.2013.02.020

Lucrezi S, Milanese M, Markantonatou V, Cerrano C, Sara A, Palma M, Saayman M (2017) Scuba diving tourism systems and sustainability: Perceptions by the scuba diving industry in two Marine Protected Areas. Tour Manag 59:385–403. https://doi.org/10.1016/j.tourman.2016.09.004

Mowery L (2017) Will the sport of Scuba Diving end by 2050? Retrieved from https://goo.gl/S2vDEd

Musa G, Seng WT, Thirumoorthi T, Abessi M (2011) The influence of scuba divers' personality, experience, and demographic profile on their underwater behavior. Tour Mar Environ 7(1):1–14

Nicholls S (2006) Climate change, tourism and outdoor recreation in Europe. Manag Leisure 11(3):151–163. https://doi.org/10.1080/13606710600715226

Obenaus S (2005) Ecotourism—sustainable tourism in national parks and protected areas. Banff National Park in Canada and National park Gesause in Austria—a Comparison. Diplomarbeit an der Universität Wien. Wien, 171 S, p 171. Retrieved from https://www.nationalpark.co.at/images/Forschung/2260/Dokumente//Obenaus_2005_Ecotourism_Sustainable_Tourism_in_National_Parks_and_Protected_Areas.pdf

PADI (Professional Association of Diving Instructors statistics (2017) The business of women in diving. Retrieved from: https://padiprosamericas.com/2017/06/14/the-business-of-women-in-diving/

PADI (Professional Association of Diving Instructors statistics) (2019) Worldwide corporate statistics. Retrieved 18 Feb, Retrieved from https://goo.gl/TLJATm

Rangelm MO, Pita CB, Goncalves JMS (2014) Developing self-guided scuba dive routes in the Algarve (Portugal) and analyzing visitors' perceptions. Mar Policy 45:194–203

Sauter R, ten Brink P, Withana S, Mazza L, Pondichie F with contributions from Clinton J, Lopes A, Bego K (2013) Impacts of climate change on all European islands, a report by the Institute for European Environmental Policy (IEEP) for the Greens/EFA of the European Parliament. Final Report. Brussels

Scalkos G, Strigas A, Moudakis C, Stergioulas A (2009) Mapping of the content state of diving tourism in Greece. J Appl Sci 9(21):3829–3835

Scott D, Simpsons MC, Sim R (2012) The vulnerability of Caribbean coastal tourism to scenarios of climate change related sea level rise. J Sustain Tour 20:883–898. https://doi.org/10.1080/09669582.2012.699063

SDL (Scuba Dive Life) (2017) Making sustainable tourism the standard in the dive industry. Retrieved from https://scubadiverlife.com/making-sustainable-tourism-standard-dive-industry/

Scubandros (2019) Scuba diving in Greece. Retrieved from https://scuba-andros.gr/scuba-diving-in-greece/

Siddiqui S, Imran M (2018) Impact of climate change on tourism. In: Sharma R, Rao P (eds) Environmental impact of tourism in developing nations. IGI Global, Herseey PA, USA, pp 68–84. https://doi.org/10.4018/978-1-5225-5843-9.ch004

Skanavis C, Kounani A (2017) Diving ecotourism as an educating tool for encouraging sustainable development: the case of Skyros Island, innovation Arabia 10: health and environment conference, Dubai, pp 188–197, 6–8 Mar 2017

Tapsuwan S, Rongrongmuang R (2015) Climate change perception of the dive tourism industry in Koh Tao Island, Thailand. J Outdoor Recreat Tour 11:58–63

Townsend C (2008) Dive tourism, sustainable tourism and social responsibility: a growing agenda. In: Garrod B, Gössling S (eds) New frontiers in marine tourism, pp 139–152. Elsevier. https://doi.org/10.1016/b978-0-08-045357-6.50010-3

UNICEF (2013) Climate change and environmental education—a companion to the child friendly schools manual. Retrieved from https://goo.gl/u27U4J

UNWTO (United Nations World Tourism Organization) (2008) Climate change and tourism-responding to global challenges. Madrid, Spain. Retrieved from https://sdt.unwto.org/sites/all/files/docpdf/climate2008.pdf

USEPA (United Stated Environmental Protected Agency) (2019) Climate impacts on coastal areas. Retrieved from https://19january2017snapshot.epa.gov/climate-impacts/climate-impacts-coastal-areas_.html

Wabnitz CCC, Cisneros-Montemayor AM, Hanich Q, Ota Y (2018) Ecotourism, climate change and reef fish consumption in Palau: benefits, trade-offs and adaptation strategies. Mar Policy 88:323–332

Wirth V, Prutsch A, Grothmann T (2014) Communicating climate change adaptation—state of the art and lessons learned from ten OECD countries. GAIA 23(1):30–39

Yazdanpanah H, Barghi H, Esmaili A (2016) Effect of climate change impact on tourism: a study on climate comfort of Zayandehroud River route from 2014 to 2039. Tour Manag Perspect 17:82–88

Chapter 26
A Model to Integrate University Education Within Cultural Traditions for Climate Change Resilience

Keith Morrison

Abstract A community-based learning-system model is outlined. The model has been constructed to frame university education and research as interacting with cultural traditions, so as to understand how to better use university education to enhance climate change adaptation within the South Pacific region. The model features the learning systems involved, including the functions and purpose of the components of the learning systems. The function and purpose of education and research in the context of climate change adaptation is shown to be the critical development of innovations, so as to enhance resilience through maintaining and gaining flexibility within social-ecological systems. The adaptations can be technological or institutional, including policy development, development of institutional arrangements, and development of cultural traditions. But because of the relevance of cultural traditions to the communities of the South Pacific region, the model focuses on adaptations that involve cultural traditions interacting with education that universities can provide. The role of cultural traditions of the South Pacific region for climate change adaptation is shown to be important, and even of global significance, because South Pacific nations are at the forefront of climate change adaptation. Living, and hence developing, cultural traditions provide high adaptive capacity through facilitating the questioning of the goals and assumptions of development processes. In particular they facilitate clarification of what is of highest importance and in needing of being maintained with the highest priority. This enables non-traumatic and hence civil adaptation to climate change to proceed, through maintaining what is most valuable, so as to avoid the arising of trauma; with adaptation only changing what is of lesser importance. Flexibility is maintained by having multiple non-traumatic options to choose from. The model explores how clarification of what is of highest importance enables the 'letting-go' of fixation on any particular views of development; views of development that may not actually be those of the communities. The model also clarifies how letting go of ideological fixations about development goals frees up greater sensitivity to what is essential for civil society, which is care and concern for the well-being of others and the natural environment. Simultaneously the model

K. Morrison (✉)
Sustainable Community Development Research Institute, Christchurch, New Zealand
e-mail: keithdmorrison59@gmail.com

© Springer Nature Switzerland AG 2019
W. Leal Filho and S. L. Hemstock (eds.), *Climate Change and the Role of Education*,
Climate Change Management, https://doi.org/10.1007/978-3-030-32898-6_26

outlines how the South Pacific region's cultural traditions provide a resilient self-reinforcing system of civil adaptation to climate change. Finally the model explores what is essential for a university pedagogy to contribute to the resilience of climate change adaptation by the South Pacific region's communities.

Keywords Pedagogy · Climate change · Resilience · Cultural tradition · University

Introduction

Many countries most vulnerable to climate change are developing nations, where cultural traditions still play a major role within their communities. The island nations of the South Pacific are examples. This is relevant because anthropogenic climate change is linked to unsustainable 'development' that seeks to 'modernise' nations. While it is hoped that achieving sustainable development can avert runaway anthropogenic climate change and mitigate what is already underway, it begs the question of whether or not this will still entail the 'modernisation' of nations. In the meantime however, adaptation to climate change is going to be necessary, and is indeed already underway in the South Pacific region where the effects of climate change are already being experienced. Moreover, the adaptation to climate change already underway is not new, but rather an extension of a well proven traditional resilience to climate variability. Furthermore, because these traditional adaptation processes to climate variability are not part of the unsustainable development linked to anthropogenic climate change, they are plausibly a feature of what is required for sustainable development as well as climate change adaptation. If so, the traditional learning processes involved in the cultural systems successfully adapting to climate variability are also relevant to questions about the role of education in climate change adaptation.

Based on research and university teaching within the South Pacific region, an argument is proffered here that the adaptive processes inherent to endogenous[1] cultural traditions can indeed help ensure development is sustainable, and moreover, also ensure resilient adaptation to climate change.

The development of the argument begins by exploring how neither resilient adaptation to climate change nor sustainable development simply emerge out of a singular focus on economic growth to 'modernise' nations (Jackson 2017). It explores how they rather require an understanding of what is required for well-being, and how to maintain adaptive capacity and flexibility for resilience (Raworth 2017, Hone 2017). The essence of the argument proffered here is that the rich complexity of endogenous cultural traditions provide the multi-dimensional focus necessary to garner such

---

[1] Endogenous cultural traditions refers to those that have co-evolved with a particular natural environment. Endogenous cultural traditions are predominantly indigenous cultures, but not all aspects of indigenous cultures can be assumed to be endogenous, and endogenous cultural traditions are not restricted to indigenous cultural traditions.

understanding. The multi-dimensional focus is explored, following early leads provided by Weil (1952) and Rappaport (1979), as a complex yet coherent and resilient set of obligations, from rules through to worldviews and horizons.

The obligations constituting the foundations of these traditional societies are however open-ended processes, neither one-off events nor of fixed form, because adaptive capacity is a key feature of them. The obligations are inseparable from a continuous open-ended processes of learning. Therefore the learning in these societies founded on cultural traditions is not restricted to what is currently contained in their cultural traditions. Education is highly prized by these communities, including university education. The task is to ensure that the openness to learning through integration of what can be exogenous[2] traditions, is carried out as a continuation of adaptation by endogenous traditions, through co-evolution with the natural environment.

The chapter here develops a community-based learning-system model to outline how the traditional (mostly indigenous) societies of the South Pacific are able to integrate university education with cultural traditions to nurture leadership in climate change adaptation, and culminates in an appropriate pedagogy for university education to assist in resilient climate change adaptation.

The components of the community-based learning system can all be linked to sustainable development, as well as climate change adaptation; with each component primarily associated with either development or sustainability (Morrison and Singh 2009). The community-based learning system model seeks to integrate sustainability and development, to reveal important aspects of the dynamics of climate change adaptation. In particular, the model reveals the complementary roles of cultural traditions and university education, and hence the complementary contributions that cultural traditions and university education can make to climate change adaptation, when they collaborate.

Background is provided by first reviewing relevant literature on development, and second by reviewing relevant literature on sustainability, to provide a detailed focus on the dynamics of sustainable development and how they accord with traditional and systems theoretical models of adaptation to climate variability (and change). This is then used to review relevant literature on learning systems as they pertain to adaptation, including those of cultural traditions, to reveal the role of education, including that provided by universities.

The conceptual development of the community-based learning system model for climate change adaptation is then outlined, followed by discussion of the specific role and features of university education that are appropriate to focus on. Finally, practical comments are made on the unique opportunity that universities have to contribute to appropriate climate change adaptation globally for both 'developed' as well as 'developing' nations vulnerable to climate change.

---

[2]Exogenous traditions are those that originate outside of the cultural tradition of interest. Modern scientific disciplines and paradigms are exogenous to most cultures.

Background

Development is one pole of sustainable development. Sustainability is the other. To have an adequate model of sustainable development, development and sustainability must be shown to be in dynamic interaction rather than contradictory (Morrison and Singh 2009). This is not straightforward. It requires careful and challenging conceptualisation of both development and sustainability. In particular it requires avoidance of a commonplace equating of development with economic growth. But even this step is fraught with challenges. Raworth (2017) provides a contemporary very readable traverse of how the ideology of economic growth emerged, highlighting how it is now becoming exposed by reality, with anthropogenic climate change being one of the main expositors. Raworth (2017) also expresses a perspective on an emerging economics and development paradigm that replaces the exposed ideology of economic growth. Another acclaimed contemporary expression of the same emerging paradigm, by Jackson (2017), provides a more detailed technical analysis of the logical need for sustainable development to decouple economic growth from material resources use. He does so in relation to the challenge provided by climate change. He considers the proposed "green growth" method of how to decouple, and shows conclusively that it is unrealistic. Then, critically extending yet another acclaimed contemporary work of macro-economics, by Piketty (2014), Jackson points out the possible ways to avoid collapse. In the second edition Jackson (2017) argues for a recovery of "wisdom traditions" and associated greater reliance on service 'industries', which are enhanced by low rather than high productivity; 'industries' that are enhanced by the spending of time for community engagement to strengthen relationships and belonging. Jackson (2017) points out that this can solve a productivity dilemma in 'modernisation', to potentially ensure full employment and hence sharing of prosperity, without the need for continuous exponential economic growth to desperately attempt to "trickle" further prosperity down.

These three acclaimed contemporary authors, Jackson (2017), Raworth (2017) and Piketty (2014) all agree that development is an ideal that seeks to fulfil the needs required for the well-being of all. Raworth (2017: p 44) proffers an appealing humorous metaphor of a doughnut defining how the holistic and requisite fulfilment of multiple needs are threatened by both too little (the inner edge of the doughnut) and by too much provision of them (the outer edge of the doughnut). Raworth proposes 12 needs: energy, food water, health, education, income and work, peace and justice, political voice, social equity, gender equality, housing, and networks.

Raworth's model builds on an emerging paradigm going back into the recent past at least as far as Weil (1952). Weil defined 14 "needs of the soul" to ensure civility, emphasizing that the rights to having these needs met were constituted by the obligations on everyone to provide them for all others. Weil's list is: order, liberty, obedience, responsibility, equality, hierarchism, honour, punishment, freedom of opinion, security, risk, private property, collective property, and truth.

Another expression of the paradigm was by a team (Max-Neef et al. 1991) who proposed a matrix of needs, all of which must all be fulfilled to avoid poverty, and when fulfilled, provide wealth. They argued that there are as many different types of poverty and wealth as there are cells in their matrix (see a summary simplification in Table 26.1). Like Weil (1952), Max-Neef et al. (1991) argued that the purpose of a society is to achieve well-being for all through the obligations on all to overcome all the various poverties that threaten everyone. For Max-Neef et al. (1991) this obligation defined development. As for Weil (1952), Max-Neef et al. (1991) defined needs as universally valid, but able to be fulfilled in multiple ways, which they defined as "satisfiers". Max-Neef et al. (1991) argued that there is not a one-to-one correspondence between needs and "satisfiers", and so there are always multiple "satisfiers", as well as always potentially more appropriate "satisfiers" to discover, hence giving an intrinsic adaptive movement within development. They recognise an intrinsic flexibility to well-being and hence development. Moreover they pointed out that the ideal for well-being and hence development is to achieve "synergistic satisfiers", which efficiently fulfil multiple needs within any particular context, which is also always in a process of change. They therefore implicitly refer to endogenous development, occurring inevitably through societal co-evolution with the local context. Furthermore, Max-Neef et al. (1991) explicitly point out that cultural practices can mistakenly instead provide "pseudo-satisfiers", which claim to be "satisfiers" but are actually over satiating one need to mask the failure to satisfy others—that is, addictive practices.

Another proponent of the paradigm was Bossel (1998) who took a systems approach and mapped the various functions (defined as "orientors") of living systems onto the various types of interacting systems within human communities; all of which need to be sustained for well-being (see Table 26.2). These components map well onto the typology of needs defined by Max-Neef et al. (1991). They also map well onto the model proffered by Raworth (2017), and are also largely similar to those outlined by Weil (1952). Moreover, the "functions" of a system refer in effect to the obligations the system components have to the whole system: there is in effect, in agreement with Weil (1952), an inter-dependence of mutual obligation

Table 26.1 Matrix of needs

Needs	Being (qualities)	Having (things)	Doing (actions)	Interacting (setting)
Subsistence				
Protection				
Affection				
Understanding				
Participation				
Leisure				
Creation				
Identity				
Freedom				

A summary adapted from Max-Neef et al. (1991)

Table 26.2 Orientors of living systems

Orientor	Social system		Support system		Natural system	
	Subsystem	Total	Subsystem	Total	Subsystem	Total
Existence						
Effectiveness						
Freedom						
Security						
Adaptability						
Co-existence						
Psych. needs						

A summary adapted from Bossel (1998)

between all components of a system, which is a process of co-evolving endogenous embeddedness, and what Weil (1952) describes metaphorically as the "growing of roots".

Bossel's (1998) work is particularly pertinent because the definition of living system components imply a link between development and sustainability. This is because living systems to be sustained must maintain well-being through maintaining all systems components as well as the larger social and natural systems they are part of. Bossel's (1998) work provides some detail to the overall 'doughnut' structure of Raworth's (2017) model. Moreover, because all living systems exist within environments that are changing, and indeed co-evolve with them, as do systems with system components, a necessary function of system components is to ensure adaptive capacity—they have an obligation to provide it for 'their' system. Just as a human cultural system has an obligation to ensure the resilience of 'their' natural environment as well as other social systems they participate within.

Prior to Bossel's (1998) work, Rappaport (1979) had already applied systems theory to anthropological systems; to human cultures. Rappaport focused on the role of adaptation and how cultural systems maintain adaptive capacity. A point he mentions without fully developing it is that evolutionary adaptation, defined as adaptation that involves change in the structure of a system, is driven counter-intuitively; namely, it is driven by what cannot change in the system. He argues that what cannot change, or in other words, what is essential within a particular environment for the survival of any particular living system, drives adaptation because what is not essential is then left free to adapt and diversify. The implication is that what is essential for a system within a changing environment may also change. Therefore, ecological resilience is created due to a diversity of potentially apt features being present to choose from. Later Rappaport (1999) applied this dynamic of evolutionary adaptation to cultural systems and the emergence of distinct traditions, exploring what this means for the challenges facing contemporary modern cultures. Rappaport (1999) argued that the conceptualisation of natural systems that humans participate within enables them to be modified; cultural traditions are a synergy of natural processes and human construction, and the potentiality for human construction is part of a natural process. Weil (1952) had already recognized the synergy, in terms of the need for "roots", where obligations required toward the natural processes people participate within take precedence over those artificially constructed by people; society has the overarching obligation to fulfil the needs of natural processes and not vice versa.

Rappaports (1979, 1999) work is relevant because its "environmental anthropology" effectively equates cultural traditions with endogenous cultures: endogenous cultural traditions are by definition, cultural practices that are sustained within any particular environment. Therefore the structure of cultural traditions proffered by Rappaport (1999) provides an adequate definition of the essence of cultural traditions in the concern here to analyse how cultural traditions assist sustainable development and climate change adaptation. Such a definition therefore, as well as detailing features of endogenous cultural traditions, goes part of the way to provide discernment of the difference between unsustainable (maladaptive, producing ill-being and lack of resilience) cultural practices, and sustainable resilient cultural traditions.

Once again akin to Weil (1952), Rappaport (1999) proposed that cultural systems have an intrinsic hierarchy, expressed by different types of knowledge, and where each level in the hierarchy has a different function (obligation) necessary for adaptation. He argues that there are four levels to the structure of cultural systems, and that only when adaptive pressure becomes high enough does adaptation at the next higher level 'kick- in'. For ease of presentation, this aspect of Rappaport (1979) work has been tabulated (see Table 26.3).

Just knowing what the four levels are, the type of adaptation that occurs, and when they are triggered, is however insufficient. What is also necessary is to be able to recognise what is successful adaptation and what is mal-adaptation. This further piece of knowledge is also intrinsic to a cultural system. As indicated above, it is only partially explored in Rappaport's (1999) work. I have already previously attempted to make it explicit through a critical adaptation of Rappaport's (1999) work (Morrison 2016). Figure 26.1 is my previous attempt to define the difference between successful adaptation and mal-adaptation explicit in a conceptual model, comprised of three levels, and applied it to disaster risk reduction, and to migration as a climate change adaptation.

A poignant paradox emerges here to reveal tension within a commonplace assumption found in sociology and development theory, that "development" requires "modernisation" to replace "traditional" societies (see for example Giddens 1990). Even

Table 26.3 Hierarchy of knowledge

Level	
4 (highest)	Invariant principle(s)
3	Worldviews
2	Rules
1 (lowest)	Application of rules

Summarised from Rappaport (1979)

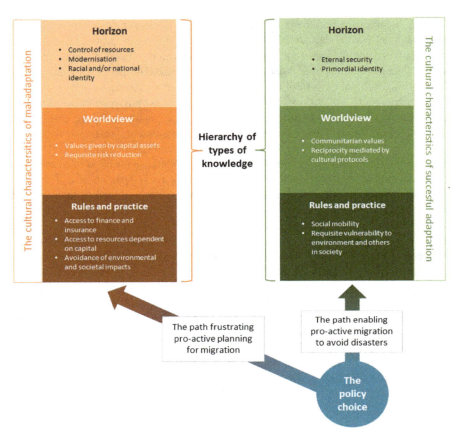

Fig. 26.1 Conceptual model for the role of cultural traditions to guide migration as climate change adaptation (Morrison 2016: p 256)

Raworth (2017: p 268) appears to condone it. It is to the credit of Jackson (2017) in his second edition that he has started to tentatively question the assumption. I am explicitly rejecting the assumption. I argue instead, that to sustain the benefits of risk-taking, through supposedly 'freeing' itself from so-called 'traditional' societies' natural embeddedness, 'modernisation' needs to be limited by sensitivity to negative feedback from all systems as to whether all needs are still being met, as well as sensitivity to all obligations to maintain the resilience of all systems (natural and social). I am arguing that contra to a common implicit misconception, cultural traditions (as defined here, based on Rappaport's work, as endogenous cultures) are the source of adaptive capacity and flexibility: cultural traditions maintain adaptive capacity to naturally co-evolve diverse "synergistic satisfiers" to ensure resilience, through being able to freely move up and down as necessary the hierarchy of knowledge, as defined by Rappaport. I am arguing that in contrast to the supposed 'freeing' from natural embeddedness of 'traditional' societies, 'modernisation' actually becomes prone to fixate on "pseudo satisfiers" of needs, through having an horizon that closes off questioning the higher levels of knowledge. Rappaport (1999) argued in his tentative exploration of the topic that this is driven by the use of money as a cultural symbol for exchange. In other words, modernisation has a fixed worldview, with only superficial flexibility, which is provided by enhanced exchanges at the lower levels of rules and the application of rules, and facilitated through the use of money. Rappaport (1999) argued that this creates in reality a sort of evolutionary and structural rigidity and brittleness. There is by contrast far more diversity and flexibility between and within (endogenous) cultural traditions at the level of worldview, than there is in the monolith of global 'modernisation'. This is significant because evolutionary adaptation requires the ecological resilience provided by a diversity of worldviews, because they maintain a diversity of sets of "satisfiers", which may become necessary for the transformation necessary for survival as the environment changes. Empirical evidence for this is found in how 'modernisation' is found to decrease risk of low impact events, but to increase the risk of disasters (Lauer et al. 2013; Maru et al. 2014; Morrison 2016).

Others have also modelled a hierarchy in adaptation, and the need to be conversant with it. The 'school' of Transition Management (Geels 2002; Foxon et al. 2009) articulated a hierarchical typology for adaptation. Here the hierarchy refers to the scope of adaptation, rather than to the knowledge processes involved in cultural systems. The typology of Transition Management uses ecological metaphors. The lowest level is that of 'niche' and refers to changes wrought by individuals and their inter-personal interactions. The middle level is that of 'ecosystem' and refers to institutional changes. The highest level is that of 'landscape' and refers to large-scale political changes.

Geels (2002) argues, as does Rappaport (1999), that adaptation occurs "bottom-up". Institutions adapt when small scale community innovations force them to, and institutional changes eventually bring about political and cultural adaptation. But it is surely actually more complex than that. Political changes can often leave institutions reeling; with a legal and policy vacuum, requiring institutions to catch up. Institutions in such a situation drag and slow down what political change has already decreed. Also it is commonplace to recognise that regulatory changes can occur without community acquiescence.

The actual complexity can nevertheless still be characterized as "bottom-up" by recognising that it is still individuals or think-tanks and task forces of innovative individuals who form the catalyst for political and institutional transformations. It is actually individuals or small groups who act within institutions to catalyse their change and then impose regulation on communities. Also it is individuals and small groups who catalyse innovative institutional forms and herald them into political life, and which are then imposed on all institutions in the 'landscape'. Foxon et al. (2009) coined the term "shadow spaces" to refer to how adaptations are catalysed in this hidden way. The term "shadow spaces" is however inadequate, because it refers to the space of adaptation in relation to institutional structures. It is also necessary to conceive of it in terms of what individuals experience, which is rather a space of inspiring enlightenment with a hopeful horizon of "eternal security" and "primordial identity"(see Fig. 26.1). As Weil (1952) argued, there is need for "rootedness" in natural processes, which is where the hopeful horizon of "eternal security" and "primordial identity" is known, and which has priority over the structures of, and roles within, society. In a review and synergy of the literature of psychological resilience with the psychology of grief, Hone (2017) defines this hopeful horizon as the need to "see the good", in the natural environment and human societies.

Recognizing the operation of inspiring "shadow spaces" to enlighten grounded adaptive capacity to catalyse adaptation, enables the integration of Rappaport's (1999) work with Transition Management theory (Geels 2002; Foxon et al. 2009). Every individual can potentially act creatively and critically to constructively catalyse adaptation at any of the three levels defined by Transition Management, and they do so by using the appropriate level of adaptation in the hierarchy of knowledge defined by Rappaport (1999), but always toward, and in the light of, the hopeful horizon of the highest level. This means that adaptation is something that potentially every member of a community can constructively catalyse and participate in; and indeed has an obligation to do so, as well as to empower and enable others to be able to do so as well. Moreover as Bossel's (1998) model emphasizes, it is a necessary and continuous process, and indeed intrinsic to all living systems, including therefore human cultural systems. Next we look at how the adaptive process has been conceived of as learning, and hence we explore the role of education for adaptation within cultural systems.

Cultural learning-systems.

One portrayal of cultural learning systems comes directly from indigenous academics who themselves culturally participate within traditional cultures. Ironically, their analysis has come to fore due to a misinterpretation and misappropriation of cultural knowledge. None other than Maslov's (1954) famous five step 'hierarchy of needs'; with "physiological needs" at the base, rising to "safety and security", then "love and belonging", to the penultimate need for "self-esteem", and with "self-actualization" at the apex, was informed from the Blackfoot Indians in Canada, to Maslov as a young researcher. Maslov's (1954) work has however recently been interrogated as to whether or not it correctly recorded the Blackfoot cultural knowledge (Blackstock 2011). There are indigenous researchers who have as a result questioned the validity of the supposed "hierarchy of needs" Maslov (1954) proposed, and suggested a corrected version (Blackstock 2011). The proposed corrected model of needs and well-being is actually rather more like what Weil (1952), Rappaport (1979, 1999), Max-Neef et al. (1991), Bossel (1998) and Raworth (2017) have proffered.

Instead of a hierarchy of needs there is are levels of intensity in what needs to be focused upon for well-being. The highest or centremost intensity of focus is defined as "spirituality", the middle intensity of focus defined as "love, belonging and relationships", and the lowest and outermost intensity of focus, defined as "shelter, safety, water and food". The three intensities of focus correspond to the levels of knowledge defined by Rappaport (1999) as I have previously critically adapted them (Morrison 2016). The difference between Blackstock (2011) and Rappaport (1999) is that Rappaport appears to consider that a person and society will only focus on what Blackstock considers to be of centremost importance, when they are pressured to do so by a loss of a more superficial fulfilment of needs. The differences between Blackstock and Rappaport are however best seen as constituting a complementarity rather than a contradiction. Rappaport was probably simply realistically describing how there is a tendency to put off adaptations that require a greater intensity focus, until they become necessary, and not implying that continual awareness of the higher levels of knowledge is not necessary. Blackstock by contrast is focusing on well-being, and therefore emphasizing that to ensure well-being through holistic fulfilment of all needs, there has to be continual effort to enculturate the highest intensity of focus for spirituality. Hone's (2017) work on psychological resilience is in implicit agreement with Blackstock—a focus on "seeing the good" has to be continuous—and can be seen to implicitly challenge Rappaport's view that it needed to be clearer about how all adaptation, at any level, requires a continuous focus on spiritual and psychological needs, even if the adaptations are not of the psychological and spiritual realms, but to do with how to fulfil physiological needs.

This correction of Maslov's (1954) 'hierarchy of needs' is significant, because of the huge influence of Maslov's work. Maslov's (1954) model of a supposed 'hierarchy of needs' was initially published at about the same time as the first English translation of Weil's (1952) work outlined the "needs of the soul". But up till now, Maslov's (1954) work has had a far greater influence. The reason is probably due in part to how Maslov's (1954) work can be used to rationalize the equating of development with economic growth. Maslov's (1954) model can be interpreted to mean that the initial focus of development has to be to ensure physiological needs are provided by material goods, and hence met by economic growth to allow increased consumption of material goods. Moreover, Maslov's (1954) model implies that there is no use focusing on fulfilling other needs until physiological needs are first met. And because physiological needs continue to be unmet, Maslov's (1954) model can be used to rationalize a focus on development that is singularly on economic growth to increase consumption. Therefore the central thrust of Weil's (1952) work, which is an emphasis on the obligations on all to seek to fulfil all the needs necessary for well-being for all, is almost completely undermined, as is Bossel's (1998) argument for the obligation to ensure the resilience of all systems, including the natural environment. It also makes it clear why a focus on economic growth can lead to the prevalence of "pseudo-satisfiers" or addictive practices: instead of explicitly addressing all needs, not only those not able to be fulfilled by material consumption, over-satiation of material consumption can mask the actual poverties of unmet psychological and spiritual needs. Moreover it shows how 'modernisation' with its comprehensive use of monetary valuation creates a rigidity and loss of adaptive capacity: flexibility when restricted to monetary valuation is limited to being only between commensurate needs, that is, between those able to be replaced by others through the exchange of money. Therefore there is only flexibility to fulfil tradeable needs, and hence an intrinsic pressure to create "pseudo-satisfiers" whenever 'other' needs become unable to be fulfilled, which gives rise addiction. Evolutionary adaptation is required to bring about the transformation in society necessary to get beyond such an impasse that gives rise to "pseudo-satisfiers". Not surprisingly therefore, the loss of transformative or evolutionary adaptation in 'modernisation' is empirically manifest in an epidemic of ill-being (mental illness) (Hone 2017).

Far from being idle speculation, such analyses gives understanding as to why the so-called 'modernised' and 'developed' world has an empirically recognised crisis of mental illness, ill-being and loss of psychological resilience (Hone 2017), as well as being the cause of anthropogenic climate change. Further insight can be gained by also considering what the same psychological resilience literature has recognised empirically as providing psychological resilience. Hone (2017) emphasizes four things. One is the need to be positive, as already mentioned. This is explored primarily along the lines of learning to "seeing the good" but can also take such forms as "hunting for the good stuff", or in other words to seek out the good—it is an active as well as passive pursuit of spirituality. The second is the need for relationships, or community. The third is the need to recognise that suffering is normal and to be expected. Challenges requiring adaptation are to be expected, prepared and planned for. The fourth is that resilience and well-being are not static events, but a continuous process of learning. Hone's (2017) work is therefore in agreement with that of Blackstock (2011), namely that there is a need for cultivating, through teaching and learning, the necessary intensity of focus required to maintain spirituality, love, belonging and relationships, as well as material needs. What Hone (2017) implicitly adds to the work of Blackstock (2011) is that the learning requires sensitive openness to challenges, and suffering. The negative feedback provided through sensitivity to challenges and suffering provides metaphorically an inoculation and strengthening of the mental health immune system (my metaphor, not Hone's).

Given the need for continuous learning, the central role of education comes to fore. Continuous process of learning to ensure adaptation for well-being and resilience is required, which is the role of education. For successful adaptation, a strong argument can be made for a focus on teaching (facilitating the learning of) adaptive capacity, which requires nurturing openness to the ethical orientation of a hopeful horizon of "primordial identity" and "eternal security"—spirituality (Blackstock 2011, Morrison 2012, 2016; Hone 2017). A focus on education nurturing spirituality and facilitating adaptive capacity, is what actually enables all other needs to be met, including physiological needs.

There are models in learning-system theory that have come to the same conclusion. These models include cultural learning as a third loop within a Triple-loop learning process (Argyris and Schon 1996; Morrison and Singh 2009; Tosey et al. 2011; Shinko et al. 2019: p 98). A feature of the Triple-loop learning model is that it successfully models, to multi-dimensionally integrate, the adaptation levels proposed by both Rappaport (1999) and Transition Management (Geels 2002; Foxon et al. 2009). Triple-loop learning makes explicit reference to personal development constituting creative innovation, as a third learning loop necessary for transformative or evolutionary adaptation (Morrison and Singh 2009; Morrison 2012, 2016). The third learning loop therefore is what catalyses adaptation within inspiring "shadow spaces". Triple-loop learning recognises the necessity (obligation) of individuals to engage in all types of learning at all levels of adaptation; namely where worldviews

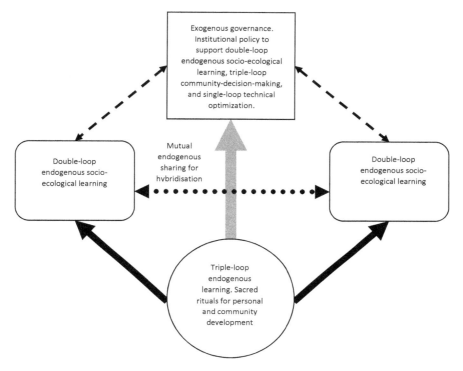

Fig. 26.2 Structure of endogenous socio-ecological learning (Morrison 2012: p 191)

and paradigms are questioned (double-loop learning), and where learning occurs within paradigms and disciplines (single-loop learning). The focus however, out of which all other learning and adaptation occur, is the third learning loop.

I have previously modelled (see Fig. 26.2) relationships between Triple-loop learning and cultural traditions (Morrison 2012). The model addresses the interactions between endogenous (local community) and exogenous (external—for example universities within a global context) learning to maintain adaptability.

The model emphasizes how hybridisation is a key feature of adaptation. "Shadow spaces" are not only or usually solitary works of inspiration, but rather usually places of creative dialogue with other persons, who transcend and even transgress their social roles, disciplinary expertise, and even cultural traditions.

Conceptual Model

The conceptual model constructed here is of a community-based learning system, which maintains traditional cultural learning systems whilst participating in university education and carrying out university-based research (see Fig. 26.3). It uses

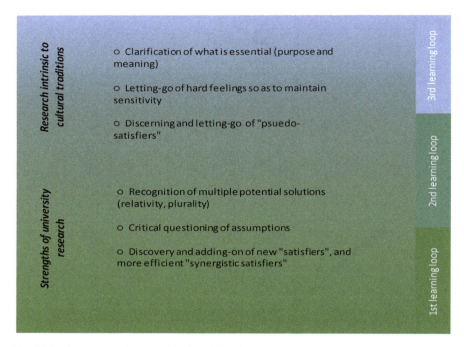

Fig. 26.3 Components of community-based learning system

Triple-loop learning as a meta-frame. This enables: (i) clarification of what are the significant interacting system components, as modelled by Bossel (1998); (ii) incorporation of the multi-dimensional levels of adaptation, as explored by Rappaport (1999) and Transition Management (Geels 2002; Foxon et al. 2009); (iii) the requisite and holistic fulfilment of all needs for well-being and resilience argued for by Blackstock (2011), Hone (2017) and Raworth (2017), along with the fulfilment of them through "synergistic satisfiers" as developed by Max-Neef et al. (1991); and (iv) the overall need for "roots" for civility argued for by Weil (1952), and that this overall need for "roots" is provided by "wisdom traditions" as previously argued for by myself (Morrison 2012), to solve the dilemma of how to decouple development from unending growth, as postulated by Jackson (2017).

The model does not separate university education and research from learning and research intrinsic to cultural traditions. Rather it implies, in agreement with the model of endogenous socio-ecological learning (see Fig. 26.2) that they interact and interpenetrate (Morrison 2012). But at the same time it clarifies the contributions and strengths of university education. In general, university education and research facilitates double-loop learning and excels in single-loop learning.

Through defining where third-loop learning occurs, the model also clarifies that the personal capacity to be a catalyst for adaptation; whether in a community, institution or politically, it is due to learning in, and being able to carry out research which is intrinsic to cultural traditions. Therefore the values and process of cultural traditions enhance university education and research to ensure it is innovative. In particular, the learning and research that is intrinsic to cultural traditions ensures adaptation is successful, by focusing on how to make sure needs are met for well-being. Firstly, cultural traditions clarify what is essential within a particular context to ensure adaptation is successful: what has to drive adaptation to ensure it is successful is clarified. Secondly, cultural traditions ensure all needs are truly met by adaptation, including clarifying what can and needs to be changed: cultural traditions discern "pseudo-satisfiers" of needs, to let-go of them and to avoid them.

Clarification of the significant components of the community-based learning system does not however reveal how the components interact to successfully carry out climate change adaptation. To do this requires an understanding of vulnerability within climate change adaptation. Vulnerability is intrinsic to living systems. Vulnerability refers to the stress necessary to drive adaptation within social-ecological systems. But beyond a certain level it becomes a cause of trauma and hence system dysfunction. Therefore it is usually necessary to avoid increases in vulnerability so as to avoid system dysfunction and mal-adaptation. But without a certain level of vulnerability there is no adaptive response. A degree of negative feedback is also necessary for system function. Therefore, within cultural systems, requisite vulnerability is that which provokes adequate adaptive capacity to create the necessary flexibility of "satisfiers" to choose from, but without the stress creating trauma and hence breakdown in civil society (Morrison 2016). Requisite vulnerability ensures the emotional bonds maintaining a social-ecological system are kept flexible, but not broken.

Not surprisingly however, given the near hegemony of modernisation development ideology, the need for requisite vulnerability is not yet well appreciated (Morrison 2016). There have been a few heralding the need for requisite vulnerability in other contexts, for example Butler (2004) in face of terrorism, and it is also foundational to psychological resilience (for example Hone 2017), but it has not yet become included into the definitions and analyses of vulnerability to climate change by the Intergovernmental Panel on Climate Change (IPCC 2014). I have previously already broached this lacunae in relation to vulnerability within the field of climate change adaptation (Morrison 2016). Here I seek to detail how to do so in practice, which is to understand how requisite vulnerability is maintained through cultural traditions. A lead into this is given by social vulnerability theory.

An equation defining vulnerability (Eq. 26.1) can be deduced from social vulnerability theory. Social vulnerability theory clarifies the significant things that need to be considered within cultural systems, and defines their relation to vulnerability (Fischer et al. 2013).

$$Vulnerability = Fn\left(\frac{sensitivity,\ exposure}{adaptive\ capacity}\right) \qquad (26.1)$$

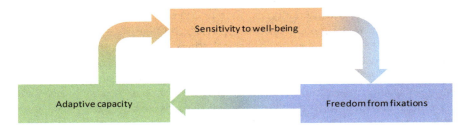

Fig. 26.4 Resilient self-reinforcing system

Equation 26.1 implies that there is a self-reinforcing system of positive feedback within cultural systems, which gives them potentially high adaptive capacity, and which potentially makes them very resilient (see Fig. 26.4).

The adaptive learning system within cultural traditions is self-reinforcing because it maintains both high sensitivity and high adaptive capacity. This occurs through the Triple-loop learning intrinsic to cultural traditions.

Cultural systems are able to maintain high *adaptive capacity* because Triple-loop learning avoids fixation on any particular set of "satisfiers" to fulfil needs, through maintaining clarity about what is essential for a community. Triple-loop learning maintains a process of 'letting-go', to question specific goals, so as to regain vision of the bigger picture of the overall purpose for communities.

But the third loop within Triple loop learning also maintains a vision of what is essential for a community due to *sensitivity* to care for the well-being of all others, including non-human beings in the environment. And moreover, this also provides great flexibility, because sensitivity to care for others is not restricted to any particular structure or set of "satisfiers". So, even though the vulnerability equation shows that sensitivity increases vulnerability, because cultural systems at the same time increase adaptive capacity when sensitivity increases, vulnerability can be maintained at a requisite level even when sensitivity increases. Therefore through Triple-loop learning, cultural systems can maintain both high adaptive capacity and high sensitivity to care for the well-being of the whole community and environment. Therefore the flexibility of the social-ecological system of a cultural tradition can be bootstrapped to whatever level of adaptation is necessary, without increasing vulnerability and so without creating the trauma that can break down civil society; just as long as by doing so it does not increase *exposure* to what is being adapted to (climate change in our case).

However, because the Triple-loop learning processes are what nurture the positive feedback in the self-reinforcing system of cultural systems, and because learning involves openness to negative feedback, cultural traditions therefore operate with both negative feedback and positive feedback. Indeed, the positive feedback in cultural traditions can be considered to be the engine that gives the energy and momentum for the sensitivity and adaptive capacity required for cultural traditions to continue to carry out the learning and adaptation in response to negative feedback. There has to be sensitivity and adaptive capacity to be able to learn for adaptation to occur.

But if the energy and momentum to maintain the intensity of focus to continually sustain sensitivity and adaptive capacity are not there, then developing the higher levels of knowledge through Triple-loop learning will also not occur. The energy and momentum sustained by the positive feedback in the self-reinforcing system, is the engine for the concerted effort and intensity of focus required to ensure openness to the highest levels of knowledge, so that transformative or evolutionary adaptation occurs when it is necessary. Without the intensity of focus, the learning is likely to be superficial and to varying degrees fixated because it is unable to be transformative (Shinko et al. 2019; p 98). As already mentioned, when only relying upon lower levels of knowledge and the flexibility to transfer between commensurate values able to be symbolised by money, there is a lack of capacity to engage in transformative or evolutionary adaptation.

There has to therefore always be a concerted effort to ensure sensitivity to negative feedback, even when denial or of it would be possible in the short term. Low levels of irritation or discomfort and challenges, even suffering, constitute the requisite vulnerability that give the necessary preparation to initiate the necessary planning for resilience. The concerted effort ensures that "seeing the good" (spirituality) is able to be continually focused upon, whilst recognising and processing all the negative things that arise because everything else is continuing to change. The stable balance to accept the bad whilst maintaining focus on the good is required for resilience, and is nurtured by cultural traditions.

The proviso however in the analysis, namely that the self-reinforcing system can only be maintained if exposure does not increase, is not something that the third loop of Triple-loop learning, and therefore cultural traditions focus on. Cultural traditions do sensitise people to respond to negative feedback so as to stay out of trouble (Maru et al. 2014; Mercer 2010; Morrison 2016). To this extent they are able to go part of the way to ensure that exposure does not increase. But what is required to stay out of trouble is contextual. It will depend on the degree of change occurring. What may be required is to simply the making of a bigger water tank, or it could be to participate in international migration. Now, because the adaptive process within cultural traditions operates only to let-go of fixations on any particular sets of "satisfiers" of needs, it only works to stay out of trouble if the cultural tradition has adequate diversity of "satisfiers" to choose from, so that there is at least still one that is going to work. This is not always the case. Indeed, a lack of options can give rise to an apparent lack of flexibility in cultural traditions. Letting-go can only go so far. Adding-to is also necessary. Eventually there may be nothing left to let-go, and so than adaptation fails because an increase in exposure to climate change cannot then avoided, and so the positive feedback driving the openness to a positive horizon and familial and community relationships can be lost. Therefore, even though mal-adaptation is implicitly avoided by cultural traditions, this is inadequate if the rate of climate change and therefore the rate of necessary adaptations is greater than what an unfocused increase in "satisfiers" by cultural traditions can produce. Cultural traditions can indeed lack flexibility in face of climate change. It is this fact that is probably behind the commonplace assumption implicitly condoned by Raworth (2017; p 268) that the first step of "development" is to take leave from "traditional" society. Even though cultural traditions do tend to contain a relatively high diversity of "satisfiers", as for example seen in organic farming techniques, increasing the diversity of "satisfiers" does not appear to be an explicit focus of cultural traditions, in the same way that increasing the diversity pertaining to higher levels of knowledge does. The focus on increasing the diversity of "satisfiers" appears to be a peculiarity of 'modernisation' and university education. Therefore a contribution university learning and research contributes to community-based climate change adaptation, is to explore innovative ways to decrease exposure to climate change, so as to increase flexibility by increasing the options of available potential "satisfiers". However the contributions of university education and research for climate change adaptation go far beyond this, even though this is a prominent outcome. Everything that helps optimise the self-reinforcing highly resilient system of cultural traditions is also a contribution. Specifically, university education facilitates double-loop learning through emphasizing the value of being critical, and hence the ability to recognize the relativity of perspectives and the need for pluralism to respect different perspectives: university education and research helps avoid rigid thinking. University education therefore increases flexibility to climate change adaptation by recognising that there are always multiple possible solutions and always further assumptions that could still be questioned to explore yet further possible "satisfiers" of needs. University education can help increase openness to exploring other ways of doing things. But also, once there is requisite flexibility through having enough options to choose from, yet not superfluous choice, university research excels at optimizing single-loop research to discern efficient "synergistic satisfiers" of multiple needs.

But for university education to makes its unique contributions to increase flexibility in climate change adaptation, it has to be couched within cultural traditions of endogenous social-ecological systems. If the innovation nurtured by 'modernisation', and ably pursued by university education, is to lead to increased flexibility in climate change adaptation, it has to be: (i) directed for the explicit purpose of enhancing well-being and the resilience of natural systems through "seeing the good"; and (ii) couched within cultural traditions that are "rooted" in natural processes and relationships giving rise to belonging within communities. Otherwise it is prone to become merely a superficial flexibility to increase consumption, which increases the risk of disasters (Lauer et al. 2013; Maru et al. 2014; Morrison 2016).

These features of university education can be generalized into an integrative conceptual model of pedagogy for university education, to help guide how university education can be is effective within community-based learning systems for climate change adaptation (see Fig. 26.5).

There are three key facets to the proposed pedagogy; innovative thinking, critical thinking and community engagement. All three facets exist within the same three dimensions, which are defined by the three nested learning systems of Triple-loop learning. The model reflects models of traditional cultural knowledge pertaining to needs and well-being. The focus is on what is most essential, which is transcendence of fixation on any particular "satisfier", to instead be open to a horizon of opportunities by "seeing the good", to discover new "satisfiers" to nurture well-being—termed

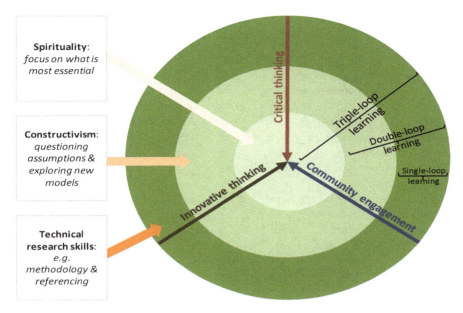

Fig. 26.5 A pedagogy for university education within community-based learning

spirituality by cultural traditions. It is what the third loop of Triple-loop learning provides. All three facets of university education require to have this focus if they are to remain open, and to hence to have clarity about purpose (about what is most essential). This openness intrinsically comes with questioning of assumptions, through seeing the relativity of all constructions of "satisfiers", including conceptual models or paradigms and disciplines used to frame them, which is double-loop learning. This is equally true for community engagement as it is for innovative thinking and critical thinking, because otherwise inclusivity to maintain an authentic community perspective is not able to be maintained.

Finally, it is not possible to engage in any research activity if disciplinary techniques and expertise are not utilised and upheld, which is single-loop learning. Without specific disciplinary skills, research is empty and ineffective, no matter how focused it is on good purpose and aware of its relativity. This is once again equally true for community engagement as it is for innovative thinking and critical thinking, because otherwise there would be nothing to actually assist a community with, just a good approach. This highlights clearly the important contribution that university education can provide for climate change adaptation.

Discussion

Even though the community-based learning system model was developed to assist in carrying out university teaching and research on sustainable development and climate change adaptation in South Pacific universities and communities, general features of community-based learning, and specific features of how cultural systems maintain their resilience, and an appropriate pedagogy for university education to assist in climate change adaptation, were also clarified through the process. Importantly, because systems components are defined, this led to the clarification of where there are leverage points to enhance the operation of university education within the learning system.

One leverage point occurs around the need for university education and research to participate in the learning and research intrinsic to cultural traditions. Or to put it better, university research needs to be carried out within the community context where cultural traditions are lived, if they are to be effective catalysts for adaptation by and within the community. Far from these being needing to be imposed on people, what is required is to rather nurture their intrinsic presence in the idealism of youth. For example the global climate action initiative among youth. By the 2020s this generation of high school climate activists will be receiving university education and will be expecting universities to have a correct focus about what is most essential; so that their idealism can be nurtured and made effective to achieve the transformations that they already intuit are necessary. The ethical ideals of professional responsibility and integrity need to explicitly nurtured in students to extend and give adult life to youthful climate activism. Moreover these future students can expect to be guided so that they also have their own (psychological) resilience personally nurtured, to help prepare them for the challenges coming in their professional life in relation to climate change. An holistic approach to the university education of students' youth is necessary.

A second leverage point is the need for university education and research to fulfil its role to complement the research and learning that is intrinsic to cultural traditions, namely to exert a critical approach, whereby assumptions are always open to be actively questioned, to ensure multiple and new options are always explored and made explicit. The idealism of youth needs to be sharpened and widened. This has traditionally been the role of the humanities in universities. The role has to be recovered and recognised as essential for all students. Traditional university humanities curricula may need to be adapted, but this could also breathe new life into the humanities, and may itself be well overdue. A feature of it could be to explicitly engage with communities, to hear their stories and to create new ones with them through community-based action research. Not only would this give a context in which students can truly learn how to express ethical professional ideals, it would forge a bridge between the communities and universities, to ensure that the overarching purpose for university education remains firmly focused upon.

A third leverage point is the need for university research to use disciplinary techniques and expertise to construct in an unprejudiced way, multiple realistic and practical and efficient solutions—to provide flexibility and opportunity by adding-on "satisfiers" to the mix of options. But for the "satisfiers" to be innovative and synergistic options to enhance well-being and resilience, they need to be inspired by the "shadow spaces" of sincere dialogue, positively open to opportunities, whilst realistic about the continual challenges. The dialogue has to therefore be inter-disciplinary as well as pluralistic. Triple-loop learning has to be lived, so the full range of disciplines are discussed together to how they can be used together for common purpose, and multiple cultural traditions are be shared from and learnt from simultaneously. Otherwise, the continual challenges in society are matched by the inability to coherently use rationally intelligence to do anything about them. Instead, out of desperation, irrational denial of what can be avoided in the short-term is chosen instead, in an attempt to mask and to forget about the challenges. And so "pseudo-satisfiers" breeding addiction and the epidemic of ill-being are added-on as the default activity, instead of the constructive yet challenging task of discovering new synergistic "satisfiers". And so the system becomes yet more brittle, and climate change resilience lessens even further. It is not an exaggeration to say that this is the most pervasive and most pernicious climate change denial. But to confront it at universities requires a coherent and comprehensive trans-disciplinary pedagogy. Actually the very name university surely points to this being what is expected of universities. Triple-loop learning provides one possibility, and so can be considered as a starting point for such pedagogical development.

There is nothing startling in these discoveries. They could be seen as underwhelming, except that it is precisely because they emphasize an holistic approach to university education, which makes them interesting. It is their ordinariness that is their greatest strength. They are what students entering university expect to be provided for them in their university education. Universities are expected to nurture their future leadership to bring about the necessary transformation of society in face of climate change. The time for strategic cynicism toward youthful idealism is well past. Indeed there is a danger that if universities do not live up to what is expected of them, the idealism behind the current youthful climate activism could become misdirected, and so the greatest potential resource of universities wasted.

Conclusion

An ambitious hypothesis has been articulated: that universities working within communities guided by cultural traditions can help overcome a prime cause of anthropogenic climate change, namely a cynical fixation on economic growth, and by doing so, put in place missing steps in the quest for sustainable development and resilience to climate change. To the extent the community-based learning system model proffered can provide a framework for further exploration of these crucial issues, then it will be a hypothesis that can potentially be useful.

At the very least, the model has clarified how traditional cultural systems can potentially maintain high resilience, and also, what is helpful within university pedagogy to amplify the resilience of cultural traditions, when universities collaborate with them. This can be relatively straightforward in regions like the South Pacific where cultural traditions are extant and vibrant, and where education is prized, especially if there can develop global collaboration to share knowledge and university resources. There is hope also however that this synergy, which has been developed in a region that is very vulnerable to climate change yet remains "rooted" in cultural traditions, can potentially help developed countries that have lost their "roots" in cultural traditions. It may potentially inspire from out of a "shadow space" of dialogue in the South Pacific between academics, postgraduate students and community members, some helpful insights. It could well be that the current climate action youth are an akin emergent expression of such an inspiring "shadow space" in the 'developed' and 'modernised' world. There is even the hope, found previously glimmering in the work of Weil during an equally challenging and fraught global challenge in early 1940s Europe, that synergies between universities and communities can help maintain and strengthen civilisation. This includes obligations toward those most vulnerable to climate change, who will be forced into migration and may even become refugees. The 'developed' countries, need to find in their traditional roots a way to prioritise hospitality and gift the future immigrants and refugees the opportunity of higher education. But it also includes obligations towards the members of their own 'developed' countries communities who are facing an epidemic of ill-being (mental sickness) and lack of psychological resilience. A bridge between universities and these broken communities needs to be forged for the sake of both of them, the 'developed' world, and the world generally. This is required so as to retain the hope that opportunities will never cease from being on the horizon, just as long as the present challenges are realistically addressed. Currently there is however often a tendency to reverse this; the easier option being short-term denial to enjoy what remains in the present, whilst being in fearful gloom about what the future holds. For meaningful resilience to be gained by university students there is nothing more urgent than to correct this inverted narrative of terror, so that the youthful idealism surging hope and expectation, can be directed instead to fulfil its natural purpose. Universities have an obligation to help construct this narrative of realistic hope.

References

Argyris C, Schon DA (1996) Organisational learning II: theory, method and practice. Addison-Wesle, Reading
Blackstock C (2011) The emergence of the breath of life theory. J Soc Work Val Eth 8(1):1–16
Bossel H (1998) Earth at a crossroad: paths to a sustainable future. Cambridge University Press, Cambridge, UK
Butler J (2004) Precarious life: the powers of mourning and violence. Verso, London

Fischer AP, Paveglio T, Carroll M, Murphy D, Brenkert-Smith H (2013) Assessing social vulnerability to climate change in human communities near public forests and grasslands: a framework for resource managers and planners. J For 111(5):357–365

Foxon T, Reed M, Stringer L (2009) Governing long-term socio-ecological change: What can resilience and transitions approaches learn from each other? Environ Policy Govern 19:3–20

Geels FW (2002) Technological transitions as evolutionary reconfiguration processes: a multi-level perspective and a case study. Res Policy 31:1257–1274

Giddens A (1990) The consequences of modernity. Stanford University Press, Stanford

Hone L (2017) Resilient grieving: how to live with loss that changes everything. Allen & Unwin, Auckland

IPPC (2014) AR5 climate change 2014: impacts, adaptation, and vulnerability. Cambridge University Press, Cambridge

Jackson T (2017) Prosperity without growth: foundations for the economy of tomorrow. Routledge, New York

Lauer M, Albert S, Aswani S, Halpen BS, Campanella L, La Rose D (2013) Globalisation, pacific Islands, and the paradox of resilience. Glob Environ Chang 23(1):40–50

Maru YT, Smith MS, Sparrow A, Pinho PF, Dube OP (2014) A linked vulnerability and resilience framework for adaptation pathways in remote disadvantaged communities. Glob Environ Chang 28:337–350

Maslov A (1954) Motivation and personality. Harper & Brothers, New York

Max-Neef MA, Elizalde A, Hopenhayn M (1991) Human scale development: conception, application and further reflections. The Apex Press, New York

Mercer J (2010) Disaster risk reduction or climate change adaptation: are we reinventing the wheel? J Int Dev 22(3):247–264

Morrison KD (2012) The promise of orthodox christianity for sustainable community development. In: Williams L, Roberts R, McIntosh A (eds) Radical human ecology: intercultural and indigenous approaches, pp 179–203. Ashgate, Farnham, Surry

Morrison KD (2016) The role of traditional knowledge to frame understanding of migration as adaptation to the "slow disaster" of sea level rise in the South Pacific. In: Sudmeier-Rieux K, Jaboyedoff M, Fernandez M, Penna IM, Gaillard JC (eds) Identifying emerging issues in disaster risk reduction, migration, climate change and sustainable development, pp 249–266. Springer

Morrison KD, Singh SJ (2009) Adaptation and indigenous knowledge as a bridge to sustainability. In: Lopes Priscila, Begossi Alpina (eds) Current trends in human ecology. Cambridge Scholars Publishing, Newcastle upon Tyne, pp 125–155

Piketty T (2014) Capital in the twenty-first century. The Belknap Press of Harvard University, Cambridge, Massachusetts

Rappaport RA (1979) Ecology, meaning and religion. North Atlantic Books, Berkeley

Rappaport RA (1999) Ritual and religion in the making of humanity. Cambridge University Press, Cambridge

Raworth K (2017) Doughnut economics: seven ways to think like a 21st-century economist. Random House Business Books, London

Shinko T, Mecher R, Hochrainer-Stigler S (2019) The risk and policy space for loss and damage: integrating notions of distributive and compensatory justice with comprehensive risk management. In: Mecher R, Bouwer LM, Shinko T, Surminski S, Linnerooth J (eds) Loss and damage from climate change: concepts, methods and policy options, pp 83–110. Springer Open. Accessed 20th Mar 2019. http://doi.org/10.1007/978-3-319-72026-5

Tosey P, Visser M, Saunders MNK (2011) The origins and conceptualizations of triple-loop learning: a critical review. Manag Learn 43(3):291–307

Weil S (1952) The need for roots. Routledge Classics, London

Chapter 27
Nurturing Adaptive Capacity Through Self-regulated Learning for Online Postgraduate Courses on Climate Change Adaptation

Keith Morrison, Moleen Monita Nand and Heena Lal

Abstract The University of the South Pacific provides postgraduate courses on climate change adaptation across the South Pacific region to 12 countries. In 2014 a programme to implement research skill development was extended to the postgraduate courses. The implementation process over four semesters culminated in a pilot study to evaluate the effectiveness of a model of pedagogy for postgraduate climate change education that was critically developed over the period. A challenge facing the project was the need to increasingly use a purely online mode of delivery. The approach taken was to explore possible synergies that might be found between self-regulated learning necessary for online courses, research skill development, and the adaptive capacity required for effective climate change adaptation. Postgraduate students were nurtured to understand, and required to participate with, adaptive capacity to initiate research relevant for the climate change adaptation needed within the region they live and work. Self-regulated learning pedagogy was integrated with research skill development pedagogy, by focusing on the recursive processes involved, including the need for feedforward as well as feedback, and aligning it with the adaptive capacity required for climate change adaptation. Helpful practical developments in online delivery were discovered. In the fourth semester, to gain an initial assessment of the effectiveness of the pedagogy for postgraduate climate change education that was critically developed, a pre-post survey was conducted. A total of 21 students out of a 34 (63%) participated in the pre-survey and 16 students out of 30 (53%) participated in the post-survey. There were 19 questions in the survey. 68% (13) of the questions showed highly significant improvement ($p < 0.05$) and 90% (17) showed significant improvement ($p < 0.1$). It was concluded that it is possible to efficiently provide effective and relevant postgraduate climate change education for developing nations that are bearing the brunt of climate change.

Keywords Postgraduate · Adaptive capacity · Self-regulated learning · Pedagogy · Climate change

---

K. Morrison (✉)
Sustainable Community Development Research Institute, Christchurch, New Zealand
e-mail: keithdmorrison59@gmail.com

M. M. Nand · H. Lal
The University of the South Pacific, Suva, Fiji

© Springer Nature Switzerland AG 2019
W. Leal Filho and S. L. Hemstock (eds.), *Climate Change and the Role of Education*,
Climate Change Management, https://doi.org/10.1007/978-3-030-32898-6_27

Introduction

Climate change is already having a dramatic impact on the nations of the South Pacific. The University of the South Pacific (USP), which is a regional university providing for 12 of the South Pacific nations, has responded. A Postgraduate Diploma, M.Sc. and Ph.D. programme in Climate Change have been established. However, being a regional university, where each member nation has its own campus, delivery of these programmes has faced challenges. This is especially the case for the Postgraduate Diploma in Climate Change, because approximately half of students enrolled in it remain working while doing their study; working in positions where the study is recognized as a beneficial, if not necessary, upskilling requirement. With enrolments for the Postgraduate Diploma reaching over 30 a year, delivery of the postgraduate diploma programme had to be made available through all campuses.

The challenge USP faced with the climate change postgraduate programmes was however not new to the university. USP had already been delivering courses in three ways. There was the 'traditional' delivery where students attend lectures, tutorials, laboratories and fieldwork in person at a campus. A second type of delivery has the whole subject taught online through use of the MOODLE platform, facilitated by a course coordinator. A third delivery was a 'blending' of traditional delivery and online delivery. Whereas some courses are relevant only for one campus, where traditional delivery is entirely appropriate, some courses, like those in the postgraduate Diploma in Climate Change, that are delivered to multiple campuses, require for efficiency and consistency to be delivered by online mode across the 12 member nations of the South Pacific.

But the type of delivery is not pedagogically neutral. It is more than merely a matter of efficiency and consistency. Some aspects of what is learnt, are learnt through the mode of delivery. Learning is a process. Moreover the learning process is a dynamic interaction between concepts conveyed symbolically through language and the social and natural environment; a dynamic interaction usually termed praxis. Learning is a process of praxis. Instead of seeing the effects of the mode of delivery on education as a problem to be overcome, it is more appropriate to rather see the mode of delivery as an intrinsic feature of and participation in the social, and often also natural, context. Therefore the model of delivery is something to be developed and used for particular purpose within praxis.

Praxis has long been a focus of "critical pedagogy", heralded by Freire (1970). Freire long ago eloquently critiqued the view that pedagogy was purely about instilling conceptual content by teachers into the minds of students, metaphorically labelling it the "banking" approach to education. Freire argued instead for a focus in pedagogy on how concepts are co-constructed in the real-life social and natural environment, and then applied in the same real-life, which then changes the co-learning environment; and so on in a continual social learning and adaptive recursive[1] process, or praxis-loop. Given that the contemporary concern about climate change is explicitly concerned with raising awareness about what is happening to the natural as well as social environment, and how to mitigate and adapt to it, critical pedagogy as heralded by Freire is an appropriate starting point for explicit concern about the pedagogy of climate change education, including concern about the appropriate mode of delivery.

The explicit use of critical pedagogy began at USP at the culmination of two climate change education projects run by the Pacific Centre for the Environment and Sustainable Development, over the period 2014–2015. However it was implicit in the two projects right from their beginning.

One of the projects was funded by the Australian Aid agency, then called AUSAID. It was a project within the Climate Leaders programme (completed in 2014). The other project was funded by the European Union for a project by the Global Climate Change Alliance (EU-GCCA) (completed in 2016). The two projects focused on different aspects of the praxis of addressing climate change. The AUSAID Climate Leaders Programme project focused on the nurturing of young leaders about climate change, including about processes underway to address it among the countries of the South Pacific region. The EU-GCCA project focused on establishing linkages between the multiple potential stakeholders participating in climate change adaptation, to establish viable climate change adaptation programmes in all countries of the South Pacific region. The linkages were catalysed by "In-country coordinators" who participated in local communities as well as national and South Pacific regional wide public sector and NGO agencies, and who established pilot projects in local communities according to best practice in climate change adaptation. One place where the two projects crossed over and resonated to support each other was in the Postgraduate Diploma in Climate Change. The young "climate leaders" and the "In-country coordinators" all took the Postgraduate Diploma in Climate Change, or participated in some of the workshops run as part of it. The Climate Leaders Programme project and the EU-GCCA project also both helped fund the teaching of the Postgraduate Diploma in Climate Change.

The reason why critical pedagogy was nascent in the teaching of the Postgraduate Diploma in Climate Change run by USP before 2014–2015, and only became explicit during 2014–2015, was twofold. It was partly because of an increased focus globally on climate change adaptation, rather than only climate change mitigation, after the fifth assessment by the Inter-governmental Panel on Climate Change (IPPC 2014). It was also because USP had begun in 2014 to roll out a programme to embed research skill development in all postgraduate courses at USP.

A focus on climate change adaptation is appropriate for South Pacific nations. South Pacific nations can do little directly to mitigate climate change, even though they are indirectly doing so, as leading voices in international forums to argue that

---

[1]Recursive processes are ones that fold back onto themselves through feedbacks, resulting in repeated similar processes. Therefore patterns emerge as a resonance of underlying looped processes.

larger and industrialised countries need to mitigate climate change because of the effect it is having on countries in the South Pacific, and also to other small developing countries elsewhere.[2] South Pacific nations are therefore mainly focusing on their own need for climate change adaptation, for example the EU-GCCA project. The IPPC (2014) fifth assessment detailed best practice in adaptation and highlights regional issues that need to be addressed. The IPPC (2014) therefore became the new bench mark for what had to be taught in the Postgraduate Diploma in Climate Change. With a clearer focus on adaptation, the postgraduate courses became more concerned with praxis, and so critical pedagogy became more pertinent. But it was the embedding of the RSD framework into the courses that was the catalyst for making explicit use of critical pedagogy.

The research skill development (RSD) framework that USP required to be embedded into courses, was based on Bloom's taxonomy (Bloom et al. 1956), as developed by Willison and O'Regan (Willison and O'Regan 2006, 2013). It was a comprehensive framework. To seriously embed this RSD framework into courses, rather than to just tack it on, required conscientious reflection on the pedagogy behind the whole of each of the courses. This process reinforced how facilitating the learning of climate change adaptation implies the need to understand praxis, and hence requires a critical pedagogy. Two related concepts also emerged, cutting across all aspects of the courses. One was the need to facilitate "transformative" leadership capacity in students (Kuhnert and Lewis 1987). The other was the need to explicitly facilitate self-directed learning.

"Transformative" leadership is necessary for climate change adaptation, because the adaptations required have to be innovative. The other recognised type of leadership, "transactional" leadership, is only appropriate when processes are clearly laid out and needing only to be implemented. But innovation also requires self-regulated learning. Innovation requires the adaptive capacity to create something new from out of one's thinking. And creativity requires an energy or self-empowerment to initiate and maintain (to regulate) one's learning. Moreover, self-regulation of learning implies a certain type of engagement with the learning process, or praxis, and so implies a certain type of mode of delivery. Therefore the logistical need for online delivery of the postgraduate climate change courses had to be critically juxtaposed with the need for self-regulated learning within the courses.

Addressing this challenge of what is the appropriate mode of delivery became forced, when the Climate Leaders Programme project's funding of the transportation and accommodation (with per diems) of all students of the Postgraduate Diploma for a week-long intensive workshop at the Laucala campus in Suva, Fiji, ceased in 2014. Efficiency in the delivery of the climate change courses also became a critical issue, and therefore there was increased pressure to provide purely online courses. An explicit pedagogy was required that allowed for purely online delivery, whilst enabling and, if possible, enhancing self-regulated learning, along with embedding of the RSD framework. The pedagogy of online delivery had to become focused on. It was however fortuitously noticed that a feature, though also challenge, of online

---

[2] The political leaders of Tuvalu, Kiribati and Fiji have been very vocal.

courses is that they require a relatively greater reliance on self-regulated learning. Moreover it was noticed that the pedagogy of self-regulated learning is similar to that of the RSD framework. These commonalities gave a coherent starting point for the necessary multi-faceted integration.

A literature review of self-regulated learning in relation to research skill development and adaptive capacity is presented in the next section, followed by a conceptual model of the pedagogy that was developed for postgraduate online courses in climate change. Next the methodology is outlined, followed by the results of the research. Finally there is discussion of the results of the research, and then concluding remarks about the process of utilizing the pedagogy for climate change adaptation and education.

Literature Review

Even though all education, including postgraduate university education, requires a degree of self-regulated learning, online delivery of a course demands self-regulated learning to a high degree. But, even though online delivery of courses requires a high degree of self-regulated learning, what is necessary to facilitate self-regulated learning is itself challenged by online delivery. The challenge revolves around how self-regulated learning involves a "bootstrapping" process, where students need to pro-actively develop "feedforward" within a recursive process.

Bootstrapping refers to self-initiated processes, which then over time take on a particular form, which is the pattern that emerges from an underlying looped recursive process. The pattern emerges because of resonance that occurs due to the feedback loops. Bootstrapping is the creative process, and is what adaptive capacity enables. But while being necessary, creativity is by itself insufficient for a pattern to emerge. Creativity has to also be tested. Creativity can only create hypotheses, which then have to be repeatedly tested by engagement with reality, through the recursive process of praxis, until some stable pattern emerges as vindication of what are adequate, but only ever contingently validated, hypotheses. Hypotheses are a projecting (feeding) forward of a symbol describing the potentially resonating pattern emerging from the recursive process of praxis.[3] The truth emerges through the praxis, as a stable symbolic representation of the reality that is repeatedly engaged with. Praxis is therefore experimentation. The reality is discovered (successfully symbolically represented) by the engagement with it through the recursive process of praxis. Discovery occurs as "feedforward" (a hypothesis) is vindicated by the feedback in the recursive process of the praxis. It is how research occurs, as well as how adaptation occurs. Therefore, to bootstrap, is to gain the adaptive capacity to initiate research, as well as to initiate adaptation. And to facilitate the bootstrapping process is to facilitate research skill development, as well as to nurture adaptive capacity.

---

[3]The epistemology proffered here is a type of critical realism. A reality is assumed to objectively exist, but how it is to be described remains relative to the observer.

This is however only what initiates adaptation. Once some particular feedforward (hypothesis) has been vindicated through research, the truth that emerged as a symbolic representation of the pattern emerging from praxis, becomes a guide for future engagement with that particular aspect of reality. It becomes a guide or map for leadership to implement the policy and to develop of programmes and projects concerned with the particular aspect of reality. The leadership required to implement policy is "transactional "leadership". But to get to that point there has to first be the "transformative" leadership engaged in anticipatory learning (Tschakert and Dietrich 2010), to discover and establish the emergent truth about the aspect of reality. Transformative leadership creates feedforward within a process of anticipatory learning, which is then implemented incrementally by transactional leadership.

Therefore there is also a recursive process to adaptation, in both research and leadership. There is initially (i) a creative transformative stage, followed by (ii) a stage of incremental transactional development, which eventually levels off into (iii) a relatively steady state, until some event will trigger (iv) a collapse, and then the re-emergence of another creative transformation, and so on. The recursive process has been called the "adaptive cycle" (Holling and Gunderson 2002), and modelled by the infinity symbol (∞). Modelling of the adaptive cycle came out of research on adaptive environmental management (Holling 1978), which developed to define two types of resilience: "engineering resilience", and "ecological resilience" (Gunderson et al. 2003).

"Engineering resilience" refers to what transactional leadership seeks to maintain. It refers to the ability to engineer the implementation of policies into programmes and projects. The adaptation required for engineering resilience involves the fine-tuning of the processes of learning to ensure the successful implementation of policies.

"Ecological resilience" refers to what transformative leadership seeks to maintain, by considering the whole adaptive cycle, where the four stages in the adaptive cycle are repeated. It refers to the need to attune to natural and social processes, which are not able to be controlled (engineered). Ecological resilience refers to the ability to anticipate the limits to steady development, as well as to anticipate the limited period of any steady state before eventual collapse. The adaptation maintaining ecological resilience requires anticipatory learning; to be prepared for the levelling off of development and eventual collapse, by having transformative leadership ready to research new policies that will take root to start new forms of development as and when necessary.

There is a broad inter-disciplinary field of many researchers applying and developing the model of the adaptive cycle. The researchers have grouped themselves as the Resilience Alliance.[4] Natural conservation, adaptive governance, economy, indigenous and traditional knowledge, as well as adaptive environmental management, are all researched. It is in the concern for indigenous and traditional knowledge that pedagogical concerns are often raised. In particular, the role for "spirituality" in pedagogy is explicitly addressed.

---

[4]The Resilience Alliance publishes an online journal Ecology and Society (http://ecologyandsociety.org).

The recursive process of adaptation in both research and leadership, as explored by the Resilience Alliance, is also recognised in critical pedagogy. The critical pedagogy initiated of Freire predated the work of the Resilience Alliance, but, one feature of the adaptive cycle, the principle of renewal emerging out of collapse, is nevertheless present in Freire's work. It is indeed the central focus of it: there is the possibility of transformative leadership in face of inflexible and "oppressive" policies and practices. Freire emphasizes that the anticipatory learning for transformative leadership emerges firstly amongst those who are "oppressed". A clear example can be seen in the global concern for climate change. Some of the most vocal international leadership seeking to ensure climate change mitigation by the larger developed and industrialised nations, is coming from under-developed low-lying micro-nations who are bearing the brunt of the onset of climate change, even though they have been an infinitesimal contributor to it.[5] Even though Freire's work remains therefore highly relevant, it nevertheless only deals with the pedagogy concerned with the stage of transformation, the first stage in the adaptive cycle. What is not considered by the critical pedagogy of Freire is the pedagogy concerned with wider recursive process of the adaptive cycle, and the need therefore to anticipate and to be prepared for continual renewal, including when steady incremental progress is possible and appropriate.

The wider concern of continual renewal was first addressed in pedagogy by Argyris and Schon (1978), where a Triple-loop learning process was proposed. Three nested loops of learning were postulated. Triple-loop learning does not explicitly refer to the pedagogy concerning the adaptive cycle, but it does nevertheless cover the same processes found in the adaptive cycle. The first loop refers to steady development through implementation of policy by optimising variables, often referred to as key performance indicators, to maintain engineering resilience. To recap, this is where transactional leadership is appropriate. The second loop refers to recognition that diversity adds to ecological resilience, and so multiple viable developmental processes are simultaneously maintained. This still only involves transactional leadership, but flexible transactional leadership. The third loop refers to recognition of what are the most appropriate of the viable development processes to focus upon, as well as to recognise the need for creation of new ones in preparation for when they may become necessary. This is where transformative leadership is required, to maintain ecological resilience. The third loop therefore has an implicit ethical dimension to it—the ethic of maintaining developing processes.

Argyris and Schon's (1978) pedagogy can be contrasted to Freire's (1970) critical pedagogy, because unlike Freire, Argyris and Schon do not refer explicitly to inflexibility, collapse, "oppression" and injustice. Nevertheless they do refer implicitly to such negative processes, but in the general and neutral terminology of negative feedbacks that need to be learnt from. It is up to the user of the Triple-loop learning model to recognise the "spiritual", ethical, sociological, economic, political and ecological dimensions, and to learn from negative events that are experienced within them. They

---

[5]This is the thrust of what the political leaders of Tuvalu, Kiribati and Fiji have regularly spoken out strongly in international forums.

are all experienced in climate change. Postgraduate climate change education has to therefore recognize and also facilitate the experience of critical engagement with the challenging complexity of climate change, and to be able to learn from it.

The bootstrapping involved in self-regulated learning is how the critical engagement with the complexity of climate change is experienced. It involves creative and critical exploration, and experimentation. But at USP, and especially from 2014 onwards, there was also pressure to do so through a purely on-line mode of delivery. This was problematic because there are two facets of the bootstrapping of self-regulated learning that are particular challenging for online courses. The first is to do with facilitating the initiation and maintenance of bootstrapping. The second is to do with the role of feedforward in bootstrapping.

Firstly, it is recognized that normally, explicit facilitation of self-regulated learning is required. Self-regulated learning does not normally spontaneously emerge. Moreover, there are two parts to what needs to be facilitated. The first part is the initiation of bootstrapping through mutually supportive roles provided by peers, the supportive role of teachers, and the support provided by the community cultural context (Nicol and Macfarlane-Dick 2006; Winne 2010a). The second part is the need for explicit supportive guidance of the self-regulated learning process once it is underway (Kirschner et al. 2010). Even though this may bear little resemblance to what is facilitated through a traditional lecture room delivery mode, it bears even less resemblance to an isolated student in front of a laptop. Clearly there has to be special design of online delivery to enable the types of and degrees of interaction with peers, teachers and communities, which are necessary to lead to successful bootstrapping of self-regulated learning—both its initiation and continued facilitation.

Secondly, it is recognised that there is the need to ensure students develop feedforward to sustain their bootstrapping of SRL. Feedforward is intrinsic to self-regulated learning, because self-motivation to take responsibility to plan and execute study is intrinsic to it. But it is not a one-off action. Rather it is a recursive process where feedforward from the learner repeatedly changes the context in which learning occurs (Winne 2010b)—the praxis loop. Therefore the learner has to also be flexible. But they need to be flexible with not only the external context of the supportive roles provided by peers, teachers and communities, but also to be flexible in themselves toward their reliance on tacit knowledge of meta-cognitive processes—of how their learning occurs (Murtagh and Baker 2004; Sadler 2010; Zimmerman 2002). Both create feedforward from reflection on feedback perceived through the lenses of their tacit knowledge (Quinton and Smallbone 2010). Students have to be intuitive and creative to pro-actively explore new possible avenues for further learning; both potentially new supportive contexts, and potentially new ways of thinking. It is a complex process whereby a student has feedforward inspired by supportive interactions, which provide feedback. Therefore there is the challenge to ensure that online delivery facilitates, in a flexible manner, repeated feedback from peers, teachers and the community. But it has to be of the sort that is provided by supportive interactions. If students are going to be inspired to pro-actively explore through creative feed-forward and to critically test it, they have to feel empowered to explore, rather than to feel constrained or judged by the feedback they receive. Only if they feel empowered, is the bootstrapping of their self-regulated learning sustained.

Conceptual Model of the Pedagogy for Postgraduate Courses in Climate Change

The conceptual model proposes a way in which to visualise the key facets needing to be reconciled and integrated together. The challenge is to communicate nested dynamic interactions occurring in multiple dimensions. The conceptual model is an expression of feedforward, aimed to nurture a steady development and implementation of postgraduate climate change education. The mode of delivery is not referred to in the model. The model is applicable for all possible modes of delivery, including a purely online mode of delivery. It is expected that the model is implemented by considering first the feasible modes of delivery, and to then implement the model with that chosen mode of delivery.

The integrative concept that describes all of the dynamic interactions is bootstrapping, both its initiation as well as its maintenance. This covers all four phases of the adaptive cycle. But, neither bootstrapping nor the adaptive cycle are explicitly referred to in the model, because the conceptual model refers to pedagogy and not the dynamics that the pedagogy has to facilitate learning about.

The overarching skill facilitated to be learnt is that of adaptive capacity, and it is to be learnt through self-regulated learning. The pedagogical theories that have informed the model are: critical pedagogy, Triple-loop learning, adaptive learning, and constructivism.

Constructivism refers to theory that emphasizes the creative process of hypothesising symbolic representations of reality. However even though there are multiple possible and viable forms of constructed truth, and they are relative to the intentions of the agent actively engagement with reality, they are not arbitrary; they do not constitute reality. Rather, reality provides negative feedback to correct the inadequacies of the relative truth discovered and constructed, provides positive feedback to affirm them, and moreover inspires the feedforward of creativity, as well as nurturing rational construction of conceptual understanding. Note that reality inspires, corrects and affirms; it does not dictate or determine a one-to-one correspondence between itself and truth. Moreover it cannot be presumed that learning results in the discovery of truth that iteratively converges toward a singular form. The importance of constructivism for our purposes is that it is an understanding that gives space for creativity and plurality, and hence is consonant with transformative adaptation, bootstrapping and self-regulated learning. It clarifies that an adequate pedagogy has to facilitate openness and inspired creativity, as well as rationality.

The various facets of learning implicit in constructivism are found empirically to occur in cultural traditions. Rappaport (1999), working in environmental anthropology, defined a hierarchy of types of knowledge production found within cultures, which he characterised by four levels. The three lower levels are covered by constructivism. The lowest two levels are the rational construction of rules or policies (second level) and the application of rules or policies (first level). The first level refers to the implementation of policies by transactional leadership, which is where truth is fine-tuned by (both positive and negative) feedback within the first loop of Triple-loop learning. The second level refers to the latter part of the bootstrapping process, where feedforward (hypotheses) are tested by feedback (both positive and negative) to select appropriate rules or policies. It is maintained by transformative leadership within the second loop of Triple-loop learning. The third level is the rational construction of worldviews, which includes integrative conceptual models, such as that proposed here. It also refers to the bootstrapping process carried out by transformative leadership within the second loop of triple-loop learning. But it is still only the middle stage of the bootstrapping process, where rational development of hypotheses and new policies is occurring, though without them yet being tested through interaction with reality. The fourth level refers to the "invariant principles" or "horizon" that inspires the creative process. It refers to the initiation of the bootstrapping process, also carried out by transformative leadership, but within the third loop of Triple-loop learning. It is the stage of creative inspiration, where creativity has yet to begin to rationally construct coherent hypotheses. This fourth stage is not covered by constructivist theory. It is the creative inspiration that "spirituality" refers to.

Constructivism is concerned with rational concept-making processes within praxis. The critical interaction with reality through feedbacks within praxis means that constructivism recognises that the context influences the knowledge that is constructed. The context is social as well as natural. Therefore it recognises that knowledge is "socially constructed", by which is meant, the construction is socially influenced. For example by cultural traditions, educational institutions' syllabi, and global organisations, for example the IPPC. But because there is a hierarchy of knowledge, where the highest level is beyond the knowledge construction process, the construction of knowledge is not determined by either social or natural influences. There is "free-play" in the knowledge construction process (McGowan 1991). Free-play refers to how the learning process cannot be determined, for example by forcing economic and/or institutional change. It can only ever be influenced, and only to a relative degree manipulated. The free-play is human agency; both inspired creativity and rational construction. It is adaptive capacity. It is also the flexibility in sociocultural institutions. It is therefore essential for the resilience of society. Even the strongest analysts of the influence of context to create and subconsciously maintain ideology (Foucalt 1977) and hegemony (Gramsci 1992), also hold the ethic of seeking to increase human agency or adaptive capacity. And moreover, they hold that increasing human agency is a rational process of social learning, gained through become aware of the contextual influences on the constriction of knowledge.

An ethic to enhance human autonomy or adaptive capacity, gives the purpose and direction to social learning and anticipatory learning. Moreover it is developed through a process of rational discourse, or dialogue with others. Rational discourse and co-learning through dialogue is the pathway for resilient development. What is necessary is to have a horizon that is open to the possibility of transformative change when it is necessary, and incremental progress when it is possible. This is openness to the higher levels of cultural knowledge, where not only policies, but also worldviews can become changed. The autonomy or adaptive capacity of humans, and the flexibility in socio-cultural, including economic, institutions, is an openness to what is ethical and beautiful, which transcends conceptual or discursive knowledge. It is the realm of ethical intuitions, aesthetics and spirituality. These are not less than rationality. They are not expression of irrationality, but rather of trans-rational exploration of alternative threads of rationality, which then allow the weaving together of new interdisciplinary and pluralistic tapestries of rational thought – the ever continuing co-evolution of socio-cultural institutions and traditions between each other, and their ever continuing co-evolution with their local and global natural environments.

There are many way that a conceptual model can be rationally constructed. There is no singularly true model to be discovered. Intentions about what needs to be covered and focused upon gives the overarching perspective and flavour, and the degree of detail pertaining to already existing models and theories of pedagogy gives its texture. To outline how to enhance adaptive capacity to create flexibility and resilience is the overarching intention, and the terrain that is covered has to show all the significant aspects of self-regulated learning, whilst detailing key features of Triple-loop learning and the levels of knowledge construction, and what they involve. In particular, the metaphor of horizon has been used to indicate what the ethic and purpose is, by placing it at the extremity of the model. What is placed centrally is by contrast the lowest level of knowledge construction, to give the sense that it is the "nuts and bolts" of the process. The centre is portrayed with the lightest tone to give the sense that it is where there is the greatest precision, though also the greatest rigidity. As one moves outwards toward the horizon, to ensure ethical purpose is maintained, darker tones are used because things become fuzzier: things are increasingly imprecise, but at the same time more flexible. The paradoxical tonal movement is to indicate that all levels are equally necessary and complement each other, and must be used together as a whole. The model is outlined in Fig. 27.1.

Adaptive capacity is simultaneously individuation or personal development, and community engagement. Or in terms of research skills, it is simultaneously the gaining of autonomy and the gaining of the ability to dialogue, so as to contribute meaningfully with originality within disciplines and traditions. The complementary movement paradoxically heads towards the same horizon and purpose, to emphasize that adaptive capacity is not due to convergence onto a singular solution, but rather never-ending inspired, creative and rational process of dialogue and engagement. To the extent creativity is gained through autonomy or agency, there is simultaneously the aesthetic gaining of respect with a sense of awe and wonder at others and all of reality. To the extent autonomy develops critical thinking, there is simultaneously

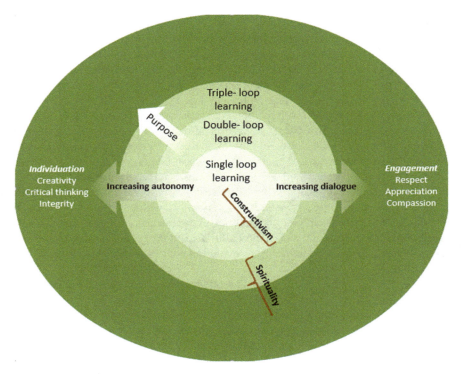

Fig. 27.1 Pedagogy for postgraduate climate changed education

the development of appreciation for the intrinsic value of others and all of reality. To the extent ethical integrity is maintained as autonomy and power is gained, there is simultaneously increase in compassion for others who likewise are facing the same or similar challenges, or worse.

The rational clarity sought by constructivism is the appropriate vehicle for implementing carefully the 'nuts and bolts' of climate change adaptation. However it requires an inspirational horizon informed by spirituality, and this is what also inspires continual striving to gain enhanced skills to better fulfil the purpose and aim of climate change education.

Figure 27.1 is a pedagogical tool to be used directly by students. It is an expression of feedforward that has coalesced from participation in the praxis of educating postgraduate students about climate change, to give students some means by which to begin to become aware of their tacit knowledge of their cognitive processes, and then to expand on it to help bootstrap their self-regulated learning, and hence adaptive capacity and research skills.

Methodology

The aim of the research was to develop an explicit pedagogy for purely online delivered postgraduate courses in climate change. There were two key prompts for the research. One was the resource pressure to have purely online courses. The other was the requirement to embed an explicit research skill pedagogy, the RSD framework.

There were three steps to the research. The first step had the objective (i) to postulate a coherent conceptual model for an appropriate pedagogy. The pedagogy was required to critically incorporate the RSD framework, whilst simultaneously ensuring successful outcomes from a purely online mode of delivery.

The second step had the objective (ii) to implement the conceptual model over several semesters, to obtain feedback. The feedback was to be informally obtained through all interactions that occurred when teaching the courses that implemented the pedagogy.

The third step had the objective (iii) to formally and statistically evaluate the success of the implementation of the pedagogy.

Objective 1

The methodology used for the first objective was literature review. It is assumed that the methodology of literature review is well known, and so will not be commented upon, except to point out that the literature review was required to be interdisciplinary. There was the need to juxtapose several different disciplinary perspectives in the attempt to find sufficient common ground to enable their integration. Whenever plausible common ground was found, it became then the object of further literature review to explore how others had previously explored that route. Multiple steps were taken in this manner until cross-cutting concepts were found across all the fields needing to be integrated. Because the review was interdisciplinary, the cross-cutting concepts were not necessarily the same word or phrase, but rather required to be analysed within the context of their development, to ascertain if they were actually meaning the same thing. A trans-disciplinary frame was used to assist in this, namely systems theory. For example, terms such as feedback, feedforward, resilience and flexibility had similar systems theoretical meaning, but it was nevertheless required to determine how they were actually used in the context of the different disciplines.

Objective 2

The methodology for the second objective was action research.

The essence of the methodology of action research is that it is recursive. It involves repeated experiments, where the results of the previous one directly guide the production of the new hypothesis. The methodology is appropriate when seeking to gain clarity in messy complex situations. In this case there were four semesters of teaching where each course became its own informal experiment. The pedagogy implemented was informed informally from the feedback obtained from all possible sources during the previous semester's teaching, along with further literature review over the semester. For the first semester of the process, in 2014, there was feedback recorded in various forms from previous semesters, which was utilised in the construction of the initial conceptual model for a pedagogy, along with the initial literature review.

Objective 3

The methodology used for the third objective was a pre-post statistical test to determine if there was significant improvement in adaptive capacity. Because it was a formal research project, it required ethics approval. It was not run by the course coordinator (teacher), but rather by the USP Research Skill Development programme coordinator. All provision of information and interaction with students about the pilot study were carried out by the Research Skill Development programme coordinator. All participation by students was voluntary. Neither the course coordinator nor the teaching assistant knew which of the students participated in the pilot study, nor discussed it with them. Students were informed of this. Their confidentiality was strictly maintained.

The methodology for pre-post formal statistical tests is standard. First the appropriate indicators had to be ascertained. Then the appropriate test statistic used. The appropriate test statistic could only be decided upon once the normality or otherwise of the data obtained was ascertained. If the data was not sufficiently normal, a non-parametric measure would be required, otherwise an appropriate parametric measure used. The methodology did not contain any experimental design. This was because it was a single pilot study evaluation, with no control nor replication. The indicators were derived from the research skill development pedagogy (RSD framework), which was embedded into the course. Even though the pedagogy of the course had become critically developed during the course of the research, the RSD framework nevertheless provided an appropriate comprehensive range of indicators that could be used.

The pedagogy of the RSD framework models a repeated sequence of "six research facets" or meta-cognitive processes: from (i) "embark and clarify—curious" to (ii) "find and generate—determined", to (iii) "evaluate and reflect—discerning", to (iv) "organize and manage—harmonising", to (v) "analyse and synthesize—creative", to (vi) "communicate and apply ethically—constructive". The process is explicitly modelled as a spiral, to recursively bring ever fuller expressions of each of the meta-cognitive processes during each turn, starting from: Level 1—"prescribed research"; to Level 2—"bounded research"; to Level 3—"scaffolded research"; to Level 4— "student initiated research"; to Level 5—"open research". Moreover, Level 1 is designated to be what is expected for the first year of university study, Level 2, to the second year of study and so on, up to Level 5, which is designated as research for a thesis. Postgraduate courses are designated to require "student initiated research".

The range that the set of indicators has to cover is what the six "facets of research" cover. But the details of what the sets of indicators are depends on the level of research. Some of the indicators covered what was required for transformative leadership, and others, what was required for transactional leadership. 19 indicators were selected.

The 19 indicators were evaluated by 19 questions. Pre-test and post-test questions were identical (see Table 27.1). There was however no expectation that student

27 Nurturing Adaptive Capacity Through Self-regulated Learning ...

Table 27.1 *P* values for the comparison of pre-test and post-test of the embedding of the RSD framework into PC425 in 2015

Question	*P* value
1. I am able to define the overall aim and specific objectives for both Environmental Impact Assessment (EIA) and Strategic Environmental Assessment (SEA) report	0.057
2. I know how to and the types of research questions required to be asked for an EIA and SEA	0.0
3. I know where and how hypotheses are used for the research carried out for within EIA and SEA	0.003
4. I know how to and am able to access all information required for an EIA and SEA	0.010
5. I am always able to distinguish between a reputable and legitimate source of information form a dubious one	0.048
6. I know all of the types of methodology appropriate for EIA and SEA, as well as how to choose and integrate the most appropriate ones for any particular EIA and SEA	0.002
7. I am able to reference all types of sources correctly	0.030
8. I am able to determine the multiple perspectives needing to be considered for an EIA and SEA	0.002
9. I am able to determine the appropriate discipline for information needed for each perspective required for an EIA and SEA	0.0
10. I know how to appropriately structure an EIA and SEA report	0.003
11. I construct tables well	0.170
12. I construct figures well	0.168
13. I am able to create themes according to the appropriate disciplines used in EIA and SEA	0.0
14. I am able to construct cross cutting themes for EIA and SEA and know why they are necessary for a successful EIA and SEA	0.001
15. When new knowledge is found through carrying out an EIA or SEA I know how to incorporate it into to extend the appropriate disciplines	0.0
16. I know how to structure and argue key points in an EIA and SEA based on what has been found in relation to the hypotheses that have been tested	0.0
17. I know what is verbose or waffle language and know how to avoid using it	0.058
18. I know if there is missing information and do not try to cover it up	0.025
19. I know how to address all ethical issues that arise in EIA and SEA	0.098

would have to participate in both, or either, of the pre-test and post-test surveys. It was decided that the most effective way to evaluate this was to carry out a pre and post evaluation of students' skill in relation to each of the six research facets of the RSD framework, as they pertain specifically to the material relevant to the course. Questions were therefore formulated to indirectly seek information. It was the relative change in the skills questioned about, which was evaluated. Each question was actually a statement that respondents were asked to respond to. Ordinal values were assigned through use of a five-point Likert scale, ranging from 1 to 5. The data were statistically evaluated.

Questions 1 and 2 were concerned with the first facet of research in the RSD framework, "Embark & clarify—curious". Questions 3-5 were concerned with the second facet of research "find & generate—determined". Questions 6-7 were concerned with the third facet of research, "evaluate and reflect—discerning". Questions 8–12 were concerned with the fourth facet of research, "organise and manage—harmonising". Questions 13 -16 were concerned with the fifth facet of research "analyse and synthesize—creative". Question 17–19 were concerned with the sixth facet of research "communicate and apply ethically—constructive".

The questions that provided indicators of transformative leadership were those covering the first, third and sixth facets of research, namely questions: 1, 2, 6, 7, 17–19. The questions that provided indicators of transactional leadership were those covering the second, fourth and fifth facets of research, namely questions: 3–5, 8–16.

Results

Objective 1: *Literature review*

The literature review started at the beginning of 2014, before the course was first taught with the explicitly developed pedagogy, and continued throughout the whole period, of four semesters. It also continued afterwards. Key finding of the literature review are found in the introduction, literature review and conceptual model sections of this chapter. The literature review covered a very complex field of very active research. There is the need to continue it unabated, including now to look out for other explicit pedagogies that might have been constructed to deal with the issue of postgraduate education on climate change, and which also focus on nurturing adaptive capacity, the use of self-regulated learning, leadership, research skill development and online modes of delivery.

Objective 2: *Action research*

The conceptual model of the pedagogy developed iteratively over four semesters. However right from the beginning, in the first semester of the process in 2014, a conceptual model was explicitly constructed and provided to students. The reason is because it was initially intuitively assumed that explicit reference to pedagogy helps students reflect upon their own learning processes, to improve them. Later literature review vindicated this intuitive assumption. As outlined previously, student awareness of their "meta-cognitive" processes assists the bootstrapping of student feed-forward within self-regulated learning.

The pedagogy that was communicated to students during the first semester of the research, revolved around the RSD framework. Three models were added to complement the RSD framework. They were all communicated to students in their first online learning module, along with an introduction to the RSD framework. Students were required to develop their own syntheses of the complementary frameworks, through reflection on their "meta-cognitive processes", to guide how they developed their own self-regulated learning.

One complementary model was actually a critical application of the RSD framework. It was the construction of sets of rubrics to provide feedback for each of the first four levels of research in the RSD framework. The reason it was a critical application and provided a complementary view was because it was in tension with the supposed spiralling up through the levels of research through the years of education at university, as portrayed by the RSD framework. Whereas the course was supposedly to require Level 4, "student initiated research", and to be evaluated accordingly, the approach taken was to instead recursively nurture students from Level 1 right up to Level 5, each semester. The rubrics were also designed to be repeatedly utilised, through the online MOODLE platform MARKSHEET application, for ease of provision of feedback to students. Moreover this was explicitly done in a supportive empowering manner, rather than judgemental manner. The repetitive assessments, and constructive supportive feedback allowing resubmission, aimed to bootstrap self-regulated learning, adaptive capacity, and research skill development.

It was explicitly stated to students that the purpose of the rubrics and evaluation was for learning to occur, so students could resubmit if they wished to; if they considered that they had learnt from the feedback. And that this could continue until the final exam. The final exam itself was explicitly stated to be seen as a "prompted opportunity to communicate what had been learnt". Assessments at Level 1 started in week two and continued weekly throughout the course. There were two assessments at Level 2 starting in week 4, and one each at Levels 3 and 4 near the end of the course. The exam was set to evaluate Level 5 learning. The process aimed to gently nurture momentum in the bootstrapping of self-regulated learning, adaptive capacity and research skill development, up the levels of research.

The assumption of the assessment model was that authentic research skill development requires authentic increase in autonomy, in agreement with the pedagogy of the RSD framework, but that because this equated to heightened self-regulated learning, it paradoxically also required and depended upon increasing flexible authentic and supportive engagement with others. Therefore care was taken to avoid opportunities for individuals to 'game' assessment through making it an end in itself. For example, participation in MOODLE discussion pages with peers and the coordinator and teaching assistant was not evaluated. Rather, participation was encouraged as a means to learn from and with others as the proper end in itself. Likewise, participation in group projects (assessed at Level 3) was not evaluated according to participation.

The second complementary pedagogy was the model of Triple-loop learning. It was communicated to describe the various recursive processes of learning involved in the course. Explicit cross references to the "six facets of research" in the RSD framework were made. It was explained how the first learning loop deals with technical skill development, for example methods and referencing; the second level deals with critical questioning and juxtaposing of multiple perspectives, models and paradigms; and the third level deals with the purpose for the process of learning, in this case climate change adaptation and leadership.

The cross-referencing between the RSD framework and Triple-loop learning also questioned the supposed spiralling adaptive process modelled by the pedagogy of the RSD framework. This is because the supposed spiralling sequence in which research skills are learnt operates simultaneously at multiple nested levels according to the pedagogy of Triple-loop learning. The process is complex and surely unique to each learner/researcher and their particular task, depending on how they enter in and out of the three nested levels to accomplish what is required.

The third complementary pedagogy was that of critical pedagogy, as initiated by Freire (1970). Once again, cross references to research skills within the RSD framework were made, which like The triple-loop learning model, questions the model of spiralling progression. But in this case it was because the contexts of the external social and natural environment were explicitly evoked. Moreover, it was emphasized that there is a praxis loop of interactive learning between thinking and engagement with the external world, which is not normally a steady spiralling progression, but rather jumpy, involving discontinuous innovations according to the adaptive cycle. But once again, how to incorporate this challenge was left to students to synthesize with their own feed-forward visualisations.

During the first semester a "blended" mode of delivery was used. There was a fully funded workshop that brought all students from across the South Pacific region to Laucala campus in Fiji for one week. Up until 2014 the workshop had been a training workshop. But to implement the research skill development pedagogy, the training programme was replaced with facilitation of "student initiated research". Only the initial exercises involved training, and then steadily moved on and up to higher levels of research skill development, where greater autonomy and expectation for engagement was required.

The second semester threw up the challenge of a change in the blended mode of delivery. There was no longer to be a fully funded week-long workshop. The blended mode of delivery could not extend to the whole region, but only to Laucala campus in Fiji, to fund a local workshop. This raised the challenge of how to facilitate Level 4 "student initiated fieldwork" when only a minority fraction (about one third) could attend. It was decided to make projects contextual so every student had to choose their own research project, and that the local workshop would instead be where the meta-cognitive processes involved could be discussed. Therefore the challenge became one of how to communicate the process as best as possible online, so that everyone could participate at least to a degree.

The situation was discussed with students and their feedback resulted in transforming the weeklong workshop into multiple shorter weekend and daily events. The need to recognise the change in context and to be innovative was itself an important learning process that students co-authored. The result was actually a far more realistic research process for students to participate within, especially as it was required to be initiated by themselves. There was far more opportunity for praxis to occur, where there was an event and then reflection and critical adaptation to create the next event and so on. Students themselves worked out when further fieldwork or meetings with stakeholders were required.

As well as being careful to allow and support the innovative research process, care was taken to ensure that students reflected on the processes involved, and communicated them for the benefit of all students. Students across the region were required to develop their own research projects as teams. Assistance was given to ensure everyone had a project and team to work on and to belong to.

During the third semester the blended mode of delivery changed once again, because there was no longer any funding at all for any workshops. Funding for the whole region was for only a purely online mode of delivery. Given the challenge of how to carry out research in such a difficult financial environment, students once again leapt to address the challenge and to find an innovative solution. They quickly managed to obtain their own external funding from interested sponsors and took the lead, at Levels 4 and 5 of research skill capability, when given the opportunity.

There were however not surprisingly many unexpected twists and turns to the process. One was how an initial invitation to several community members to participate in a weekend workshop to scope a project involving the community, resulted in jealousy among others in the community who were not invited. Therefore shortly afterwards, the students had organised a large and high profile one-day workshop for the whole community, which had government and media involvement, and also funded by external sponsors.

Students were given the opportunity to explicitly engage in feed-forward to develop their "student-initiated" research, and responded well. The guidance given by the course coordinator and teaching assistant was to ensure the creation of teams, with functional division of labour for students across the region. Unlike the previous semester where each student could develop their own project, this time there was only one project with the explicit requirement to create a functional team where everyone was given a place. For example, those local to the field site could do the primary data collection and communicate it effectively to all others. Others carried out literature review and others sought secondary data, and so on.

Students were allowed and explicitly enabled to authentically carry out student initiated research, as well as being prompted to critically reflect on the process, to become aware of the importance of their flexibility and innovation and coordination for the research process they were engaged within. It was actually inspiring for the course coordinator and teaching assistant to observe the motivation of students and momentum of the self-regulated learning and research skill development emerge spontaneously and grow among the students to innovatively develop appropriate climate change adaptation.

By the fourth semester the assessment structure and use of rubrics had taken on a stable format. It appeared to be successful, so no change was made, though care was exercised to ensure that those charged with providing the feedback understood that their role was to be supportive and encouraging through their feedback, rather than to knock back or judge the students. It had been learnt increasingly over the prior three semesters, how to facilitate the emergence or bootstrapping of empowering innovative processes of self-regulated learning and research skill development through repeated interaction with students who authentically strove to learn together.

Based on the relative success of the coordination of all students across the region to work on one project, it was decided once again to make a single team, but to seek to better use online media. It was sought to go beyond the online discussion blogs and uploading of lectures. In particular it was sought to utilise the use of audio and video files of recorded events, which could be uploaded, as well as real-time participation in audio and visual interactions across the region. Feedback has shown that there continued to be a need to ensure the fullest participation as possible by all across the region in the course activities.

A challenge was the relatively slow internet services and often minimal services available in some parts of the South Pacific region, especially for real-time participation. Generally there was however always the opportunity to attend a local USP campus to retrieve files of recorded events. Nevertheless several region-wide Skype discussions were also organised by students.

The role of the course coordinator and teaching assistant was mainly to participate as members of the team (often as consultants with expertise in the matter), learning along with the students, as well as assisting with communication processes. The aim was to bootstrap the momentum to get to this stage. It occurred easily.

Objective 3: *Statistical evaluation*

A total of 21 students out of 34 students (62%) participated in the pre-survey while 16 students out of 30 students participated (53%) in the post-survey.

Ordinal values were assigned to the responses with 1 for strongly disagree, 2, for disagree, 3 for neutral, 4 for agree, and 5 for strongly agree.

The statistical test on pre and post responses for each question was:

$$H_o : \left(\mu_{post} - \mu_{pre}\right) \leq 0$$
$$H_a : \left(\mu_{post} - \mu_{pre}\right) > 0$$

The P values for the single-tailed analysis are shown in Table 27.1. P values less than 0.1 were considered to be significant enough to question the null hypothesis, while P values less than 0.05 were considered to be highly significant and therefore allowing for the rejection of the null hypothesis.

The data were found to not be adequately normally distributed for parametric analysis, and so a non-parametric measure (Mann-Whitney test) was carried out, using SPSS software.

90% (17 out of 19) of questions showed a significant ($p < 0.1$) improvement in research skill. The two that did not (questions 11 and 12) were both technical issues, namely how to construct tables and figures.

68% (13 out of 19) of questions showed a very significant ($p < 0.05$) improvement. The questions for which there as only a significant and not very significant increase in research skill were found in the sixth facet of research, "communicate and apply ethically—constructive", and the first facet of research, "embark and clarify—curious". The second, third and fifth facets of research, "find and generate—determined", "evaluate and reflect—discerning", and "analyse and synthesize—creative", all had very significant improvements in research skill in all questions associated with them, and thus allowed the null hypotheses to be rejected for these research facets.

Discussion

The research indicates that it is possible to develop a coherent pedagogy in a very complex interdisciplinary field, and that appropriate feedback is available from multiple sources and types, including formal statistical evaluation. But it also revealed that feedback needs to be coupled with rational yet creative forward-feeding innovation from ongoing thorough and critical literature review. Action research methodology was found to be an appropriate overarching methodology in which to do this. Ongoing critical literature review was able to be incorporated into it, along with the inclusion of formal statistical evaluation to provide indications of the correctness of the direction, but without the need to determine causality. In such a messy complex interdisciplinary field as climate change, trying to plumb an abstract ontological subterranean causal matrix is counter-productive. It is also inappropriate to endeavour to do so in face of the urgency to address the complex messy phenomenon of climate change. Continuous evaluation and seeking of feedback from supportive colleagues and communities is however necessary. It is empowering, to maintain the momentum of adaptation. Notwithstanding the inability to make conclusions about the causes of the improvement in research skill development, self- regulated learning and adaptive capacity, it can nevertheless be concluded that the pedagogy and its implementation was successful, because the formal statistical evaluation showed that there was significant research skill development for the majority of all aspects, for each of the research facets portrayed in the RSD framework.

Several lessons can be taken from the research. The most important is that it is possible to facilitate the bootstrapping of self-regulated learning within online postgraduate courses, and that this simultaneously results in research skill development. Another is that postgraduate students can be trusted not to 'game' assessment, and therefore to authentically develop adaptive capacity, if given the opportunity. Moreover, transformative leadership skills were comprehensively enhanced. The aspects of transformative leadership least enhanced were those to do with ethical skills, found in the "communicate and apply ethically—constructive" facet of research (questions 17 and 19). Possibly the pedagogy and its implementation needs to focus more on explicitly informing about the need for, and to nurture openness to, an ethical horizon and clear sense of purpose.

It was only a few transactional leadership skills that were not improved. Even though there was a successfully implemented emphasis on developing teamwork in the pedagogy, the use of the team spirit was not comprehensively nurtured. What was lacking in the implementation of the pedagogy was a careful and detailed focus on the procedural "nuts and bolts" of climate change adaptation. Even though this "single loop learning" is at the centre of the pedagogy, it was not adequately implemented over the four semesters of the research.

But both transactional and transformative skills are required, because both "engineering resilience" and "ecological resilience" are required. Participation in continual transformation risks making a person a "rebel without a cause", and unable to implement (engineer) any innovative solutions. Creativity that is not carefully and rationally tested with sober purpose in an ethical horizon, can become irrational expression of frustration, and even lead to violence. There is certainly a risk of this in climate change. The protests of Pacific Climate Warriors[6] among youth in the South Pacific region is possibly a foretaste of it.

The leadership required for climate change adaptation has to include transactional leadership. Without it, awareness of climate change and frustrated expression about it will not lead to successful adaptation. Authentic transformative leadership has to be expressed with the aim of enabling transactional leadership to emerge to implement policies that have become rationally discussed and vindicated. Policies are then able to be progressively implemented. An important aspect of adaptive capacity is to be able to know when to shift from a transformative mode of leadership to a transactional mode. "Psychological resilience" requires it (Hone 2017). "Psychological resilience" requires to be able to know when it is necessary to see the bigger picture and to consider what is necessary for innovations required by "ecological resilience", but equally when it is necessary to focus carefully on the details of how to maintain "engineering resilience", so that the innovations are actually implemented. An important feature of successful transactional leadership is a necessary part of "psychological resilience", namely the need to forge and maintain interpersonal relationships (Hone 2017). Such flexible working relationships are essential for functional implementation of policies, especially if they involve working across multiple institutions and agencies and nations, as found in climate change adaptation. Another necessary feature of "psychological resilience" is to have the balance to accept and to learn from the negative feedback that continually occurs, even, or especially, while soberly maintaining ethical purpose in an ethical horizon (Hone 2017). Without such balance, successful policies cannot be successfully implemented, to become successful innovations for climate change adaptation.

But, even though there was this limitation in how the pedagogy was implemented, the Triple-loop learning framework for the pedagogy is appropriate. Recent work by the German development agency, GIZ, to implement the pedagogy of Triple-loop learning, has vindicated that Triple-loop learning is able to integrate when necessary the need for transformation for climate change adaptation, but also to maintain incremental (first loop) change, and to enable fundamental (second loop) change, as appropriate (Shinko et al. 2019). Moreover, their work on evaluating how to best avoid loss and damage from disasters brought by 'development', indicates that the pedagogy of Triple-loop learning is able to help forge transformative adaptation that is common to both climate change adaptation and sustainable development. By doing so, their work is beginning to address what is simultaneously required for both climate change adaptation and climate change mitigation. Innovative proposals about how to do so are already being floated, and in need of being tested through careful

---

[6]The Pacific Climate Warriors is a group active within 350.org. (http://350.org).

and sober rational and ethical adaptive implementation. For example, proposals for a New Green Deal (Elliot et al. 2008). Whether or not the innovative proposal for a New Green Deal proves to be successful or not is not important. Rather it is important that the adaptive capacity to create and explore the potential viability of such innovations is able to be carried out. It will be those with postgraduate education and the associated research skill development who will be able to do so, and who are likely to take the lead.

Conclusion

The pedagogy proposed here sought to nurture adaptive capacity, so that the necessary balance between transformative leadership and transactional leadership was nurtured in those taking postgraduate courses in climate change. Some insights have been obtained. First, adaptive capacity for postgraduate students is research capability. Second, it can be facilitated by online delivery of postgraduate courses by focusing on how to facilitate self-regulated learning. To do so involves people using what online media enables, whilst also using the freedom from not being tied to a physical campus, to engage pro-actively and more fully with the local natural and social context. Online media can enhance the global network of communication, so important for science, as well as for national, regional and international policy-making and planning, whilst paradoxically putting pressure on students to become more proactive in engaging with their local context. If the pressure is used constructively to prompt self-regulated learning, then the use of purely online delivery of courses can potentially enhance the nurturing of adaptive capacity and research skill development. For this to occur however there is need for enhanced capacity for self-regulated learning. Even though on the face of it, purely online delivery of postgraduate courses makes self-regulated learning more difficult, the research found that it can nevertheless be easily overcome if there is care to facilitate supportive constructive feedback. It can then paradoxically open up new opportunities for an increase in adaptive capacity, leadership and resilience. It was found to add momentum and enthusiasm. The only caveat is that care is required to ensure that sober purpose and a rational focus on the practical "nuts and bolts" of climate change adaptation and sustainable development toward an ethical horizon, does not become overlooked in the enthusiasm.

The research indicates that transformative leadership was nurtured well by the proposed pedagogy, but that a greater focus was required on the details of implementing policies, to enable transactional leadership. But with continuation of the action research methodology, the feedback can be incorporated, and a better balance obtained.

The statistical evaluation of the courses indicates that it is possible to facilitate adaptive capacity in postgraduate students, but that it is not likely to be ever able to determine what aspect(s) of the pedagogy and its implementation cause it to be successful. It would be unrealistic to carry out a statistical design to determine it.

Moreover it would be inappropriate because of the urgency to do the best that is realistically possible straight away. It should be accepted comfortably that there is always the need for the pedagogy to be experimental to explore pro-actively possible improvements—to explore the "free-play" in the educational system at any one time, to seek ways in which it can successfully adapt. Ideological purity has to be avoided, and to accept instead the inevitable messy and pluralistic complexity, but which nevertheless, and also because of it, always has opportunities for improvement. The messy diversity gives flexibility and resilience.

The conceptual model outlines a process broadly enough to allow for continual adaptation, and is also structured adequately to establish what constitutes the dynamics of adaptation. Hence the conceptual model provides requisite detail to ensure resilience. Notwithstanding this, it is not claiming to be the singular and correct truth. Alternative equivalent or better models are recognised to be certain to able to be constructed, and that this will always be the case. Moreover, what is appropriate to include in the model may change as the context changes. For example, already the reference to "spirituality" may be unhelpful. Also, the constructivist epistemology may well become replaced by an improved theory that subsumes it. Finally, the Triple-loop learning pedagogy is surely to become more nuanced and fleshed out as neuro-biological understanding of cognitive, including affective, processes increases. But it can nevertheless serve at the present moment of crisis and urgent need, as a launching pad for bootstrapping how to facilitate climate change leadership among postgraduates. Moreover, because online mode of delivery has been found to be not only plausible, but to also open up opportunities for both access to information, networking and local engagement, the efficiency of postgraduate course provision can be expected to increase. That the students themselves innovatively responded to address the challenge and to find opportunities in face of dire resources cuts, is suggestive that it is likely to be possible under most situations. This bodes well for effective and efficient provision of postgraduate courses on climate change for poorly resourced nations that are already threatened by disastrous effects from climate change.

References

Argyris C, Schon D (1978) Organizational learning: a theory of action perspective. Addison-Wesley, Reading, MA

Bloom BS, Engelhart MD, Furst EJ, Hill WH, Krathwohl DR (1956) Taxonomy of educational objectives: The classification of educational goals. Handbook 1: cognitive domain. David Mckay Company, New York

Elliot L. Hines C., Juniper T., Legetti J., Lucas C., Murphy R., Pettifor A., Secrett. & Simms A. 2008. A green new deal: joined-up policies to solve the triple crunch of the credit crisis, climate change and high oil prices. [online] URL http://neweconomics.org

Focault M (1977) Discipline and Punish: the Birth of the Prison. Pantheon, New York

Friere P (1970) Pedagogy of the oppressed. 30th Anniversary Edition, Bloomsbury Academic, London

Gramsci A (1992) In: Buttigieg JA (ed) Prison notebooks. Columbia University Press, New York

Gunderson L, Holling CS, Pritchard L, Peterson PD (2003) Resilience. In: Mooney HA, Canadell JD (eds) The Earth system: biological and ecological dimensions of global environmental change, of encyclopedia of global environmental change, vol 2, pp 530–531. Wiley, New York

Holling CS (1978) Adaptive environmental assessment and management (ed) Wiley & sons, London

Holling CS, Gunderson LH (2002) Resilience and adaptive cycles. In: Gunderson LH, Holling CS (eds) Panarchy: understanding transformations in human and natural systems. Island Press, Washington

Hone L (2017) Resilient grieving: how to live with loss that changes everything. Allen & Unwin, Auckland

IPPC (2014) AR5 climate change 2014: impacts, adaptation, and vulnerability. Cambridge University Press, Cambridge

Kirschner PA, Sweller J, Clark RE (2010) Why minimal guidance during instruction does not work: An analysis of the failure of constructivist, discovery, problem-based, experiential, and inquiry-based teaching. Edu Psychol 41(2):75–86

Kuhnert K, Lewis P (1987) Transactional and transformational leadership: a constructive/developmental Analysis. Acad Manag Rev 12(4):648–657

McGowan J (1991) Postmodernism and its critics. Cornel University Press, Ithaca

Murtagh L, Baker N (2004) Feedback to feed forward: student response to tutors' written comments on assignments. Pract Res High Edu 3(1):20–28

Nicol DJ, Macfarlane-Dick D (2006) Formative assessment and self-regulated learning; a model and seven principles of good feedback practice. Stud High Educ 31(2):199–218

Quinton S, Smallbone T (2010) Feeding forward: using feedback to promote student reflection and learning—a teaching model. Innov Edu Teach Int 47(1):125–135

Rappaport RA (1999) Ritual and religion in the making of humanity. Cambridge University Press, Cambridge

Sadler DR (2010) Beyond feedback: developing student capability in complex appraisal. Assess Eval High Edu 35:535–550

Shinko T, Mecher R, Hochrainer-Stigler S (2019) The risk and policy space for loss and damage: integrating notions of distributive and compensatory justice with comprehensive risk management. In: Mecher R, Bouwer LM, Shinko T, Surminski S, Linnerooth J (eds) Loss and damage from climate change: concepts, methods and policy options, pp 83–110. Springer Open. Accessed 20th Mar 2019. http://doi.org/10.1007/978-3-319-72026-5

Tschakert P, Dietrich KA (2010) Anticipatory learning for climate change adaptation and resilience. Ecol Soc **15**(2):11. [online] http://Ecologyandsociety.org/Vol15/iss2/art11/

Willison JW, O'Regan K (2006, 2013) The research skills development framework. http://www.adelaide.edu.au/rsd/

Winne PH (2010a) Bootstrapping learner's self-regulated learning. Psychol Test Assess Model 52:472–490

Winne PH (2010b) Improving measurements of self-regulated learning. Educ Psychol 45:267–276

Zimmerman BJ (2002) Becoming a self-regulated learner: an overview. Theory Into Pract 41(2):64–70

Chapter 28
Increasing Environmental Action Through Climate Change Education Programmes that Enable School Students, Teachers and Technicians to Contribute to Genuine Scientific Research

Elizabeth A. C. Rushton

Abstract School communities provide an important context for climate change education that explores the implications of climate change at a range of spatial scales and provides young people with information about how they can positively respond to the challenges posed by the Anthropocene. This research explores the experiences of secondary school teachers and technicians who, with their students, have actively participated in science research that has a climate change and/or biodiversity focus, for at least four months. The study uses reflexive thematic analysis to analyse semi-structured interviews of twenty-eight participants from a diverse geographical range across England and Scotland, U.K., and from a variety of educational and socio-economic contexts. This study suggests that some teachers and technicians are motivated to participate in research projects to enable students to engage with, and contribute to, science and research that could provide solutions to real-world challenges and problems including climate change and the loss of biodiversity. Teachers and technicians reported that participation in research projects developed some students' sense of agency in the context of global challenges and this included students who were not already active in pro-environmental groups or activities. Learning from this approach to climate change education will be useful for educators in a range of settings as well as policy makers in the fields of climate change and education.

Keywords Climate change · Education · High school students · Independent research · Environmental action

---

E. A. C. Rushton (✉)
Institute for Research in Schools, London, UK
e-mail: elizabeth.rushton@kcl.ac.uk

King's College London, London, UK

© Springer Nature Switzerland AG 2019
W. Leal Filho and S. L. Hemstock (eds.), *Climate Change and the Role of Education*,
Climate Change Management, https://doi.org/10.1007/978-3-030-32898-6_28

Introduction

The place of environmental education in UK secondary schools

There is distinct variability in the content, coherence and clarity of environmental education in the four UK National Curricula of England, Scotland, Wales and Northern Ireland. As this study includes those working in the English and Scottish secondary contexts, the place of environmental education in England and Scotland is briefly considered. In their review of environmental education in English secondary schools, Glackin and King (2018) found that provision was 'patchy and restricted' due to a 'lack of intention or ideological vision' for environmental education in English education policy (p.1). Students in English secondary schools who studied geography had far greater exposure to environmental education but, as this is a non-mandatory subject post-14 and 50% of students do not study Geography at GCSE,[1] a substantial proportion of students in English schools have a restricted and environmental education (Glackin and King 2018). In their analysis of secondary school science and geography teachers and staff from subject associations/learned societies Glackin et al. (2018) report that environmental education continues to suffer from negative stereotyping that it is a 'soft science' and that this is exacerbated due to the lack of coherence and clarity of the subject within the national curriculum and often a lack of support at a whole school level. This contrasts with the place of environmental education in Scotland, where sustainability is a core concept and 'Learning for Sustainability' is integral to the General Teaching Council for Scotland Professional Standards Framework (GTCS, 2012). WWF-Scotland (2012, p 2.) defines 'Learning for Sustainability' in the school context as:

> ...a whole school approach that enables the school and its wider community to build the values, attitudes, knowledge, skills and confidence needed to develop practices and take decisions which are compatible with a sustainable and equitable world.

The term 'sustainable development' does not appear in the English national curriculum (Glackin and King 2018) and this and the lack of a coherent national framework for environmental education presents a concerning gap in the learning of young people in English schools. Informal science education provides contexts and opportunities for young people to engage with environmental issues outside of the classroom, and this includes youth citizen science initiatives. The role of citizen science in developing environmental agency is now considered.

Citizen science and environmental agency

Over past two decades there has been a rapid expansion of citizen science projects—projects that involve members of the public in research science (Curtis 2015; Gura 2013). With this expansion, the range of approaches taken when involving the public in science research activities has also increased (Bonney et al. 2009; Wiggins and

---

[1]GCSE or General Certificate of Secondary Education, taken by students aged 15–16 years in England, Wales and Northern Ireland.

Crowston 2011). This growth has led to classifications or typologies of citizen science projects (Bonney et al. 2009; Bonney et al. 2016; Cooper et al. 2007; Wiggins and Crowston 2011; Wilderman 2007). The most frequent approach to grouping citizen science projects is focused on the varying levels of participation that members of the public have in directing or leading the scientific process, and this has created the three categories; contributory projects, collaborative projects and co-created projects (Bonney et al. 2009). Bonney et al. (2016) developed a further classification of citizen science projects, identifying four categories that are defined by the types of activities participants are engaged in, including; Data Collection, Data Processing, Curriculum-based and Community Science. Bonney et al. (2016) suggest that scientific outcomes for citizen science projects are well documented for Data Collection and Data Processing projects and, that there is growing evidence that citizen science projects develop both scientific knowledge and public understanding of scientific research. Ballard et al. (2017, p. 65) suggest that environmental science agency (ESA) is a combination of an:

> …understanding of environmental science and inquiry practices…[and]…youths' identification with those practices and their belief that the ecosystem is something on which they act.

Ballard et al. (2017, p. 65) suggest that citizen science projects can, 'foster youth participation in current conservation actions and build their capacity for future conservation actions. Calabrese Barton (2012) highlights the need to recognise the concept of place as a crucial component of a citizen science program. Calabrese Barton (2012) asks whether "citzen science" should be reframed as "citizens' science", placing importance on science expertise that is linked to societal change, and, located in a specific community (p. 3). This emphasis on place is especially pertinent to citizen science projects that seek to engage young people with environmental issues such as climate change and loss of biodiversity. Calabrese Barton (2012) has challenged the capacity of schools to provide contexts for citizen science programs that bring about social change. However, the lack of environmental education for 50% of post-14 students in English schools means that if access to environmental education, particularly that which is inquiry based, is to be democratically available, the school setting is a context that cannot be overlooked.

School student research

The Institute for Research in Schools (IRIS) is a UK-based charity that launched in March 2016 to develop an approach to school education where research is a key element of STEM (science, technology, engineering and mathematics) learning that offers opportunities for students to work on genuine problems (Parker et al. 2018; Rushton and Parker 2019). This approach resonates with the concept of authentic learning, as students and teachers are contributing to knowledge by focusing on what is not already known, as part of a disciplined inquiry that has value beyond the classroom (Newmann et al. 1996; Lombardi 2007; Bennett et al. 2018). IRIS research projects are a hybrid of the two citizen science frameworks outlined by Bonney et al. (2009) and Bonney et al. (2016) and are founded on a dialogic and participatory approach where students and their teachers co-create knowledge with scientists and other specialist partners (Rushton and Parker 2019). The learning goals are appropriate and achievable and are rooted in genuine scientific research that contributes to community genome annotation in a way that has been previously recognised as valid, reliable, accurate and useful (Elsik et al. 2006; Loveland et al. 2012). The role of the school teacher and technician is to encourage, support and facilitate their students' participation. Informal conversations between IRIS staff and teachers revealed that teachers involved in two projects with links to climate change and/or biodiversity viewed this experience as a positive approach to climate change education for both themselves and their students. These conversations provided the starting point for this current study so that the experiences could be more closely explored within the context of climate change education. The two projects are *Monitoring the environment, Learning for Tomorrow (MELT)* and *Well World*. Both projects were supported by grants from IRIS' core funding as well as external funders (UK Space Agency, Wellcome) and these grants supported the evaluation work from which this paper has developed. These two projects are briefly outlined.

Monitoring the Environment, Learning for Tomorrow (MELT) is supported by researchers at the Centre for Polar Modelling and Observation, University of Leeds and funded by the UK Space Agency. Launched in early 2018, MELT enables student to research environmental change through the Carbon Footprint Challenge and Earth Observation. The Carbon Footprint Challenge is appropriate for students aged 7 – 18 years. Students use a carbon calculator to measure the carbon footprint of their school community, develop a plan to reduce their carbon footprint over a defined period of time (e.g. 3 months) before recalculating the carbon footprint for a second time. Students work in teams to produce a research poster to present to their school communities and through IRIS conferences (Rushton et al. 2019). The Earth Observation strand is appropriate for students aged 14–18 years and many participating students study geography as well as traditional STEM subjects. Students use images from the European Space Agency's Sentinel-1 satellite and Synthetic Aperture RADAR data to measure iceberg formation and movement on the Antarctic Peninsula. Over the last twenty years there have been significant iceberg break-off or 'calving' events including at the 'Larsen-B Ice Shelf' in 2002 and the 'Larsen-C Ice Shelf' in 2017. These events may suggest that environmental conditions in the Antarctic Peninsula have changed. Teachers and students share their analysis with scientists and receive feedback and further guidance through electronic-based networking including emails and webinars. Students and teachers from twenty schools have participated in *MELT*, including schools in England, Scotland and Norway. Research findings have been presented by students at school conferences hosted by IRIS and the Royal Society, as well as international student research conferences.

Well World developed from the work of school students (16–18 years of age) and their teachers, from a school in Canterbury, Kent, United Kingdom who were supported by researchers at the University of Cambridge and funding from Wellcome and the Royal Society. The students posed an initial research question, "Does biodiversity make us happy?" inspired by a wooded area of their school site, known locally as "The Orchard". Students and teachers were initially motivated to research the links between biodiverse areas and health and well-being to raise awareness and appreciation of the value of "The Orchard" within their school community. When developing their study, students considered research that demonstrates the positive role green space can have in promoting physical and mental well-being (Barton and Pretty 2010; Burgess et al. 1988; Pretty et al. 2005; Wood et al. 2017). Green space includes a range of environments including national parks, public parks and domestic gardens (Kaplan et al. 1998). In the context of increasing global concern about the protection of the environment (including, anthropogenic global warming, deforestation, soil degradation and water and air pollution) (Millennium Ecosystem Assessment 2006) and rising levels of child and adult physical inactivity and obesity (UK Department of Health 2009) multi-disciplinary initiatives including Green Gym (Pretty et al. 2007) and Blue Gym (White et al. (2016) have sought to reconnect populations to their environments and promote positive health outcomes associated with physical activity in the natural environment.

Students (mainly aged 16–18 years, studying biology and/or psychology) designed an experiment to answer the question "Does biodiversity make us happy?" In the experiment they identified three areas in the school site with different levels of biodiversity (high, low, none) that participants were asked to walk in for a similar distance and length of time. These areas included an indoor school area (no biodiversity), the outdoor running track (low biodiversity) and the highly biodiverse wooded area known as "The Orchard." Participants' physical well-being was measured before and after walking through pulse and blood pressure measurements and their mental well-being was assessed through the shortened Spielberger State Trait Anxiety Index (STAI) questionnaire (Marteau and Bekker 1992). The students presented their research as part of an IRIS student conference in November 2017, and this led to the development of student and teacher guides and resources to support other schools to develop their own *Well World* research projects. Since the national launch of the *Well World* in 2018, 15 schools have developed student-led research projects in their communities.

Based upon the literature, the following research questions were identified:

1. What are the experiences of UK-based secondary school teachers and technicians who are research active with their students in the areas of climate change and/or biodiversity?
2. Are school-based research projects in the areas of climate change and/or biodiversity an effective approach to climate change education?

Methods

Participants

Twenty-eight participants were recruited for the study from the IRIS network of teachers and technicians. Sixteen are female, twelve are male and there are two technicians and twenty-six teachers in the participant group. All teachers teach one or more of the following subjects: biology, chemistry, physics, psychology and science at secondary school level (Table 28.1). All have been working on research projects with their students and with teachers and students from other schools for at least four months. Teachers have diverse experience, ranging from newly qualified teachers to those who had been teachings for over 30 years (Table 28.1). One technician had more than 10 years school-based experience and the other technician had two years school-based experience. Nine participants have a PhD in a science subject and seventeen are or have held school management roles, including subject leader, lead technician and curriculum leader (Table 28.1).

Participants were drawn from a wide geographic range in England and Scotland, from Cornwall to Stirling, and schools included selective and non-selective Academies, selective and non-selective Local Authority schools, and fee-paying independent schools (Table 28.2). This ensured that the study incorporated the experiences of teachers from a range of educational contexts with different ideologies. The ethnicity of participants was not requested or disclosed.

Data collection procedure

Participants were recruited during February 2018–January 2019 through email request. An interview schedule was prepared with questions in three main sections: *background* information, including information about the participant's teaching role and research role; the *impact* of the research on the participant's experience of teaching; and the *experience* of research projects for themselves and their students. The interviews were conducted during visits to the participants' schools to ensure that they were in a comfortable, familiar environment and the interviews were at a time chosen by the participant, carried out during February 2018–January 2019. The interviews were audio-recorded and transcribed shortly after each interview according to Braun and Clarke's (2006, 2019) guidelines for reflexive thematic analysis.

Results and Discussion

The interview transcripts were analysed using the six phases of Reflexive Thematic Analysis (RTA) originally outlined in Braun and Clarke (2006) and further developed in Braun and Clarke (2019). Three superordinate themes were identified from the data, each with sub-ordinate themes, and were created from several initial codes (Clarke et al. 2015), as summarised in Table 28.3. Each of these three key findings are discussed in turn before some more wider ranging consideration of the implications for environmental education.

Superordinate theme A: *Student development through research*

Participants acknowledged that the time they had to provide research project opportunities was limited and they attributed their motivation to 'make time' to the students

Table 28.1 Main subjects, experience and current management roles of participants

	Main subject					Number of years					PhD	Management role
	Bio	Che	Phy	Psy	GenSci	1-5	6-11	12-17	18+	Mean		
Female	4	4	5	0	3	6	3	2	5	**25**	5	7
Male	5	2	4	1	0	1	4	2	5	**15**	4	10
Total	**9**	**6**	**9**	**1**	**3**	**7**	**7**	**4**	**10**	**20**	**9**	**17**

Table 28.2 Geographical location and school type

School type	Academy (selective)	Academy (non-selective)	Local authority (non-selective)	Local authority (selective)	Independent
Unitary authority or local authority district of school	South Hams	Bedford City of Bristol Cotswold Sheffield Worthing	Camden City of Edinburgh Cornwall Dartford Medway Northumberland North Yorkshire Sheffield South Lanarkshire Stirling	Canterbury North Yorkshire Shepway	Canterbury Oxford Tunbridge Wells Vale of the White Horse
Total number of schools	1	5	10	3	4
Total number of teachers	1	7	11	4	5

they teach. Teacher and technicians were motivated to respond in this way because they could see the opportunities for student development, and some were responding to the enthusiasm and passion of their students. When participants discussed the contribution that working on a research project had made to the skills development of their students, communication skills and increased confidence were the most commonly identified and were identified as important development opportunities. The networks and wider connections students forge through research are also recognised by participants as valuable:

> In terms of skills development, research projects are invaluable; students are finding the skills through experience of research, communication skills, teamwork, planning, leadership … and this increases their confidence and self-belief.

Technicians and teachers made a link between student development of confidence and communication skills to the opportunities students have to forge networks through research that extend beyond their own school. This enabled students to share their experiences and findings with others and so better understand that the contribution that they have made is seen by others as valuable:

> Through research students have a chance to develop their communication skills and confidence … they can see that they have achieved something and that they need to share that with the outside world.

Participants also suggest that the confidence students gain through research is something that moves with them into other educational spaces. For students, sharing research with younger students gives them opportunities to develop communication

Table 28.3 Superordinate and sub-ordinate themes which emerged through a process of coding

Superordinate theme	Subordinate themes	Codes
(A) Student development through research	(1) Development of students' skills and wider connections (2) Students' independent engagement in research	Students doing real research, practical skills, problem solving, independence, questioning, foundation for next research stage, supporting undergraduates, university choices, student presentation and communication skills, confidence
(B) Societal development through research	(1) Contribution to science and the world of research (2) Providing solutions to problems that impact the wider world	Real science, real solutions, students enthused, wider world relevance and importance, connecting with science/research, science in and for the community, student-led
(C) Increasing student environmental agency and pro-environmental behaviour	(1) Providing a space to grow and develop students' passion for the environment (2) Positive contribution to solutions (3) Environmental advocacy within school community	Shared passion for the environment, students' capacity to contribute, research project approach to teaching environmental issues, encouragement and support, environmental advocacy, behaviour change, space to develop ideas and solutions, contrast in student response, biodiversity, climate change, global challenges, school assemblies

skills that went beyond their teachers' own expectations of their abilities. Another teacher identified that students develop their evaluation skills through research and are more able to understand the quality of information they are presented with:

> I think this research project has got them thinking about the quality of the science a little bit more.

Through research projects, participants suggested that students implicitly learn about careers in science that is therefore likely to have an impact upon their future choices:

> … with the research project they [the students] are experiencing something that is more like the reality of science and so they are getting a better understanding of what a career in science might be like, so we are not explicitly telling them, they are learning it for themselves and I think that ultimately that is a more powerful way.

Teachers and technicians also suggested that post-14 students included those who were no longer studying geography and post-16 students were drawn from across subject choices, including those who were no longer studying geography or STEM subjects.

Superordinate theme B: Societal development through research

As well as the opportunities for students' development provided by research projects, some teachers were motivated to participate to enable students to engage with, and contribute to, science and research that could provide solutions to real-world challenges and problems. Participants identified that some students value the opportunity to contribute to wider science research, and this is related to research projects providing students with the ability to forge wider connections through the contribution they make:

> [The students] also feel ... that they are contributing to research, there are some of them that regard themselves as researchers having taking part in this project.

> The students come from across the sixth form subjects and they create a research community in school, and it was so important to be connected with researchers and make a contribution to that world.

As well as contributing to research, some students were also motivated to participate in research projects if they made a connection with the real-life implications of the work and identified with the wider story:

> One of the things that really appealed to the students was that this ... has a narrative behind it with real life implications ... because it was real they [the students] were prepared to sit down and work through the theory and relearn how to apply it.

One teacher described how students as a group repeatedly, verbally connected their work with making a wider positive contribution:

> The students develop such enthusiasm and confidence, in that they know what they have to do, and they know how to do it and they are able to explain it to people, what they are doing and why they are doing it and they have this little mantra, "because we are saving the world!".

Students' recognition and value of the wider societal link was not universally observed by participants, with one technician suggesting that sometimes the wider context was not visible to all students:

> I think sometimes the societal link to the research project, the contribution that students can make gets a bit lost on them...some of the students were so focused on the task and on getting it right they sometimes lost sight of the wider context and significance which was a shame.

Participants suggested that the societal link was an important way to encourage students to initially participate, but that as they encountered the challenge of research this link could diminish. Some participants identified that it was only after working on the research project in a sustained way for at least two-three months that the wider context and societal links were firmly embedded in students' discussions of research. Once those links were made as part of research, they did appear to increase environmental agency and shape pro-environmental behaviour, as is reported in the third superordinate theme.

Superordinate theme C: Increasing student environmental agency and pro-environmental behaviour

Participants described their role in the research projects as one of 'encouragement and support' and saw that research projects such as *Well World* and *MELT* were an opportunity to develop and harness the passion many young people have for the environment. One teacher identified a need to embed the research project approach into classroom teaching of environmental issues:

> To really contribute to the environmental understanding of all students you would need this type of research project to be part of the way that students are taught.

Teachers and technicians reported that participation in research projects developed some students' sense of agency in the context of global challenges and that this increase in agency was found in school contexts which had previously provided school clubs and projects with an environmental focus:

> Over the years we have done lots of different clubs and projects that have tried to help students learn about their environment and to be responsible citizens, and these have been good for the students but when we got them involved in research we really saw them take the lead with the project and I think that has translated into them having a sense of ownership of the problems their generation faces.

One teacher described how she saw research as a successful and positive way of connecting students to problems that impact their lives, now and in the future:

> Through research, students have become enthusiastic about biodiversity and have seen the real benefits that biodiversity brings them ... it is that personal link, that tangible link that people have with biodiversity and research has created that link for the students ... research can be a mechanism to develop a personal link with a subject for a student.

Students' environmental agency was developed through research projects for some students, but not all. One teacher described how different students responded to the *MELT* research project:

> Some of the students who worked on *MELT* had no interest in the environmental aspect of the project when they started, for them it was an interesting project, a puzzle to get their teeth into and try and solve, but for some of these students, during the course of the project they started to talk about the environmental aspect to me and their peers. I think the consequences and the wider context of what they were studying in an academic way in the research project started to creep into their wider thinking.

There was a recognition from participants that environmental challenges, including biodiversity loss and climate change, were often communicated in a negative way to students and that students discussed with them that being part of research projects gave them a way to contribute in a constructive way:

> Getting young people to think positively about the environment can be a challenge as in their lessons they are getting the accurate message that the vast majority of scientist view climate change as a global challenge, but this can be reported in a very doom-laden way, with lots of shouty headlines. When students were involved in the *MELT* project, they were part of the research team, part of the solution, they had a sense there was a positive way forward and they were part of it.

Four participants explicitly described how students who had participated in the *MELT* and Well World projects were motivated to consider their own activities and behaviours in the context of global warming and biodiversity loss and also identified ways in which these personal changes could be shared with their wider school community. One teacher described this change:

> The *Well World* project had only been running in my school for a few months and some of the students who were part of it were in my form and they asked if they could lead an assembly to talk about the project to their year group. When they presented, I was surprised that they included lots of information about biodiversity in their local area, research that they had done independently and they included a section at the end of the assembly that gave the audience some actions they could take to look after their environment e.g. not mowing all of the lawn, creating habitats for bees, checking for hedgehogs before lighting a bonfire but also, reducing energy use. The project really inspired them to think about their actions and then they wanted to share that with their peers.

This is a clear example of how, from the perspective of a teacher, students have identified the actions they and their peers can take to change the impact they have on their local and global environments and, how students demonstrate increased agency in communicating these issues with their peers. This aspect of climate change research projects is discussed in the wider context of environmental education.

Implications for environmental education in England and Scotland

This study draws on the experiences of teachers and technicians at a range of career stages, management and research experience. The range of levels of experience suggests that with support, teachers can embed research projects with a climate change focus at varied points in their career and do not seem to need threshold levels of teaching or research experience. Teachers and technicians reported that climate change research projects provide an opportunity for students to engage with environmental issues in a way that can develop students' agency and pro-environmental behaviours. This is consistent with previous findings from youth citizen science projects with a conservation focus (Ballard et al. 2017). When considering how to expand and enhance the provision of secondary school environmental education it is important to note that the findings from this study are formed from the perspectives of teachers and technicians who are working as part of school-based science and psychology provision. Geography teachers did not contribute to this study because, to date, *Well World* has not been implemented by geography teachers, and *MELT* has only had limited uptake in school geography departments. Glackin et al. (2018) suggest that environmental education is well served in the view of geography teachers but, science teachers reported that environmental education was often overlooked and suffered from the negative stereotype as a less academically rigorous subject area. It is possible that geography teachers may have a different perspective on the importance or otherwise of research projects such as *MELT* and *Well World* in developing student environmental agency and pro-environmental behaviours, and future work would look to consider the perspectives of geography teachers.

Environmental education, and sustainability in particular, is embedded in the national curriculum and teaching standards framework of Scotland (WWF 2012).

This is in sharp contrast to the patchy and, in some contexts, absent provision in England. This study included the perspectives of three teachers working in schools in Scotland as well as twenty-three teachers working in English schools. From this small sample size, it was not possible to determine any difference in experience of research projects between these two groups of teachers. Future work would look to further compare and contrast the experiences of teachers in Scotland and England in light of their different contexts. There are also opportunities to consider the experiences of teachers in Wales and Northern Ireland. The Welsh government is currently reviewing the National Curriculum for Wales during a five-year period of reform, 2017–2022. The Welsh framework divides the curriculum into six Areas of Learning and Experience (AoLEs). Each of the AoLEs presented as a series of 'What Matters' statements and environmental education features in the sixth of eight statements included in the Science and Technology AoLE:

> There are ethical and ecological impacts and implications of science and technology at personal, local and global levels and beyond. (CAG, 2017b p. 6)

Environmental education and ideas around sustainability are also found in the fourth of five 'What Matters' statements from the Health and Well-being AoLE:

> Our physical, social and cultural environments are connected to our health and well-being. (CAG 2017a p. 5)

As with the Scottish national curriculum, environmental education is clearly embedded in the planned Welsh framework and this would provide a further dimension when considering the role of national educational policies on the efficacy of climate change education. Given the variety of approach to environmental education in the National Curricula of the UK, further research could explore in more detail the similarities and differences in experiences of environmental education from the perspectives of teachers, technicians and students to better understand provision across England, Scotland, Wales and Northern Ireland.

Drawing on the findings from this study, it is possible to make three distinct recommendations for environmental education which incorporates research in the areas of climate change and/or biodiversity:

1. *Environmental education that has a focus on rigour, reproducibility and enables students to make a genuine contribution.* Research projects such as *MELT* and *Well World* are rooted in the context of authentic science where students, supported by their teachers and technicians, contribute to science in partnership with university-based scientists. Teachers and technicians who were already experienced in delivering after school clubs and activities with an environmental focus recognised that research projects gave students a way to engage with climate change in a positive way that developed their agency and, in some cases, pro-environmental behaviours.

2. *Environmental education that has a purposeful cross-curricular approach.* Climate change research projects were predominantly supported by science teachers and technicians and student participants were drawn from a range of subjects

including biology, chemistry, general science, geography, physics and psychology. Some teachers and technicians reported that students who no longer studied geography or science subjects at A-level were drawn to participate in the project because they had a passion for the environment. This cross-curricular approach enables all secondary students to access climate change education which is especially important when geography is a non-mandatory subject. This study also suggests that young people want to engage with climate change issues as part of their school experience and research projects are one way to achieve this.

3. *Environmental education which connects students to personal and societal development.* Student participation in research projects was widely recognised to develop a range of skills and provided students with opportunities to extend and enhance their learning. This is consistent with previous studies of secondary school student participation in independent research projects (Bennett et al. 2018; Rushton et al. in review). Research projects with an explicit link to societal development provided some students with additional motivation to persevere through challenge periods of research although some students did lose sight of the societal component. Rushton and Reiss (in review) suggest that the elucidation of societal links is more likely to happen after students and teachers have been working as part of research projects in a sustained and intensive way, for at least one academic year. Participants in this study had been working as part of *MELT* or *Well World* for between four and fifty months, so the societal link may become more apparent over time.

Conclusion

Teachers and technicians who are research active in climate change projects with their students are positive about the contribution they can make to the development of their students and the wider contribution to society through research. This enhanced professional experience is consistent with the findings of Rushton and Reiss (2019). As has been found with some youth citizen science programs with a conservation focus (Ballard et al. 2017), widening participation in school-based research projects will enable more young people to see the value, importance and relevance of climate change education that is linked to increased environmental agency. Embedding opportunities for schools to genuinely contribute to research concerning climate change and/or biodiversity enhances environmental education in three ways. Firstly, research gives a focus on rigour and reproducibility. Secondly, research engenders a purposeful, cross-curricular approach and, thirdly, research connects students to personal and societal development. This research suggests that environmental education in England would particularly benefit from this approach due to the patchy nature of current coverage (Glackin and King 2018). Further consideration of the experiences of teachers actively involved in MELT and Well World in Scotland would provide an opportunity to explore in greater detail the contrasting approaches to sustainability in English and Scottish national curricular, and how these are perceived by teachers and students. Future studies considering the role of research projects in environmental education could consider students' experiences of research. This would provide a better understanding of challenges and opportunities this approach to climate change education presents for young people. Research should initially focus on understanding the nature of participation so that all young people see research activities as opportunities open to students like themselves.

Funding Acknowledgements This study developed from the evaluation of two projects that were funded by the UK Space Agency (*MELT*) and Wellcome (*Well World*). Both projects were also supported through core grant funding from the Institute for Research in Schools.

References

Ballard HL, Dixon CG, Harris EM (2017) Youth-focused citizen science: examining the role of environmental science learning and agency for conservation. Biol Cons 208:65–75. https://doi.org/10.1016/j.biocon.2016.05.024

Barton J, Pretty J (2010) What is the best dose of nature and green exercise for improving mental health? a multi-study analysis. Environ Sci Technol 44(10):3947–3955. https://doi.org/10.1021/es903183r

Bennett J, Dunlop L, Knox KJ, Reiss MJ, Torrance-Jenkins R (2018) Practical independent research projects in science: a synthesis and evaluation of the evidence of impact on high school students. Int J Sci Educ. https://doi.org/10.1080/09500693.2018.1511936

Bonney R, Cooper CB, Dickinson J, Kelling S, Phillips T, Rosenberg KV, Shirk J (2009) Citizen science: a developing tool for expanding science knowledge and scientific literacy. Bioscience 59(11):977–984. https://doi.org/10.1525/bio.2009.59.11.9

Bonney R, Phillips TB, Ballard HL, Enck JW (2016) Can citizen science enhance public understanding of science? Public Understand Sci 25(1):2–16. https://doi.org/10.1177/0963662515607406

Braun V, Clarke V (2006) Using thematic analysis in psychology. Qual Res Psychol 3(2):77–101. https://doi.org/10.1191/1478088706qp063oa

Braun V, Clarke V (2019) Reflecting on reflexive thematic analysis. Qual Res Sport Exerc Health 11(4):589–597

Burgess J, Harrison CM, Limb M (1988) People, parks and the urban green: a study of popular meanings and values for open spaces in the city. Urban Stud 25:455–473. https://doi.org/10.1080/00420988820080631

Curriculum Assessment Group (2017a) Health and well-being AoLE. Welsh Government. Available at https://beta.gov.wales/sites/default/files/publications/2018-07/health-and-well-being-aole-december-2017.pdf

Curriculum Assessment Group (2017b) Science and technology AoLE. Welsh Government. Available at https://beta.gov.wales/sites/default/files/publications/2018-07/science-and-technology-aole-december-2017.pdf

Calabrese Barton AM (2012) Citizen (s') science. A response to "The Future of Citizen Science." Democr Educ 20(2):12. Academic one file. Retrieved from https://democracyeducationjournal.org/cgi/viewcontent.cgi?referer=https://scholar.google.co.uk/&httpsredir=1&article=1044&context=home

Clarke V, Braun V, Hayfield N (2015) Thematic analysis. In: Smith JA (ed) Qualitative psychology: a practical guide to research methods. Sage, London, pp 222–248

Cooper C, Dickinson J, Phillips T, Bonney R (2007) Citizen science as a tool for conservation in residential ecosystems. Ecol Soc 12(2). http://www.ecologyandsociety.org/vol12/iss2/art11/

Curtis V (2015) Motivation to participate in an online citizen science game: a study of Foldit. Sci Commun 37(6):723–746. https://doi.org/10.1177/1075547015609322

Elsik CG, Worley KC, Zhang L, Milshina NV, Jiang H, Reese JT, Gibbs RA (2006) Community annotation: procedures, protocols, and supporting tools. Genome Res 16(10):000. https://doi.org/10.1101/gr.5580606

General Teaching Council for Scotland (2012) The standards for career-long professional learning. GTCS, Edinburgh. Available at http://www.gtcs.org.uk/web/FILES/the-standards/standard-for-career-long-professional-learning-1212.pdf

Glackin M, King H (2018) Understanding environmental education in secondary school in England. Report 1: perspectives from policy. King's College London. Available at https://kclpure.kcl.ac.uk/portal/files/101862531/EnvironmentalReport1_2018pdf.pdf

Glackin M, King H, Cook R, Greer K (2018) Understanding environmental education in secondary school in England. Report 2: the practitioners' perspective. King's College London. Available at https://kclpure.kcl.ac.uk/portal/files/101862558/EnvironmentalReport2_2018.pdf

Gura T (2013) Citizen science: amateur experts. Nature 496(7444):259–261. https://doi.org/10.1038/nj7444-259a

Kaplan R, Kaplan S, Ryan RL (1998) With people in mind. Design and management of everyday nature. Island Press, Washington, DC

Lombardi MM (2007) Authentic learning for the 21st century: an overview. Educause Learn Initiat 1:1–12. Available at https://library.educause.edu/~/media/files/library/2007/1/eli3009-pdf.pdf

Loveland JE, Gilbert JG, Griffiths E, Harrow JL (2012) Community gene annotation in practice. Database 2012. https://doi.org/10.1093/database/bas009

Marteau TM, Bekker H (1992) The development of a six-item short-form of the state scale of the Spielberger State—Trait Anxiety Inventory (STAI). Br J Clin Psychol 31(3):301–306. https://doi.org/10.1111/j.2044-8260.1992.tb00997.x

Millennium Ecosystem Assessment (2006) Ecosystems and human well-being. Island Press, Washington, D.C

Newmann FM, Marks HM, Gamoran A (1996) Authentic pedagogy and student performance. Am J Educ 104(4):280–312. https://doi.org/10.1111/j.2044-8260.1992.tb00997.x

Parker B, Fox E, Rushton EAC (2018) IRIS—promoting young peoples' participation and attainment in STEM and reigniting teachers' passion for science education. Impact 2. Available at https://impact.chartered.college/article/parker-iris-stem-students-teachers-participation-research/

Pretty J, Peacock J, Hine R, Sellens M, South N, Griffin M (2007) Green exercise in the UK countryside: effects on health and psychological well-being, and implications for policy and planning. J Environ Plann Manag 50(2):211–231. https://doi.org/10.1080/09640560601156466

Pretty J, Peacock J, Sellens M, Griffin M (2005) The mental and physical health outcomes of green exercise. Int J Environ Health Res 15(5):319–337. https://doi.org/10.1080/09603120500155963

Rushton EAC, Charters L, Reiss MJ (2019) The experiences of active participation in academic conferences for high school science students. Res Sci Technol Educ. https://doi.org/10.1080/02635143.2019.1657395

Rushton EAC, Parker B (2019) Empowering young people to develop STEM careers through active participation in genuine scientific research. In: Hiller SE, Kitsantas A (eds) Citizen science programs: guidelines for informal science educators in enhancing youth science motivation and achievement. Nova Science Publishers, Inc, Hauppage, NY, pp 97–127

Rushton EAC, Reiss MJ (2019) From science teacher to 'teacher scientist': exploring the experiences of research-active science teachers in the UK. Int J Sci Educ 41(11):1541–1561. https://doi.org/10.1080/09500693.2019.1615656

UK Department of Health (2009) Be active and healthy. A plan for getting the nation moving. London, England. Available at http://www.laterlifetraining.co.uk/wp-content/uploads/2011/12/DoH-Be-Active-Be-Healthy-2009.pdf

White MP, Pahl S, Wheeler BW, Fleming LEF, Depledge MH (2016) The "Blue Gym": What can blue space do for you and what can you do for blue space? J Mar Biol Assoc United Kingdom 96(1):5–12. https://doi.org/10.1017/S0025315415002209

Wiggins A, Crowston K (2011) From conservation to crowdsourcing: a typology of citizen science. In: 2011 44th Hawaii international conference on system sciences (HICSS), pp 1–10,IEEE Computer Society, Washington D.C. Available at https://dl.acm.org/citation.cfm?id=1956101

Wilderman CC (2007) Models of community science: design lessons from the field. In: McEver C, Bonney R, Dickinson J, Kelling S, Rosenberg K, Shirk JL (eds) Citizen science toolkit conference. Cornell Laboratory of Ornithology, Ithaca, NY

Wood L, Hooper P, Foster S, Bull F (2017) Public green spaces and positive mental health–investigating the relationship between access, quantity and types of parks and mental wellbeing. Health Place 48:63–71. https://doi.org/10.1016/j.healthplace.2017.09.002

WWF-Scotland. (2012) Learning for sustainability: the report of the one planet schools working group. Scottish Government, Edinburgh, Scotland. Available at http://assets.wwf.org.uk/downloads/1planetschools_web2.pdf

Chapter 29
Teenagers Expand Their Conceptions of Climate Change Adaptation Through Research-Education Cooperation

Oliver Gerald Schrot, Lars Keller, Dunja Peduzzi, Maximilian Riede, Alina Kuthe and David Ludwig

Abstract Unlike previous generations, today's youth is directly affected by global anthropogenic climate change (CC), and its increasing consequences throughout their lifetimes. However, both the educational strategies to prepare them for CC adaptation, and their conceptions of CC adaptation, remain insufficiently understood. This study sets out to investigate the CC adaptation conceptions of 120 students from four high-schools in Austria and Italy. The influence of a year-long research-education cooperation between students and 28 CC adaptation experts is examined. In the educational design, the focus lies on moderate-constructivist theories, and the transdisciplinary dialogue between students and experts. A mixed-methodologies approach is applied, which combines content analysis to study students' conceptions of CC adaptation and test statistics (*chi-square* and *t*-test) to assess the impact of the educational intervention. The results show that students' conceptions differ in degree of sophistication, and also include misconceptions. Some students relate adaptation to limiting disadvantages due to CC, others confuse adaptation with mitigation or environmental protection. After the educational intervention, most students have expanded their CC adaptation conceptions and overcome misconceptions, and their performance to

---

O. G. Schrot (✉) · L. Keller · D. Peduzzi · A. Kuthe
Institute of Geography, University of Innsbruck, Innsbruck, Austria
e-mail: Oliver.Schrot@uibk.ac.at

L. Keller
e-mail: Lars.Keller@uibk.ac.at

D. Peduzzi
e-mail: Dunja.Peduzzi@uibk.ac.at

A. Kuthe
e-mail: Alina.Kuthe@uibk.ac.at

M. Riede
alpS-GmbH, Innsbruck, Austria
e-mail: riede@alps-gmbh.com

D. Ludwig
Knowledge, Technology and Innovation (KTI), Wageningen University and Research,
Wageningen, The Netherlands
e-mail: david.ludwig@wur.nl

© Springer Nature Switzerland AG 2019
W. Leal Filho and S. L. Hemstock (eds.), *Climate Change and the Role of Education*,
Climate Change Management, https://doi.org/10.1007/978-3-030-32898-6_29

differentiate between adaptation and mitigation increased significantly. This paper will be useful to researchers and teachers interested in utilizing education as a means to adapting to CC.

Keywords Climate change · Adaptation · Conceptual change · Constructivist learning · Collaborative research

Introduction

Article 7 of the 2015 Paris Agreement leaves no doubt: adaptation to man-made climate change (CC) is a global challenge that is faced by everyone (UNFCCC 2015). The evidence of consequences of warming is alarming (Hansen and Stone 2016), and adaptive solutions employed across the globe show that actions are required to protect people, livelihoods and ecosystems (Adger et al. 2003; Ford et al. 2011). In the decades ahead, effective adaptation will be indispensable: (i) to cope with climate-related risks and (ii) and to (re-)build resilience in human-environment systems (Berrang-Ford et al. 2011; Lesnikowski et al. 2017; Runhaar et al. 2018).

Against this background, CC educators are observed taking up the concept of adaptation (Anderson 2012). Efforts in CC education aim to increase public concern about climate-related changes, and intend to improve understanding of climate response among different groups of people (Lutz et al. 2014; Rumore et al. 2016). Besides promoting a citizenship that is knowledgeable about the causes of change (Shepardson et al. 2011), learners should also be gaining scientifically informed insights into documented and expected CC effects. This should also be combined with training on how to adapt to these changes (Anderson 2012). In this study, a special emphasis is given to the enhancement of teenagers' cognitive capabilities for adaptation response through educational intervention.

Knowledge on climate change, its effects and the skills necessary for making informed adaptation decisions are particularly critical for young people, since they play a unique role in CC response (Ojala and Lakew 2017). Given that they already suffer from harmful CC impacts, depending on location and socio-cultural status (UNICEF 2015), they will be challenged to deal with unprecedented CC effects throughout their lifetimes (Hansen et al. 2013). Capacity building by educational approaches therefore is essential to prepare young people for upcoming environmental and development issues (Körfgen et al. 2017). Furthermore, it will be increasingly important for the next generations to balance the aftereffects (such as greenhouse gases or degraded ecosystems) of former generations (Hansen et al. 2017), while seeking opportunities for a climate-resilient living (Folke et al. 2010).

While there is ample literature on the nexus between young people and CC education (Corner et al. 2015), comparatively little is known on how to teach CC adaptation (Anderson 2012). Knowledge, such as being able to put cause-effect relations of CC into societal context, determines how people appraise possible response options to changes in social-ecological systems (Gorddard et al. 2016), and therefore, is a socially limiting factor for adaptation (Adger et al. 2009). Next to awareness or willingness to act, knowledge influences choices regarding response options (Kroemker and Mosler 2002), and refers to a critical determinant of adaptive capacity (Grothmann and Patt 2005). In this light, two approaches on knowledge acquisition appear useful for teaching CC adaptation successfully. Fazey et al. (2007) argue that without an adequate knowledge base for adaptation, societies may not be capable to build social-ecological resilience. Consequently, flexible learning for CC adaptation should focus on knowledge about (1) CC impacts, (2) the direction of change, and (3) how to achieve change. Frick et al. (2004) express similar ideas. They argue that (1) system knowledge (knowing what), (2) action knowledge (knowing how) and (3) effectiveness knowledge (knowing the effectiveness of actions) shape environmental behaviour. Detailed descriptions of adolescents' system, action and effectiveness knowledge about CC adaptation are insufficiently covered by contemporary literature on CC education (Bofferding and Kloser 2015). This research gap appears particularly striking for two reasons. Societal responses to CC suffer from constraints as uninformed perceptions of response strategies prevail (i.e. examples of maladaptation) (Adger et al. 2009). On the other hand, adolescents and adults are found to hold misconceptions about human interference with the Earth's system (Kuster and Fox 2017), which may not be resolved without appropriate CC education. For example, Bofferding and Kloser (2015) show that adolescents' (between 11 and 18 years) understanding of CC adaptation strategies is limited, and that there is severe confusion between adaptation and mitigation.

In this paper, it is proposed that the educational approach of moderate-constructivist learning (Krahenbuhl 2016) offers useful insights on how to prepare young learners for CC adaptation. This is due to the fact that this approach believes learning to be active, self-determined, and be a part of the social process (Terhart 1999). In moderate-constructivist learning, individuals enter teaching with personal conceptions of a specific phenomenon and educational intervention may influence how individuals construct their own conceptions (Riemeier 2007). The meaning of conception is as follows: people hold subjective conceptions (i.e. mental representations that are tacit) about a specific concept (i.e. CC adaptation) at a particular point in time. Conceptions are built around individuating knowledge, and they reflect an individual's sociocultural experiences (Ezcurdia 1998; Shepardson et al. 2011). The degree to which learners actively reshape their pre-instructional conceptions towards more scientific concepts can be assessed within the framework of conceptual change. Here, it is important to realise that learners test the new (scientific) concepts, introduced to them by educational means. If a new concept leaves a learner dissatisfied with his or her prior conception, if it appears plausible and intelligible to them, and if is useful to solve problems, learning leads to strong (fundamental knowledge restructuring), weak (marginal knowledge assimilation) or no conceptual change (Duit and Treagust 2003).

Framed as a research-education cooperation (Riede et al. 2017), in which young learners, academic and non-academic experts collaborate on the topic of CC adaptation, moderate-constructivist learning offers several benefits. It encourages learners to perform research on their own in authentic learning settings, and allows individuals to contribute with their own conceptions, capacities or views (Wamsler 2017). Furthermore, research-education cooperations create space for deliberative discussion between learners and trusted communicators, while offering platforms for knowledge production. When interacting with other people, social learning may occur, which is a prerequisite for solving adaptation issues (Phuong et al. 2017). Another distinct advantage of research-education cooperations in CC education, is the applicability of the format outside of the school context to generate community outreach (Monroe et al. 2017). Ultimately, this should assist young learns in building a range of capabilities needed for CC adaptation; such as, appraising the social acceptability of adaptation measures or dealing with uncertainty. Furthermore, teenage learners can be motivated to share and co-create adaptation knowledge that could be relevant for their wider communities (Moser 2014).

Focusing on the issue of CC adaptation and embedding it into the context of CC education, the purpose of this study is two-fold. First, by empirical means, this study investigates the personal conceptions that high-school students hold in order to make sense of the strategy of CC adaptation. Second, students' ability to use their conceptions in an exercise in CC education will be studied. The two research questions are as follows:

1. Which conceptions do high-school students hold about CC adaptation and what is the influence of a year-long research-education cooperation?
2. What is the influence of a year-long research-education cooperation on students' ability to differentiate between CC adaptation and CC mitigation?

Methods

Participants and Data Collection

Overall, 120 participants between 16 and 18 years ($M = 16.85$; SD 0.83; 52% female) were part of this study. All of them were high-school students from two schools in North Tyrol (Austria) and two schools from South Tyrol (Italy), all German-speaking. They were grouped into one intervention group (pre-test: $n = 60$, post-test: $n = 60$), and one control group (pre-test: $n = 40$, post-test: $n = 40$). Missing data can be explained by student absences on the day of pre- or post-tests. A pre-test—intervention—post-test design was chosen, (i) to study the conceptions of CC adaptation and, (ii) to assess students' ability to differentiate CC adaptation and CC mitigation. Data was collected through a web-based survey directly before and after the intervention. Furthermore, it was assessed whether CC adaptation content had been part of the school curricula prior to the study.

Educational Design: The Research-Education Cooperation Generation F^3—Fit for Future

The educational design reflects a research-education cooperation based on the principles of moderate-constructivist learning (Bardsley and Bardsley 2007). This setting was applied in the context of CC education and combines two issues, i.e. (i) learning about adaptation (Fazey et al. 2007) and (ii) transdisciplinary collaboration (Cundill et al. 2018). Basically, *Generation F^3—Fit for Future* followed two main objectives: (1) raising awareness of CC effects in Tyrol (Austria and Italy), and (2) increasing understanding of applied and potential adaptation actions. The setting was chosen to reduce cognitive adaptation barriers, such as misconceptions about CC, and possible strategies responding to it (Moser and Ekstrom 2010).

During a period of eight months, intervention group students were asked to autonomously perform a research project on a local adaptation problem. For example, a group of three students investigated snow-farming as a strategy to adapt Tyrol's winter tourism to CC, while another group examined adaptation options for Tyrolean apple farmers in response to CC-induced apple diseases, like apple scab. Two researchers assisted the students in this process during seven in-school workshops. The in-class trainings were structured as follows: (1) identifying a research question of interest, (2) finding a suitable research method in order to answer the research question, (3) collecting data, (4) analysing, interpreting, and discussing data, and (5) disseminating the results.

In *Generation F^3—Fit for Future*, transdisciplinary collaboration and dialogue (Park and Son 2010) between 120 high-school students, 4 teachers, and 28 adaptation scientists and practitioners was established in the first year. During two consecutive expert workshops, the teenagers collaborated with those scientists and practitioners from multidisciplinary backgrounds. In a final poster session, the intervention group students presented their research outcomes to each other, to scientists, societal and political stakeholders and the media. The expert workshops were designed to facilitate the exchange of knowledge about the issue of CC adaptation between intervention group students and experts. Control group students did not participate in the research-education cooperation and received lessons according to their standard curriculum (see Fig. 29.1).

Data Analysis

First, conventional content analysis was chosen to study the different conceptions that students hold about CC adaptation. Here, a systematic procedure was followed to elicit meaning from data and to draw conclusions from it (Bengtsson 2016). The analysis started by reading through the students' responses to obtain immersion with data. Then, meaningful units from the responses were derived and an inductive coding process was started to group the meaningful units into several categories.

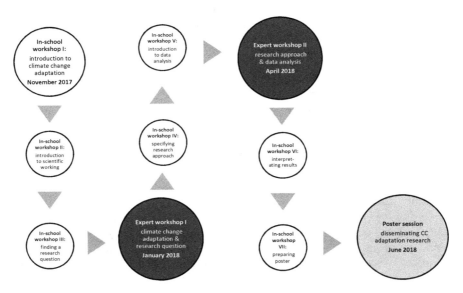

Fig. 29.1 Overview of all interventions received by intervention group students during the research-education cooperation *Generation F³—Fit for Future*. White colours indicate a low transdisciplinary level (two researchers during in-school workshops). Dark colours indicate a high transdisciplinary level (18 experts during expert workshop I and 10 experts during workshop II). Grey colours indicate a moderate transdisciplinary level (less than 10 experts during the poster session)

After 50% of the data was categorized, the coding was revised to make sure that each meaningful unit could be related to a distinctive category; no response was left un-coded (Mayring 2010). At this point, inter-coder reliability was calculated to increase trustworthiness in the content analysis (Krippendorff 2004). A simple percent agreement approach was followed to express inter-coder reliability among three different coders. The mean agreement among coders was 81%, which means that 19% of the coding was erroneous. For each category, one sampling example from the original data is presented, and absolute frequencies of categories (n) from the intervention and control group students are shown. Later, the identified categories were investigated in a quantitative manner and absolute frequency differences (n) between pre- and post-test were assessed performing a statistical *chi square* test ($p < 0.05$).

Second, a multiple-choice instrument was designed to check whether students could differentiate between adaptation and mitigation after engaging in the research-education cooperation. The control instrument asked students to separate CC adaptation from CC mitigation (four answers were correct for adaptation, three for mitigation, and two fitted neither to adaptation nor mitigation) and followed the IPCC's AR5 WGII glossary (IPCC 2014) (see Appendix). Here, absolute frequency differences between pre- and post-test were assessed performing a statistical *t*-test ($p < 0.05$).

Results

Conceptions of CC Adaptation Held by Intervention Group Students

Before CC Education Intervention (Pre-test)

High-school students from the intervention group were asked to explain what CC adaptation meant for them before and after engaging in the research-education cooperation (see Appendix).

Data revealed that about 12% of students held no conception of CC adaptation at all. Either students gave no response, or a statement was given that explained their lack of understanding. Another 20% of students showed a flawed conception of CC adaptation since they confused adaptation with mitigation. Misconceptions of this kind mainly related CC adaptation to mitigation goals (*respondent A*), or specific mitigation options. Another common misconception was the confusion between CC adaptation with environmental protection (*respondent B*), and nature conservation (see Table 29.1). Examples that reflect these categories are as follows:

Respondent A: *'[CC adaptation] means to undertake countermeasures against climate change with the aim to slow its rate down.'*

Respondent B: *'[CC adaptation] means to reduce environmental pollution.'*

In addition, about 48% of students understood CC adaptation as a process of adjustment. These conceptions strongly differed in degrees of sophistication. Some students explained that adaptation would be necessary in order to respond to a changing climate in general (*respondent C*), while others related adaptation to specific effects of CC (*respondent D*). Only a small number of students could link adaptation with the need to moderate, or avoid harm from actual, or expected CC effects (*respondent E*), or to exploit beneficial opportunities (*respondent F*) (see Table 29.1). Examples that reflect these categories are as follows:

Respondent C: *'Adaptation refers to how mankind adjusts to a new climate.'*

Respondent D: *'Humans try to cope with the impacts resulting from CC.'*

Respondent E: *'CC adaptation means to avoid potential harm due to changes.'*

Respondent F: *'[CC adaptation] means that humans adjust to climate change and are trying to get the best out of it or to antagonize in positive manner.'*

Finally, 20% of students related CC adaptation to a major transformation of lifestyles or argued for the development of new coping strategies and measures (*respondent G*) (see Table 29.1). One example is given below:

Respondent G: *'[CC adaptation means] to develop new technologies and methods for a life with CC.'*

Table 29.1 Conceptions of CC adaptation from intervention group students

Pre-test			Post-test			Comparison pre-post (%)
Category	Sub-category	Sampling example	Category	Sub-category	Sampling example	
1. No conception of adaptation (n = 7)	Answer sheet was left blank		1. No conception of adaptation (n = 4)	Answer sheet was left blank		−43
	No knowledge was indicated	*'I don't know much about [CC adaptation]'*		No knowledge was indicated	*'I am not very familiar with CC adaptation'*	
2. Misconception of adaptation (n = 12)	Adaptation confused with mitigation	*'[CC adaptation means] to undertake countermeasures against climate change, with the aim to slow it down'*	2. Misconception of adaptation (n = 3)	Adaptation confused with mitigation	*'[CC adaptations are] measures to reduce CC'*	−75*
	Adaptation confused with environmental protection	*'[CC adaptation means] to reduce environmental pollution'*		Adaptation confused with environmental protection	*'[CC adaptations] refers to waste separation, ban of diesel fuels, abandonment of coal plants, ban of nuclear plants, reduction of plastic bottles and more resistance against tropical clearings'*	

(continued)

Table 29.1 (continued)

Pre-test			Post-test			Comparison pre-post (%)
Category	Sub-category	Sampling example	Category	Sub-category	Sampling example	
	Adaptation mistaken for an organization	*'I think that adaptation refers to an organization that informs about CC and how to adapt to it'*				
3. Adaptation is adjustment (n = 29)	To a changing climate	*'Adaptation refers to how mankind adjusts to a new climate'*	3. Adaptation is adjustment (n = 24)	To a changing climate	*'Adaptation means to adjust to changing weather and climatic conditions in all aspects in economy and society''*	−17
	To expected climate and its effects	*'Humans try to cope with the impacts resulting from CC'*		To expected climate and its effects	*'[CC adaptation refers to] adjustment to changing conditions due to climate change'*	
	To moderate or avoid harm	*'CC adaptation means to avoid potential harm due to changes'*		To moderate or avoid harm	*'[CC adaptation means to] adjust to new conditions that the globe will stay in equilibrium'*	

(continued)

Table 29.1 (continued)

Pre-test			Post-test			Comparison pre-post (%)
Category	Sub-category	Sampling example	Category	Sub-category	Sampling example	
	To exploit beneficial opportunities due to CC	'[Adaptation] means that humans adjust to a climatic change and are trying to get the best out of it or to antagonize in positive manner'		To exploit beneficial opportunities due to CC	'Adaptation means to identify also positive side-effects of CC and to strengthen them instead of fighting it'	
4. Adaptation is transformation (n = 12)	Transforming life-styles	'Adaptation to CC means that animals and humans adjust to impacts, which means to change life-styles'	4. Adaptation is transformation (n = 9)	Transforming life-styles	'[CC adaptation means] to change the personal life-style towards increasing environmental friendly behaviour and to cope with impacts due to CC as best as possible'	−25
	Developing major coping strategies or measures	'[CC adaptation means] to develop new technologies and methods for a life with CC'		Developing major coping strategies or measures	'[CC adaptation refers to] new technology for agriculture'	

(continued)

Table 29.1 (continued)

Pre-test			Post-test			Comparison pre-post (%)
Category	Sub-category	Sampling example	Category	Sub-category	Sampling example	
			5. Adaptation is learning (n = 4)	To get informed about adaptation	'[CC adaptation means] that young people but also adults should inform themselves about how to adapt'	+100
				To reflect upon adaptation choices	'[CC adaptation means] to identify possible changes due to CC in first place before taking solutions'	
			6. Adaptation is climate action (n = 16)	Implementing adaptation measures in different contexts	'[CC adaptation means] to start actions that are helpful to protect ourselves from negative impacts due to CC, to recognize beneficial opportunities and to prepare for a life with CC'	+100

*between group significant differences from pre- and post-test based on chi-square standard residuals, $p < 0.05$, n number of responses

After CC Education Intervention (Post-test)

After the students engaged in the research-education cooperation and collaborated with scientists and practitioners during the period of eight months, the content of their CC adaptation conceptions changed fundamentally. Students used more words to explain the concept of CC adaptation, and in contrast to the pre-test, they could describe adaptation to CC more accurately. Most misconceptions were replaced by more valid representations of a scientific-informed perception of adaptation, and interestingly, two new categories emerged during the inductive coding process.

The proportion of students that held no conception or a flawed misconception were reduced by 43 and 75%. Most intervention group students successfully expanded their conceptions of CC adaptation, however, certain misconceptions persisted. Some students still confused CC adaptation with mitigation goals, mitigation measures or environmental protection (*respondent H*) (see Table 29.1). One example is given below:

Respondent H: *'[CC adaptation] refers to measures to reduce climate change.'*

Those students, who believed CC adaptation to be an adjustment in the pretest were found to expand their conceptions of adaptation as well. Some students gained the insight that the concept of vulnerability refers to CC adaptation, while others showed a preliminary understanding of the concept of resilience, since they used an equilibrium approach to describe how CC effects may affect human-climate relationships. Furthermore, a larger number of students explained that CC adaptation should be addressed in a context-specific manner (*respondent I*), and that different sectors of society depend upon different adaptation measures (see Table 29.1). One example is given below:

Respondent I: *'Adaptation means to adjust to changing weather and climatic conditions in all aspects of economy and society.'*

About 15% of students perceived CC adaptation as a process of life-style transformation and as the development of new strategies or technologies. No qualitative changes for this category compared to the pre-test situation were identified (see Table 29.1).

After CC education, a new category emerged from the responses of the intervention group students. Adaptation conceptions were identified that framed CC adaptation as a learning and reflection process. Students argued that all people should inform themselves about CC adaptation (*respondent J*). Furthermore, they realised that scientific assessment should inform adaptation choices before they are carried out (see Table 29.1). One example that reflects this new category is as follows:

Respondent J: *'[CC adaptation means] that young people, but also adults should inform themselves about how to adapt.'*

Another new category that emerged from the post-test responses only referred to climate action. This means that 27% of students understood adaptation as a measure that needs to be implemented in different contexts of managed systems, with the aim to effectively respond to climate changes. These perceptions differed from other CC adaptation conceptions in the way that students argued explicitly for adaptation actions for achieving a more resilient state in society (see Table 29.1). One example that reflects this new category is as follows:

Respondent K: '[CC adaptation] means to start actions that are helpful to protect ourselves from negative impacts due to CC, to recognize beneficial opportunities and to prepare for a life with CC.'

Conceptions of CC Adaptation Held by Control Group Students

Before Standard Curriculum Intervention (Pre-test)

Like intervention group students, students from the control group were also asked to explain what CC adaptation meant for them.

A total of 28% of the control group either gave no answer or stated a flawed conception of CC adaptation. The detected misconceptions were similar to those of the intervention group students and referred to a confusion between CC adaptation and CC mitigation (*respondent L*), or CC adaptation and environmental protection (see Table 29.2). One standard example that reflects this category is as follows:

Respondent L: 'Humans need to adapt to CC by emitting less fossil fuels and by eating less meat.'

Most control group students (58%) recognized adaptation as adjustment to a new state in climate, or to expected CC effects (*respondent M*). Control group students were unaware that adaptation could involve the exploitation of beneficial opportunities that may result from CC (see Table 29.2). One example is given below:

Respondent M: '[CC adaptation] means that people adjust to a new climate and the resulting conditions.'

About 15% of the control group students related CC adaptation either to the transformation of life-styles, or to the development of new strategies or measures relevant for adaptation (*respondent N*) (see Table 29.2). One standard example is given below:

Respondent N: 'I think that adaptation means to adjust to CC and to find new technologies, measures as well as to support our Earth when adapting.'

After Standard Curriculum Intervention (Post-test)

Interestingly, the share of control group students who gave no answer or stated a misconception of CC adaptation increased to 45% after the post-test. This is surprising,

Table 29.2 Conceptions of CC adaptation from control group students

Pre-test			Post-test			Comparison pre-post (%)
Category	Sub-category	Sampling example	Category	Sub-category	Sampling example	
1. No conception of adaptation (n = 2)	Answer sheet was left blank		1. No conception of adaptation (n = 7)	Answer sheet was left blank		+250
	No knowledge was indicated	*'I don't know what CC adaptation is about'*		No knowledge was indicated	*'I don't know what CC adaptation means'*	
2. Misconception of adaptation (n = 9)	Adaptation confused with mitigation	*'Humans need to adapt to CC by emitting less fossil fuels and by eating less meat'*	2. Misconception of adaptation (n = 11)	Adaptation confused with mitigation	*'[CC adaptations refers to] measures to slow down or stop CC'*	+22
	Adaptation confused with environmental protection	*'I think that for a positive climate it is important to reduce using chemtrails and to keep the landscape clean'*		Adaptation confused with environmental protection	*'Under CC adaptation, I understand a more environmental friendly way to interact with nature'*	
3. Adaptation is adjustment (n = 23)	To a changing climate	*'[CC adaptation means] that people all over the globe need to adjust to a different climate'*	3. Adaptation is adjustment (n = 20)	To a changing climate	*'Adaptation is the adjustment to a new climate'*	−13

(continued)

Table 29.2 (continued)

Pre-test			Post-test			Comparison pre-post (%)
Category	Sub-category	Sampling example	Category	Sub-category	Sampling example	
	To expected climate and its effects	'[CC adaptation means] that people adjust to a new climate and the resulting conditions'		To expected climate and its effects	'[CC adaptation means] to adjust to those changes that are caused by CC'	
	To moderate or avoid harm	'Adaptation includes measures to reduce the impact of CC'		To moderate or avoid harm	'Households but also companies need to inform themselves about CC and should undertake measures to avoid negative impacts'	
				To exploit beneficial opportunities due to CC	'[CC adaptation means] to tackle CC in positive manner'	
4. Adaptation is transformation (n = 6)	Transforming life-styles	'Under adaptation, I understand that people change their everyday habits to adjust to CC'	4. Adaptation is transformation (n = 2)	Transforming life-styles	'Adaptation to climate change is the change of recent life-styles according to CC effects'	−67

(continued)

Table 29.2 (continued)

Pre-test			Post-test			Comparison pre-post (%)
Category	Sub-category	Sampling example	Category	Sub-category	Sampling example	
	Developing major coping strategies or measures	*'I think that adaptation means to adjust to CC and to find new technologies, measures as well as to support our Earth when adapting'*		Developing major coping strategies or measures	*'One has to develop certain measures in order to adapt to CC'*	

n number of responses

given that similar percentages of students holding misconceptions that were shown in the pre-test were expected. Most commonly, misconceptions were a confusion between CC adaptation and CC mitigation (*respondent O*) (see Table 29.2). One example is given below:

> Respondent O: '*[CC adaptation refers to] measures to slow down or stop CC.*'

Similar to the pre-test, nearly every second control group student related adaptation to CC as a process of adjustment. Only 5% of students understood CC adaptation as a life-style change or a development of adaptation strategies (see Table 29.2).

High-School Students' Ability to Differentiate Between Adaptation and Mitigation

The data suggest that intervention group students could differentiate between CC adaptation and CC mitigation much better after they had participated in the research-education cooperation. The decrease in total answers given for the mitigation section (answers that were correct for mitigation) between pre- and post-test is statistically significant (t-value $(3) = 4.587, p < 0.05$) (see Fig. 29.2). No significant differences in answer frequencies for the control group during pre- and post-test were detected (see Fig. 29.3).

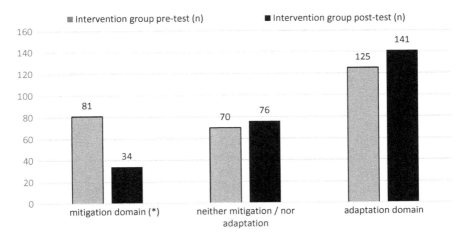

Fig. 29.2 Absolute frequencies of intervention group students' performance to distinguish adaptation from mitigation, (* = between group significant differences from pre- to post-test based on t-test standardized residuals, $p < 0.05$, n = number of responses)

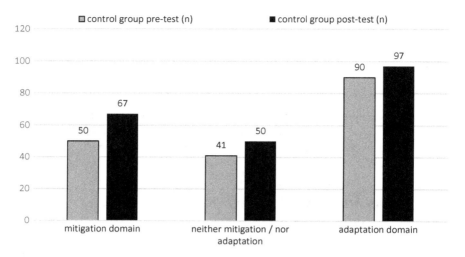

Fig. 29.3 Absolute frequencies of control group students' performance to distinguish adaptation from mitigation, (n = number of responses)

Discussion

This study set out to explore different conceptions that high-school students hold on the issue of CC adaptation, before and after educational intervention. Moreover, it was assessed whether a research-education cooperation, following the principles of moderate-constructivist learning, could serve as a tool to successfully expand knowledge about CC adaptation among high-school students.

As already highlighted in the introduction, young people specifically struggle to make sense of adaptation as a strategy to cope with anthropogenic CC (Corner et al. 2015). Filling this gap, the results from this study provide in-depth insights into teenagers' conceptions of CC adaptation. For one, it was shown that before educational intervention, a significant proportion of students hold none (or rather highly generalized conceptions) of CC adaptation. Adaptation was largely confused with mitigation, or with ideas of environmental protection. This confirms Bofferding and Kloser's (2015) findings, who found that middle and high-school students were likely to mistake mitigation and adaptation behaviours. In addition, this is in line with the findings of Davidson and Lyth (2012) and Mochizuki and Bryan (2015), who argue that recent CC education lacks to address the challenge of CC adaptation appropriately.

The results also go beyond confirming these literature results, as they indicate that a few students—from intervention and control group by equal terms—could make sense of the scientific concept of adaptation before engaging in CC education. However, most students of the intervention group demonstrated a more sophisticated understanding of CC adaptation after participating in the research-education cooperation. It is interesting that these students developed an incremental view of CC adaptation, and related it to a general process of collective response to observed or expected CC impacts. However, other student responses suggested that CC adaptation would involve societal transformation. It was also shown that after the collaboration many students gained the perception that CC adaptation is context-specific. Such views may have evolved, since the expert workshops in which intervention students collaborated with 28 scientists and practitioners were organized according to socio-economic sectors where CC adaptation is implemented. For instance, the students that worked on CC in Tyrol and the abundance of apple scab could discuss adaptation options with agricultural researchers or consultants.

In addition to the finding that intervention group students developed more elaborated conceptions of CC adaptation, one interesting contribution of this study is that students grasped other scientific principles, like vulnerability, resilience, or transformation thinking due to collaboration with CC experts in the research-education cooperation. All of these principles are crucial to understanding the strategy of CC adaptation in terms of its objectives and barriers (Folke et al. 2010). This is exemplified by the conceptions of the intervention group students, who argued that different sectors in society are not equally vulnerable to CC and that fundamental changes in everyday-habits are needed in order to cope with climate risk.

It is important to note that most personal conceptions about CC adaptation from the sample are flexible and were successfully opened-up by CC education. This insight adds evidence to literature findings that, next to the provision of factual knowledge, the creation of empowering learning settings (Hayden et al. 2011) and deliberate dialogue with trusted communicators (Corner et al. 2015) support teenagers to develop increased levels of CC literacy. In contrast, the data also clearly show that it was not possible to expand all CC adaptation conceptions of intervention group students. Nevertheless, this does not limit the effectiveness of the educational design, as students' conceptions should not be completely replaced by scientific principles. Teenagers decide by themselves whether their deep-rooted adaptation conceptions will be restructured or not (Duit and Treagust 2003).

Several limitations of this study are acknowledged. First, qualitative data (the students' conceptions) left room for interpretation, and were translated from German into English. Unless consistency between the three coders were assessed, it would be difficult to draw meaning from the data in each case, and therefore the coding may add simplified logic to the complex qualitative data. Furthermore, standard examples that were selected to illustrate students' conceptions cannot address the samples' full richness (Shepardson et al. 2011). It is also understood that framing CC education as a research-education cooperation has some influence on students' conceptions on its own. While the intervention group students argued for the importance of learning for CC adaptation, the critical role of learning was not mentioned from any control group student. Control group students did not gain the advantage of working with experts. To partly alleviate this disadvantage, they will benefit from improved in-class climate change education by the experimental group's teachers in the second year of *Generation F³—Fit for Future* (Bournot-Trites and Belanger 2005). Finally, the study did not investigate affective factors, such as motivation for, or interest in,

CC education. Both factors could influence whether a learner replaces, restructures or sticks to his or her personal conception of CC adaptation (Monroe et al. 2017).

Notwithstanding these limitations, this study's empirical findings indicate that research-education cooperations are promising programmes of CC education. Such settings create deliberative dialogue between learners, teachers and experts, and it was possible to convince academic and non-academic professionals to participate in this transdisciplinary dialogue with teenagers by establishing an output-oriented framing. As Tolppannen and Aksela (2018) claim, research-education cooperations give teenagers the opportunity to answer their own questions on CC adaptation, and offer orientation for those teachers, who aim to cross the borders of their classrooms.

Conclusion

In conclusion, it is shown that a research-education cooperation like *Generation F³—Fit for Future:* (i) supports adolescent students to expand their conceptions of CC adaptation and (ii) assists them to better differentiate between adaptation and mitigation. The results encourage CC educators working on the issue of CC adaptation to create similar learning settings, and to offer space for close collaboration between students and experts. Since this study focused on CC adaptation on a conceptual level, it is recommended that future research about CC education should investigate what determines teenagers' understanding of adaptive solutions in more detail.

Funding Research funding was received from 3. Forschungswettbewerbsausschreibung Autonome Provinz Bozen-Südtirol (Abteilung 34. Innovation, Forschung und Universität).

Appendix

(1) **'Adaptation to climate change'** means:
   (Several answers possible)

   - to reduce greenhouse gas emissions. *(M)*
   - to avoid harm from CC effects. *(A)*
   - to moderate harm from CC effects. *(A)*
   - to live with CC effects. *(NOR)*
   - to avoid global warming. *(M)*
   - to reduce the risks of CC. *(NOR)*
   - to exploit beneficial opportunities due to CC. *(A)*
   - to moderate harmful CC effects on creatures and ecosystems. *(A)*
   - to limit the causes of CC. *(M)*

(2) What does **adaptation to climate change** mean to you?

Please describe shortly.

References

Adger WN, Huq S, Brown K, Conway D, Hulme M (2003) Adaptation to climate change in the developing world. Prog Dev Stud 3(3):179–195. https://doi.org/10.1191/1464993403ps060oa

Adger WN, Dessai S, Goulden M, Hulme M, Lorenzoni I, Nelson DR, Wreford A (2009) Are there social limits to adaptation to climate change? Clim Change 93:335–354. https://doi.org/10.1007/s10584-008-9520-z

Anderson A (2012) Climate change education for mitigation and adaptation. J Educ Sustain Dev 6(2):191–206. https://doi.org/10.1177/0973408212475199

Bardsley DK, Bardsley AM (2007) A constructivist approach to climate change teaching and learning. Geogr Res 45(4):329–339. https://doi.org/10.1111/j.1745-5871.2007.00472.x

Bengtsson M (2016) How to plan and perform a qualitative study using content analysis. NursingPlus Open 2:8–14. https://doi.org/10.1016/j.npls.2016.01.001

Berrang-Ford L, Ford JD, Paterson J (2011) Are we adapting to climate change? Glob Environ Change 21(1):25–33. https://doi.org/10.1016/j.gloenvcha.2010.09.012

Bofferding L, Kloser M (2015) Middle and high school students' conceptions of climate change mitigation and adaptation strategies. Environ Educ Res 21(2):275–294. https://doi.org/10.1080/13504622.2014.888401

Bournot-Trites M, Belanger J (2005) Ethical dilemmas facing action researchers. J Educ Thought 39(2):197–215

Corner A, Roberts O, Chiari S, Völler S, Mayrhuber ES, Mandl S, Monson K (2015) How do young people engage with climate change? The role of knowledge, values, message framing, and trusted communicators. Wiley Interdisc Rev Clim Change 6(5):523–534. https://doi.org/10.1002/wcc.353

Cundill G, Harvey B, Tebboth M, Cochrane L, Currie-Alder B, Vincent K et al (2018) Large-scale transdisciplinary collaboration for adaptation research: challenges and insights. Glob Challenges 1700132. https://doi.org/10.1002/gch2.201700132

Davidson J, Lyth A (2012) Education for climate change adaptation—enhancing the contemporary relevance of planning education for a range of wicked problems. J Educ Built Environ 7(2):63–83. https://doi.org/10.11120/jebe.2012.07020063

Duit R, Treagust DF (2003) Conceptual change: a powerful framework for improving science teaching and learning. Int J Sci Educ 25(6):671–688. https://doi.org/10.1080/09500690305016

Ezcurdia M (1998) The concept-conception distinction. Philos Issues 9:187–192. https://doi.org/10.2307/1522969

Fazey I, Fazey JA, Fischer J, Sheeren K, Warren J, Noss RF, Dovers SR (2007) Adaptive capacity and learning to learn as a leverage for social-ecological resilience. Front Ecol Environ 5(7):375–380. https://doi.org/10.1890/1540-9295(2007)5

Folke C, Carpenter SR, Walker B, Scheffer M, Chapin T, Rockström J (2010) Resilience thinking: integrating resilience, adaptability and transformability. Ecol Soc 15(4). http://www.ecologyandsociety.org/vol15/iss4/art20/

Ford JD, Berrang-Ford L, Paterson J (2011) A systematic review of observed climate change adaptation in developed nations. Clim Change 106(2):327–336. https://doi.org/10.1007/s10584-011-0045-5

Frick J, Kaiser FG, Wilson M (2004) Environmental knowledge and conservation behaviour: exploring prevalence and structure in a representative sample. Pers Individ Differ 37(8):1597–1613. https://doi.org/10.1016/j.paid.2004.02.015

Gorddard R, Colloff MJ, Wise RM, Ware D, Dunlop M (2016) Values, rules and knowledge: adaptation as change in the decision context. Environ Sci Policy 57:60–69. https://doi.org/10.1016/j.envsci.2015.12.004

Grothmann T, Patt A (2005) Adaptive capacity and human cognition: the process of individual adaptation to climate change. Glob Environ Change 15(3):199–213. https://doi.org/10.1016/j.gloenvcha.2005.01.002

Hansen G, Stone D (2016) Assessing the observed impact of anthropogenic climate change. Nat Clim Change 6:532–537. https://doi.org/10.1038/NCLIMATE2896

Hansen JP, Kharecha P, Sato M, Masson-Delmotte V, Ackerman F, Beerling DJ et al (2013) Assessing "dangerous climate change": required reduction of carbon emissions to protect young people, future generations and nature. PLOS ONE 8:e81648. https://doi.org/10.1371/journal.pone.0081648

Hansen J, Sato M, Kharecha P, von Schuckmann K, Beerling DJ, Cao J, Ruedy R (2017) Young people's burden: requirement of negative CO_2 emissions. Earth Syst Dyn 8:577–616. https://doi.org/10.5194/esd-8-577-2017

Hayden M, Houwer R, Frankfort M, Rueter J, Black T, Mortfield P (2011) Pedagogies of empowerment in the face of climate change uncertainty. J Activism Sci Technol Educ 3(1):118–130

IPCC (2014) Climate change 2014: impacts, adaptation, and vulnerability. Part B: regional aspects. In: Barros VR, Field CB, Dokken DJ, Mastrandrea MD, Mach KJ, Bilir TE et al (eds) Contribution of working group II to the fifth assessment report of the intergovernmental panel on climate change. Cambridge University Press, Cambridge

Körfgen A, Keller L, Kuthe A, Oberrauch A, Stötter H (2017) (Climate) change in young people's minds—from categories towards interconnections between the anthroposphere and natural sphere. Sci Total Environ 580:178–187. https://doi.org/10.1016/j.scitotenv.2016.11.127

Krahenbuhl KS (2016) Student-centered education and constructivism: challenges, concerns, and clarity for teachers. Clearing House J Educ Strat Issues Ideas 89(3):97–105. https://doi.org/10.1080/00098655.2016.1191311

Krippendorff K (2004) Reliability in content analysis—some common misconceptions and recommendations. Hum Commun Res 30(3):411–433. https://doi.org/10.1111/j.1468-2958.2004.tb00738.x

Kroemker D, Mosler HJ (2002) Human vulnerability—factors influencing the implementation of prevention and protection measures: an agent based approach. In: Steininger K, Weck-Hannemann H (eds) Global environmental change in alpine regions, impact, recognition, adaptation and mitigation. Edward Elgar, Cheltenham

Kuster EL, Fox GA (2017) Current state of climate education in natural and social sciences in the USA. Clim Change 141(4):613–626. https://doi.org/10.1007/s10584-017-1918-z

Lesnikowski A, Ford J, Biesbroek R, Berrang-Ford L, Maillet M, Araos M, Austin SE (2017) What does the Paris agreement mean for adaptation? Clim Policy 17(7):825–831. https://doi.org/10.1080/14693062.2016.1248889

Lutz W, Muttarak R, Striessnig E (2014) Universal education as a key to enhanced climate adaptation. Science 346(6213):1061–1062. https://doi.org/10.1126/science.1257975

Mayring P (2010) Qualitative Inhaltsanalyse. In: Flick U, von Kardoff E, Keupp H, von Rosenstiel L, Stephan W (eds) Handbuch qualitative Forschung: Grundlagen, Konzepte, Methoden und Anwendungen. Verlag für Sozialwissenschaften, München

Mochizuki Y, Bryan A (2015) Climate change education in the context of education for sustainable development: rationale and principles. J Educ Sustain Dev 9(1):4–26. https://doi.org/10.1177/0973408215569109

Monroe MC, Plate RR, Oxarart A, Bowers A, Chaves WA (2017) Identifying effective climate change education strategies: a systematic review of the research. Environ Educ Res 1–22. https://doi.org/10.1080/13504622.2017.1360842

Moser SC (2014) Communicating adaptation to climate change: the art and science of public engagement when climate change comes home. WIREs Clim Change 5(3):337–358. https://doi.org/10.1002/wcc.276

Moser SC, Ekstrom JA (2010) A framework to diagnose barriers to climate change adaptation. PNAS 107(51). https://doi.org/10.1073/pnas.1007887107

Ojala M, Lakew Y (2017) Young people and climate change communication. In: Oxford research encyclopedia of climate science. Oxford University Press, Oxford

Park J-Y, Son J-B (2010) Transitioning toward transdisciplinary learning in a multidisciplinary environment. Int J Pedagogies Learn 6(1):82–93. https://doi.org/10.5172/ijpl.6.1.82

Phuong LTH, Biesbroek GR, Wals AEJ (2017) The interplay between social learning and adaptive capacity in climate change adaptation: a systematic review. NJAS Wageningen J Life Sci 82:1–9. https://doi.org/10.1061/j.njas.2017.05.001

Riede M, Keller L, Oberrauch A, Link S (2017) Climate change communication beyond the 'ivory-tower': a case study about the development, application, and evaluation of a science-education approach to communicate climate change to young people. J Sustain Educ 12:1–24

Riemeier T (2007) Moderater Konstruktivismus. In: Krüger D, Vogt H (eds) Theorien in der biologiedidaktischen Forschung - Ein Handbuch für Lehramtsstudenten und Doktoranden. Springer, Heidelberg

Rumore D, Schenk T, Susskind L (2016) Role-play simulations for climate change adaptation education and engagement. Nat Clim Change 6(8):745–750. https://doi.org/10.1038/NCLIMATE3084

Runhaar H, Wilk B, Persson Å, Uittenbroek C, Wamsler C (2018) Mainstreaming climate adaptation: taking stock about "what works" from empirical research worldwide. Reg Environ Change 18(4):1201–1210. https://doi.org/10.1007/s10113-017-1259-5

Shepardson DP, Niyogi D, Choi S, Charusombat U (2011) Students' conceptions about the greenhouse effect, global warming, and climate change. Clim Change 104(3–4):481–507. https://doi.org/10.1007/s10584-009-9786-9

Terhart E (1999) Konstruktivismus und Unterricht - Gibt es einen neuen Ansatz in der Allgemeinen Didaktik? Zeitschrift für Pädagogik 45:629–647

Tolppanen S, Aksela M (2018) Identifying and addressing students' questions on climate change. J Environ Educ 49(5):375–389. https://doi.org/10.1080/00958964.2017.1417816

UNFCCC (2015) Adoption of the Paris agreement—report No. FCCC/CP/2015/L.9/Rev.1. UNFCCC, Bonn. Retrieved from http://unfccc.int/resource/docs/2015/cop21/eng/l09r01.pdf

UNICEF (2015) Unless we act now—the impact of climate change on children. UNICEF, New York. Retrieved from https://www.unicef.org/publications/files/Unless_we_act_now_The_impact_of_climate_change_on_children.pdf

Wamsler C (2017) Stakeholder involvement in strategic adaptation planning: Transdisciplinarity and co-production at stake? Environ Sci Policy 75:148–157. https://doi.org/10.1016/j.envsci.2017.03.016

Chapter 30
Engaging and Empowering Business Management Students to Support the Mitigation of Climate Change Through Sustainability Auditing

Kay Emblen-Perry

Abstract In spite of growth in specialist courses and modules, and integration of some sustainability content into business management curricula, engaging business management students in sustainable business practices continues to lag behind needs of graduates in the workplace; most students do not understand the role businesses can play in supporting the mitigation of climate change and alleviation of impacts. As the learning environment is an important determinant of behaviour, new means and methods of learning, teaching and assessment are required to enhance education for sustainability for business management students to improve their knowledge of climate change mitigation, adaption and impact reduction and empower future graduate employees to change businesses from within. In response, an innovative learning, teaching and assessment approach has been designed for a 3rd year undergraduate module to equip business management students with sustainability knowledge and tools to develop and communicate climate change mitigation activities through the completion of a Global Reporting Initiative sustainability audit of a bespoke case study company. This conceptual paper presents the means and methods employed in this innovative module. It describes the theoretical and practical contexts of the module, its experiential learning, teaching and assessment means and methods designed for active, collaborative learning and outlines the opportunities and challenges experienced in designing the module. It will help members of the sustainability community seeking to build new means and methods of generative education for sustainability through active, experiential learning. It builds on existing pedagogic discourse on innovative means and methods for learning, teaching and assessment of opportunities for sustainable business futures and climate change mitigation and contributes to research into participatory approaches to education for sustainability.

Keywords Sustainability audit · Learning, teaching and assessment · Education for sustainability · Business management students · Active, experiential learning

---

K. Emblen-Perry (✉)
Worcester Business School, University of Worcester, WR1 3AS Worcester, UK
e-mail: k.emblenperry@worc.ac.uk

© Springer Nature Switzerland AG 2019
W. Leal Filho and S. L. Hemstock (eds.), *Climate Change and the Role of Education*,
Climate Change Management, https://doi.org/10.1007/978-3-030-32898-6_30

549

Introduction

As most future business managers and decision-makers are educated by universities (UNESCO 2011) there is a growing expectation that tertiary education systems should contribute to developing a sustainable society (United Nations 2015). Cortese (2003) goes further and argues that universities bear a profound moral responsibility to develop the knowledge, skills and values required to empower students to act to achieve a just and sustainable future including alleviation of impacts of climate change. United Nations (2017) suggest this can be achieved within tertiary education curricula through the provision of pedagogic approaches that develop skills and attributes to help mitigate the impacts of climate change. Such learning, teaching and assessment will encourage individuals to take responsibility to resolve challenges and contribute to creating a more sustainable world (UNESCO 2017). For this reason, education underlies the goals and targets of all 17 Sustainable Development Goals (United Nations 2017).

As tutors are powerful change agents able to influence students in the classroom (Bourn 2013) the learning environment becomes an important determinant of behaviour (Genn 2001). Therefore, to maximise the opportunities this offers new means and methods of learning, teaching and assessment are required to enhance education for sustainability for business management students (Figuero and Raufflet 2015; UNESCO 2017). These are required to improve students' knowledge of climate change mitigation, adaption and impact reduction and to empower future graduate employees to act as active pro-sustainability citizens (Seatter and Ceulemans 2017) to change businesses from within. However, pedagogic literature does not provide a consensus on what learning, teaching and assessment approaches should be used to facilitate sustainability thinking and behaviour (Seatter and Ceulemans 2017); rather pedagogic approaches rely on the interpretation of policy recommendations and intuition of educators (Boeve-de Pauw et al. 2015).

However, waiting for this consensus before introducing new, innovative means and methods of sustainability learning, teaching and assessment into the business curriculum will not adequately equip business management students (who will shortly graduate and move into their graduate workplaces) with the appropriate knowledge, skills and values to act as the change agents able to address the causes and consequences of climate change and promote sustainable business futures. Consequently, the author designed a 3rd year undergraduate module that aligns pedagogic means and methods to an assessed Global Reporting Initiative sustainability audit (GRI 2016) of a bespoke case study company.

A sustainability audit is a methodical examination of a business's procedures and practices that determine or influence environmental, social or economic impacts; they have long been recognised as a method of improving sustainability performance and prioritising improvement actions (Hillary 2004). Adopting a sustainability audit as the focus of a business sustainability module therefore presents an opportunity to engage business management students in real world learning of business sustainability practice and encourage learners to construct a personal, evidence-based view of sustainability whilst delivering specific, general and transformative employment skills and values. As a means and method of learning, teaching and assessment an audit requires students collect, collate, synthesise and evaluate information provided in the case study to form credible judgements of the case study company's performance. This promotion of sustainability, employment and higher order cognitive skills is designed to equip students with the necessary competences to promote sustainable futures and help alleviate climate change impacts from within their future workplaces.

This conceptual paper presents this 3rd year business management undergraduate module introduced in academic year 2016–17. It introduces the author's innovative pedagogic approach that may provide business management students the freedom to explore sustainability issues, investigate climate change mitigation opportunities and experience the inherent complexities and challenges that sustainability practitioners face on a daily basis. As Seatter and Ceulemans (2017) consider a key goal of tertiary education is to promote the understanding of the complexity of sustainability, sharing this module's innovative approach with other educators may support the wider enhancement of education for sustainability. The author describes the theoretical and practical contexts of the module, its' pedagogic means and methods designed immerse students in the complexities of business sustainability through their preferred means and methods of active, collaborative learning (Oblinger and Oblinger 2005; Abdel Meguid and Collins 2017) and shares the opportunities and challenges experienced in designing the module. Although the paper does not attempt to validate the impacts and outcomes of the pedagogic approach presented the paper may be of practical use to other educators considering experiential learning, teaching and assessment.

In addition, this paper contributes to the research into participatory approaches to education for sustainability recommended by Vare and Scott (2007) and Mader and Mader (2012) and develops current pedagogic discourse of innovative means and methods for education for sustainability. These are vital for developing sustainability literacy to inform behaviour change (Disterheft et al. (2015) and stimulating more intelligent and strategic decision making (Craik 1943) to promote sustainable business futures and the necessary skills to monitor, analyse and communicate climate change impacts.

Education for Sustainability in the Business School Context

Climate change cannot be seen as an isolated issue; rather it should be treated as an integral part of developing sustainable futures (Baumgartner 2014). As universities play a key role in promoting these sustainable futures through learning and teaching (Rieckmann 2011; Sterling et al. 2013; HEFCE 2013; Higher Education Academy 2014; United Nations 2017), developing students with appropriate knowledge and skills (Chalkley 2006; Rieckmann 2011; Quality Assurance Agency for Higher Education 2014) and engaging with business to solve real-life problems for both businesses and society (Molthan-Hill 2014) the author incorporates the causes and consequences of climate change alongside social and environmental responsibility and economic accountability within the business curriculum to provide an holistic business education for sustainability.

However, in the massified and marketised environment of universities (Lynch 2006), in which students are treated as consumers of educational output (Vanderstraeten 2004), equipping business management students with the appropriate knowledge and skills to promote sustainable futures and the alleviation of climate change impacts from within their future workplaces in increasingly challenging. This challenge emanates from the need to deliver the expectations of the three main actors within business sustainability learning, teaching and assessment: students, employers and educators. Each of these actors has different and evolving needs, hopes and demands which combine to affect students' perceptions, dispositions and approaches to learning (Pennington et al. 2012). Students expect good grades and employment skills delivered through participatory user interactions (Conole and Alevizou 2010; Abdel Meguid and Collins 2017); Employers demand employment-ready graduates who possess appropriate sustainability knowledge and employment skills such as negotiation, collaboration and influencing (Drayson 2015); Educators hope for student engagement, sustainability literacy and values for advocacy and action. In addition, although not a main consumer of tertiary education, the Government's strategic hopes and expectations of increasing the skills base of the UK and enhancing the employability of graduates (Harvey 2000) also indirectly influence the three main actors' behaviours. Together these have combined to create a complex learning context and environment that challenges business school academics to instigate new means and methods for business sustainability learning, teaching and assessment to prepare students to respond to the environmental, social and ethical dilemmas that increasingly confront organisations (Navarro 2008). Pedagogic approaches to business sustainability must therefore transform learning systems to promote competencies and capabilities that emphasise the need for thinking holistically and envisioning change (Ferreira et al. 2006; UNECE 2012) whilst helping to close the employment skills gaps identified by Culpin and Scott (2011) by assisting students to acquire skills required for the 21st Century workplace such as collaboration, critical thinking and communication (Buck Institute for Education 2017).

Despite the growth in specialist courses and modules and integration of some sustainability content into business management curricula, learning, teaching and assessment has not adequately engaged business management students in sustainable business practice including potential climate change alleviation actions. Consequently, sustainability within the business curriculum continues to lag behind needs of graduates in the workplace such that the majority of students are unprepared to deal with sustainability issues they will face on a daily basis (Waddock 2007; Govender 2016). Many graduates continue to perpetuate professional practices that exploit the environment and people (Mula et al. 2017) as they have not been exposed to the appropriate knowledge, skills and values required for developing sustainable futures such as critical and reflective thinking (UE4SD 2013; Howlett et al. 2016) or questioning skills to challenge unsustainable behaviours (Ryan and Tilbury 2013). It is now widely accepted that education for sustainability has fallen behind the internal and external sustainability interests of businesses and change agents (Lozano et al. 2013; Environmental Audit Committee 2017). This has resulted in a growing sustainability skills gap (Benn and Dunphy 2009; Lambrechts and Ceulemans 2013; Edie 2015; Sadler 2016 and Laurinkari and Tarvainen 2017) that has left businesses struggling to respond to the material but underestimated risks to business posed by the wide ranging and varied effects of climate change (UNFCCC 2007; KPMG 2008).

Although universities are playing an important role in closing the first of these two recognised skills gaps by increasing investment in resources to enhance graduate employment skills (Cashian et al. 2015), the academic community remains less focused on two critical issues. Firstly, how students, who will be the managers and leaders of the future, can be educated most effectively to become change agents for business sustainability (Hesselbarth and Schaltegger 2014; Edie 2015; Laurinkari and Tarvainen 2017). Secondly, how to integrate business sustainability into management education or where sustainability should fit into the curriculum (Figuero and Raufflet 2015). Even though many graduates are not aware, the knowledge and capabilities to manage corporate sustainability have become significant components of graduate employment (Hesselbarth and Schaltegger 2014; Docherty 2014): 85% of graduate jobs in the UK now require a knowledge of sustainable business practices and the impacts of climate change (Drayson 2015). Sustainability skills are therefore critical to graduates' future career success.

The capability of universities to address sustainability issues of environmental accountability and social and environmental responsibility through research is well recognised (Nicolaides 2006; HEFCE 2013; Higher Education Academy 2014). However, there is now a growing expectation that universities should also contribute to and act as role models for a sustainable society through learning and teaching (Sterling et al. 2013; HEFCE 2013; Higher Education Academy 2014; United Nations 2017). Despite this, no specific best practice for this learning and teaching has been agreed upon. Ongoing pedagogic discourse suggests effective approaches to business education for sustainability are to engage students in realistic, relevant and authentic experiences in contexts and learning environments they find meaningful (Crossthwaite et al. 2006; Matthew and Butler 2017). To encourage the best student outcomes, this environment should be immersive (Staniskis and Katiliute 2016), take a constructivist pedagogical approach (Howlett et al. 2016), incorporate the examination of potential scenarios for a sustainable future in a reflexive and participative process (Rieckmann 2011) and engage students in authentic inquiry (Hensley 2017) to empower learners to transform the way they think and act (UNESCO 2017). Although Rieckmann (2011) considers this environment should have a problem orientation and link formal and informal learning to facilitate development of key competencies that are needed to deal with unsustainable development, Wiek et al. (2014) argue the pedagogic approach should include real-world settings so that students are educated in relevant environments and able to develop workable solutions to promote sustainable futures.

To maximise the benefit of these recommendations, the author designed the business sustainability module that is the focus of this study to offer an immersive learning environment (completion of an assessed sustainability audit), active, participative, reflexive learning, teaching and assessment (participative in-class activities aligned to the audit) with a problem solving focus (reflection on the audit findings) in a real-world setting (the case study) and collaborative learning culture of environmental, social and economic responsibility (pedagogic practice).

Although this range of recommendations for effective education for sustainability has been proposed, a broad range of functional barriers are also recognised. These include a lack of awareness among university educators and researchers, the disciplinary structure of business schools, or a lack of funding (Lambrechts and Ceulemans 2013) and declining student engagement (Leach 2016). To overcome these, Molthan-Hill (2014), Figuero and Raufflet (2015) and UNESCO (2017) advise educators need to radically rethink means and methods of management education to encourage students to think in new ways. Consequently, the author has designed the 3rd year undergraduate sustainability module for business management students to overcome these barriers and promote sustainability and employment skills to develop employment ready graduates.

Aligning Learning, Teaching and Assessment to a Sustainability Audit

A sustainability audit of a case study company provides a focus for learning, teaching and assessment in the module presented here: the audit and analysis of findings forms the assessment, all taught subjects and in-class activities are aligned to a topic within the audit. Aligning learning, teaching and assessment to the audit is designed to offer learning by stealth and overcome preferences for just-in-time learning as it presents students a reason for attending and engaging with the topics presented in-class, i.e. contribution to their assignment. In turn this may help students achieve the aspirations of a good grade and development of employment skills.

Theoretical context for using a sustainability audit for learning, teaching and assessment

A sustainability audit is a methodical examination of a business's procedures and practices that determine or influence environmental, social or economic impacts; they have long been recognised as a method of improving sustainability performance. Audits add value to businesses as they are a voluntary but essential management procedure that allows an organisation to detect problems before they affect operations (Hillary 2004), provide a benchmark from where to measure subsequent change (Clark and Whitelegg 1998) and enable the development of a systematic approach to improving sustainability performance whilst improving economic performance (Viegas et al. 2013). As such, an audit may therefore be considered an effective tool to engage students in collecting, collating, synthesising and evaluating information, which is both outcome and process of learning (Corcoran and Wals 2004). The author has adopted this tool for learning, teaching and assessment as it may stimulate students' sustainability literacy and higher order cognitive and employment skills that can feed forward into future workplaces to support the development of sustainable futures and alleviation of climate change impacts.

Learning based on completing a sustainability audit combines three learning styles: firstly, 'learning by doing', which can generate student engagement (Dewey 1916; Drayson 2015) and result in knowledge being retained for longer than knowledge gained in passive learning experiences (Gardiner and D'Andrea, 1998); secondly, 'project-based learning', which can equip students with softer employment skills such as collaboration, negotiation and influencing and transferable knowledge such as enquiry, problem solving and critical analysis (Shepherd 1998). Thirdly, 'active learning' in a real world setting which can help students find their place in the world by doing what they learn (Bardati 2006). Together these can stimulate students to be active learners rather than simply consumers of knowledge (Juarez-Najera et al. 2006).

A review of pedagogic literature suggests that sustainability audits are infrequently employed in learning, teaching and assessment. A small number of environmental audits have been used to offer extracurricular learning (Ferreira et al. 2006; Alshuwaikhat and Abubakar 2008) or adopted for an environmental module (Bardati 2006); there is little evidence to suggest sustainability audits have been adopted as a formal learning, teaching and assessment tool for a business curriculum. Research does agree, however, that audits can offer a valuable pedagogic approach as they can stimulate students to be active learners instead of knowledge consumers as the audit process requires them to collect, analyse and synthesis information.

In addition, an audit may offer valuable double loop learning as it can expose students to a variety of possible future scenarios through a critical evaluation of their own decisions (Beckett and Murray 2000). Wrestling with empirical data for themselves to collect, collate, synthesise and evaluate information, rather than receiving pre-digested analyses from lecturers or secondary sources, supports deeper learning as it empowers students to examine issues with greater insight (Centre for Teaching 2017). This independent learning process is also valuable as it can enhance personal meaning (Meyer et al. 2008) which is required to develop the sustainability values that educators aspire to engender and to empower students to act to alleviate the impacts of climate change in their future workplaces.

Practical context of using a sustainability audit for learning, teaching and assessment

Since academic year 2016–17 students taking the Level 6 business sustainability module have completed an assessed sustainability audit of a case study company. Over the course of 12 weeks (the module period) the students act as Sustainability Consultants whose role is to undertake a modified GRI Audit (GRI 2016) of a case study company and design a sustainability strategy to report and address the positive and negative issues identified utilising the case study material provided. This mirrors the real-life roles and responsibilities of internal and external sustainability auditors.

To achieve the module learning outcomes (and to achieve a good grade), students are required to explore, critically evaluate and synthesise the mixed media case study material to make evidence-based judgements on the case study company's (un)sustainability performance in social, environmental and economic sustainability, i.e. conduct the audit. The taught content and in-class activities are designed to support students in this as they are all aligned to the audit.

The module is designed to engage students in a week-by-week development of sustainability knowledge, skills and values that aims to challenge current preferences for just-in-time learning whilst providing the continuous support and extensive scaffolding that Ertmer and Simons (2006) and Kandiko and Mawer (2013) argue is expected by students in the current university environment. Each week's taught session explores an area of sustainability practice within a business. This is aligned to a topic from the GRI audit such as raw material use and recycling, employee benefits, procurement practices or investment agreements and related to the case study company through in-class activities which are undertaken in an engaging and fun environment of games, quizzes and creative activities. Aligning learning, teaching and assessment in this way may provide a constant reinforcement of the case study and ongoing opportunities for reflection on the case study company's behaviours. It may also allow students to see how sustainability theory can be used in practice and develop their self-efficacy which in turn may generate sustainability knowledge, skills and values and deliver expected learning outcomes. To assist other members of the sustainability community the module learning and teaching plan is shown in Table 30.1.

The audit-based assignment

The assignment for this business sustainability module is divided into three sections. Firstly, a sustainability audit of the case study company utilising the case study material and following the GRI process and modified template. Secondly, a critical situational analysis of the case study company's (un)sustainability performance using the audit findings, and thirdly, the design of an evidence-based sustainability strategy to address the issues and enhance opportunities identified in the audit.

To undertake this assignment, students act as Sustainability Consultants and operate within a real world processes of the GRI audit; to achieve a successful assignment outcome, i.e. a good grade, students are expected to show a detailed and holistic knowledge of sustainability issues and resolutions and apply them at a business level. This requires them to consider the most appropriate improvement processes or strategic responses within a business that must continue to operate on a day to day basis with conflicting and uncertain existing systems which Beckett and Murray (2000) highlight as a key objective for organisations. The assignment may therefore represent a valuable employment skill learning opportunity.

The audit, combined with the situational analysis and proposed strategic response presents a different style of assignment to the more usual essay or exam for business management students. The author is aware that this may provoke nervousness in some students as it may take them out of their assignment comfort zone of exams and essays. However, to engage future sustainability advocates and develop new ways of thinking through the new learning approaches advocated by Ferreira et al. (2006) and UNECE (2012) the managers and leaders of the future must be challenged to

Table 30.1 Module learning and teaching plan

Module week	Lecture title	In-class activities	Link between lecture and assignment This lecture will…
1	Sustainability: the current management challenge	Living lab research: sustainability knowledge survey The Oxfam inequality quiz	Review the module outline and introduce the students to the case study
2	The business case for sustainability	The banana split game (CAFOD 2018) Debate—is there a benefit from adopting sustainability as a business approach?	Introduce potential benefits and challenges for the case study company which is required for the students' assignment
3	Mitigating sustainability challenges: taking action on energy, waste, water, air and land	The sustainability in music challenge Drawing the site plan of the case study company. Environmental Impact Assessment of the case study company	Enables the students to undertake an environmental impact assessment of the case study company to understand the impact of the business on the environment/local communities that is required for their assignment
4	Auditing sustainability performance: taking action to improve assurance and reporting	Meet the Managers: discuss the case study company's sustainability performance with the company's sustainability manager and consultant	Enable the students to conduct an audit meeting with managers from the case study company. The students can test their understanding of the case study, which will provide evidence for their assignment
5	Ecomedia: taking action to improve communication to customers	Create a 2-minute film to communicate the case study company's sustainability performance	Enable the students to identify techniques to communicate sustainability to stakeholders, which is required for their assignment
6	Managing waste: taking action to rethink resources	The plastic bottle challenge Audit practice using the campus as the audit scope	Enables the students to identify environmental social and economic impacts of waste management practices, which is required for their assignment

(continued)

Table 30.1 (continued)

Module week	Lecture title	In-class activities	Link between lecture and assignment This lecture will...
7	Sustainable supply chains: taking action to manage the supply chain	Mapping the case study company's supply chain; identifying the impacts of businesses on people and the environment within the supply chain and proposing improvements	Enables the students to identify social and environmental impacts of the case study company within the supply chain, which is required for their assignment
8	Adding social value: taking action to improve social sustainability	The orange trading game	Enable the students to identify the risks of social exposure for the case study company which is required for their assignment
9	Sustainability in practice	Meet the managers: audit meetings with the case study company's management team	Allows the students to discuss the social, environment and economic performance of the case study company to support their audit data collection
10	Eco-efficiency: taking action to improve environmental sustainability	Affinity diagram and interrelationship diagraph to explore cause and effect of potential improvements (root cause analysis)	Enables the students to identify environmental opportunities that will contribute to their recommendations to improve the case study company's performance, which is required for their assignment
11	Developing sustainability strategies: taking action on improving corporate performance	The sustainable strategies game	Enable the students to identify potential improvement strategies, which are required for their assignment
12	Assignment support: 1:1 student-tutor meetings	This session gives the students an opportunity to discuss their assignment with module tutor to obtain advice and guidance on their performance	

develop the higher order cognitive academic and life skills such as critical thinking, reasoning, synthesising, evaluating and communicating advocated by Sadler (2016).

The sustainability audit

To complete the first part of their three-part assignment effectively and achieve the modules' learning outcomes, students should collect, collate, synthesise and present evidence from their analysis of the case study on the modified GRI Audit template (an extract is shown in Table 30.2), i.e. conduct an audit. The processes of audit preparation and completion are designed to promote an understanding of the complexity of business sustainability practices, advocated by Richert et al. (2017), knowledge of improvement opportunities and higher order cognitive skills such as critical thinking, reasoning, synthesising, evaluating and communicating that Sadler (2016) considers lacking in graduates.

To undertake their audit students must filter the information presented in the case study to decide on what is important and what is not required. Students should then synthesise and evaluate this spotlighted information to form credible judgements regarding the case study company's (un)sustainability performance related to the topics within the GRI audit. Engaging students' in the filtering of information which is required to complete their audit may also promote understanding of what is important and what can be discarded (Sadler 2016) which can cultivate long-term thinking and prioritisation of actions as students decide how to frame appropriate solutions for themselves (Taylor 1974). The process and practice of the audit may therefore help students recognise their own learning and perhaps, more importantly learn what they need to learn.

The students are required to complete their audit on a modified GRI Audit template. This is provided to students as a Word document which can be completed and embedded as a PDF in their final assignment script. The author recommends that students submit their completed audit within an appendix of their assignment script to enhance the flow of their report and facilitate marking.

To complete the audit template students are required to write audit questions (Column B in Table 30.2) to address the audit topic presented (Column A in Table 30.2) and utilise information collected, collated and synthesised from the case study and associated in-class activities to answer them. This process creates the audit findings (Column C in Table 30.2). As in a real sustainability audit, the students must justify their findings with observations of (un)sustainability performance; for the case study audit they do this by referencing the document, photograph, or report used (Column D in Table 30.2). Students' initial thoughts on improvement interventions to address the issues identified (Column E in Table 30.2) with potential owners and a timeframe (Column F in Table 30.2) are included in the template as would be required in a real-world audit.

Throughout the module the students are encouraged to behave as auditors would in a real-world audit to gain employment skills. This includes conducting an assessment of the processes being explored rather than the person implementing it and looking for processes that are working well. Based on experience of real-world audits the author

Table 30.2 Extract of modified GRI audit (GRI 2016) utilised for the Audit-based Learning assignment

Audit Ref.	Audit topic (A)	Audit questions (B)	Audit answer (sustainability performance—current situation observed during audit including historical actions impacting on current sustainability) (C)	Evidence used (D)	Post audit recommendation (E)	Action by and recommended timing (F)
Social sustainability						
EC1	Direct economic value generated and distributed, including revenues, operating costs, employee compensation, donations and other community investments					
Transport						
EN27	Significant environmental impacts of transporting products and other goods and materials used for the organisation's operations, and transporting members of the workforce					
Diversity and equal opportunity						
LA14	Ratio of basic salary of men to women by employee category					

suggests this gathers the most detailed responses as assessing the person generally leads to weak disclosure of audit information and fear of reprisals. By looking for positive performance, positive reinforcement can be applied; whilst looking for good practice, the negative issues will emerge automatically.

The module is designed to encourage students to analyse the case study in groups although students must submit an individual audit, situational analysis and strategy to complete their assignment. The sharing of the data collection through in-class activities, Meet the Manager sessions and independent research aims to stimulate students to be resource efficient and engage in group work to practice their employment skills of collaboration, negotiation and influencing. This may engender reflexive, self-directed learning within the safe environment of the classroom and facilitate peer-to-peer learning, formative feedback and the face-to-face interactions for learning and support advocated by Kandiko and Mawer (2013).

The case study

The bespoke case study presenting the company to be audited was written by the author in 2015–16 in preparation for its' introduction in 2016–17. It was created as part of the development of teaching materials for the module and as such elicited no funding.

The case study is hosted on a WordPress website to offers a novel format for business management students who are used to traditional paper-based case studies. It presents them with information regarding the (un)sustainability performance of the company to be audited within a series of emails, photographs, corporate reports, purchase orders and invoices, job advertisements, wage sheets, HR records, letters to the company from local residents and the company's responses, etc. This mixed media case study incorporates interrelated sustainability good practice and issues that emerge through the collation and synthesis of the material available. This format was chosen for several reasons. Firstly, it replicates the types of information encountered by an auditor and evaluated in a real life sustainability audit. Secondly, it engages students in generative sustainability, i.e. they must generate their own views and opinions rather than downloading their responses from the internet. Thirdly, it allows sustainability topics to be introduced week by week, with some relevant documentation released each week to support the taught session and in-class activities. Fourthly, it enables the weekly taught sessions and in-class activities to be linked to a specific subject within the assessed audit to promote students' engagement with the topic of sustainability and encourage lecture attendance.

In-class activities

Two types of in-class activity are incorporated into the classroom sessions: starter activities and audit-linked activities. A starter activity (see Table 30.1) is implemented at the beginning of a session to give time to students to settle down and engage them in sustainability thinking. As students' engagement can be increased when their competitive spirit is engendered (Emblen-Perry 2018) these starter activities are designed to be fun but foster competition between groups. Activities include a Plastic Bottle Challenge, Environmental Impact Spot the Difference, Sustainable Clothes Challenge (Molthan-Hill 2014), drawing sustainability, the Banana Split Game (Cafod 2018), etc.

Audit-linked activities are designed to expose students to a variety of learning, teaching and assessment approaches to keep taught sessions novel, engaging, fun and most importantly different. These individual and group based activities include case study quizzes to engage students with case study company early in the module, the Sustainability Strategies Game that explores conflicts of natural resource use in profit driven organisations, 'Meet the Manager' audit meetings to give students an opportunity to discuss their audit findings with members of the case study company's management team role played by colleagues, greenwash video making, etc. An overview of the activities, their aims and their link to the taught session is shown in Table 30.3.

The in-class audit-linked activities are planned to be practical and experiential but cognitively demanding so that students are encouraged to develop the problem-solving and decision-making skills advocated by Fabricatore and Lopez (2012) by exploring case study information through participative and reflexive learning processes individually and within groups. Reflection on the values, actions and activities encountered within the case study company and audit processes may promote the new ways of learning recommended by Rieckmann (2011) and Molthan-Hill (2014) to support and empower students to transform the way they think and act (UNESCO 2017).

The approach of aligning in-class activities to the audit aims to overcome students' preferences for just-in-time learning and embed sustainability practice knowledge and skills through active, collaborative and relevant learning tasks. Relevant activities are important; students will only engage if they consider the task worthy of investing their time and effort (Robertson 2013). The relevance of in-class activities in this module is enhanced through their alignment to the students' assignment, which may in turn promote engagement, generate peer-to-peer learning and develop confidence with the case study, sustainability theory and good practice as well as building sustainability knowledge, skills and hopefully values, along with softer employment skills.

To provide ongoing support and guidance each in-class activity includes a post-activity debriefing and discussion. Debriefings may also enable students to provide and receive peer reflection and feedback and share and explore individual and group generated information and feelings which Kuh et al. (2006) suggest promotes student engagement. Post-activity reflection also allows the author to monitor students' progress and evaluate their perception of the effectiveness of activities. This feedback is used to inform research into means and methods of enhancing students' engagement with education for sustainability.

30 Engaging and Empowering Business Management Students … 563

Table 30.3 In-class activities designed to promote sustainability knowledge, skills and values ad employment skills

Activity	In lecture(s)	Details of the activity	Activity designed to
Drawing sustainability	1 and 10	In groups students discuss their experience of sustainability and draw their groups interpretation of "sustainability"	• Establish students' knowledge of sustainability • Help students recognise their learning throughout the course of the module
Case study quiz	1	In groups, students use the case study to answer 20 questions on the people, facts, features of the case study company	• Engages students in case study content, format and inter-connectivity of the information presented. • Emphasise need for students to explore the case study early in the module • Highlight some key issues to be followed up as independent study
Drawing a site plan	3	In groups, students utilise the information in the case study to draw a site plan (Drawn on flip chart paper)	• Engage students in the details of the case study • Develops mental model of how the case study company's physical location and position of factory and facilities in relation to communities, employees, water, biodiversity etc. • Encourages early exploration of how the case study company operates e.g. production operations, logistics, raw material storage, employee housing, water sources/consumption, energy sources and uses, emissions to atmosphere, site biodiversity, etc. • Exposes students to simulated real-world view of sustainable/unsustainable businesses practices • Encourages investigation of the interconnections within and between aspects of the case study
Environmental impact assessment	3	Students extend their site plan to annotate it with the environmental impacts identified from the case study	• Engages students in positive and negative impacts of operations, people and activities carried out within the case study company's site and exploration of potential improvement opportunities

(continued)

Table 30.3 (continued)

Activity	In lecture(s)	Details of the activity	Activity designed to
Meet the managers sessions	4 and 9	Students hold audit meetings with members of how the case study company's newly appointed sustainability manager and sustainability consultant. These are role played by an hourly paid lecturer and a local energy consultant Week 4: Class meeting Week 9: Class discussions and individual one-to-one meetings	• Provide opportunities for students to explore sustainability issues and opportunities in more detail with experts in a simulated real-world environment in the safe environment of the lecture room • Allow students the opportunity to explore their perceptions of the case study company, request additional information and check understanding and obtain feedback on their improvement recommendations • Replicate real world audit practice and process to enable student to gain employment skills • Enhance student learning by introducing guest speakers to extend the boundaries of the classroom, which engages students in perceiving the course as part of a larger network of ideas and conversations (Centre for Teaching 2017)
Making a sustainability communication film	5	Groups of students create short films to greenwash the case study company's sustainability performance utilising the information presented in the case study Films can comprise students acting, pictures with voiceovers or slides with music depending on confidence of students	• Encourage students to explore sustainability performance of the case study company • Engages students in thinking in different ways about sustainability • Engages students in communicating sustainability in different ways • Encourage the development of sustainability values • Offer students opportunity to act within their preferred learning style and environment
Audit practice	6	Students choose 3 questions from the modified GRI template and apply them to campus sustainability	• This allows students to practice audit skills, particularly looking and questioning

(continued)

Table 30.3 (continued)

Activity	In lecture(s)	Details of the activity	Activity designed to
Mapping the supply chain	7	In groups, students use the case study information to map the supply chain using the Six Sigma SIPOC model	• Engage students in analysis of the movement and role of goods and people through the supply chain and the potential sustainability impacts created • Provide sustainability skills, assignment knowledge and employment tools
Orange trading game (Traidcraft 2017)	8	In groups, students act as orange growing families	• Engage students with the opportunities presented by Fair Trade. • Connect students with issues raised within the game with employee and community issues within the case study
Affinity diagram and interrelation-ship diagraph	10	Students filter case study information to evaluate what evidence is valuable and what can be discarded	• Engage students in differentiating between causes and symptoms of issues in the case study • Use Six Sigma tools to engage students in root cause analysis; a key skill for business professionals • Encourage the development of sustainability values
Sustainable strategies game	11	In groups, students play a role-based game that explores the challenge of sustainability decision	• Engage players in the challenges of decision making for sustainability that students will encounter in their future careers and • Support development of sustainability values

Discussion

The module was created to provide innovative means and methods of management education that may encourage students to think in the new ways demanded by HEFCE (2013) and address sustainability skills gaps recognised by Benn and Dunphy (2009), Lambrechts and Ceulemans (2013), Edie (2015), Sadler (2016) and Laurinkari and Tarvainen (2017). The author's pedagogic approach of immersing business management students in distinctive, memorable academic challenges through a sustainability audit aims to expose learners to the process-based approaches used by the sustainability sector (Ferriera et al. 2006). In essence the approach to education for sustainability presented in this conceptual paper is designed to develop employment-ready, sustainability literate students equipped with the tools and techniques to become sustainability advocates who can act to deliver sustainable business futures and alleviate the impacts of climate change. The sustainability audit and audit linked in-class activities, which form the core of this novel approach to learning, teaching and assessment, may consequently deliver the expectations of the three key actors in education for sustainability: students, employers and educators.

The transformative and innovative active learning, teaching and assessment approaches advocated by Oblinger and Oblinger (2005) and Abdel Meguid and Collins (2017) are utilised in this module to provide the participative and collaborative learning styles students prefer (Stubbs 2011; Armier Shepherd and Skrabut 2016), and deliver the extensive support for learning students have come to expect (Ertmer and Simons 2006; Kandiko and Mawer 2013). In addition to meeting students' preferences for participative, collaborative activities and expectations of scaffolded learning the author's innovative learning, teaching and assessment pedagogy may deliver the learning experiences advocated by Molthan-Hill (2014), Figuero and Raufflet (2015) and UNESCO (2017) to encourage the wider perspectives on sustainability that United Nations (2015) consider are required to contribute to sustainable societies and support the mitigation of climate change.

In-class activities may be relevant students as they are aligned to the assessed sustainability audit; they can support the achievement of successful module outcomes (i.e. good grades) and the development of employment skills which are key expectations of students. Through this alignment the module may stimulate a number of effective module outcomes. Firstly, it may stimulate enhance engagement in sustainability as participation in alignment linked activities may incentivise business management students to attend lectures. This in turn may help to overcome the declining student engagement recognised by Leach (2016). Secondly, the active approach to learning, teaching and assessment adopted in the module may facilitate peer-to-peer learning, ongoing formative feedback and face-to-face interactions to engender the reflexive, self-directed learning advocated by Kandiko and Mawer (2013). Thirdly, the generative learning approach designed for the module may provide the learning environment advocated by Kuh et al. (2006) in which students can develop personal attitudes and sustainable behaviours through reflection on the values, actions and activities encountered. Finally, active learning processes and the (un)sustainable behaviours encountered in the in-class activities may stimulate students to develop the individual and collective sense of responsibility required for sustainability advocacy and the alleviation of climate change impacts.

Throughout the module business management students are supported to collate, synthesise and evaluate the online case study and information gained through in-class activities to complete their sustainability audit and reflect upon their audit findings to create an improvement strategy. The actions required to complete the audit provide a generative learning process that may promote the construction of viable solutions to real-world problems of climate change thus providing the learning approach Crossthwaite et al. (2006), Partnership for 21st Century Skills (2007) and Weik et al. (2014) suggest students find relevant, engaging and meaningful. By analysing and evaluating complex sustainability scenarios within the case study, completing the GRI audit and undertaking independent research to inform a sustainability strategy, students may be helped to recognise and take ownership of their learning needs; Savery and Duffy (1995) recognise this ownership is an essential component of learning.

Immersing students in a critical analysis of the audit findings may also provide them with an opportunity to engage in double-loop learning. This may enhance students' ability to identify their own learning needs (Argyris 1982) and consider a variety of possible future alternatives through a critique of their own decisions (Beckett and Murray 2000); skills which are required for the development of sustainable business futures.

Engaging students with a variety of learning, teaching and assessment means and methods that challenge them to collaborate, communicate and practice enquiry, problem solving and critical thinking in a real-world sustainability context (the assessed GRI audit is designed to mirror a real-world business audit process) may advance appropriate sustainability knowledge and employment skills required by future employers (Drayson 2015) by promoting a holistic understanding of sustainability issues within an organisation in a format that involves students in real world learning. This may contribute to the closure of students' sustainability and higher order cognitive skills gaps recognised by Sadler (2016) and to the future-proofing their careers. In turn this may promote personal learning and demonstrate students' capabilities and knowledge development thus promoting academic and workplace confidence along with sustainability literacy. This module may therefore contribute to achieving sustainable business futures and support the alleviation of climate change impacts by developing employment-ready graduates whilst closing higher order cognitive skills gaps by promoting reasoning, synthesising, evaluating and communicating (Sadler 2016).

Equipping business management students with appropriate knowledge (Chalkley 2006; Rieckmann 2011), enhancing learners' capacity for critical and reflexive evaluation (Howlett et al. 2016) and providing them with experience of key sustainability skills such as recognition of the inherent complexity of issues and strategic responses may promote valuable employment skills that allow the cross-communication of ideas and detection of problems before they effect operations (Beckett and Murray 2000). Linking sustainability theory and good practice to the case study and audit may therefore engage students with the sustainability knowledge and employment skills expected by employers (Docherty 2014) and, in turn, enhance students' employment prospects which Drayson (2015) considers key challenges for tertiary education.

Implementing a module learning structure that progressively builds the sustainability knowledge and academic skills through the alignment of learning and teaching and in-class activities to a sustainability audit may offer both the outcome and process of learning advocated by Corcoran and Wals (2004). The process of learning may be provided by the audit itself and participation in in-class activities whilst the outcome of learning including awareness rising of business (un)sustainability and potential resolutions may emerge from the learning, teaching and assessment approaches created. Alongside these skills for academic success the necessary skills of environmental and social impact monitoring and analysis may also emerge as the outcome of learning to empower students to support the mitigation of climate change and communicate the need for change within their future workplaces.

Emergent Issues and Future Opportunities

This paper presents a conceptual analysis of an innovative learning, teaching and assessment approach designed by the author to enhance the outcomes of education for sustainability for 3rd year undergraduate business management students, their future employers and current educators. It contributes to the research into participatory approaches to education for sustainability recommended by Vare and Scott (2007) and Mader and Mader (2012) and to the dialogue on education for sustainability in universities which is vital for the development of new ways of thinking to provoke policy and behaviour change (Disterheft et al. 2015) and stimulate the mitigation of climate change. Through learning, teaching and assessment means and methods focused on the completion of a sustainability audit future business managers and decision-makers may be helped to understand the complex problems of business sustainability and climate change (Richert et al. 2017) thus encouraging them to change businesses from within.

Although the impact of student engagement, learning outcomes and employment skills is not validated here, the learning, teaching and assessment methodology presented may be of interest to others in the sustainability community building the capacity of staff who are looking for new ways to engage students in sustainable knowledge, skills and values. The author recommends that further research is undertaken to test students' responses to the module, learning processes, teaching methodology and audit experience to validate a module focused around a sustainability audit as an effective tool for business sustainability learning, teaching and assessment in tertiary education.

Conclusion

The author combined previous business experience and literature validating active, experiential learning for business sustainability and supporting the value of audits as extracurricular learning opportunities to create the learning, teaching and assessment approach to the business sustainability module for 3rd year business management undergraduates presented in this paper. Although further research is required to corroborate the value of learning, teaching and assessment aligned to a sustainability audit for enhancing student engagement, learning outcomes and employment skills, the author suggests this innovative pedagogic approach may provide students the freedom to explore sustainability issues and gain a sense of the inherent complexities and challenges that business sustainability practitioners face on a daily basis. In turn, this may engage learners in generative sustainability and challenge them to develop their own perspectives rather than simply searching online for others' established answers.

In-class activities, on-going learning support and formative assignment guidance, audit meetings as well as the process of undertaking the audit and reflecting on findings may allow students to develop sustainability knowledge, skills and values; attributes required if future leaders are to communicate the need for change and generate the desperately required sustainable business futures that can contribute to the alleviation of climate change impacts. The real-world tools and learning environment created are designed to encourage students to construct a personal, evidence-based view of sustainability and the alleviation of climate change whilst delivering specific, general and transformative employment skills and values that are required to future proof their careers and promote sustainable business futures.

References

Abdel Meguid E, Collins M (2017) Students' perceptions of lecturing approaches: traditional versus interactive teaching. Adv Med Educ Pract 8:229–241

Alshuwaikhat H, Abubakar I (2008) An integrated approach to achieving campus sustainability: assessment of the current campus environmental management practices. J Clean Prod 16:1777–1785

Argyris C (1982) Reasoning, learning, and action: individual and organizational. Jossey-Bass, San Francisco, CA

Armier D, Shepherd C, Skrabut S (2016) Using game elements to increase student engagement in course assignments. Coll Teach 64:64–72

Bardati D (2006) The integrative role of the campus environmental audit: experiences at Bishop's University, Canada. Int J Sustain High Educ 7:57–68

Baumgartner R (2014) Managing corporate sustainability and CSR: a conceptual framework combining values, strategies and instruments contributing to sustainable development. Corp Soc Responsib Environ Manag 21:258–271

Beckett R, Murray P (2000) Learning by auditing: a knowledge creating approach. TQM Mag 12:125–136

Benn S, Dunphy D (2009) Action research as an approach to integrating sustainability into MBA programs: an exploratory study. J Manag Educ 33:276–295

Bourn D (2013) Teachers as agents of social change. Int J Dev Educ Glob Learn 7(3):65

Buck Institute for Education (2017) Why project based learning (PBL)? Retrieved from https://www.bie.org/about/why_pbl

Cafod (2018) Banana split game. Retrieved from https://www.google.co.uk/?gfe_rd=cr&ei=EMBjV9eUDfHS8AfDvaKACw&gws_rd=ssl

Cashian P, Clarke J, Richardson M (2015) Perspectives on: Employability. Is it time to move the employability debate on? Retrieved from https://charteredabs.org/wp-content/uploads/2015/06/Employability-Debate1.pdf

Centre for Teaching (2017) Teaching sustainability. Retrieved from https://cft.vanderbilt.edu/guides-sub-pages/teaching-sustainability/

Chalkley B (2006) Education for sustainable development: continuation. J Geogr High Educ 30:235–236

Clark G, Whitelegg J (1998) Maximising the benefits from work-based learning: the effectiveness of environmental audits. J Geogr High Educ 22:325–334

Conole G, Alevizou P (2010) A literature review of the use of Web 2.0 tools in higher education. High Educ Acad. Retrieved from http://www.heacademy.ac.uk/

Corcoran P, Wals A (2004) Higher education and the challenge of sustainability: problematics, promise, and practice. Kluwer Academic Publishers, Boston, MA

Cortese A (2003) The critical role of higher education in creating a sustainable future. Plan High Educ 31:15–22

Craik K (1943) The nature of explanation. Cambridge University Press, Cambridge

Crossthwaite C, Cameron I, Lant P, Litster J (2006) Balancing curriculum processes and content in a project centred curriculum: In pursuit of graduate attributes. Educ Chem Eng 1:39–48

Culpin V, Scott H (2011) The effectiveness of a live case study approach: increasing knowledge and understanding of 'hard' versus 'soft' skills in executive education. Manag Learn 43(5):565–577

Dewey J (1916) Democracy and education; an introduction to the philosophy of education. Macmillan, New York, NY

Disterheft A, Caeiro S, Azeiteiro U, Leal W (2015) Sustainable universities—a study of critical success factors for participatory approaches. J Clean Prod 106:11–21

Dochety D (2014) Universities must produce graduates who are ready for any workplace. Retrieved from https://www.theguardian.com/higher-education-network/2014/may/22/universities-must-produce-graduates-who-are-ready-for-workplace

Drayson R (2015) Student attitudes towards and skills for sustainable development, executive summary: employers. Retrieved from www.heacademy.ac.uk/system/les/executive-summary-employers.pdf

Edie (2015) Minding the gap: Developing the skills for a sustainable economy. Retrieved from https://www.edie.net

Emblen-Perry K (2018) Enhancing student engagement in business sustainability through games. Int J Sustain High Educ 19(5):858–876

Environmental Audit Committee (2017) Sustainable development goals in the UK. Retrieved from https://publications.parliament.uk/pa/cm201617/cmselect/cmenvaud/596/59602.htm

Ertmer P, Simons K (2006) Jumping the PBL implementation hurdle: supporting the efforts of K-12 teachers. Interdiscip J Probl-Based Learn 1:40–54

Fabricatore C, López X (2012) Sustainability learning through gaming: an exploratory study. Electron J E-Learn 10(2):209

Ferreira A, Lopes M, Morais J (2006) Environmental management and audit schemes implementation as an educational tool for sustainability. J Clean Prod 14:973–982

Figuero P, Raufflet E (2015) Sustainability in higher education: a systematic review with focus on management education. J Clean Prod 106:22–33

GRI (2016) GRI standards. Retrieved from https://www.globalreporting.org

Gardiner V, D'Andrea V (1998) Teaching and learning issues and managing educational change in geography. Cheltenham and Gloucester College of Higher Education, Cheltenham, UK

Genn J (2001) AMEE Medical education guide no. 23 (part 1): curriculum, environment, climate, quality and change in medical education–a unifying perspective. Med Teach 23(4):337–344

Govender I (2016) Evaluating student perceptions on the development management curricula to promote green economy. Environ Econ 7:1–10

Harvey L (2000) New realities: the relationship between higher education and employment. Tert Educ Manag 6:3–17

Hensley N (2017) The future of sustainability in higher education. J Sustain Educ, 03. Retrieved from http://www.jsedimensions.org/wordpress/content/the-future-of-sustainability-in-higher-education_2017_03/

HEFCE (2013) Sustainable development in higher education: consultation on a framework for HEFCE. Retrieved from http://www.hefce.ac.uk/workprovide/Framework/

Hesselbarth C, Schaltegger S (2014) Educating change agents for sustainability—learnings from the first sustainability management master of business administration. J Clean Prod 62:24–36

Higher Education Academy (2014) Education for sustainable development: guidance for UK higher education providers. Retrieved from http://www.qaa.ac.uk/en/Publications/Documents/Education-sustainable-development-Guidance-June-14.pdf

Hillary R (2004) Environmental management systems and the smaller enterprise. J Clean Prod 12:561–569

Howlett C, Ferreira J, Blomfield J (2016) Teaching sustainable development in higher education: building critical, reflective thinkers through an interdisciplinary approach. Int J Sustain High Educ 17:305–321

Juarez-Najera M, Dieleman H, Turpin-Marion S (2006) Sustainability in Mexican Higher Education: towards a new academic and professional culture. J Clean Prod 14:1028–1038

KPMG (2008) Climate change your business: KPMG's review of the business risks and economic impacts at sector level. KPMG International, Netherlands

Kandiko CB, Mawer M (2013) Student expectations and perceptions of higher education. King's Learning Institute, London, UK

Kuh G, Kinzie J, Buckley J (2006) What matters to student success: a review of the literature. Retrieved from http://nces.ed.gov/IPEDS/research/pdf/Kuh_Team_Report.pdf

Lambrechts W, Ceulemans K (2013) Sustainability assessment in higher education. Evaluating the use of the Auditing Instrument for Sustainability in Higher Education (AISHE) in Belgium. In: Caeiro S, Leal Filho W, Jabbour C, Azeiteiro U (eds) Sustainability assessment tools in higher education institutions. Mapping trends and good practice around the World. Springer, Switzerland

Laurinkari J, Tarvainen M (2017) The policies of inclusion. EHV Academic Press, London, UK

Leach L (2016) Exploring discipline differences in student engagement in one institution. High Educ Res Dev 35:772–786

Lozano R, Lukman R, Lozano FJ, Huisingh D, Lambrechts W (2013) Declarations for sustainability in higher education: becoming better leaders, through addressing the university system. J Clean Prod 48:10–19

Lynch K (2006) Neo-liberalism and Marketisation: the implications for higher education. Eur Educ Res J 5:1–17

Mader C, Mader M (2012) Innovative teaching for sustainable development—approaches and trends. In: Global university network for innovation (ed) Higher education in the world 4: higher education's commitment to sustainability: from understanding to action. Palgrave Macmillan, Basingstoke, UK

Matthew A, Butler D (2017) Narrative, machinima and cognitive realism: constructing an authentic real-world learning experience for law students. Australas J Educ Technol 33:148–162

Meyer B, Haywood N, Sachdev D, Faraday S (2008) What is independent learning and what are the benefits for students? Retrieved from http://www.curee.co.uk/files/publication/[site-timestamp]/Whatisindependentlearningandwhatarethebenefits.pdf

Molthan-Hill P (2014) The business student's guide to sustainable management: principles and practice. Greenleaf Publishing, Sheffield, UK

Mulà I, Tilbury D, Ryan A, Mader M, Dlouhá J, Mader C, Benayas J, Dlouhý J, Alba D (2017) Catalysing change in higher education for sustainable development: a review of professional development initiatives for university educators. Int J Sustain High Educ 18:798–820

Navarro P (2008) The MBA core curricula of top-ranked US business schools: a study in failure? Acad Manag Learn Educ 1:108–123

Nicolaides A (2006) The implementation of environmental management towards sustainable universities and education for sustainable development as an ethical imperative. Int J Sustain High Educ 7:414–424

Oblinger D, Oblinger J (2005) Educating the net generation. Retrieved from http//www.educause.edu/educatingthenetgen. https://www.educause.edu/ir/library/PDF/pub7101.PDF

Partnership for 21st Century Skills (2007) Framework for 21st century learning. Retrieved from http://www.p21.org/storage/documents/docs/P21_framework_0816.pdf

Pauw J, Gericke N, Olsson D, Berglund T (2015) The effectiveness of education for sustainable development. Sustainability 7(11):15693–15717

Pennington R, Joyce T, Tudora J, Thompson J (2012) Do different learning contexts, processes and environment affect perceptions, dispositions and approaches to learning? Retrieved from https://www.raeng.org.uk/publications/other/factors-that-affect-learning

Quality Assurance Agency for Higher Education (2014) Education for sustainable development: guidance for UK higher education providers. Retrieved from http://www.qaa.ac.uk/en/Publications/Documents/Education-sustainable-development-Guidance-June-14.pdf

Richert C, Boschetti F, Walker I, Price J, Grigg N (2017) Testing the consistency between goals and policies for sustainable development: mental models of how the world works today are inconsistent with mental models of how the world will work in the future. Sustain Sci 12(1):45–64

Rieckmann M (2011) Future-oriented higher education: which key competencies should be fostered through university teaching and learning? Future 44:127–135

Robertson R (2013) Helping students find relevance. Retrieved from https://www.apa.org/ed/precollege/ptn/2013/09/students-relevance

Ryan A, Tilbury D (2013) Uncharted waters: voyages for education for sustainable development in the higher education curriculum. Curr J 24:272–294

Sadler D (2016) Three in-course assessment reforms to improve higher education learning outcomes. Assess Eval High Educ 41:1081–1099

Savery J, Duffy T (1995) Problem based learning: an instructional model and its constructivist framework. Educ Technol 35:31–38

Seatter C, Ceulemans K (2017) Teaching sustainability in higher education: pedagogical styles that make a difference. Can J High Educ Rev 47(2):47–70

Shepherd H (1998) The probe method: a problem-based learning model's effect on critical thinking skills of fourth and fifth-grade social studies students. North Carolina State University, Raleigh, NC

Staniskis J, Katiliute E (2016) Complex evaluation of sustainability in engineering education: case and analysis. J Clean Prod 120:12–20

Sterling S, Maxey L, Luna H (2013) The sustainable university: progress and prospects. Earthscan, London, UK

Stubbs W (2011) Addressing the business-sustainability nexus in postgraduate education. Int J Sustain High Educ 14:25–41

Taylor R (1974) Nature of problem ill-structuredness: implications for problem formulation and solution. Decis Sci 5:632–643

Traidcraft (2017) Orange trading game. Retrieved from https://www.traidcraft.co.uk/resources

UE4SD (2013) University educators for sustainable development. Retrieved from https://www.ue4sd.eu/images/leaflets/UE4SD-eng.pdf

UNECE (2012) Learning for the future: competences in education for sustainable development. Retrieved from http://www.unece.org/leadmin/DAM/env/esd/ESD_Publications/Competences_Publication.pdf

UNESCO (2011) Definition of education for sustainable development. Retrieved from http://www.unescobkk.org/fr/education/esd-unit/definition-of-esd/

UNESCO (2017) Education for sustainable development goals learning objectives. Retrieved from http://unesdoc.unesco.org/images/0024/002474/247444e.pdf

UNFCCC (2007) Uniting on climate: a guide to the climate change convention and the kyoto protocol. UNFCCC Secretariat, Bonn, Germany

United Nations (2015) Sustainable development goals, 17 goals to change our world. Retrieved from http://www.un.org/sustainabledevelopment/sustainable-development-goals/

United Nations (2017) Sustainable development knowledge platform: sustainable development goals. Retrieved from https://sustainabledevelopment.un.org/sdgs

Vare P, Scott W (2007) Learning for a change: exploring the relationship between education and sustainable development. J Educ Sustain Dev 1:191–198

Vanderstraeten R (2004) Education and society: a plea for a historical approach. J Phil Educ 38(2):195–206

Viegas C, Bond A, Duarte Ribeiro J, Selig P (2013) A review of environmental monitoring and auditing in the context of risk: unveiling the extent of a confused relationship. J Clean Prod 47:165–173

Waddock S (2007) Leadership integrity in a fractured knowledge world. Acad Manag Learn Educ 6:543–557

Wiek A, Xiong A, Brundiers K, van de Leeuw S (2014) Integrating problem-and project-based learning into sustainability programs. Int J Sustain High Educ 15:431–449

CPSIA information can be obtained
at www.ICGtesting.com
Printed in the USA
LVHW082104141220
674148LV00001B/9

9 783030 329006